D1743155

Nitrate:
Processes, Patterns
and Management

Nitrate: Processes, Patterns and Management

Edited by

T.P. BURT

School of Geography,
University of Oxford,
UK

A.L. HEATHWAITE

S.T. TRUDGILL

Department of Geography,
University of Sheffield,
UK

JOHN WILEY & SONS

Chichester · New York · Brisbane · Toronto · Singapore

Other Wiley Editorial Offices

John Wiley & Sons, Inc., 605 Third Avenue,
New York, NY 10158-0012, USA

Jacaranda Wiley Ltd, G.P.O. Box 859, Brisbane,
Queensland 4001, Australia

John Wiley & Sons (Canada) Ltd, 22 Worcester Road,
Rexdale, Ontario M9W 1L1, Canada

John Wiley & Sons (SEA) Pte Ltd, 37 Jalan Pemimpin #05-04,
Block B, Union Industrial Building, Singapore 2057

Library of Congress Cataloging-in-Publication Data

Nitrate : processes, patterns, and management / edited by T.P. Burt, A.L.
  Heathwaite, and S.T. Trudgill.
      p. cm.
    Includes bibliographical references and index.
    ISBN 0-471-93476-3
    1. Nitrogen cycle. 2. Nitrates — Environmental aspects.
  3. Nitrogen in agriculture. I. Burt, T.P. II. Heathwaite, A.L.
  (A. Louise) III. Trudgill, Stephen T. (Stephen Thomas), *1947–*
  QH344.N515   1993
  574.5′222—dc20   92—23877
                                                        CIP

British Library Cataloguing in Publication Data

A catalogue record for this book is available from the British Library

ISBN 0-471-93476-3

Typeset in 10/12pt Times by MHL Typesetting Ltd
Printed and bound in Great Britain by Bookcraft (Bath) Ltd

For Lorna and Bob Troake
who started the nitrate research at Slapton

Contents

List of Contributors

A.C. Armstrong
ADAS Soil and Water Research Centre, Anstey Hall, Trumpington, Cambridge CB2 2LF

S. Ball
Faculty of Law, University of Sheffield, Crookesmoor Building, Conduit Road, Sheffield S10 1FL

G. Bentham
School of Environmental Sciences, University of East Anglia, Norwich NR4 7TJ

T.P. Burt
School of Geography, University of Oxford, Mansfield Road, Oxford OX1 3TB

B.T. Croll
Anglian Water Services Ltd, Yare House, 62−64 Thorpe Road, Norwich NR1 1SA

M.T. Downes
Taupo Research Laboratory, Tuwharetoa St, PO Box 415, Taupo, New Zealand

T. Hall
WRc Swindon, Frankland Road, Blagrove, PO Box 85, Swindon SN5 8YR

N.E. Haycock
Department of Agricultural Water Management, Silsoe College, Bedford MK45 4DT

A.L. Heathwaite
Department of Geography, University of Sheffield, Sheffield S10 2TN

C. Howard-Williams
Taupo Research Laboratory, Tuwharetoa St, PO Box 415, Taupo, New Zealand

P.J. Johnes
Department of Environmental and Evolutionary Biology, University of Liverpool, PO Box 147, Liverpool L69 3BX

D. Johnson
Biological Science Centre, Desert Research Institute, University of Nevada, Reno, Nevada 89506, USA

T. O'Riordan
School of Environmental Sciences, University of East Anglia, Norwich NR4 7TJ

N.J.P. Owens
The Plymouth Marine Laboratory, Prospect Place, The Hoe, Plymouth PL1 3DH

R.J. Parkinson
University of Plymouth, Seale-Hayne Department of Agriculture, Newton Abbot, Devon TQ12 6NQ

K.A. Smith
SAC, School of Agriculture, West Mains Road, Edinburgh EH1 3JG

S.T. Trudgill
Department of Geography, University of Sheffield, Sheffield S10 2TN

H. van Miegroet
Environmental Sciences Division, Bldg 1505, MS-6038, Oak Ridge National Laboratory, Oak Ridge, TN 37831-6038, USA

A.J.A. Vinten
SAC, School of Agriculture, West Mains Road, Edinburgh EH1 3JG

R.P. Wayne
Physical Chemistry Laboratory, University of Oxford, South Parks Road, Oxford OX1 3QZ

Preface

We decided to edit this book for a number of reasons, but perhaps the most important was that we were unaware of any other text which brought together the various facets of the 'nitrate issue' at a research level that would be of interest and use to a wide range of environmental and agricultural scientists working in this general area. It is true that other books are available which deal with this issue at a popular level such as Dudley's *Nitrates* (1990) or which make particular reference to the agricultural industry (Addiscott *et al.*'s *Farming, Fertilisers and the Nitrate Problem*, 1991) However, recent changes in government policy have confirmed what we have argued for a number of years: the nitrate issue is not merely a question of the relationships between farming, fertilisers and drinking water, important though these matters are. Large losses of nitrate (and other forms of nitrogen) from farmland which can have widespread environmental consequences, often far away from the source, show that once one starts to consider the fate of nitrate leached below the root zone, a broader catchment-based hydrological view is in order. Such a view can encompass the multiple sources of nitrogen compounds and indicate possible viable options for management. This view thus develops from the Royal Society's *The Nitrogen Cycle of the United Kingdom* (1983). A geographical perspective, stressing especially the importance of the drainage basin as the fundamental unit and scale of analysis, is therefore a central focus of this book.

Ignoring the prologue and epilogue, our book is divided into three parts. In the first, the authors examine the processes of nitrogen cycling in particular environments. As one might expect, two chapters are devoted to soils, though only one deals with agricultural soils. However, it was a deliberate decision not to concentrate on soil at the expense of other systems such as wetlands, marine waters, and so on. An integrated approach to the nitrate issue requires this broad appraisal, in our view. The central part deals with the hydrology of nitrate, the delivery of nitrate to streams, the movement of nitrate in groundwater systems, and the patterns of nitrate in surface waters. The third part considers management responses to the nitrate problem, with discussions of both preventative and curative solutions, and a detailed examination of the legal requirements surrounding the issue.

Much of our own nitrate research has been focused within the catchments draining into Slapton Ley, a small lake in south west England whose status as a nature reserve has been threatened over the last few decades because of eutrophication. We have been particularly interested in the links between land use and nitrate loss and in the hydrological pathways by which nitrate moves from soil to stream. In our research we have certainly not avoided nitrate leaching from the soil profile (at least we hope not!) but we have sought to place such studies within the wider context of the drainage basin and to relate our work with that of ecologists working in the area. As noted above, this hydrological theme and integrated approach have provided the basis for this book.

Like all field science, our research has been enjoyable as well as interesting. We conceived of this project while working at Slapton and there could be no more proper location for the composition of this preface.

Tim Burt
Louise Heathwaite
Steve Trudgill

The Tower Inn, Slapton
7 May 1992

Part I

NITROGEN CYCLING AND NITRATE PRODUCTION IN CATCHMENT ECOSYSTEMS

1 Overview — the Nitrate Issue

A.L. HEATHWAITE

Department of Geography, University of Sheffield

T.P. BURT

School of Geography, University of Oxford

and

S.T. TRUDGILL

Department of Geography, University of Sheffield

1.1 THE NITRATE ISSUE: A QUESTION OF SCALE

There is strong evidence to suggest that, particularly over the last 20 years, 'the nitrate issue' has shifted in scale from what was once a local pollution problem to what is now one of regional dimensions. Indeed, there is evidence to suggest that unless measures are taken to reduce the rate and magnitude of nitrate input to receiving waters, the scale of the nitrate issue may shift towards a continent-scale problem involving widespread pollution of both fresh and marine waters. Figure 1.1 illustrates the increasing scale of water pollution problems since the Roman period. Although nitrate pollution has not yet reached the 'continental' scale, Figure 1.2 shows the current extent of water pollution from nitrate sources. Furthermore, the importance of nitrogen in controlling the trophic status of freshwaters means that measures to control nitrate pollution in aquatic systems are urgently needed.

On a global scale, Meybeck (1982) suggested that the average concentration of nitrogen for world rivers in a pristine state is to 0.015 mg NH_4-N l^{-1}, 0.001 mg NO_2-N l^{-1} and 0.1 mg NO_3-N l^{-1}. A report published jointly by the United Nations Environment Programme (UNEP) and the World Health Organisation (WHO) in 1988 found that in some regions of the world, particularly Europe, less than 10% of the river stations may be classified as pristine (Table 1.1). The median concentration for nitrate in Global Environment Monitoring Systems (GEMS) stations outside Europe was 0.25 mg NO_3-N l^{-1} whereas for European rivers the median concentration was 4.5 mg NO_3-N l^{-1} (WHO/ UNEP, 1987). Over 10% of European rivers had a nitrate concentration ranging from 9 to 25 mg NO_3-N l^{-1}, thus a number of rivers exceed the EC maximum admissible concentration (MAC) for nitrate in drinking water of 11.3 mg NO_3-N l^{-1} (EC Directive on Drinking Water 80/788). In the UK, 125 groundwater sources supplying 1.8 million people in 1983/84 exceeded 11.3 mg NO_3-N l^{-1} (Lean, 1990). This is compared to 90 groundwater supplies in 1980 and 60 supplies in 1960 (DoE, 1986). Furthermore, the amount of nitrate by which the EC limit is exceeded is increasing in a number of areas in England, notably Norfolk, Cambridgeshire, Lincolnshire, Hereford and Worcestershire. In Germany, where the drinking-water supply comes mainly from borehole sources, 5−10% of boreholes exceed the EC limit and the average

Nitrate: Processes, Patterns and Management. Edited by T.P. Burt, A.L. Heathwaite and S.T. Trudgill
© 1993 John Wiley & Sons Ltd

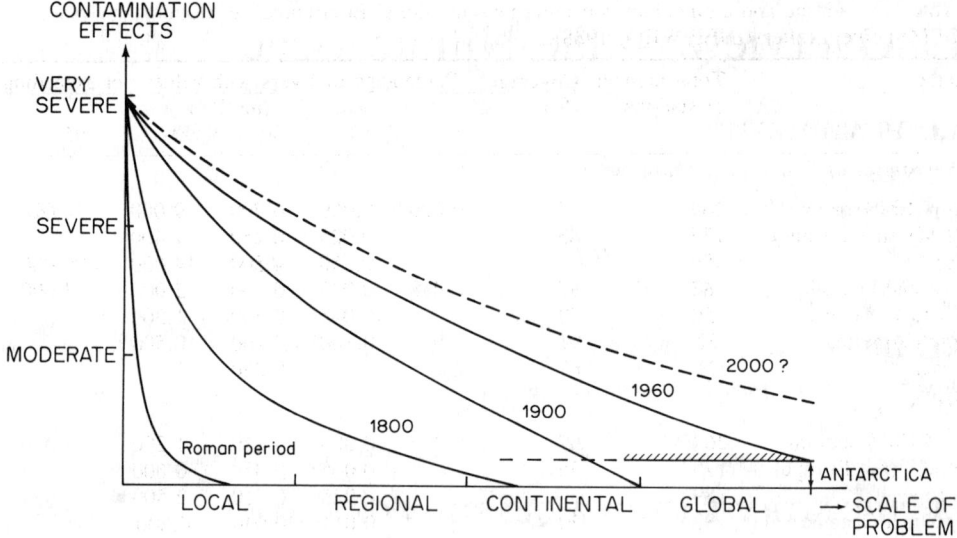

Figure 1.1 The evolution of water quality problems (after UNEP/WHO, 1988)

concentration of nitrate is increasing by $1-2$ mg NO_3-N l^{-1} per year in intensively cultivated areas (Saull, 1990).

It is generally accepted, therefore, that the concentration of nitrate-nitrogen in surface waters has increased in recent decades (see, for example, Burt *et al.*, 1988; Heathwaite and Burt, 1991; Meybeck, Chapman and Helman, 1989). However, as nitrogen reaches surface waters from a primarily diffuse source area it is difficult to clearly isolate the assumed link between land use, fertiliser application and stream water quality (Dermine and Lamberts, 1987). This is despite the fact that there is strong evidence that the scale of the nitrate pollution problem

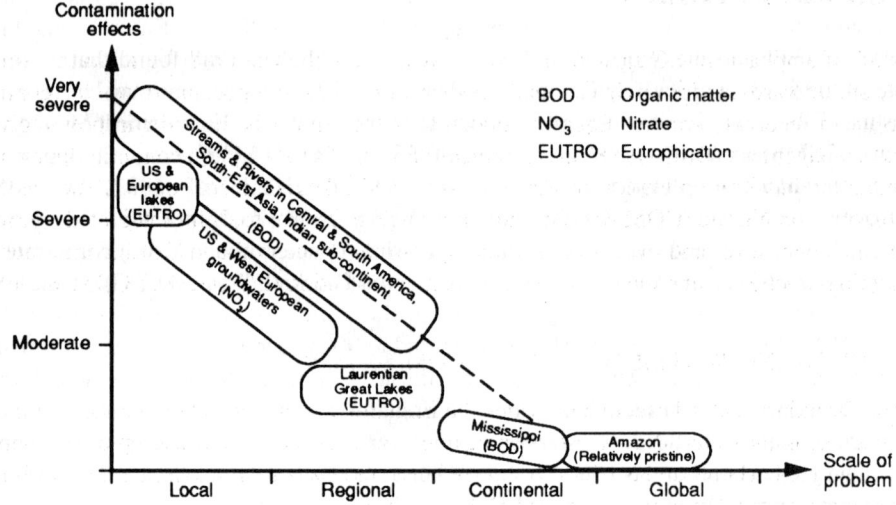

Figure 1.2 The occurrence and importance of organic water pollution (after UNEP/WHO, 1988)

Table 1.1 Nitrate and ammonium concentrations in Global Environmental Monitoring System (GEMS) rivers (after UNEP/WHO, 1988)

Region	Total number of stations	Coverage[a] (%)	Percentage of rivers with values not exceeding (mg l^{-1})				
			5	10	50	90	95
NO$_3$ N mg l^{-1}							
All GEMS stations	264	87	0.020	0.050	0.700	9.000	14.000
GEMS (not Europe)	175	83		0.025	0.250	1.400	
Europe	89			0.250	4.500	14.000	20.000
Asia and Oceania	82	87	0.018	0.035	0.350	2.000	4.000
N and C America	50	82		0.045	0.300	1.300	
South America	22	91		0.100[b]	0.200	0.500	
Africa	21	67			0.250		
NH$_4$-N mg l^{-1}							
All GEMS stations	264	67	0.006	0.009	0.110	1.200	3.000
GEMS (not Europe)	175	56		0.005	0.070	0.600	
Europe	89			0.025	0.210	2.500	
Asia and Oceania	82	50		0.011	0.040	0.800	
N and C America	50	44		0.005*	0.015	0.200[b]	
South America	22	76		0.020	0.080	0.250	
Africa	21	53			0.020		

[a] Proportion of the stations effectively monitoring the variable.

[b] Approximate values.

is increasing (see, for example, Foster et al., 1985; Roberts and Marsh, 1987). The processes of nitrogen transfer between atmosphere, land and receiving waters will be examined in more detail in this book and will form a basis for the interpretation of the scale and magnitude of the nitrate issue and the mechanisms for its control.

1.2 SHIFTS IN POLICY

The shift of emphasis in EC agricultural policy away from the Common Agricultural Policy (CAP) and towards wider environmental concerns provides an opportunity for the control of nitrate in receiving waters. Equally important is the emphasis this shift in policy gives to land management (DoE, 1988) at an appropriate scale as a viable, economic option for the reduction of nitrate pollution in surface waters and lakes. Current policies, such as the EC Directive on Nitrate (COM 88/708), add more weight to proposals for integrated controls at the catchment level and even stretch to compensating farmers for loss of income caused by adopting practices compatible with environmental protection (EC Directive COM 90/366).

1.2.1 DRINKING WATER DIRECTIVE (COM 80/788)

The EC Drinking Water Directive provides the basis for national legislation concerning the quality of drinking water. New legislation has been introduced by, for example, Germany, France, the Netherlands and Spain to bring national standards more closely in line with the requirements of the Directive (Agg and Derick, 1991).

Nitrate concentrations in all public water supplies must be kept below the maximum admissible concentration (MAC) permitted by the EC of 11.3 mg l^{-1} NO_3-N (equivalent to 50 mg NO_3^- l^{-1}). The UK MAC for other nitrogen species is 0.1 mg NO_2^- l^{-1}, 0.5 mg NH_4^+ l^{-1} and 1 mg Kjeldahl N l^{-1}. The MAC for nitrate has major financial implications for the UK water industry (DoE, 1986). In order to achieve this level of control, strategies aimed either at *preventing* nitrate inputs to water sources or *treating* nitrate inputs once they have reached watercourses, must be devised. The DoE concluded in 1988 that land-use control would prove to be the cheaper option (DoE, 1988),

There remain two loopholes in the EC Drinking Water Directive (COM 80/788) which deal with derogations and delays. Article 9(1) allows member states to apply for derogations from MAC where high nitrate concentrations result from 'situations arising from the nature and structure of the ground' that is, natural rather than anthropogenic sources. In 1985 the UK government applied for derogations on 52 water supplies affecting over 900 000 people in England, primarily in the Anglian and Severn-Trent water areas (DoE, 1988). A number of environmental groups have stated that the high nitrate concentrations in these areas result from anthropogenic rather than natural sources and, therefore, should not be exempt from control (Dudley, 1986). The second key loophole in the EC Directive is found in article 20 of COM 80/788, which allows time extensions for nitrate control of up to 10 years from 1985 'in exceptional cases from geographically defined population groups'; in practice, ten years' grace has been given for nitrate control in a number of areas rather than requiring immediate action.

1.2.2   NITRATE DIRECTIVE (91/676)

The EC Nitrate Directive (91/676) 'concerning the protection of waters against pollution caused by nitrates from agricultural sources' recognises that 'excessive use of fertilisers' constitutes an environmental risk and goes further than previous Directives on water quality in suggesting that this risk applies to both inorganic and organic (slurry and farmyard manure) fertilisers. Unlike the previous legislation, the Nitrate Directive includes the use of mandatory orders for nitrate control in rivers and groundwaters used for the supply of drinking water, in order to achieve two objectives. First is the prevention of nitrate concentrations in freshwaters (surface and subsurface) reaching levels that may interfere with the 'legitimate' uses of these waters. This may create real difficulties for water supply companies in some areas of the UK. Second, is the prevention of the eutrophication of fresh and coastal waters, an objective that does not apply solely to nitrate but represents a departure from previous Directives which have focused on human health rather than environmental concerns.

The EC Nitrate Directive will compel member states to designate all zones vulnerable to water pollution from nitrogen compounds. Vulnerable zones are defined as areas of land which drain, either directly or indirectly, into one or more of the following waters: surface freshwaters and groundwaters, intended for drinking water and which would contain nitrate concentrations in excess of 11.3 mg l^{-1} if protective action was not taken; and natural freshwater lakes, other natural freshwater bodies, estuaries, coastal waters and seas which are already eutrophic or which may become eutrophic if preventative action is not taken. Within designated 'vulnerable zones' farming practices will be restricted so that the timing and amount of fertiliser applied is controlled, taking into account the characteristics of the vulnerable zone concerned and preserving consistency with good agricultural practice. However, the final form of the Directive does not go nearly as far as the earlier draft (88/708) in laying down detailed requirements; in particular, the definition of terrain on which fertilisers may or may not be applied is left to the voluntary code of good agricultural practice (see Section 1.2.3).

Middleton (1990) stated that 'if one attempts to control nitrate by agricultural means alone, substantial areas in vulnerable zones that are to be designated will have to come out of production altogether'. The role of land-use management through the use of, for example, vulnerable zones will no doubt have widespread social and economic repercussions within the farming sector.

A number of problems remain with the designation of vulnerable zones such as the size and extent of the zones and where and when nitrate testing will take place in order to determine the extent. Lord Middleton, speaking to the House of Commons (*Science in Parliament*, February 1990), argued that the draft EC Nitrate Directive (COM 88/708 draft) was too inflexible in its insistence of an absolute standard of 11.3 mg NO_3-N l^{-1} as the upper limit for the designation of vulnerable zones. He suggested that switching to an average rather than an absolute limit for nitrate could reduce the area of land included in vulnerable zones by up to 30%. See Chapters 11 and 12 for further discussion of these issues. Ball argues (Chapter 14) that vagueness in Directive 91/676 is likely to give rise to debate as to its scope and spatial coverage.

1.2.3 CODE OF GOOD AGRICULTURAL PRACTICE

It is widely accepted that plant productivity responds to increases in the supply of available nitrogen (Cooke, 1982) and, as a result, nitrogen fertiliser use in the UK increased a hundred-fold between 1950 and 1980 (Chapter 3, this volume). Fertiliser 'wastage' through denitrification, volatilisation and leaching of nitrogen compounds, and the impact of such wastage on receiving waters, has become a major 'nitrate issue'. Nitrogen equivalent to 30—50% of the fertiliser nitrogen applied may be lost to the atmosphere through denitrification, while up to 30% may be lost through leaching (Frissel and Van Veen, 1982; Jenkinson, 1982; Stewart and Rosswall, 1982).

The *Code of Good Agricultural Practice* (MAFF, 1991) is primarily concerned with instigating *voluntary* controls on manure, slurry and dirty water to agricultural land but comments briefly on sewage sludge application and milk products. It is a statutory code under Section 116 of the Water Act 1989; additionally Section 107 of the Act makes it an offence to knowingly discharge pollutants into receiving waters. Five stages are identified in the Code which should be followed prior to the application of organic fertilisers to land:

(1) Identification of areas of land on which organic fertilisers should not be spread at any time;
(2) Undertaking measures to match the land area available for spreading to the nitrogen load of the waste;
(3) Estimating the risk of pollution from spreading;
(4) Choosing the size of storage facilities to contain organic wastes at times when spreading is not feasible; and
(5) provisions for the design and building of storage facilities for organic waste (see Chapter 11).

The Code is essentially an attempt to control the timing, amount and application method associated with the utilisation of organic wastes on agricultural land. An application limit of 250 kg ha total nitrogen per year is proposed which does not include nitrogen directly voided onto the land by livestock. A lower limit may be introduced in Nitrate Sensitive Areas (NSAs) (see Section 1.4.2).

In a similar manner to the EC Directive on Nitrate (Section 1.2.2) which identifies 'vulnerable zones' for nitrate, the *Code of Good Agricultural Practice* attempts to control land use by prohibiting the application of organic fertilisers within 10 m of a watercourse or 50 m from a spring, borehole or well. Indeed, the Code will need to be amended in order to comply with the requirements of Directive 91/676 (see Section 14.9.4), which requires a code of good agricultural practice to be implemented as a voluntary basis by farmers with the aim of providing for all waters a general level of protection against pollution.

Organic fertilisers contain appreciable quantities of nitrogen in soluble forms and should, therefore, only be applied to agricultural land when the crop can use the nitrogen supply, which suggests that their use should be restricted in autumn. Organic wastes with a high soluble nitrogen fraction include cow slurry, pig slurry, poultry manure and liquid sewage sludge (MAFF, 1991). Some organic wastes contain a small proportion of total nitrogen in readily available forms (for example, farmyard manure and sewage sludge cake). Restrictions on the timing of applications of these organic wastes do not need to be as stringent.

1.3 NITRATE: THE ISSUES

1.3.1 HEALTH

The potential link between nitrate and human health has focused on (1) the element of risk and (2) the provision of evidence to support claims. There are two key areas of concern, methaemoglobinaemia and stomach cancer (see Chapter 15). Opinion is divided over the human health concerns and nitrate usage (Jollans, 1983; Shuval and Gruener, 1976). In 1983, the *Municipal Journal* (Anon, 1983) called for action on nitrate in water, while Wild (1977) stated that a health hazard was unlikely.

Methaemoglobinaemia ('blue-baby' syndrome) is a well-recognised medical condition and is thought to arise where ingested nitrate impairs the oxygen-carrying capacity of the blood. Ingested nitrate is converted to nitrite in the gut of infants. High nitrite concentrations result in the oxidation of haemoglobin (which occurs more readily in infants than in adults), ultimately restricting the ability of haemoglobin to carry oxygen and resulting in a bluish-tinged or oxygen-starved baby. Furthermore, infants do not have a sufficiently well-developed enzyme-reduction system to enable the re-transformation of methaemoglobin to haemoglobin. The condition does not affect adults, who are able to convert any nitrite formed back to nitrate which is then excreted.

While the risk from infant methaemoglobinaemia or cyanosis is clear, the incidence is low. The last case of methaemoglobinaemia in Britain was reported in 1972. The World Health Organisation (WHO) reported 2000 cases worldwide between 1945 and 1986; 160 of these infants died as a result of consuming water with a nitrate concentration greater than 25 mg NO_3-N l^{-1} from unsterilised feeding bottles. The greatest risk is likely to be in rural agricultural areas where nitrate fertiliser application may be increasing and water quality may be less closely monitored (Pretty and Conway, 1988).

The link between nitrogen and stomach cancer is not well understood (Kornberg, 1979; Beresford, 1985; Forman, Al-Dabbagh and Doll, 1985; and see Chapter 15). It is known that nitrosamine compounds are carcinogenic (Magee, 1982; Hill, Hawksworth and Tattersall, 1973; Winnerberger, 1982; Pocock, 1985) but the link between nitrate ingestion and nitrosamine formation is less clear (Magee, 1982). Nitrosamines may be formed from food amines and nitrite in saliva. Indeed nitrate in vegetables (even without fertiliser nitrogen application) may be an equal or greater source of nitrate ingestion than water (Kornberg, 1979). On the other hand, the consumption of fresh vegetables may provide protection against stomach cancer through the anti-oxidant action of their high vitamin C content (Doll and Peto, 1981; OPCS, 1981).

Attempts to link the spatial distribution of the incidence of stomach cancer with that of high nitrate concentrations in surface waters have been largely unsuccessful. In the UK, stomach cancer is highest in western and northern upland areas whereas high nitrate concentrations occur in sourthern and eastern arable regions. However, as cancer takes a number of years to develop and nitrate concentrations have increased significantly only in the past 20 years, such geographic correlations may be misleading (Dudley, 1986).

1.3.2 FARMING PRACTICE

In heavily urbanised catchments, wastewater from domestic and industrial sources may result in large nitrogen loads in receiving waters (see, for example, Meybeck, Chapman and Helman, 1989). These high nitrogen loads comes primarily from *point* sources. In rural catchments, however, agricultural sources are the most significant contribution to the nitrogen load of receiving waters and arrive largely from *non-point* or diffuse sources (see Chapter 5).

Land-use practices influence the amount of nitrogen reaching surface waters and lakes. For example, cultivation alters soil structure and the rate of microbial breakdown of organic matter, releasing a potentially leachable nitrogen source (Skjemstad, Vallis and Myers, 1988; McEwen *et al.*, 1989; Dowdell, Crees and Cannell, 1983). The *Code of Good Agricultural Practice* (Section 1.2.3) suggests that ploughing grassland, leaving ground bare after an arable crop has been removed and excessive cultivations (for example, reseeding grass leys) make a significant contribution to nitrate leaching from agricultural land. Dairy and pig farming also have potentially high nitrogen loads. Dairy farming, which uses large volumes of water and has a high production of organic waste, results in a large volume of slurry which must be stored when application to land, particularly during winter months, is restricted. MAFF (1991) suggest that a 100-herd dairy produces the organic waste equivalent to a large village of 400 houses. Pig farming produces highly organic waste and, because it is usually confined to smallholdings, has the added problem of the disposal of organic wastes on a restricted area of land. Between 1987 and 1989, slurry, silage effluent and dirty water accounted for 28%, 25% and 19% of the serious pollution incidents recorded in England and Wales (MAFF, 1991).

The addition of inorganic and organic fertilisers to agricultural land is another key factor controlling the amount of nitrogen reaching streams and rivers. Powlson *et al.* (1986) found that soils with a long history of nitrogen fertiliser application generally had a large mineralisable nitrogen fraction which could form a potentially large leachable nitrate source given the right environmental conditions (warm, moderately wet soils). There is, however, no simple relation between nitrogen fertiliser application and nitrogen leaching loss. Factors such as climate, soil type, soil drainage, fertiliser timing and application rate and the presence of grazing animals all influence nitrogen leaching losses (Armstrong, 1984; Armstrong and Garwood, 1991; Garwood, Salette and Lemaire, 1980; Ryden, Ball and Garwood, 1984). Heathwaite, Burt and Trudgill (1990) showed that nitrogen losses in surface runoff from bare ground can exceed that from arable or temporary grassland, although losses of nitrogen from heavily grazed permanent grassland were greater than those from bare ground. Most of the nitrogen lost ($>80\%$) was in the inorganic (NH_4-N) fraction.

The increasing intensity with which land is used for crop production and the extent of land-use change in lowland Britain in particular is reflected in long-term increases in the nitrate concentration of many rivers and lakes. The link between land-use change and river-nitrogen concentrations has been demonstrated in, for example, the south Midlands (Johnes and Burt, 1991) and south-west England (Heathwaite and Burt, 1991). Although most research on nitrogen pollution and agriculture has focused on temperate agricultural land under a wide range of geology, soil, climate and cropping conditions there is evidence to suggest that similar problems may be occurring in intensively irrigated land in, for example, Sri Lanka (Lawrence and Kuruppuarchchi, 1986) where nitrate concentrations in shallow aquifers may reach 45 mg NO_3-N 1^{-1} in some seasons.

Only a few studies have examined nitrogen losses in small drainage basins of less than 10 km^2 (for example, Lowrance, Todd and Asmussen, 1984; Jacobs and Gilliam, 1985). Burt and Haycock (in press) suggest that this is the most important scale of investigation if the aggregation of plot studies covering variations in land use and topography are to be adequately linked to patterns of nitrogen loss at the drainage basin outlet. Burt and Arkell (1987) studied the spatial pattern of nitrate runoff from different land uses in the 1 km^2 Slapton Wood catchment in south-west Devon. They suggest that it may not be just 'near-stream' or riparian land to which stream water quality is sensitive: valley-side slopes and plateaux which drain through hillslope hollows, may also require land-use management for nitrogen control.

1.3.3 EUTROPHICATION OF FRESHWATERS

Nutrient enrichment of surface waters (primarily nitrogen and phosphorus) is undesirable both as a result of the changes in the freshwater or marine ecosystem (Henderson-Sellars and Markland, 1987) and the implications for drinking-water supplies affected by toxic algal growth (Collingwood, 1977; Johnson et al., 1977). In May 1990 in the UK, the National Rivers Authority (NRA) detected blue-green algal blooms in 90% of their water regions at a total of 30 locations. At least 70% of these blooms were toxic to humans (Robbins, 1990). The early formation of algal blooms in 1990 was compounded by poor water management, a mild winter and low rainfall. Algal prevention measures include the provision of granular activated carbon substrates and jetting or destratification measures to stop the stratification of the water column and the formation of stable conditions in which algae can thrive.

Control of eutrophication depends on which nutrient (usually N or P) is growth-limiting (see Chapter 5). Although phosphorus is commonly held to be the limiting nutrient, nitrogen may become limiting as the balance between these two key nutrients in a water body shifts on a monthly basis. The nature of the limiting nutrient in the eutrophication of freshwaters is also largely determined by the catchment of the water body, particularly its land use and the characteristics of the water body itself such as flushing rate (Lack and Johnson, 1983), water depth and quantity of nutrients stored in the sediments.

1.3.4 NO_X AND AIR QUALITY

Wet and dry atmospheric deposition is an important source of nitrate and ammonium, often becoming relatively important in remote areas unaffected by direct pollution. On a global basis, soils and, to a lesser extent, the oceans are major sources of nitrogen emissions to the atmosphere (Jenkinson, 1990). The combustion of fossil fuels, particularly vehicle emissions, may produce locally important sources of NO_X in the atmosphere. For example, in south-east England, Goulding (1990) suggests that atmospheric nitrogen deposition is of the order 35−40 kg ha^{-1} each year. Assuming the average annual runoff in south-east

England is around 200 mm per year and that the atmospheric nitrogen input is balanced by a nitrogen leaching loss, Goulding calculated that the resultant nitrate concentration in drainage waters would be approximately 20 mg l^{-1} NO_3-N, which is almost double the EC limit for nitrate in drinking waters. Atmospheric nitrogen sources may, therefore, limit the control of nitrogen by agricultural restrictions in such areas (see Chapter 16).

1.4 NITRATE: THE STRATEGIES

The debate on the control of nitrogen pollution in receiving waters still centres on whether farmers who (arguably) need to use nitrogen fertilisers in order to maximise productivity should bear the cost of their polluting actions (the 'polluter pays' principle) which may lead to *preventative* measures for nitrogen control being developed, or whether the water companies (and, ultimately, the taxpayer) should take action against the increasing nitrate concentration in surface waters and thus enact largely *curative* measures of nitrogen control. Dudley (1986) argues that the taxpayer in effect pays three times for nitrate pollution: by subsidising farming, by paying to clean up the resulting pollution, and through any possible effects on human health. More recently, measures have been introduced to compensate farmers for loss of income (the Nitrate Sensitive Areas (NSA) scheme, see Section 1.4.2) when using nitrogen-prevention measures.

1.4.1 CURE

The use of nitrogen 'cures' largely ignore the ecological implications of raised nitrogen concentrations in receiving waters. The nitrogen 'cures' available are applicable to drinking water treatment only and include (1) water substitution from low-nitrogen concentration sources, (2) blending of high and low nitrate concentration water and (3) nitrate removal through chemical treatment or storage (Wilkinson and Greene, 1982). Despite the potentially high expense, the Royal Commission on Environmental Pollution (Kornberg, 1979) concluded that it would be more cost-effective for water companies to install plant for nitrate removal from drinking waters than to impose restriction on nitrogen fertiliser use.

Water substitution can result in high cost, especially if the water supply frequently exceeds the EC MAC for nitrate or the extent of the high nitrate concentration area is increasing. Blending is dependent on a readily available supply of low nitrogen concentration water otherwise extensive piping is required (White, 1983). Anglian Water, for example, spent £4 million on pipework to blend a low nitrate concentration bunter sandstone water supply with high nitrate concentration Lincolnshire Limestone water (Wilkinson and Greene, 1982). The *Hatton Catchment Study* (Severn-Trent Water, 1988) concluded that unless land use in the catchment area changes, groundwater supplies would exceed the EC standard for nitrate by the twenty-first century. Severn-Trent suggest that blending water supplies could reduce the impact on groundwater but that such a solution could only be a short-term measure for nitrogen control.

Reservoir *storage* may be an effective mechanism for removing nitrate from water 'naturally' through denitrification. However, at least 6 months' storage is necessary for effective nitrate loss to take place (Lack and Johnson, 1983) so demands for water use often conflict with the need for water storage.

Biological denitrification appears to be one of the most favoured 'cures' for nitrogen enrichment of surface waters and lakes. This may be largely a result of its relative cheapness. Biological denitrification requires the addition of a carbon source (for example, methanol, ethanol or acetic acid) to a fluidised sand bed to stimulate the growth of nitrate-removing bacteria. The process relies on the reduction of nitrate to N_2 or N_2O in the absence of oxygen. These processes are reviewed further in Chapter 13. Biological denitrification is inhibited in groundwaters by the lack of an adequate carbon source. Ion exchange may be an effective nitrogen 'cure' for groundwater treatment, although there are problems of disposal of the regenerant used to remove nitrate in the resin of the exchange column (Greene, 1978).

1.4.2 PREVENTION

Recently, attention has focused on nitrogen-prevention measures as a result of the EC Nitrate Directive (see Section 1.2.2), which insists that nitrogen control should be by prevention at source and gives recommendations for changes in agricultural practice and land use. Similar strategies include recent policies such as set-aside and extensification (Nychas, 1990) which are discussed further in Chapters 12 and 14. It should be noted, however, that these earlier policies were as much a response to a changing economic climate in the case of, for example, EC agricultural subsidies as to the perceived need to aim for nitrate 'prevention' rather than nitrate 'cure' strategies. The key changes to farming practices which are proposed include:

(1) Accurate determination of crop nitrogen requirements;
(2) Application of nitrogen fertilisers in late spring when warmer weather will help maximise crop uptake of the supplied nitrogen;
(3) Split applications of fertiliser rather than one or two major dressings;
(4) Encouraging the use of slow-release nitrogen fertilisers;
(5) Use of late summer/autumn cover crops to reduce the area of bare ground; and
(6) Provision of buffer zones between arable land and the drainage network.

In 1989, the Ministry of Agriculture, Fisheries and Food (MAFF) produced a consultative document on the Nitrate Sensitive Areas (NSA) Scheme which included compensation measures for the loss of agricultural production. This scheme became operational in autumn 1990 and has *voluntary* agricultural restrictions in ten *groundwater* catchments together with nine Nitrate Advisory Areas in which a campaign will promote 'good farming practices'. The voluntary agricultural restrictions include: avoiding fertiliser, slurry or manure application in autumn, not exceeding recommended maximum rates of fertiliser application, planting a winter cover crop in the autumn before spring-sown crops, and avoiding ploughing grasslands in autumn. Full details of the NSA scheme are given in the relevant Statutory Instruments issued by the UK government, the Nitrate Sensitive Areas (Designation) Order S1 1990/1013. The NSA scheme is discussed further in Chapters 11 and 14 of this book.

More recently the Agricultural Development and Advisory Service (ADAS) of MAFF published a *Code of Good Agricultural Practice for the Protection of Water* (see Section 1.2.3) aimed at encouraging farmers to minimise the loss of nutrients such as nitrogen from their land by making allowances for any residual nitrogen left in the soil after cropping and reducing subsequent nitrogen fertiliser applications (see, for example, Beaton, 1988). It is still unclear whether a reduction in fertiliser nitrogen application will reduce nitrate concentrations in receiving waters. Addiscott (1988), for example, demonstrated that unmanured soils could sustain nitrogen losses over a 40-year period. There is pressure from the farming body to weaken the measures contained in the Code, and these include removing both the absolute limit on fertiliser application to crops and the restriction on ploughing grassland that is greater than 3 years old (Lean, 1990).

Currently, attention is focused on the construction or utilisation of *buffer zones* in riparian areas to reduce nitrate concentrations entering receiving waters by creating an environment in which nitrate losses through denitrification will be high. This attention is largely the result of the EC Directive on Nitrate, which requires member states to identify 'vulnerable zones' within water supply catchments where controls on land-use practices may be made (see Section 1.2.2). The nitrogen transformation processes in buffer zones are summarised below and discussed further in Chapters 6 and 12.

Lord Middleton, speaking to the House of Commons in February 1990, stated that curative measures of nitrogen control will have to continue owing to the long time delay in, for example, high concentrations of nitrate in the chalk reaching water supplies. He also said that the cost of preventative measures to the agricultural sector will be high and will result in extensive changes in the social and economic structure of rural communities. He called for maximum flexibility to be given to member states in their incorporation of the EC Nitrate Directive as policy and an allowance for keeping options open to develop combined curative/preventative measures of nitrogen control (Middleton, 1990).

1.5 NITROGEN IN DRAINAGE BASINS: PROCESS AND PATTERN

There is a wealth of information on recent trends in nitrate concentrations in surface and groundwaters. British rivers are considered by, for example, Rodda and Johnes (1983), Betton, Webb and Walling (1991) and are reviewed further in Chapters 5 and 10 of this book. For UK groundwaters, Foster, Cripps and Smith-Carington (1982), Geake and Foster (1989) and Parker, Young and Chilton (1990) provide salient information. A large number of publications refer to individual UK catchments (Walling and Webb, 1985; Webb and Walling, 1985; Burt *et al.*, 1988; Heathwaite, Burt and Trudgill, 1989; Heathwaite and Burt, 1991; Johnes and Burt, 1991). Burt and Haycock (1992) note, however, that information for small drainage basins (< 10 km^2), with the exception of Lowrance, Todd and Asmussen (1984) and Jacobs and Gilliam (1985), is generally lacking. This is unfortunate because only at this small scale can clear links be established between the hillslope and the stream channel for the hydrological pathways involved in nitrogen transport.

1.5.1 CATCHMENT-WIDE POLICIES FOR NITROGEN CONTROL

The catchment is both a clearly defined topographical unit and may be thought of as the fundamental basis for hydrological evaluation. As such, it is an ideal focus for integrated measures aimed at nitrogen control through linking the terrestrial and the aquatic systems. Despite this, Newson (1991) suggests that land management in the UK has only recently shifted towards the catchment scale, and here only on to small catchments, typically less than 100 km^2. In order to adequately respond to the requirements of the EC Nitrate Directive for nitrogen control through preventative channels it is clear that more attention must be focused on the management of land-use and farming practices for environmental as well as economic purposes. To be at all effective, such control can only operate at the catchment scale.

1.5.2 SCALE

At the catchment scale, it may be possible to model the sensitivity of the quality of receiving waters to variations in agricultural runoff. Such variation may occur as a result of land-use change but at the catchment scale, predicting the outcome would not require detailed knowledge of soil nitrogen processes and hillslope hydrological transport mechanisms. Huang and Ferng (1990a,b) for example, designed a classification based on land use to model stream water quality. Heathwaite, Burt and Trudgill (1989, 1990) adopted a plot experimental approach to examine nitrogen export from different land uses under controlled conditions. Johnes (1990) worked at the catchment scale to examine the sensitivity of river water quality to catchment land use using an export-coefficient approach first identified by Vollenwider and applied to lake systems. An alternative to this essentially 'black-box' modelling approach at the catchment scale is to scale up from the experimental soil profile scale, although it is difficult to validate mechanistic models at the field scale (MAFF, 1988).

1.5.3 DELIVERY OF NITROGEN TO SURFACE AND GROUNDWATERS

Nitrogen delivery occurs primarily in nitrate form owing to the high solubility of this ion. As a result, it is intimately linked with the hydrological pathways controlling nutrient transport from the land to the stream. Figure 1.3 illustrates some of the pathways of nitrogen transport from hillslope to stream. Nitrogen delivery is further controlled by: (1) soil structure and type, (2) rainfall, (3) the amount of nitrate supplied in fertilisers and (4) plant cover and root activity.

1.5.3.1 Nitrogen losses through surface runoff and leaching

Nitrogen in runoff from different land uses may vary in both amount and form. It is commonly held that inorganic N will be the dominate nitrogen species in runoff from arable land, whereas organic nitrogen and ammonium-nitrogen are the major nitrogen species in grassland runoff (Webb and Walling, 1985; Heathwaite, Burt and Trudgill, 1990).

Most studies of nitrogen leaching have focused on losses of nitrate-nitrogen (Wild and Cameron, 1980; Powlson et al., 1986; McGill and Myers, 1987). Nitrogen loss can also

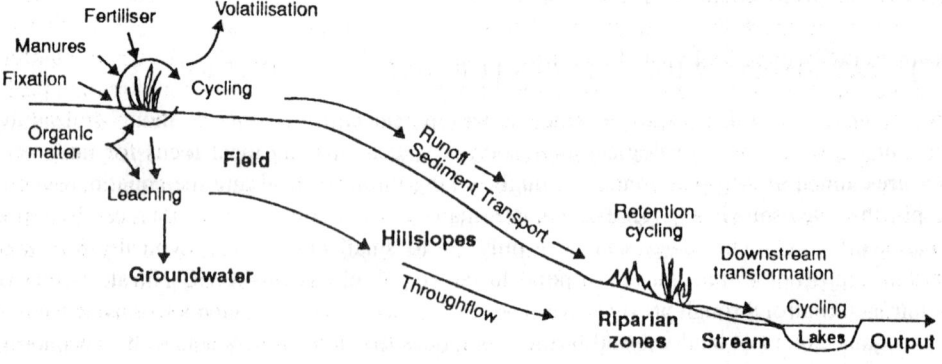

Figure 1.3 The pathways of nitrogen delivery from hillslope to stream

occur in organic and ammonium-nitrogen form, particularly from heavily grazed grassland (Heathwaite, Burt and Trudgill, 1990). Ryden, Ball and Garwood (1984) showed that nitrate leaching below a grass sward grazed by cattle was 5.6 times greater than that from a cut sward and exceeded nitrate losses normally observed on arable land. Johnes (1990) and Johnes and Burt (1991) studying nitrogen speciation in the river Windrush, a tributary of the Thames, found that the annual nitrogen load was composed of nitrate (60%) and organic nitrogen (40%), which suggests that in considering the impact of agriculture on receiving waters the focus should be on the total nitrogen load rather than nitrate alone and that variations in catchment land use will result in the spatial variation of nitrogen export.

Research into nitrogen leaching losses can be roughly grouped into modelling (Addiscott and Wagenet, 1985; Burns, 1980; Rose et al., 1982; Kaluarachchi and Parker, 1988; Selim and Iskandar, 1981a,b) or experimental (Smettem, Trudgill and Pickles, 1983; White, Wellings and Bell, 1983) approaches. The latter have tended to focus on the leaching of nitrate below the root-zone and as a result, most studies are one-dimensional investigations of nitrate loss from soil profiles, commonly using lysimeters (see, for example, Dowdell and Webster, 1980; Barraclough, Hyden and Davies, 1983). Some investigations (for example, Addiscott, 1988) include lateral hydrological pathways by using bounded plots. The potential for making generalisations are limited using a plot-based approach, although replication of results is feasible. Models are often forced to ignore or simplify potentially important solute transfer mechanisms. There have been few studies at the hillslope or catchment scale which is probably the most appropriate scale for the management of nitrogen inputs to receiving waters. Burt et al. (1983) and Burt and Arkell (1987) inferred soil hillslope processes from an analysis of outflow drainage at the base of hillslopes in the Slapton catchment in south Devon. On a longer timescale, Heathwaite and Burt (1991) attempted to link catchment land-use change together with information on nitrogen export from different land uses to increases in stream nitrate concentration using long-term (20-year) water quality records. Nitrate losses from agricultural systems are reviewed in Chapter 9.

1.5.3.2 Nitrogen in groundwater

Thirty per cent of public water supplies in Britain come from groundwater sources, roughly 50% of this fraction from the chalk aquifer and 30% from triassic sandstones (Wyndle-Taylor, 1984). In areas with a shallow water table, the upper part of the saturated groundwater zone may have concentrations of nitrate exceeding 100 mg NO_3 l^{-1} (Young, Hall and Oakes, 1976). Tracing the rate of movement of this zone of high nitrate concentration is difficult owing to the heterogeneity of the bedrock. Isotope and solute profiling in Chalk aquifers have recently been reviewed by Geake and Foster (1989). The preservation of tritium peaks at many sites shows that nitrate transport within the unsaturated zone of the Chalk is slow. The authors suggest that nitrate concentrations in (and nitrate export from) Chalk soils have increased steadily since the 1960s, although it is still uncertain whether tritium behaves in the same way as nitrate in groundwaters (Kornberg, 1979; Foster, 1975). The key assertion is that the downward movement of interstitial (pore) water in the unsaturated (vadose) zone carries with it high nitrate concentrations primarily derived from fertiliser application (Foster, Cripps and Smith-Carington, 1982) and with a relatively short association with land use (see Figure 1.4, and Young, Hall and Oakes, 1976 and also Young and Hall, 1977; Cameron and Wild, 1984). Parker, Young and Chilton (1990), showed that nitrate leaching to aquifers was strongly related to ploughing episodes.

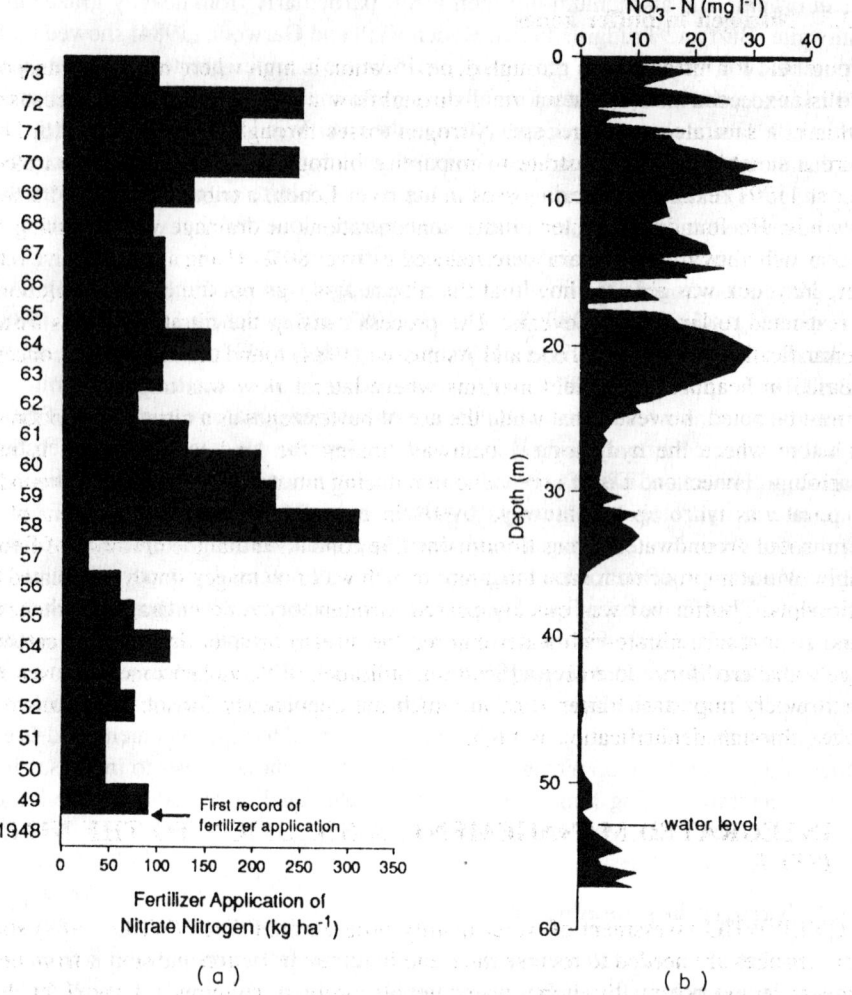

Figure 1.4   (a) Land use and fertiliser application changes (1948−74); (b) nitrate in extracted borehole water on Chalk, Winchester, Hampshire (modified from Young, Hall and Oakes, 1976)

Water abstracted from boreholes for drinking purposes is expected to continue to show an increase in nitrate concentration, even if fertiliser applications cease (Young, Hall and Oakes, 1976). However, bacterial utilisation of nitrate by, for example, ammonium-oxidising bacteria in the vadose zone, could counter, to some extent, this predicted increase (Whitelaw and Rees, 1980).

It is likely that land-use management may be used to control nitrate concentrations in thin reactive aquifers, but for deep aquifers like the Chalk, these options will not be sufficient to reverse rising nitrate concentrations, except after many decades.

1.5.3.3 Nitrogen in buffer zones

The potential for nitrogen loss through denitrification is high where nitrogen transport from the hillslope to the stream occurs via a throughflow hydrological pathway because this is essentially a saturated-zone process. Nitrogen losses through denitrification do, however, require a suitable carbon substrate to maximise biological activity in the saturated zone. Haycock (1991) examined nitrate losses in the river Leach, a tributary of the Thames in the Cotswolds. He found that winter nitrate concentrations in drainage water passing through a carbon-rich alluvium floodplain were reduced by over 80%. Using a conservative (chloride) tracer, Haycock was able to show that the nitrate loss was not the result of dilution which was restricted to large storm events. The process causing the nitrate loss was assumed to be denitrification. Lowrance, Todd and Asmussen (1984) found that the nitrate concentration decreased in headlands and field margins where lateral flow was important.

It must be noted, however, that while the use of buffer zones as a nitrate-prevention strategy is valuable where the hydrological pathway linking the land to the stream is mainly a throughflow connection, it is of little value in reducing nitrate concentrations in groundwaters. This point was taken up by Haycock (1991) in connection with the movement of nitrate-contaminated groundwater across floodplains. He concluded that the drainage of floodplains for agricultural improvement meant that nitrate-rich water no longer slowly percolated through the floodplain buffer but was quickly passed through this zone in drainage ditches or tile drains. As a result, nitrate-rich water entered the stream or river draining the catchment in a largely unaltered form. Intensive agricultural utilisation of floodplain zones removes a natural and extremely important buffer zone in which the opportunity for nitrogen transformation and loss through denitrification is high.

1.6 INTEGRATED MANAGEMENT: A SOLUTION TO THE NITRATE ISSUE

The UNEP/WHO assessment of water quality programme (UNEP/WHO, 1988) states that radical changes are needed to reverse the trend in nitrate pollution and stop it from becoming a large-scale and potentially global water quality problem (Figures 1.1 and 1.2). Foster *et al.* (1985) suggest that significant losses in productivity may occur as a result of nitrogen control measures and that recovery will be a long-term process. A key problem in the future is likely to be the unsuitability of the current focus on changing agricultural practices for nitrogen control in the developing world where the institutional and financial framework differs.

Management of nitrogen concentrations in receiving waters is very much a question of getting right the scale of investigation and interpretation. Most management decisions need to be made at the catchment scale, which requires an integrated approach and a simplified but accurate interpretation of the key hydrological pathways and soil processes involving nitrogen. It is clear that the long-term increases in nitrate concentrations in many UK rivers and lakes is strongly related to land use and farming practices (Burt and Haycock, in press). A number of studies have attempted to link land-use change with river water quality trends (for example, Smith and Stewart, 1989; Heathwaite and Burt, 1991; Johnes and Burt, 1991). Most of these studies have a historical perspective. In order to be of value for integrated catchment management of nitrogen in receiving waters, such approaches also need to be able to predict the impact on water quality of future changes in land use. This is particularly important where EC Directives are beginning to turn to land-use control as a *preventative* water quality strategy. The use of an simple export coefficient model (see, for example, Johnes, 1990) which enables the construction of stream water quality before the period of observation, and has the ability to predict the impact on water quality of future changes in land use, may be the way forward for water quality management at the catchment scale. These models are discussed further in Chapters 5 and 10 of this book.

Such models are of particular value when attempting to predict the effect of shifts in policy such as those described in Section 1.1 on nitrogen in receiving waters. Johnes (1990) combined the recommendations contained in the NSA scheme with her export coefficient model for the catchment of the river Windrush, a tributary of the Thames. She was able to demonstrate that the changes in land use encompassed in the NSA scheme could produce considerable reductions in the nitrogen load by altering the nitrogen export coefficient of land in 'sensitive' areas (see also Chapters 10 and 12 of this book).

REFERENCES

Addiscott, T.M. (1988) A simple computer model for leaching in structured soils. *Journal of Soil Science*, **28**, 554−63.

Addiscott, T.M. and Wagenet, R.J. (1985) Concepts of solute leaching in soils: a review of modelling approaches. *Journal of Soil Science*, **29**, 305−14.

Agg, A.R. and Derick, M. (1991) *Progress in Implementing the EC Drinking Water Directive in Six Member States*. Water Research Centre Technical Report No. FR 0208, Foundation for Water Research.

Anonymous (1983) Excess nitrates in water — action needed now. *Municipal Journal*, **91**, 789.

Armstrong, A.C. (1984) The hydrology and water quality of a drained clay catchment, Cockle Park, Northumberland. In Burt, T.P. and Walling, D.E. (eds), *Catchment Experiments in Fluvial Geomorphology*, Geobooks, Norwich, 153−68.

Armstrong, A.C. and Garwood, E.A. (1991) Hydrological consequences of artificial drainage of grassland. *Hydrological Processes*, **5**, 157−74.

Beaton, D. (1988) ADAS charts the easier way of knowing N needs. *Farmers Weekly*, 19 February, 35.

Barraclough, D., Hyden, M.J. and Davies, G.P. (1983) Fate of fertiliser nitrogen applied to grassland. I: FIeld leaching results. *Journal of Soil Science*, **34**, 483−98.

Beresford, S.A.A. (1985) Is nitrate in drinking water associated with the risk of cancer in the urban UK? *International Journal of Epidemiology*, **14**, 57−63.

Betton, C., Webb, B.W. and Walling, D.E. (1991) Recent trends in NO_3-N concentration and loads in British rivers. *IAHS publication 203*, 169−80.

Burns, I.G. (1980) A simple model for predicting the effects of leaching of fertiliser nitrate during the growing season on the nitrogen fertiliser need of spring crops. *Journal of Soil Science*, **31**, 187−202.

Burt, T.P. and Arkell, B.P. (1987) Temporal and spatial patterns of nitrate losses from an agricultural catchment. *Soil Use and Management*, **3**, 138−43.

Burt, T.P. and Haycock, N.E. (in press) Catchment planning and the nitrate issue: a UK perspective. *Progress in Physical Geography*.

Burt, T.P., Arkell, B.P., Trudgill, S.T. and Walling, D.E. (1988) Stream nitrate levels in a small catchment in south west England over a period of 15 years, 1970−1985. *Hydrological Processes*, **2**, 267−84.

Burt, T.P., Butcher, D.P., Coles, N. and Thomas, A.D. (1983) The natural history of Slapton Ley Nature Reserve XV: Hydrological processes in the Slapton Wood catchment. *Field Studies*, **4**, 731−52.

Collingwood, R.W. (1977) *A Survey of Eutrophication in Britain and its Effects on Water Supplies*, Water Research Centre Technical Report TR40.

Cooke, G.W. (1982) *Fertilising for Maximum Yield*, Granada, St Albans.

Department of Environment (1986) Nitrate in water. *Department of Environment Pollution Paper 26*, HMSO, London.

Department of Environment (1988) *The Nitrate Issue*, HMSO, London.

Dermine, B. and Lamberts, L. (1987) Nitrate nitrogen in the Belgian course of the Meuse River — fate of the concentrations and origins of the inputs. *Journal of Hydrology*, **93**, 91–9.

Doll, R. and Peto, R. (1981) *The Causes of Cancer*, Oxford Medical Publications, Oxford.

Dowdell, R.J. and Webster, C.P. (1980) A lysimeter study using nitrogen-15 on the uptake of fertiliser nitrogen by perennial ryegrass swards and losses by leaching. *Journal of Soil Science*, **31**, 65–75.

Dowdell, R.J., Crees, R. and Cannell, R.Q. (1983) A field study of effects of contrasting methods of cultivation on soil nitrate content during autumn, winter and spring. *Journal of Soil Science*, **34**, 367–80.

Dudley, N. (1986) *Nitrates in Food and Water*, Earth Resources Research, London.

Forman, D., Al-Dabbagh, S. and Doll. R. (1985) Nitrates, nitrites and gastric cancer in Great Britain. *Nature*, **313**, 620–25.

Foster, S.S.D. (1975) The chalk groundwater tritium anomaly — a possible explanation. *J. Hydrology*, **35**, 159–65.

Foster, S.S.D., Cripps, A.C. and Smith-Carington, A. (1982) Nitrate leaching to groundwater. *Phil. Trans. Roy. Soc. Lond.*, **296**, 477–89.

Foster, S.S.D., Geake, A.K., Lawrence, A.R. and Parker, J.N. (1985) Diffuse groundwater pollution: lessons of the British experience. *Memoirs of the 18th Congress of the International Association of Hydrogeologists*, Cambridge, 168–77.

Frissel, M.J. and Van Veen, J.A. (1982) A review of models for investigating the behaviour of nitrogen in soils. *Phil. Trans. Roy. Soc. Lond.*, **B296**, 341–9.

Garwood, E.A., Salette, J. and Lemaire, G. (1980) The influence of water supply to grass on the response to fertiliser nitrogen and nitrogen recovery. In Prins, W.H. and Arnold, G.H. (eds), *Proc. Int. Symp. Europe Grassland Federation on role of nitrogen in intensive grassland production*, Pudoc, Wageningen, 59–65.

Geake, A.K. and Foster, S.S.D. (1989) Sequential isotope and solute profiling in the unsaturated zone of British Chalk. *Hydrological Sciences Journal*, **34**, 79–95.

Goulding, K.W.T. (1990) Nitrogen deposition to land from the atmosphere. *Soil Use and Management*, **6**, 61–3.

Greene, L.A. (1978) Nitrates in water supply abstractions in the Anglian Region: current trends and remedies under investigation. *Water Pollution Control*, **77**, 478–91.

Haycock, N.E. (1991) *Riparian land as buffer zones in agricultural catchments*, Unpublished DPhil thesis, Oxford University.

Heathwaite, A.L. and Burt, T.P. (1991) Predicting the effect of land use on stream water quality. *IAHS publication* **203**, 209–18.

Heathwaite, A.L., Burt, T.P. and Trudgill, S.T. (1989) Runoff, sediment and solute delivery in agricultural drainage basins — a scale dependent approach. *IAHS publication* **182**, 175–91.

Heathwaite, A.L., Burt, T.P. and Trudgill, S.T. (1990) The effect of land use on nitrogen, phosphorus and suspended sediment delivery in lowland agricultural catchments. In Thornes, J.B. (ed.), *Vegetation and Erosion*, Wiley, Chichester, 161–79.

Henderson-Sellars, B. and Markland, H. (1987) *Decaying Lakes: origins and control of eutrophication*, Wiley, Chichester.

Hill, M.J., Hawksworth, G. and Tattersall, G. (1973) Bacteria, nitrosamines and cancer of the stomach. *British Journal of Cancer*, **28**, 562–7.

Huang, S.-L. and Ferng, J.-J. (1990a) Applied land classification for surface water quality management: I. Watershed classification. *Journal of Environmental Management*, **31**, 107–26.

Huang, S.-L. and Ferng, J.-J. (1990b) Applied land classification for surface water quality management: II. Land processes classification. *Journal of Environmental Management*, **31**, 127–41.

Jacobs, T.C. and Gilliam, J.W. (1985) Riparian losses of nitrate from agricultural drainage waters. *Journal of Environmental Quality*, **14**, 472–8.

Jenkinson, D.S. (1982) An introduction to the global nitrogen cycle. *Soil Use and Management*, **6**, 56−61.

Johnes, P.J. (1990) *An investigation of the effects of land use upon water quality in the Windrush catchment*, Unpublished DPhil thesis, Oxford University.

Johnes, P.J. and Burt, T.P. (1991) Water quality trends in the Windrush catchment: nitrogen speciation and sediment interactions. *IAHS publication* **203**, 349−57.

Johnson, D., Farley, M.R., Youngman, R.E., Yadav, N.P. and West, J.T. (1977) *Removal of algae by various unit processes*, Water Research Centre Technical Report TR45.

Jollans, J.L. (1983) *Fertiliser in UK Farming*, Centre for Agricultural Strategy, Reading, UK.

Kaluarachchi, J.J. and Parker, J.C. (1988) Finite element model of nitrogen species transformation and transport in the unsaturated zone. *J. Hydrology*, **103**, 249−74.

Kornberg, H. (1979) *Pollution and Agriculture*, Royal Commission on Environmental Pollution, 7th Report, HMSO, London.

Lack, T.J. and Johnson, D. (1983) *The effects of storage on reservoir water quality. Observations at Farmoor Reservoir, 1968−1978*, Water Research Centre Environment 547-M/1.

Lawrence, A.R. and Kuruppuarchchi (1986) *Impact of agriculture on groundwater quality in Kalpitiya, Sri Lanka*, British Geological Survey, Open File Report WD/OS/86/20.

Lean, G. (1990) Ministers weaken nitrate pollution control. *Observer*, 21 January.

Lowrance, R.R., Todd, R.L. and Asmussen, L.E. (1984) Nitrogen cycling in an agricultural watershed. *Journal of Environmental Quality*, **13**, 22−32.

Magee, P.N. (1982) Nitrogen as a potential health hazard. *Phil. Trans. Roy. Soc. Lond.*, **B296**, 543−50.

McEwen, J., Darby, R.J., Hewitt, M.V. and Yeoman, D.P. (1989) Effects of field beans, fallow, lupins, oats, oilseed rape, peas, ryegrass, sunflowers and wheat on nitrogen residues in the soil on the growth of a subsequent wheat crop. *Journal of Agricultural Science*, **115**, 209−19.

McGill, W.B. and Myers, R.K.J. (1987) Controls of dynamics of soil and fertiliser nitrogen. In Follet, R.F. *et al.* (eds), *Soil Fertility and Organic Matter as Critical Components of Production Systems*, *Soil Sci. Soc. Am. and Am. Soc.* Agron., Madison, WI.

Meybeck, M. (1982) Carbon, nitrogen and phosphorus transport by world rivers. *Amer. J. Sci.*, **282**, 401−50.

Meybeck, M., Chapman, D. and Helman, P. (1989) *Global Freshwater Quality: A first assessment*, Global Environmental Monitoring System/UNEP/WHO.

Middleton (1990) Nitrate in water. *Science in Parliament*, **47(1)**, February, 51−2.

Ministry of Agriculture, Fisheries and Food and Welsh Office Agricultural Department (1991) *Code of Good Agricultural Practice for the Protection of Water*, HMSO, London.

Newson, M.D. (1991) Catchment control and planning: emerging patterns of definition, policy and legislation in UK water management. *Land Use Policy*, **8**, 9−15.

Nychas, A. (1990) An EEC perspective on fertiliser use. *Chemistry and Industry*, 17 December, 828−31.

OPCS (1981) *Cancer Statistics; studies on medical and population subjects No. 43*, HMSO, London.

Parker, J.M., Young, C.P. and Chilton, P.J. (1990) Rural and agricultural pollution. In *A Survey of British Hydrogeology*. Royal Society, London.

Pocock, S.J. (1985) Nitrates and gastric cancer. *Human Toxicology*, **4**, 471−4.

Powlson, D.S., Pruden, G., Johnson, A.E. and Jenkinson, D.S. (1986) Recovery of ^{15}N labelled fertiliser applied in autumn to winter wheat at four sites in eastern England. *Journal of Agricultural Science*, **107**, 611−20.

Pretty, J.N. and Conway, G.R. (1988) *Cancer Risk and Nitrogen Fertilisers: evidence from developing countries*, Gatekeeper Series, **SA4**, International Institute for Environment and Development.

Robbins, J. (1990) Poison algae plague spreads nationwide. *New Civil Engineer*, 31 May, 5.

Roberts, G. and Marsh, T. (1987) The effects of agricultural practices on the nitrate concentrations in the surface water domestic supply sources of Western Europe. *IAHS publication* **164**, 365−80.

Rodda, J.C. and Johnes, G.N. (1983) Preliminary estimates of loads carried by rivers to estuaries and coastal waters around Great Britain derived from the harmonised monitoring scheme. *J. Instn, Wat. Engrs. Scient.*, **37**, 529−39.

Rose, C.W., Chichester, F.W., Williams, J.R. and Ritchie, J.T. (1982) A contribution to simplified models of field solute transport. *Journal of Environmental Quality*, **11**, 146−55.

Ryden, J.C., Ball, P.R. and Garwood, E.A. (1984) Nitrate leaching from grassland. *Nature*, **311**, 50−4.

Saull, M. (1990) Nitrates in Soil and Water. *New Scientist*, 15 September.

Selim, H.M. and Iskandar, I.K. (1981a) Modelling nitrogen transport and transformation in soils: theoretical considerations. *Soil Science,* **131**, 233−41.

Selim, H.M. and Iskandar, I.K. (1981a) Modelling nitrogen transport and transformation in soils: Validation. *Soil Science,* **131**, 303−12.

Severn-Trent Water (1988) *The Hatton Catchment Study*, Birmingham, UK.

Shuval, H. and Gruener, N. (1976) Infant methaemoglobinaemia and other health effects of nitrates in drinking water. *Proc. IAWPR Conf. Nitrogen as a Water Pollutant*, Copenhagen.

Skjemstad, J.O., Vallis, I. and Myers, R.K.J. (1988) Decomposition of soil organic nitrogen. In Wilson, J.R. (ed.), *Advances in Nitrogen Cycling in Agricultural Ecosystems*, CAB International, Wallingford, 134−144.

Smettem, K.R.J., Trudgill, S.T. and Pickles, A.M. (1983) Nitrate loss in soil drainage waters in relation to by-passing flow and discharge on an arable site. *Journal of Soil Science,* **34**, 499−509.

Smith, R.M. and Stewart, D.A. (1989) A regression model for nitrate leaching in Northern Ireland. *Soil Use and Management,* **5**, 71−6.

Stewart, W.D.P. and Rosswall, T. (1982) The nitrogen cycle. *Phil. Trans. Roy. Soc. Lond.,* **B296**, 299−576.

UNEP/WHO (1988) *Assessment of Freshwater Quality*, London.

Walling, D.E. and Webb, B.W. (1985) Local variation in nitrate levels in the Exe basin, Devon, England. *Beitrage zur Hydrologie,* **10**, 71−100.

Webb, B.W. and Walling, D.E. (1985) Nitrate behaviour in streamflow from a grassland catchment in Devon, UK. *Water Research,* **19**, 1005−16.

White, R.E., Wellings, S.R. and Bell, J.P. (1983) Seasonal variations in nitrate leaching in structured clay soils under mixed land use. *Agricultural Water Management,* **7**, 391−410.

White, R.J. (1983) Nitrate in British Waters. *Aqua,* **2**, 51−7.

Whitelaw, K. and Rees, J.F. (1980) Nitrate-reducing and ammonium-oxidising bacteria in the vadose zone of the Chalk aquifer of England. *Geomicrobiology Journal,* **2**, 179−87.

WHO/UNEP (1987) *Global Pollution and Health. Results of health related environmental monitoring*, World Health Organisation, Geneva and United Nations Environment Programme, Nairobi, Yale Press, London.

Wild, A. (1977) Nitrate in drinking water: health hazard unlikely. *Nature,* **268**, 197−8.

Wild, A. and Cameron, K.C. (1980) Soil nitrogen and nitrate leaching. In Tinker, P.B. (ed.), *Soils and Agriculture*, Blackwell, Oxford.

Wilkinson, W.B. and Greene, L.A. (1982) The water industry and the nitrogen cycle. *Phil. Trans. Roy. Soc. Lond.,* **B296**, 459−75.

Winnerberger, J.H.T. (1982) *Nitrogen, Public Health and the Environment: Some Tools for Critical Thought*, Ann Arbor Science Publications, Ann Arbor, MI.

Wyndle-Taylor, E. (1984) Nitrates in water supplies; Report to the International Standing Committee on Water Quality and Treatment. *Aqua,* **1**, 5−25.

Young, C.P. and Hall, E.S. (1977) Investigations into factors affecting the nitrate content of groundwater. Paper 14 in *Groundwater Quality, Measurement, Prediction and Protection*, Water Research Centre.

Young, C.P., Hall, E.S. and Oakes, D.B. (1976) *Nitrate in groundwater — studies on the Chalk near Winchester, Hampshire*, Water Research Centre Technical Report TR31.

2 Nitrogen and Nitrogen Compounds in the Atmosphere

R.P. WAYNE

Physical Chemistry Laboratory, University of Oxford

2.1 INTRODUCTION

The total mass of the atmosphere is about 5×10^{18}kg, which is rather less than one-millionth of the mass of the solid Earth, and roughly 300 times smaller than the mass of the oceans (Walker, 1977; Wayne, 1991). However, the atmosphere that surrounds the planet appears to have been essential to the evolution of life in the past and to its existence today. Half the mass of the atmosphere lies below an altitude of 5.5 km and 99% below 30 km, although the gaseous envelope extends, albeit tenuously, thousands of kilometres above the surface. Of the atmosphere below about 100 km nearly four-fifths by volume (and molecule for molecule) is molecular nitrogen: the *mixing ratio* is 0.78. Nitrogen is chemically rather inert in the atmosphere, mainly because of the high energy that binds two nitrogen atoms together (945 kJ mol^{-1}). However, because of the large amounts present, this gas and its compounds play a major part in atmospheric chemistry.

2.1.1 TROPOSPHERE AND STRATOSPHERE

A formal division of the atmosphere into regions is based on temperature structure. Temperatures decrease with increasing altitude through about the first $10-17$ km of the atmosphere, but then increase again through the next $30-40$ km, which is heated indirectly by the absorption of solar ultra-violet radiation in the atmospheric 'ozone layer'. The decreasing temperature of the first region leads to rather strong vertical mixing, while the 'temperature inversion' of the second results in remarkable vertical stability. Through the Greek words for 'turning' and 'layered' we obtain the terms for the atmospheric regions of *troposphere* and *stratosphere*. The troposphere is virtually in direct contact with the biosphere (a highly turbulent *boundary layer* of $0.5-2$ km thickness is often envisaged as separating the free troposphere from the Earth's surface), so that it mediates the transfer of gases and particulate material into and out of the biosphere. It is also the region of meteorological phenomena such as cloud and droplet formation, which are sources of dissolved species (including nitrate, nitrite and ammonium ions) that are precipitated to the biosphere. Most of the ozone in the atmosphere is present in the stratosphere, and life can exist on the surface of the planet because of the shield afforded by that ozone. Nitrogen compounds, both natural and anthropogenic, play a very important role in determining the detailed balance of production and loss of ozone in the stratosphere.

Nitrate: Processes, Patterns and Management. Edited by T.P. Burt, A.L. Heathwaite and S.T. Trudgill
© 1993 John Wiley & Sons Ltd

2.1.2 NITROGEN COMPOUNDS AND THEIR RELEVANCE

Among the most important compounds are the oxides NO and NO_2 (nitric oxide and nitrogen dioxide), which are precursors of the acids HNO_2 and HNO_3 (nitrous and nitric acids), N_2O (nitrous oxide), NH_3 (ammonia), and an organic nitrate $CH_3CO.O_2NO_2$ (peroxyacetyl nitrate, or PAN for short), and the free radical NO_3 (the nitrate radical). We shall explore the contributions to atmospheric chemistry of these compounds in more detail later, but it is appropriate to outline now some salient features.

NO and NO_2 are implicated in a central part of tropospheric chemistry, since they are involved in tropospheric ozone formation. In the natural atmosphere, that ozone is itself a precursor of hydroxyl radicals that initiate attack on many other compounds. Thus, the oxidising power of the troposphere and its capacity for processing chemically compounds released to it ultimately depend in part on the oxides of nitrogen (although some ozone is also transported down from the stratosphere). Human activity may result in the release of large quantities of the oxides of nitrogen, and in the presence of hydrocarbons, excess ozone may be produced photochemically: these processes lie behind pollution by photochemical smog. PAN has long been known as a component of such pollution, but the compound is now recognised as important in its own right, perhaps most as a carrier of reactive nitrogen compounds over long distances from where they are first released naturally or by human activity. Reactions of NO and NO_2 also lead to formation of the acids HNO_2 and HNO_3, so that high concentrations of the oxides may lead to elevated acidification. The nitrogen acids are components of *acid rain* or *acid deposition* that is recognised as another damaging kind of pollution.

In the stratosphere, NO and NO_2 have a role to play in the *destruction* of ozone. Some of the NO and NO_2 is produced in higher regions of the atmosphere by ionic processes and transported down, but the bulk is derived from N_2O, which is rather inert in the troposphere, and survives to reach the stratosphere.

Ammonia is, after N_2 and N_2O, the third most abundant nitrogen-containing species in the troposphere, although its concentration is rather variable. It is the only basic constituent of the atmosphere. Because of its high solubility in water, NH_3 is able to neutralise the tropospheric acids (which are present even in unpolluted environments) and it can control the pH of cloud droplets and thus the acidity of rain and snow. The loss of atmospheric NH_3 through various wet and dry deposition processes is a source of NH_4^+ (ammonium) ions to the Earth's surface.

Both N_2O and NH_3 play a significant part in the physics of the Earth's atmosphere. They are both absorbers of infra-red radiation and they are present at sufficient concentrations in the atmosphere to contribute to so-called 'greenhouse' heating (Rohde, 1990). Their presence thus has an influence on surface temperatures, and changing concentrations might affect climate (Schneider, 1989).

The nitrate radical, NO_3, has recently emerged as a key atmospheric intermediate, especially in the troposphere (Wayne *et al.*, 1991). It is the product of the interaction between NO_2 and ozone (O_3). Although it is destroyed photolytically during the day, it persists at night, when it may represent the most important oxidising species present in the troposphere. Various reactions into which it enters can also make their contribution to acidification via formation of HNO_3.

2.1.3 NO_x AND NO_y

The discussion of the previous section will have made it evident that oxidised nitrogen compounds, especially NO and NO_2, are major participants in atmospheric chemistry. It has become common practice (Wayne, 1991) to use special terms in referring to the compounds. The representation 'NO_x' refers to NO and NO_2, and if a concentration of NO_x is specified it means the sum of the concentrations of these two oxides. When 'NO_y' is written it means 'odd nitrogen': that is, the sum of NO_x and all oxidised nitrogen species that are sources or sinks of NO_x through processes that occur on relatively short timescales. Thus, besides NO and NO_2, NO_y includes N_2O_5, HNO_3, HNO_4, $ClONO_2$ and PAN. Although N_2O is a source of NO_x in the upper atmosphere, it is *not* normally considered a component of the NO_y compounds.

2.1.4 CYCLIC PROCESSES

Production and loss of atmospheric constituents have to be balanced if concentrations are not to vary. On short timescales, a steady state obviously holds approximately for a component such as molecular nitrogen. The lithosphere−hydrosphere−atmosphere system is essentially closed, though in reality some *juvenile* material is released from the Earth's interior, and the solar wind and debris left by stray bodies entering the atmosphere contribute to an inward flux. Escape from the Earth's gravitational field to the interplanetary medium constitutes an outward flux. However, within the approximation of a closed system in which material substance neither enters nor leaves, the total quantity of each element is fixed, although the distribution between the elemental and combined form can alter. This conservation of elemental quantity means that if a species appears in the atmosphere (at a rate equalled by its disappearance) then the elements involved must be passing through a series of *cyclic* chemical and/or physical transformations. One important aspect of these cycles concerns the *budgets* for different species (i.e. the distribution between different reservoirs, and the rates of physical and chemical interconversions). Among other things, an understanding of the budgets permits assessment of the relative contributions of anthropogenic and natural sources of various species and thus of the likelihood of local or even global pollution by humans. We next consider the sources and sinks of nitrogen and its compounds before trying to establish approximate budgets in the natural nitrogen cycle.

2.1.5 SOURCES AND SINKS OF ATMOSPHERIC NITROGEN SPECIES

Natural processes can *fix* nitrogen (bring it into combination) (Levine *et al.*, 1984; Franzblau and Popp, 1989). For example, lightning within the atmosphere can produce oxides of nitrogen (perhaps in amounts up to 10^{11} kg yr^{-1} of NO and NO_2) which are converted to acids (HNO_2 and HNO_3) and then precipitated dissolved in water to the planet's surface. Biological fixation is of even greater importance, at least over continental areas. Most of the nitrogen present in the atmosphere probably resulted from gases released in volcanic activity during the early history of the Earth (Wayne, 1991; Walker, 1977) although the gas is also formed biogenically through denitrification, in which NO_3^- ions are reduced back to N_2. It is the microbiological chains that maintain the enormous disequilibrium between atmospheric N_2

and O_2 concentrations and those of NO_3^- ions in sea water. The greater solubility of common nitrate minerals compared with common carbonates partly explains the dominance of N_2 over CO_2 in the Earth's atmosphere (in contrast to the situation for Venus, which is almost devoid of water, or Mars, where there is no free liquid water on the surface).

Independent micro-organisms can fix nitrogen into soils, but the greatest contribution to fixation is made by symbiotic organisms found in the root nodules of the pulses or leguminous plants. Assimilation of nitrogen is catalysed by the enzyme complex *nitrogenase*. Reduction of N_2 to NH_4^+ ions is driven in this process by the energy-rich phosphate adenosine triphosphate (ATP). In the symbiotic micro-organisms, the ATP derives from the host plant, which receives up to 90% of the fixed nitrogen in return. Other links in the soil microbiological chain involve *nitrifying* bacteria that oxidise NH_4^+ to NO_3^-, the *denitrifying* bacteria that reduce NO_3^- to N_2 and *ammonifying* bacteria that produce NH_4^+ from the wastes and remains of animals and dead plants.

Nitrous oxide (N_2O) is liberated from soils as a result of incomplete microbiological nitrification or denitrification: the biological source is probably dominated by nitrification (Anderson and Levine, 1986).

2.1.6 HUMAN CONTRIBUTION TO ATMOSPHERIC NITROGEN COMPOUNDS

Nitrous oxide is chemically inert in the troposphere, and the tropospheric residence time is about 150 years. Thus the concentration of N_2O would be expected to change to a measurable extent if release rates alter. Global mixing ratios seem, in fact, to have increased from 0.292 parts per million (ppm) in 1961 to 0.307 ppm by 1987 (Khalil and Rasmussen, 1988; Graedel and Crutzen, 1989). There seem to be several possible causes for this secular increase, most of which are related to human activity. Application of nitrogen-containing fertiliser to agricultural land leads to increased rates of nitrification (Conrad *et al.*, 1983). At low rates of application, the fractional yield of N_2O appears to increase linearly with fertiliser loading but to increase more rapidly at higher loadings. In both aquatic and soil systems, enhanced N_2O production is triggered by oxygen depletion. It seems clear, then, that fertilisation of soils with ammonium salts or organic nitrogen and the disposal of human and animal wastes strongly stimulate nitrification, and are likely to lead to a globally significant release of N_2O to the atmosphere. A recent analysis of the sources and sinks of atmospheric N_2O has been presented by Cicerone (1989).

Artificial fertilisers are, of course, often the products of fixation of atmospheric nitrogen in industrial rather than natural processes. Nitrous oxide is, indeed, also more directly a product of high-temperature combustion of fossil fuels (Linak *et al.*, 1990). However, such combustion probably makes even more impact on the atmosphere through the formation of the more highly oxidised NO_x compounds. Two damaging forms of atmospheric pollution that have already been mentioned (acid precipitation and photochemical smog formation) can be attributed largely to the burning of fossil fuels in industrial plants and internal combustion engines. While one component of acid rain involves sulphur compounds that can, at least in principle, be removed from the fuel, the NO_x species are an inevitable consequence of exposing air to high temperatures.

Biomass burning is another important way in which nitrogen compounds are released to the atmosphere (Anderson *et al.*, 1988; Lobert *et al.*, 1990; Hegg *et al.*, 1990). Both NO and N_2O are known to be formed. Part of the effect results from fixation in the high temperatures of the fires, but there seem also to be enhanced biogenic emissions of NO and N_2O from the burned land for extended periods after the fire.

2.1.7 THE NITROGEN CYCLE

Having considered the sources and sinks of nitrogen and its compounds, it is now appropriate to provide a semi-quantitative view of the nitrogen cycle. Holland (1978) showed how a suitable diagram could be constructed, and a modified version is presented in Figure 2.1.

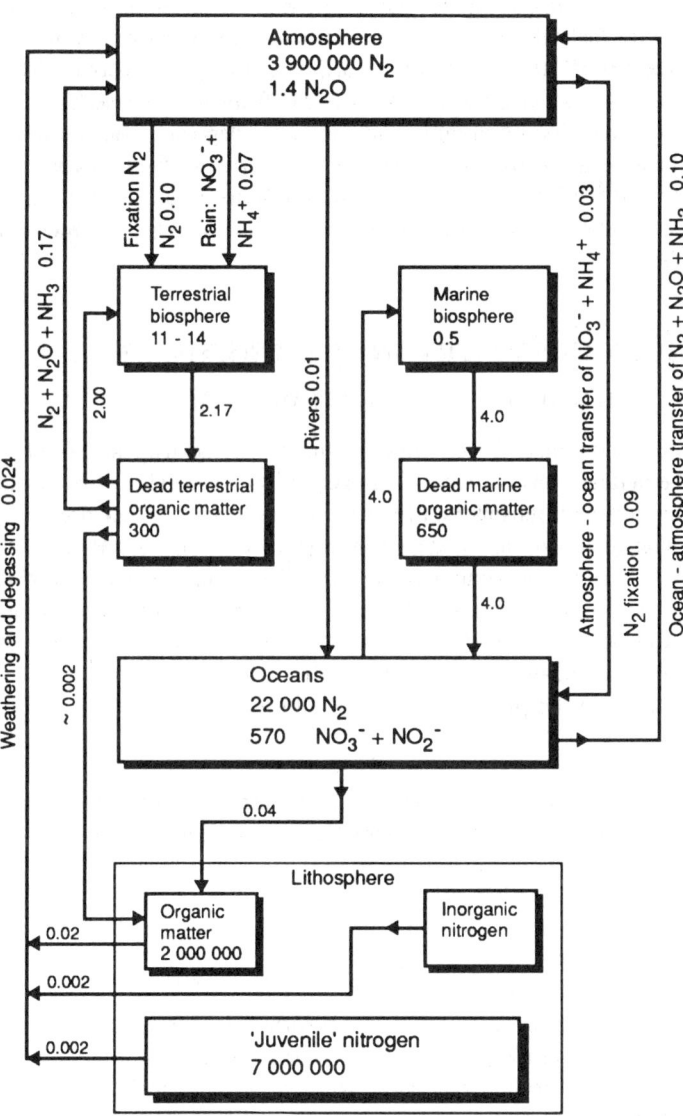

Figure 2.1 The nitrogen cycle. Arrows show the transfer of nitrogen between the different reservoirs given in the boxes. It is evident that several closed loops make up the cycle. Reservoir contents are given in units of 10^{12} kg of N (except where specified as N_2 or N_2O), and transfer rates are given in units of 10^{12} kg of N per year (reproduced from Wayne, 1991, where reference to the original data may be found, by permission of Oxford University Press)

Estimates of global budgets and reservoirs are rather imprecise for nitrogen species, although the Scientific Committee on Problems of the Environment (SCOPE) has addressed the question of the nitrogen cycle from time to time (Rosswall, 1983). Despite the lack of quantitative data, the major aspects of the nitrogen cycle are fairly clear. In both terrestrial and marine organic systems most of the nitrogen is recycled within the system, although a smaller fraction is transferred between biosphere and atmosphere. For the hydrosphere, about 0.13×10^{12} kg a^{-1} are transferred from the atmosphere as against 4.0×10^{12} kg a^{-1} cycling in the marine biosphere. Without this recycling, the lifetime of atmospheric N_2 would be about 6×10^7 a^{-1}: with the values given in the figure for atmospheric transfer, the calculated lifetime is nearer 1.6×10^6 a^{-1}. Again according to the figure, about 0.04×10^{12} kg a^{-1} of N_2 are buried as dead organic matter, mostly from the marine biosphere. Exposure, weathering and conversion of organic and inorganic deposits, and the release of juvenile nitrogen, can balance the burial rate. However, even without the restoring sources, it would take $(2 \times 3.9 \times 10^6/0.04) \simeq 2 \times 10^8$ a^{-1} to consume all atmospheric nitrogen. That is, the imbalance is small even on geological timescales because of the enormous nitrogen content of the atmosphere.

2.2 NITROGEN CHEMISTRY IN THE TROPOSPHERE

2.2.1 OXIDATION AND TRANSFORMATION IN THE TROPOSPHERE

Hydroxyl radicals dominate the daytime chemistry of the troposphere, because the high reactivity of the radical leads to oxidation and chemical conversion of most trace constituents that have an appreciable tropospheric lifetime. Free radical chain reactions oxidise hydrogen, methane and other hydrocarbons, and carbon monoxide to CO_2 and H_2O. The processes thus constitute a low-temperature combustion system. The oxides of nitrogen play a key role in the processes. They both lead to the initiation of the radical chain reactions and are involved in the propagation steps. The processes involved will now be outlined for the oxidation of methane, the most abundant of the natural hydrocarbons.

Nitrogen dioxide can be photolysed at $\lambda < 400$ nm

$$NO_2 + h\nu \rightarrow O + NO \tag{2.1}$$

and is thus photochemically active at the wavelengths that reach the troposphere. The oxygen atoms formed can then add to O_2 to form O_3, and a sequence of chemical steps yields hydroxyl radicals

$$O + O_2 + M \rightarrow O_3 + M \tag{2.2}$$

$$O_3 + h\nu \rightarrow O^* + O_2 \tag{2.3}$$

$$O^* + H_2O \rightarrow OH + OH \tag{2.4}$$

The essential feature of this scheme is that ozone, formed in step (2.2), may itself be photolysed at $\lambda < 310$ nm to yield an *electronically excited* oxygen atom represented here by O*. The excited atom, unlike the ground state atom formed in reaction (2.1), has enough energy to participate in step (2.4). Ozone photolysis is the only tropospheric source of the energy-rich atoms, so that, in the hypothetical situation where no hydrocarbon oxidation has yet occurred, hydroxyl radical production is entirely dependent on the presence of ozone. Some ozone is transported to the troposphere from the stratosphere (see Section 2.3), but *in-situ* production through NO_2 photolysis represents a major contribution to the total supply of ozone.

Hydroxyl radicals formed in process (2.4) react mainly with CO and CH_4

$$OH + CO \rightarrow H + CO_2 \qquad (2.5)$$

$$OH + CH_4 \rightarrow CH_3 + H_2O \qquad (2.6)$$

In the unpolluted atmosphere, roughly 70% of the OH reacts with CO and 30% with CH_4. The H and CH_3 fragments add to molecular oxygen

$$H + O_2 + M \rightarrow HO_2 + M \qquad (2.7)$$

$$CH_3 + O_2 + M \rightarrow CH_3O_2 + M \qquad (2.8)$$

to form the peroxy radicals HO_2 and CH_3O_2. Where NO concentrations are very low in the troposphere the peroxy radicals are consumed mainly in reactions with each other, to yield the peroxides H_2O_2 and CH_3OOH. However, if NO_x is present in the atmosphere then a quite different sequence of events follows. Both HO_2 and CH_3O_2 enter into a very fast reaction with NO

$$HO_2 + NO \rightarrow OH + NO_2 \qquad (2.9)$$

$$CH_3O_2 + NO \rightarrow CH_3O + NO_2 \qquad (2.10)$$

to produce NO_2. These reactions thus have one important effect in providing a rapid route for conversion of NO to NO_2. But, in addition, they regenerate chain-propagating free radicals. Reaction (2.9) produces the OH radical directly. The CH_3O (methoxy) radical of reaction (2.10) reacts with oxygen to generate HO_2 and formaldehyde (HCHO) that is itself photolysed in the presence of oxygen to yield two more HO_2 radicals. Three HO_2 radicals can therefore be the indirect product of reaction (2.10), and they, in turn, can yield OH through participation in reaction (2.9). Thus NO has been converted to NO_2 while preserving the active radicals capable of oxidising further molecules of CO or CH_4, or of converting more NO to NO_2.

2.2.2 OZONE FORMATION AND LOSS

As explained earlier, photolysis of NO_2 is the only known way of producing ozone in the troposphere, so that reactions (2.9) and (2.10) provide the link required for catalytic generation of O_3 through reactions (2.1) and (2.2). Ozone is destroyed by reaction with HO_2, so that this loss process and reaction (2.9) are in competition. Below a certain critical value of the ratio $[NO]/[O_3]$, ozone loss dominates over its production. Current models suggest that there is net production of O_3 for mixing ratios of $NO \geq 3 \times 10^{-11}$. Regions of the Earth characterised by extremely low concentrations of NO, such as the remote Pacific, are thus likely to constitute a net photochemical sink for odd oxygen (O and O_3), while the continental boundary layer at mid-latitudes, where concentrations of NO are relatively high, is likely to provide a net source. So long as NO_x is available, production of ozone in the troposphere is ultimately limited by the supply of CO, CH_4 and other hydrocarbons. The tropospheric abundance of ozone is obviously critically important in determining the oxidising capacity of the troposphere, because O_3 is the primary source of OH radicals as well as being an oxidising species in its own right. Future trends of oxidising capacity will thus be tied to the future atmospheric burdens of NO_x and CO, CH_4 and other hydrocarbons (Graedel and Crutzen, 1989).

2.2.3 THE ESSENTIAL ROLE OF NITROGEN OXIDES

The importance of NO_x in determining oxidation rates and ozone distributions in the troposphere has been described in the preceding sections. Figure 2.2 emphasises the influence of NO_x in a different way. The left-hand semi-circle shows the oxidation steps that do not require the presence of NO. In this case, the sequence of transformations following the heavy arrows leads from O_3 to CH_3OOH. If, however, NO is present, then the transformations of the right-hand semi-circle close the loop, with regeneration of OH and the concomitant production of two molecules of NO_2 (and thus, potentially, of O_3).

Model estimates of tropospheric ozone production obviously depend critically on the altitude−concentration profiles adopted for NO_x, which in turn are related to the atmospheric sources and sinks of the NO_x species. As explained in Section 2.1.5, natural sources include microbial actions in the soil and lightning. Oxidation of biogenic NH_3, initiated by OH radicals, would be another significant source of NO_x, while human activity, such as biomass burning and high-temperature combustion, are perhaps responsible for 50% of the present-day global budget. All these sources, with the exception of lightning, release NO_x to the boundary layer. The major loss of NO_x involves nitric acid formation in an interaction with OH radicals

$$OH + NO_2 + M \rightarrow HNO_3 + M \qquad (2.11)$$

followed by wet deposition, so that the NO_x release to the free troposphere may be appreciably smaller than the surface source strengths suggest. Injection of NO_x from the stratosphere could then provide a more important source for the upper troposphere than the surface emissions. Mixing ratios of NO in the *lower* troposphere over oceans, where there is no surface source, are very low, which would suggest that the stratospherically derived NO_x is alone important, and that ozone production would be limited to the upper troposphere in such geographical regions.

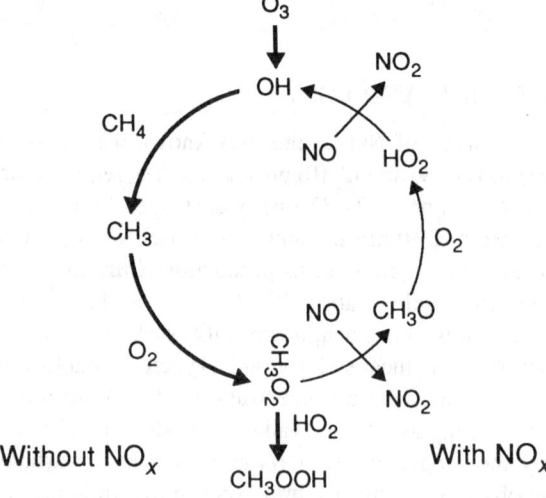

Figure 2.2 The influence of NO_x on the course of tropospheric methane oxidation according to a concept of Dr M.E. Jenkin. The heavier arrows on the left-hand side of the diagram indicate the steps that can occur in the absence of NO_x. With NO_x present, the processes on the right-hand side can close a loop, with regeneration of OH and oxidation of NO to NO_2 (reproduced from Wayne, 1991, by permission of Oxford University Press)

2.2.4 PEROXYACETYL NITRATE (PAN)

Measurements in regions remote from anthropogenic sources have generally been thought to be representative of NO_x in the natural atmosphere, because the chemical lifetime of NO_x is small compared with the times taken for geographical redistribution. The situation is rather more involved than appears at first sight, because a reservoir species for NO_x is now known that can extend the lifetime of NO_x and provide a source when others are absent (Singh, Salas and Viezee, 1986; Roberts, 1990). This reservoir species is peroxyacetyl nitrate (PAN, $CH_3CO.O_2NO_2$), and is an adduct between the peroxyacetyl radical ($CH_3CO.O_2$) and NO_2. Details of the formation of peroxyacetyl radicals do not concern us here, except to note that the radicals are analogues of the peroxymethyl radical, CH_3O_2, that we have discussed earlier, and that they are formed by the attack of OH on aldehydes that are themselves intermediates in the oxidation of hydrocarbons.

PAN has long been recognised as a component of photochemical air pollution, but more recently it was detected in the 'clean' air of remote oceanic regions (Anderson et al., 1988). Concentrations of PAN comparable with (or greater than) the concentrations of NO_2 were observed over the Pacific Ocean. PAN is in thermal equilibrium with its precursors

$$CH_3CO.O_2 + NO_2 \rightleftharpoons CH_3CO.O_2NO_2 \qquad (2.12)$$

and the equilibrium is shifted to the product side at lower temperatures. Above the boundary layer, temperatures are sufficiently low for PAN to be relatively stable, but the molecule is unstable close to the surface. PAN will thus release NO_2 as it is transferred from cooler to warmer regions. Since PAN is a compound expected to be present when hydrocarbons are oxidised in the presence of NO_2, it is now apparent that the species must be included with NO, NO_2 and HNO_3 when considering budgets and transport of NO_x and NO_y in the troposphere.

2.2.5 THE NITRATE RADICAL

Over the past two decades it has become evident that the NO_3 radical plays a significant part in chemical transformations in both the stratosphere and the troposphere of the Earth. A recent review (Wayne et al., 1991) summarises our current understanding of the radical.

In the Earth's atmosphere NO_3 is formed by the reaction

$$NO_2 + O_3 \rightarrow NO_3 + O_2 \qquad (2.13)$$

in both stratosphere and troposphere. Dissociation of N_2O_5

$$N_2O_5 + M \rightarrow NO_3 + NO_2 + M \qquad (2.14)$$

is apparently an additional source, but since N_2O_5 is formed by the reaction

$$NO_3 + NO_2 + M \rightarrow N_2O_5 + M \qquad (2.15)$$

it is ultimately dependent on the occurrence of reaction (2.13). Dinitrogen pentoxide (N_2O_5) is itself an important product, since it can react heterogeneously with H_2O to yield HNO_3 and thus contribute to atmospheric acidification (see Section 2.2.6).

During the day, the NO_3 radical is rapidly photolysed: the product channels may be NO + O_2 or NO_2 + O. In the stratosphere, therefore, reaction (2.13) followed by photolysis contributes to the detailed balance of the chemistry of odd-nitrogen compounds. Reactions of NO_3 in the troposphere present a different picture because of the multitude of organic compounds available. Although the hydroxyl radical is usually the main agent of attack on organic species during the day, the nitrate radical may be the most important oxidising component of the troposphere at night. For some compounds, such as CH_3SCH_3, the reaction at night with NO_3 may even dominate over the daytime reaction with OH. Two main kinds of initial step can be envisaged: hydrogen abstraction and addition to unsaturated bonds, as typified by the reactions

$$NO_3 + RH \qquad \rightarrow HNO_3 + R \qquad\qquad (2.16)$$

$$NO_3 + >C=C< \rightarrow >C(ONO_2)C(.)< \qquad\qquad (2.17)$$

where RH represents a hydrocarbon and R is the radical derived from it (with alkanes, for example, the radical would be an alkyl radical, analogous to the CH_3 product of reaction (2.6)). Nitric acid is thus a direct product of hydrogen abstraction by the radical. Furthermore, the radical produced in reaction (2.16) is likely to add O_2 in air to form a peroxy radical (RO_2) in a manner analogous to reaction (2.8). For aldehyde precursors, the acyl radical products of reaction (2.16) yield acylperoxy ($RCO.O_2$) and are thus potential sources of peroxyacyl nitrates.

Involvement of NO_3 in tropospheric chemistry appears to have at least four significant consequences:

(1)   Primary organic pollutants can be oxidised and removed during the night.
(2)   Nitric acid can be formed, either by hydrolysis of N_2O_5 from reaction (2.15) or as a product of the hydrogen-abstraction process (2.16).
(3)   Free radicals, probably undergoing rapid conversion to HO_2 and RO_2, can participate in further steps, so that reaction of organic compounds with NO_3 may initiate oxidation chain reactions (and provide a source of hydroxyl radicals at night).
(4)   Toxic or otherwise noxious compounds, including peroxyacyl nitrates and other nitrates and oxidised compounds, may be formed.

2.2.6   NITROGEN COMPOUNDS IN AIR POLLUTION

The chemistry outlined in the preceding sections provides the basis for explaining why elevated concentrations of nitrogen compounds can give rise to atmospheric pollution (as described by Seinfeld 1986, 1989; Finlayson-Pitts and Pitts, 1986; Harrison, 1990). *Photochemical smog* is sometimes called *Los Angeles Smog* after the city where it was first identified and where it still presents a serious nuisance, although it occurs in very many other cities. It owes its origin to the same chemistry that generates O_3 and converts NO to NO_2 in the normal atmosphere, except that the initial concentrations of NO and hydrocarbons are greatly elevated. The internal combustion engine is heavily implicated in contributing to the pollution. During the night and early morning period of heavy commuter traffic, NO and unburnt or partially oxidised hydrocarbons build up in the atmosphere. After dawn, NO becomes replaced by NO_2, and ozone is generated. By noon, there are high concentrations of O_3 and NO_2 in the atmosphere, there is a brown haze because particles are present, and the eyes run because PAN, a powerful lachrymator, is formed. Human health may be affected, primarily by the O_3, but also by NO_2, PAN and the aerosol particles. Both PAN and O_3 are phytotoxic, and the aerosols degrade visibility.

The chemistry involved is essentially that outlined in Sections 2.2.1–2.2.4. Much higher concentrations of primary species (NO and hydrocarbons) are present in the polluted atmosphere, and methane is probably no longer the dominant hydrocarbon but is replaced by a variety of saturated, unsaturated and aromatic hydrocarbons. However, the oxidation chain is still carried by OH, HO_2 and organic oxy- and peroxy-radicals, as in the natural atmosphere, and the oxides of nitrogen continue to play a critical role in ozone formation and chain propagation. A further nitrogen compound, HONO (nitrous acid), has been detected in photochemical smog. It is of particular interest since it can he photolysed at relatively long wavelengths to generate the OH radicals that are so central to the oxidation steps. HONO can be formed from NO, NO_2 and H_2O by heterogeneous and homogeneous routes. Carbonyl compounds (aldehydes and ketones) are important both as primary pollutants and as intermediates in the oxidation of hydrocarbons: they are, of course, the precursors of PAN and its analogues (see Section 2.2.4). One mechanism for the formation of the aerosol particles that contribute substantially to the undesirable properties of photochemical smog is oxidative polymerisation of hydrocarbons by ozone.

Control strategies do not concern us here, but it is worth noting that changes in NO_x emissions lead to complex responses in pollution, and it is only by understanding the underlying chemistry that effective measures can be adopted.

Acid rain, or more generally *acid deposition*, is another consequence of air pollution which is potentially damaging to the environment (Mason, 1990; Park, 1987; Schwartz, 1989). Natural precipitation is slightly acid because of the dissolved carbon dioxide. But rainwater of much greater acidity (lower pH) has been of widespread occurrence over the last few decades. The two most serious influences of lowered pH in precipitation appear to be on freshwater fish and on forest ecology. The dissolved acids are mostly the 'strong' ones such as sulphuric and nitric, and they have their origin in gaseous SO_2 and NO_2 that are a consequence of fossil fuel combustion. About two-thirds of the acidity in affected rainwater is due to H_2SO_4, but concentrations of dissolved nitrate ion have increased significantly over the last 15 years. Over the same period, both SO_2 and NO_2 emissions from fossil fuel combustion have increased sharply, but the fractional contribution of NO_x has risen even more, probably because improved desulphurisation techniques have prevented the full load of SO_2 from reaching the atmosphere.

We have already seen that an important gas-phase process for removal of NO_2 (reaction (2.11) with OH) leads directly to the formation of HNO_3. So do some reactions of the NO_3 radical, as exemplified by reaction (2.16). Reaction (2.15) is a source of N_2O_5, which is the acid anhydride of HNO_3, and which may thus be hydrolysed

$$N_2O_5 + H_2O \rightarrow HNO_3 + HNO_3 \qquad (2.18)$$

to the acid. In fact, the hydrolysis reaction proceeds more efficiently heterogeneously rather than homogeneously in the gas phase. Reaction (2.18) is likely to be one of the most important conversions that occur within the water droplets of clouds (Finlayson-Pitts and Pitts, 1986; Lelieveld and Crutzen, 1990). The gas-to-aqueous phase transfer of N_2O_5 is limited by gas-phase diffusion and transfer through the interface, while reaction (2.18) within the droplet is essentially 'instantaneous'. The dissolution of N_2O_5 is therefore irreversible.

The heteromolecular reactive condensation of gas-phase molecules on pre-existing particles is sometimes called *aerosol scavenging*. It can have an impact on bulk tropospheric chemistry by providing a sink for nitrogen and hydrogen species such as HNO_3, NO_3, N_2O_5, H_2O_2 and HO_2, as well as organic nitrates and peroxides. Clouds and raindrops have a major effect on gas-phase species through the scavenging mechanism. Rainout (removal of gases by cloud droplets) is believed to be more important than washout (removal of gases by raindrops) because of the longer lifetime and greater surface area of cloud droplets compared with raindrops. Water-soluble species, such as the acids, acid anhydrides and peroxides, are obviously particularly susceptible to removal by these mechanisms.

2.3 NITROGEN COMPOUNDS IN THE STRATOSPHERE

2.3.1 STRATOSPHERIC OZONE AND CATALYTIC CYCLES

Our brief survey of the chemistry of nitrogen compounds in the atmosphere would be incomplete without some mention of their involvement in stratospheric ozone processes (Thrush, 1989). Ozone is generated in the stratosphere by the photodissociation of molecular oxygen

$$O_2 + h\nu \rightarrow O + O \tag{2.19}$$

followed by the addition of O to O_2 in reaction (2.2). In an oxygen-only system, these formation processes are balanced by destruction of O_3 (and O) by the reaction

$$O + O_3 \rightarrow O_2 + O_2 \tag{2.20}$$

However, this reaction is rather slow at stratospheric temperatures, and if it alone were responsible for balancing the production of ozone there would be about five times as much O_3 as is really present. The key to the apparent discrepancy in calculated and measured ozone concentrations is the occurrence of reactions that destroy ozone in addition to process (2.20). The processes are chain reactions that are propagated by atoms and radicals derived from trace constituents of the atmosphere. The chain carriers are regenerated, and the overall change still corresponds to the removal of one O atom and one O_3 molecule, so that we are dealing with a catalytic destruction of O_3. The idea is summarised by the scheme

$$X + O_3 \rightarrow XO + O_2 \tag{2.21}$$

$$O + XO \rightarrow O_2 + X \tag{2.22}$$

$$O + O_3 \rightarrow O_2 + O_2 \quad \text{Net} \tag{2.20}$$

For the purposes of writing this scheme, the catalyst has been called simply 'X'. Several catalytic families have been identified. Some of the most important cases are X = OH, the hydroxyl radical, in which case XO = HO_2, the hydroperoxyl radical; X = NO (nitric oxide) so that XO = NO_2 (nitrogen dioxide); and X = Cl (atomic chlorine) with XO = ClO, the chlorine monoxide radical. Side reactions break (or *terminate*) the chains, but, even so, one 'X' can destroy very large numbers of O_3 molecules. All three cycles, especially the first two, occur in the natural atmosphere, dominating the loss of ozone and controlling its concentration in the stratosphere. Precursor molecules are converted into the catalytically active radicals: for example, methane and water vapour can be converted to OH, and nitrous oxide (N_2O) can be converted to NO.

2.3.2 NO$_x$ AS A CATALYST FOR OZONE REMOVAL

Below about 40 km altitude in the present-day atmosphere, about 70% of ozone production is balanced by loss in the NO_x cycle, which we now write out explicitly

$$NO + O_3 \rightarrow NO_2 + O_2 \tag{2.23}$$

$$O + NO_2 \rightarrow O_2 + NO \tag{2.24}$$

The NO itself is regenerated, so it is behaving as a true catalyst. It is largely produced by the reaction between excited atomic oxygen formed in the photolysis of ozone (see reaction (2.3)) and N_2O transported from the troposphere (see Section 2.1.2)

$$O* + N_2O \rightarrow NO + NO \tag{2.25}$$

The significance for ozone concentrations of changes in the tropospheric N_2O burden is now apparent. There has been increasing speculation since the 1970s that human activity could release catalytically active species that add to those already present in the atmosphere, and thin the ozone layer on which we depend for our protection (McElroy and Salawitch, 1989). One source of such increased catalyst concentrations would be the rising levels of nitrous oxide (N_2O) from increased use of nitrogen-rich fertilisers (see Section 2.1.6).

One of the first potential sources of interference with stratospheric ozone to be proposed was the SST (Stratospheric Supersonic Transport) of which the Concorde is an example. All aircraft produce quite large quantities of nitric oxide (NO) because of the high temperature of combustion in air containing both N_2 and O_2. For aircraft flying lower down, in the troposphere, the oxides of nitrogen become well mixed and ultimately converted to soluble compounds that are rained out to the surface of the Earth; extra ozone may even be produced, as discussed in the previous section. However, injection into the stratosphere could be asking for trouble. Because of the stability of the stratosphere, the lifetime against physical removal is several years, by which time chemical reaction would be likely to occur, especially since the injection is into that region where the ozone layer is found. If natural oxides of nitrogen catalyse ozone destruction, so will anthropogenic nitric oxide. Initial calculations of ozone depletion were greatly in error, both because the future Concorde fleet size was grossly overestimated and because there were substantial errors or omissions in the chemistry. However, the question has again become one of active debate (Johnston, Kinnison and Wuebbles, 1989), since plans for a second-generation SST fleet have begun to be promoted seriously. It is now recognised that both NO_x and HO_x cycles must be considered, since aircraft produce large quantities of H_2O in addition to the NO_x.

2.3.3 NO_x AND THE CFC QUESTION

Nitrogen chemistry enters in a more subtle way into what appears to be a much more serious threat to atmospheric ozone. The release of chlorofluorocarbons (CFCs) at the surface of the Earth enhances the catalytic cycle involving Cl atoms and ClO radicals. Because the CFCs are so inert in the troposphere, substantial amounts will eventually enter the stratosphere. At *stratospheric* altitudes the CFCs are *not* inert, because they can be split by the relatively short wavelength ultra-violet radiation present there

$$CF_2Cl_2 + h\nu \rightarrow Cl + CF_2Cl \tag{2.26}$$

A chlorine atom is formed in the process and it can thus initiate the ClO_x ozone catalytic destruction chain. Millions of tonnes of the CFCs are released to the atmosphere every year, and each chlorine atom can destroy as many as 100 000 ozone molecules before it is inactivated, so there is great potential for damage to the ozone layer. Indeed, were it not for another very important piece of atmospheric chemistry that involves NO_x, the effect would be much greater. Much of the chlorine released from the CFCs is not in the immediately available catalytic form (that is, chlorine atoms and chlorine monoxide) but is bound up in *reservoir* compounds, of which hydrogen chloride (HCl) and chlorine nitrate ($ClONO_2$) are the most important. The reservoir species are formed from Cl and ClO in reactions with methane and in a three-body reaction with nitrogen dioxide:

$$Cl + CH_4 \rightarrow HCl + CH_3 \qquad (2.27)$$

$$ClO + NO_2 + M \rightarrow ClONO_2 \qquad (2.28)$$

It is worth noting that the reactants that form the reservoir species are themselves important in other catalytic cycles, thus clearly illustrating interactions between the cycles and showing the complexity of atmospheric chemistry. Yet other reactions release the chlorine from the reservoirs. The important point, though, is that these reactions are slow in the normal atmosphere, so that most of the atmospheric chlorine is bound up and not available for ozone destruction. Similar reactions link the NO_x and HO_x catalytic cycles by forming HNO_3 in reaction (2.11), and HNO_4 in the analogous addition of HO_2 to NO_2.

2.3.4 NO_x AND THE POLAR OZONE 'HOLES'

There is currently much discussion about whether or not the release of the CFCs has already led to detectable global depletions of ozone (Watson, 1989). What is no longer in doubt is that the CFCs affect atmospheric ozone. The discovery of considerable depletions of stratospheric ozone in Antarctic polar regions each austral spring — the phenomenon of the 'Antarctic ozone hole' — has been clearly linked to measured greatly elevated concentrations of the ClO radical within a chemically disturbed region (Gribbin, 1988; Wayne, 1990; Stolarski, 1988; Solomon, 1990). Once again, nitrogen compounds are found to be implicated in the processes.

The explanation that has emerged for the formation of the ozone holes depends on some special features of polar stratospheric meteorology and dynamics as well as on anomalous chemistry. Temperatures are abnormally low, and polar stratospheric clouds (PSCs) are formed. The stratosphere within the disturbed region is abnormally dry (dehydrated), and highly deficient in the oxides of nitrogen (denitrified). One interpretation of these phenomena is that water condenses, and the oxides of nitrogen are converted to nitric acid, to make up the polar stratospheric clouds. The smaller cloud particles, which condense at $-80°C$, are, in fact, composed of nitric acid trihydrate ($HNO_3.3H_2O$), and the larger cloud particles are more dilute nitric acid. Sinking of the larger particles through the stratosphere thus removes both H_2O and NO_x, and it is the removal of NO_x that appears to lead to the anomalous chlorine chemistry.

We have just pointed out that in the ordinary (i.e. non-polar) stratosphere much of the inorganic chlorine is sequestered in the form of the reservoir compounds HCl and $ClONO_2$, as a result of the occurrence of reactions (2.27) and (2.28). Recent research (Moore *et al.*, 1990) has shown that these two compounds react together on the surface of the PSCs in the heterogeneous step

$$ClONO_2 + HCl \rightarrow Cl_2 + HNO_3 \qquad (2.29)$$

The nitric acid dissolves in the PSC, and is ultimately removed from the stratosphere, while free molecular chlorine is released in the gas phase. Because the vortex isolates the air within it, this part of the stratosphere can be 'pre-conditioned', with release of Cl_2, throughout the long polar night. As soon as the sun rises in spring, this Cl_2 is dissociated to atoms, which then enter into a chlorine-based catalytic chain depletion of ozone. Although this chain is itself rather unusual, the important point is that by removing the normal reservoir compound $ClONO_2$, the heterogeneous reactions allow the chlorine-chain destruction of ozone to exert far more than its usual effect. At the same time, denitrification of the stratosphere prevents re-formation of the reservoir.

The chemistry of the Arctic winter stratosphere also seems to be perturbed, although denitrification has been observed there without accompanying dehydration. No true 'holes' have *yet* been observed in the ozone layer, but it seems that the scene is set for extensive ozone depletions when the meteorological conditions are appropriate.

2.4 CONCLUSION

This brief survey has shown how nitrogen and its compounds are involved in a wide range of interesting and important atmospheric processes in the troposphere and the stratosphere. With the exception of some NO produced by lightning, almost all of the nitrogen compounds that participate in this rich chemistry are derived from the biosphere. Virtually all the nitrogen also returns to the biosphere in the form of nitric acid and nitrates. Although this book is concerned mainly with nitrogen compounds on or near the surface of our planet, this chapter will have shown what the compounds do (and what effects they have) during their excursion into the atmosphere.

REFERENCES

Anderson, I.C. and Levine, J.S. (1986) Relative rates of nitric oxide and nitrous oxide production by nitrifiers, denitrifiers, and nitrate respirers. *Appl. Env. Microbiol.*, **51**, 938.

Anderson, I.C., Levine, J.S., Poth, M.A. and Riggan, P.J. (1988) Enhanced biogenic emissions of nitric oxide and nitrous oxide following surface biomass burning. *J. Geophys. Res.*, **93**, 3893.

Cicerone, R.J. (1989) Analysis of sources and sinks of atmospheric nitrous oxide (N_2O). *J. Geophys. Res.*, **94**, 18265.

Conrad, R., Seiler, W. and Burse, G. (1983) Factors influencing the loss of fertilizer nitrogen into the atmosphere as N_2O. *J. Geophys. Res.*, **88**, 6709.

Finlayson-Pitts, B.J. and Pitts, J.N., Jr (1986) *Atmospheric Chemistry*, Wiley, Chichester.

Franzblau, E. and Popp, C.J. (1989) Nitrogen oxides produced from lightning. *J. Geophys. Res.*, **94**, 11089.

Graedel, T.J. and Crutzen, P.J. (1989) The changing atmosphere. *Scient. Am.*, **261**, September, 28.

Gribbin, J. (1988) *The Hole in the Sky*, Corgi Books, London.

Harrison, R.M. (ed.) (1990) *Pollution: causes, effects and control*, 2nd edition, Royal Society of Chemistry, London.

Hegg, D.A., Radke, L.F., Hobbs, P.V., Rasmussen, R.A. and Riggan, P. (1990) Emission of some trace gases from biomass fires. *J. Geophys. Res.*, **95**, 5669.

Holland, H.D. (1978) *The Chemistry of the Atmosphere and Oceans*, Wiley, Chichester.

Johnston, H.S., Kinnison, D.E. and Wuebbles, D.J. (1989) Nitrogen oxides from high-altitude aircraft: an update on potential effects on ozone. *J. Geophys. Res.*, **94**, 16351.

Khalil, M.A.K. and Rasmussen, R.A. (1988) Nitrous oxide: trends and global mass balance over the last 3000 years. *Ann. Glac.*, **10**, 73.

Lelieveld, J. and Crutzen, P.J. (1990) Influence of cloud photochemical processes on tropospheric ozone. *Nature*, **343**, 227.

Levine, J.S., Augustsson, T.R., Anderson, I.C., Hoell, J.M. and Brewer, D.A. (1984) Tropospheric sources of NO_x: lightning and biology. *Atmos. Environ.*, **18**, 1797.

Linak, W.P., McSorley, J.A., Hall, R.E., Ryan, J.V., Srivastava, R.K., Wendt, J.O.L. and Mereb, J.B. (1990) Nitrous oxide emissions from fossil fuel combustion. *J. Geophys. Res.*, **95**, 7533.

Lobert, J.M., Scharffe, D.H., Hao, W.M. and Crutzen, P.J. (1990) Importance of biomass burning in the budgets of nitrogen-containing gases. *Nature*, **346**, 552.

Mason, B.J. (1990) Acid rain — cause and consequence. *Weather*, **45**, 70.

McElroy, M.B. and Salawitch, R.J. (1989) Changing composition of the global stratosphere. *Science*, **243**, 763.

Moore, S.B., Keyser, L.F., Leu, M.-T., Turco, R.P. and Smith, R.H. (1990) Heterogeneous reactions on nitric acid trihydrate. *Nature*, **345**, 333.

Park, C.C. (1987) *Acid Rain*, Methuen, London.

Roberts, J.M. (1990) The atmospheric chemistry of nitrates. *Atmos. Environ.*, **24A**, 243.

Rohde, H. (1990) A comparison of the contribution of various gases to the greenhouse effect. *Science*, **248**, 1217.

Rosswall, T. (1983) The nitrogen cycle. *SCOPE*, **21**, 46.

Schneider, S.H. (1989) The changing climate. *Scient. Am.*, **261**, September, 38.

Schwartz, S.E. (1989) Acid deposition: unravelling a regional phenomenon. *Science*, **243**, 753.

Seinfeld, J.H. (1989) Urban air pollution: state of the science. *Science*, **243**, 745.

Seinfeld, J.L. (1986) *Atmospheric Chemistry and Physics of Air Pollution*, Wiley, Chichester.

Singh, H.J., Salas, L.J. and Viezee, W. (1986) The global distribution of peroxyacetyl nitrate. *Nature*, **321**, 588.

Solomon, S. (1990) Progress towards a quantitative understanding of Antarctic ozone depletion. *Nature*, **347**, 347.

Stolarski, R.S. (1988) The Antarctic ozone hole. *Scient. Am.*, **258**, January, 20.

Thrush, B.A. (1989) The chemistry of the stratosphere. *Rev. Prog. Phys.*, **51**, 1341.

Walker, J.C.G. (1977) *Evolution of the Atmosphere*, Macmillan, New York.

Watson, R.T. (ed.) (1989) *Ozone Trends Panel Report, 1988*, WMO, Geneva.

Wayne, R.P., Barnes, I., Biggs, P., Burrows, J.P., Canosa-Mas, C.E., Hjorth, J., Le Bras, G., Moortgat, G.K., Perner, D., Poulet, G., Restelli, G. and Sidebottom, H. (1991) The nitrate radical: physics, chemistry and the atmosphere. *Atmos. Environ.*, **25A**, 1.

Wayne, R.P. (1990) Punching a hole in the stratosphere. *Proc. Royal Inst.*, **61**, 13.

Wayne, R.P. (1991) *Chemistry of Atmospheres*, 2nd edition, Oxford University Press, Oxford.

3 Nitrogen Cycling in Agricultural Soils

A.J.A. VINTEN and K.A. SMITH

SAC, School of Agriculture, Edinburgh

3.1 INTRODUCTION

In most soils, well in excess of 90% of the nitrogen is present in organic forms. This organic nitrogen is made up of a vast range of compounds, derived from biological materials (roots, microflora, fauna, leaf litter and so on) and from humification processes (Stevenson, 1982). However, the relatively small amount of nitrogen present in inorganic form, as ammonium, NH_4^+, and nitrate, NO_3^-, is significantly greater in agricultural soils, especially those under intensive management, than in other ecosystems.

There has been a rapid increase in the quantities of inorganic nitrogen fertiliser used over the last few decades. In post-war Britain the desire to produce more food and be less reliant on imports stimulated the development of a very intensive agricultural system. The breeding of high-yielding varieties of cereals and other crops, and effective chemical control of pests and diseases, resulted in much higher inputs of mineral nitrogen, principally in the form of synthetic fertilisers. The amounts applied per hectare reached a plateau in the 1980s (Figure 3.1). For some crops there has been a marginal decline, and current agricultural policies are likely to cause further reductions, but overall the picture is not likely to change substantially, in the short term at least. Animal-based systems have also intensified over the same period, and large applications of nitrogen to land in the form of manures (containing up to half their nitrogen in the ammonium form) have become common.

Outputs of nitrogen from intensive agriculture are also much greater than from natural forests and grasslands. The largest outputs are in the form of crop offtake, but the quantities of readily mineralisable nitrogen in the form of crop residues are also greater, and soil disturbance through cultivation increases the rate of mineralisation compared with that prevailing in undisturbed systems. The outcome is an increased amount of mineral nitrogen in the soil profile which is vulnerable to leaching and/or denitrification. Losses of nitrogen from livestock-based agriculture have also increased with intensification, and contribute a very significant part of total losses from soils.

This chapter examines the processes involved in nitrogen transformations in agricultural soils, and attempts to estimate the magnitude of the losses to the environment in typical conditions.

3.2 MINERALISATION AND IMMOBILISATION

Mineralisation is the process by which organic compounds in the soil break down to release ammonium ion, NH_4^+, with the concurrent release of carbon as CO_2. The extent, and the

Nitrate: Processes, Patterns and Management. Edited by T.P. Burt, A.L. Heathwaite and S.T. Trudgill
© 1993 John Wiley & Sons Ltd

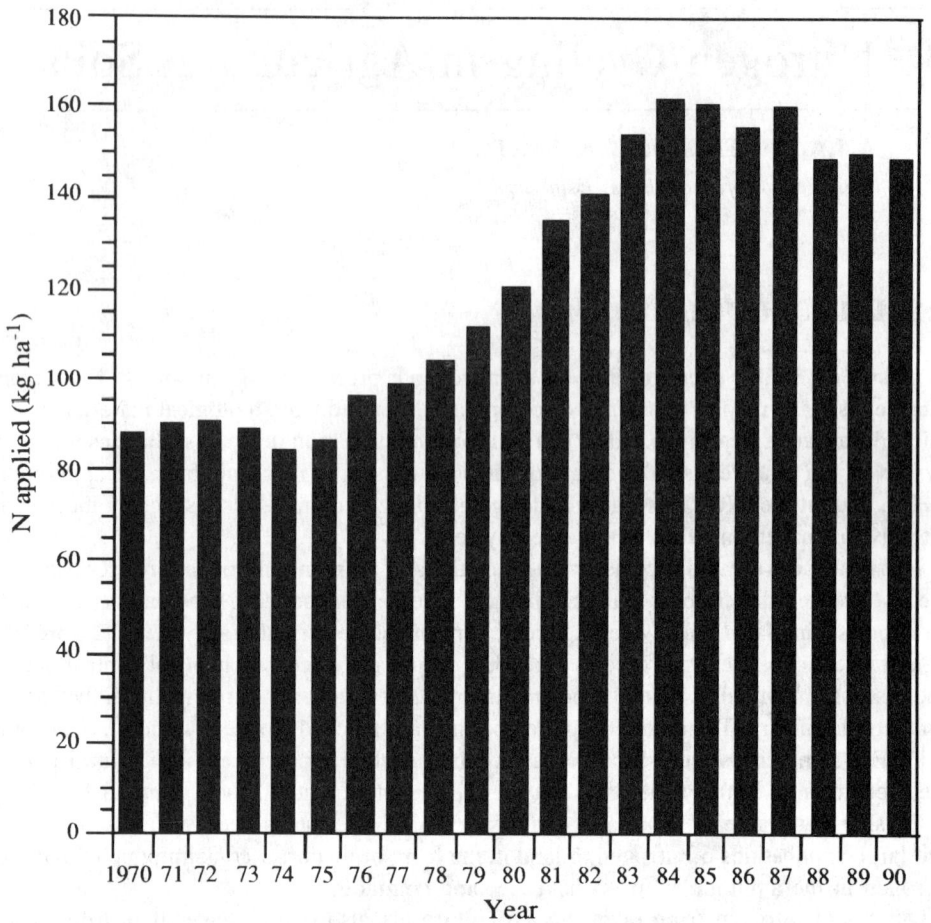

Figure 3.1 UK fertiliser use on arable crops, 1970−90

rate, of the processes are affected by many factors, including the composition of the decomposing substrate(s), soil temperature, moisture content and pH. *Immobilisation* is the reverse process, by which there is a net incorporation of mineral nitrogen, usually NH_4^+, into organic forms — effectively into microbial tissue — during the decomposition process.

A major influence on the balance between mineralisation and immobilisation is the C:N ratio in the decomposing organic substances. A low C:N ratio (high nitrogen content) generally results in net mineralisation and facilitates a high rate of decomposition. The lowest ratios (*ca.* 8:1) are usually found in microbial tissue (biomass), and this is, in consequence, relatively labile material. As for plant material resulting from cropping, the residues (roots, stems, leaves, etc.) from leguminous species such as clover, beans or lucerne have C:N ratios in the range 13:1 to 23:1; and at the other end of the scale, ratios of 60:1 to 80:1 are typical of cereal straw and other mature plant stalks (Haynes, 1986). Decomposition of this stem material following incorporation into the soil normally results in a net fall in the amount of plant-available nitrogen, for a period (Figure 3.2), even though the long-term effect of continued straw incorporation is greater mineralisation (Powlson, Brookes and Christensen, 1987).

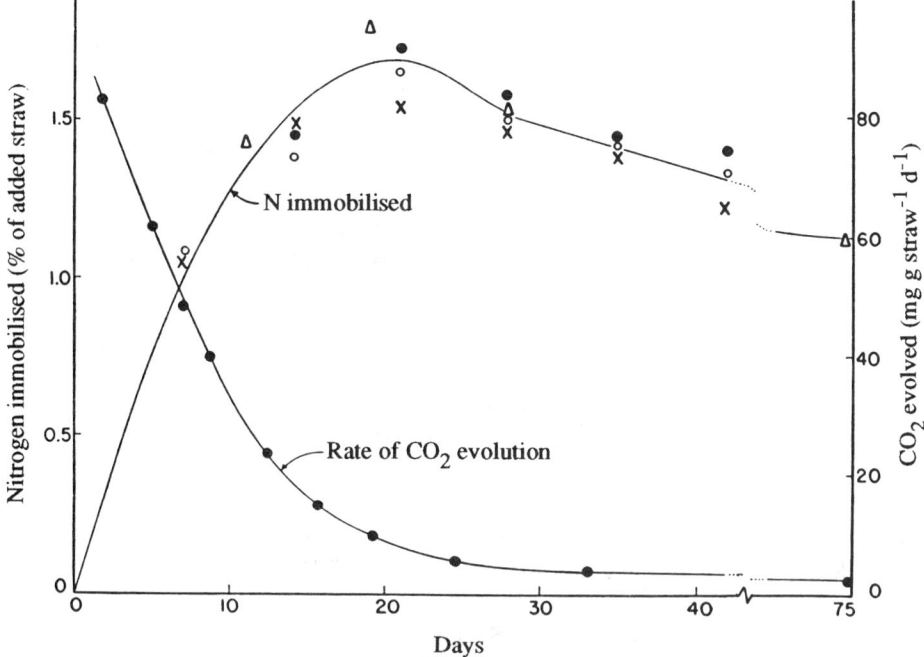

Figure 3.2 Immobilisation and release of nitrogen, and rate of carbon dioxide formation, in a soil receiving wheat straw and nitrate nitrogen (after Allison and Klein, 1962; reproduced by permission of the publishers © 1962 Williams and Wilkins)

Although there is a general trend relating net mineralisation/immobilisation to the C:N ratio, there is no precise critical value which marks the point at which reversal from immobilisation to mineralisation occurs. This is because other aspects of substrate quality, such as lignin and polyphenol content, have a major impact on rate of decomposition.

The rate of mineralisation of nitrogen from soil organic matter generally increases with increasing moisture content between permanent wilting point (-1.5 MPa) and field capacity (-5 to -10 kPa). This effect is well illustrated by the results of Stanford and Epstein (1974), who found a two-and-a-half-fold increase in the amount of mineral nitrogen accumulating over a 2-week period of incubation, as the soil moisture content was increased from <5 to about 35 g 100 g^{-1}.

As the soil moisture content is raised above field capacity, however, mineralisation rates fall because of restricted aeration. The rate of decomposition by aerobic bacteria is much greater than that brought about by anaerobic bacteria, because the former are more energy-efficient. Decomposition by fungi and actinomycetes is also predominantly an aerobic process, and thus inhibited by lack of oxygen.

It is not only moisture *content* which is important; temporal *changes* in content, i.e. cycles of drying and wetting, have a profound effect on the rate of mineralisation. There is good evidence that the rewetting of a dried soil results in a burst of microbial activity associated with an expansion in microbial populations (Campbell and Biederbeck, 1982). The substrate responsible for the stimulation of this activity is partly microbial cells killed during the drying phase, with a low C:N ratio, and partly soil organic matter newly exposed to microbial attack

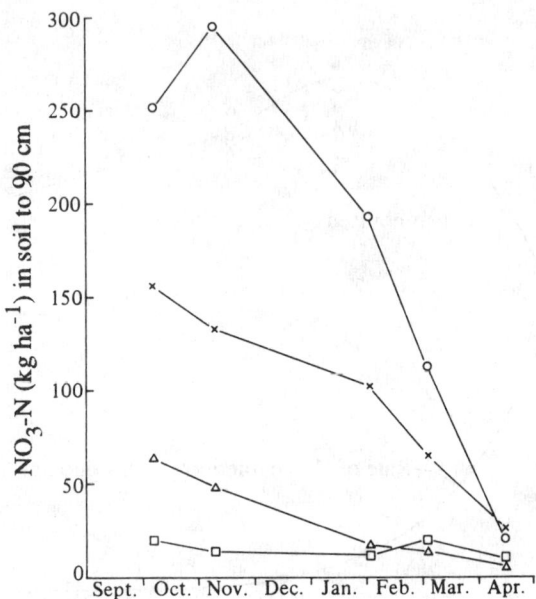

Figure 3.3 The range in the amounts of NO_3-N found in four soils under winter wheat during autumn and winter in 1981−2 (averaged oversowing dates). o sandy clay soil at Woburn, after potatoes; × silty clay soil at Rothamsted, after potatoes; △ sandy soil at Woburn, after potatoes; □ silty clay soil at Rothamsted, after oats (Widdowson *et al.*, 1987; reproduced by permission of the authors and Cambridge University Press)

as a result of physical disruption of aggregates due to swelling and shrinking of the soil (Haynes, 1986).

Each successive cycle results in a slightly smaller release of respired CO_2 and mineralised nitrogen. The size of the flush increases as the severity of drying and the duration of the dry period increase, prior to the rewetting (Birch, 1960, 1964).

In British conditions, arable soils frequently contain substantial amounts of mineral nitrogen in the rooting zone, following crop harvest, as a result of mineralisation of crop residues and the presence of unused fertiliser. Work at Rothamsted and Woburn (Widdowson *et al.*, 1987) has shown that the amounts vary greatly between soils, even after the same crop (Figure 3.3).

Freezing and thawing have comparable effects to those brought about by drying and rewetting. The freezing process kills a substantial part of the soil microbial biomass, which is then available for decomposition by the surviving population, once the temperature increases to allow the resumption of microbial activity (Biederbeck and Campbell, 1973; Edwards and Cresser, 1992). In severe conditions such as those of a Canadian winter, the effects of freezing and thawing may exceed those of drying and rewetting (Shields, Paul and Low, 1974), but in more moderate environments such as Britain, the reverse situation is more likely.

Rates of organic matter decomposition generally rise rapidly with increasing temperature, over the range normally found in soils in the field. This results in large differences in the rate of nitrogen mineralisation between typically cool conditions in early spring, especially in the north of Britain, and the rate in midsummer. This is of considerable topical interest,

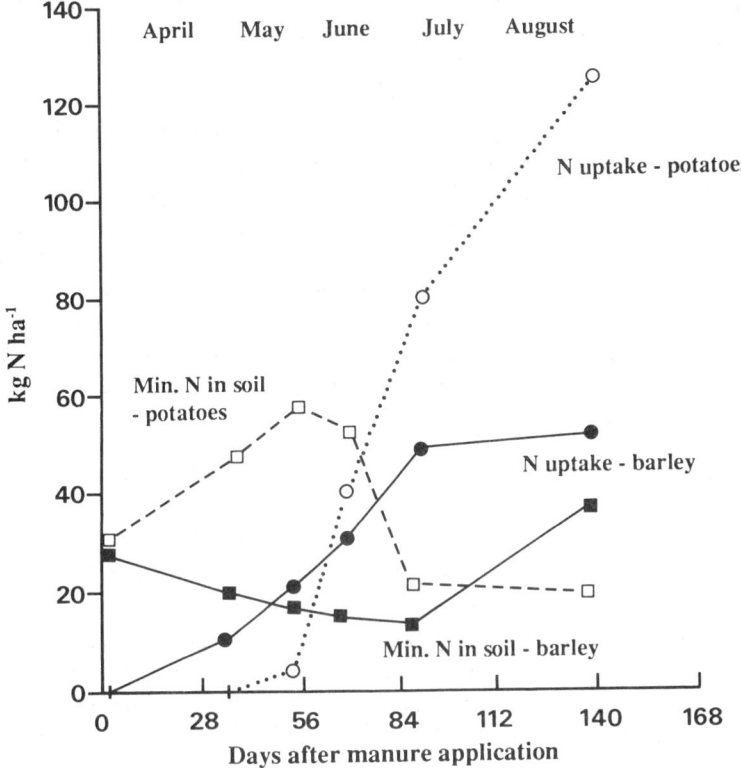

Figure 3.4 Cumulative N uptake by organically grown potato and spring barley crops at Jamesfield, Tayside, 1990. Data for potatoes are totals of haulm plus tuber N contents (Stockdale and Rees, unpublished)

because of the implications for organic farming systems. The release of mineral nitrogen from manures and soil organic matter is too slow in the spring, in the cooler parts of the UK, to provide a satisfactory supply to spring-sown cereals. In contrast, vegetable crops or potatoes whose main demands for nitrogen occur much later in the season, are better provided for by mineralisation (Figure 3.4).

The liming of acid soils is known to increase nitrogen mineralisation rates. Nyborg and Hoyt (1978) reported a doubling, on average, of the percentage of the total soil nitrogen which was mineralised over a 120-day period when a range of 40 soils were limed from their initial pH values averaging 5.0, up to 6.7. However, according to Jenkinson (1988), only the early stages of decomposition of plant residues are retarded by acidity, and differences between soils in the amounts remaining undecomposed largely disappear after five years.

Some results of research into the effects of fertiliser nitrogen on mineralisation of soil nitrogen have suggested the existence of a stimulatory or 'priming' effect on the mineralisation process. Haynes (1986), in a review, concluded that the priming effect was partially apparent and partially real. When ^{15}N-labelled inorganic fertiliser is added to the soil, this will result in some immobilisation of the labelled nitrogen and some mineralisation of unlabelled nitrogen from the soil organic matter. Apparent increased uptake of native soil nitrogen when the labelled fertiliser is applied, as compared with unfertilised controls (e.g. Westerman and Kurtz, 1973), may be due to this *pool substitution* of labelled by unlabelled nitrogen, and does not necessarily mean a net increase in mineralisation.

However, in some circumstances, the effect can be a genuine one. The fertiliser nitrogen promotes the development of a larger plant root system and this more effectively scavenges the soil volume for soil-derived mineral nitrogen (Sørensen, 1982). Work in central Scotland (Smith *et al.*, 1984) lends support to this view. The uptake of soil nitrogen by spring and winter barley in sandy soils with no restriction on rooting was constant over a range of fertiliser nitrogen applications from 0 to 150 kg ha^{-1}, whereas the uptake in clayey soils with adverse physical conditions increased with increasing amounts of fertiliser nitrogen. A contribution to the effect may also come about, under these conditions, as a result of fertiliser-stimulated growth reducing the soil water content and thus improving the aeration; this would promote the mineralisation of soil organic nitrogen, and lead to increased uptake.

3.3 PROCESSES RESPONSIBLE FOR GASEOUS LOSSES

3.3.1 DENITRIFICATION

Denitrification — the microbial reduction of nitrate to NO, N_2O and N_2 — is the major biological process by which the nitrogen cycle is completed and fixed nitrogen is returned to the atmosphere. The environment in which the greatest quantities of nitrate, the essential substrate for denitrification, are likely to be found is agricultural land receiving substantial inputs of nitrogenous fertilisers or manures. Estimates of the quantities of nitrogen lost by denitrification from agricultural land differ widely. A review by Colbourn and Dowdell (1984) produced figures of 0−20% of the applied fertiliser nitrogen from arable land and 0−7% from grassland. However, a survey by Hauck (1986) indicated losses of 20−40%. Some estimated denitrification losses during growth of crops in a number of different countries, compiled by Nieder, Schollmayer and Richter (1989) show a range of losses from 2.5% to over 50% of the applied nitrogen.

Concern about the loss of nitrogen by denitrification was originally inspired by apprehension of the agronomic and economic penalties stemming from the loss of a useful and costly nutrient. Now, however, such concerns have been supplemented by environmental problems resulting from the emission of nitrous oxide, N_2O. This gas is one of the more important contributors to the greenhouse effect and is also considered to be a partial cause of the depletion of the Earth's stratospheric ozone layer.

Subsoil denitrification as a mechanism responsible for reducing the quantity of leached nitrate in aquifers is also important in this context. If the reduction goes entirely to N_2, the outcome is a benign one, but if substantial quantities of N_2O are emitted, then all that has happened is that one potential problem has been exchanged for another.

The phenomenon of denitrification in soils, resulting in a loss of available nitrate, has been well known for a long time. Payne (1981) has cited papers at the turn of the last century reporting that the simultaneous supply of organic matter (in the form of manure) and nitrate was counter-productive, because of the stimulation of denitrification. Recent work using modern methods of analysis for the gaseous products of denitrification confirms the greater losses in the presence of manure. Increased soil carbon contents after long-term manure applications also promote the process (Webster and Goulding, 1989; Figure 3.5), and so does straw incorporation (Ball, 1990).

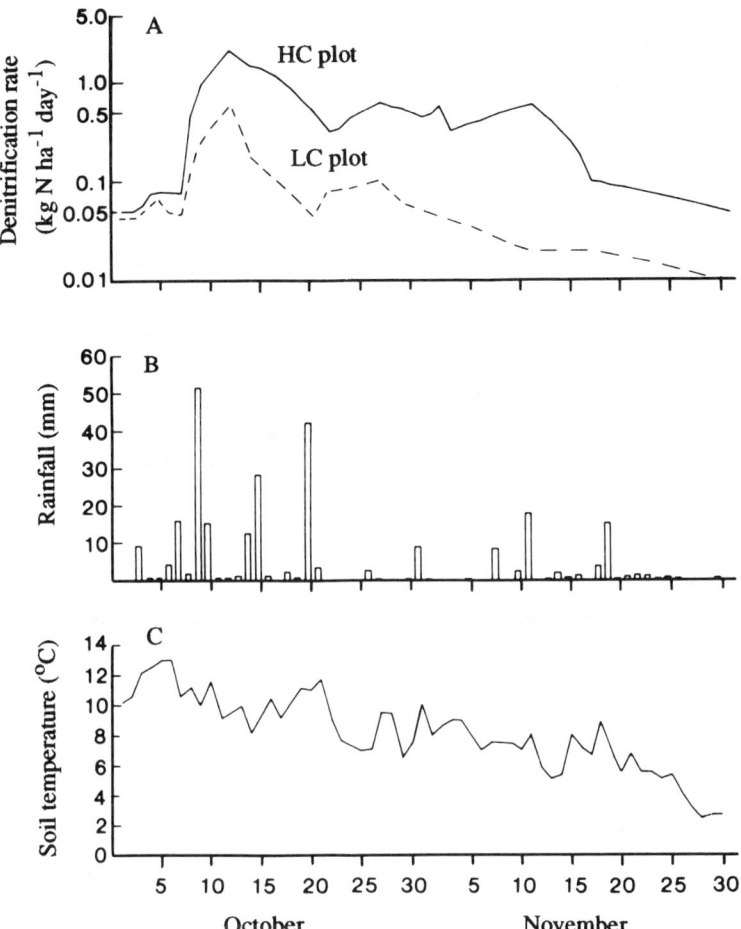

Figure 3.5 (a) Denitrification rate from high carbon content (HC) and low carbon content (LC) plots on Hoosfield, Rothamsted, during October and November 1987; (b) rainfall; and (c) soil temperature at 20 cm depth (from Webster and Goulding, 1989, reproduced by permission of the authors and the Society of Chemical Industry)

It appears to be the readily decomposable fraction of the organic matter which affects the capacity of a soil to denitrify. Bijay-Singh, Ryden and Whitehead (1988) investigated the relationship between denitrification potential (DNP) and water-soluble and anaerobically mineralisable carbon, for 32 English soils. They concluded that, in the field, denitrification may be limited by the amount of carbon susceptible to mineralisation under anaerobic conditions. The general applicability of this conclusion remains to be tested.

Denitrification rates are correlated to some extent with concentrations of nitrate in the soil. Relatively high rates (>0.2 kg N ha^{-1} d^{-1}) have been associated with soil nitrate contents exceeding 5 mg kg^{-1}. Where fertilisers containing nitrogen in the nitrate form are applied, much of the loss due to denitrification occurs in the period immediately following the application. This usually means that maximum losses from cereal-growing land and grassland occur in the spring, under UK conditions, with a tendency towards another peak in the autumn from arable land, following the release of nitrate from the mineralisation of crop residues, and an increase in soil water content.

The effect of pH on denitrification is unclear. Some effects attributed to changes in pH in the past have been shown to be due to associated changes in carbon availability.

The effects of plants on denitrification are complex. On the one hand, they can promote it by providing carbon, in the form of exudates and root cell material, that supports an increased microbial population in the rhizosphere and acts as an electron donor for nitrate reduction when oxygen availability is low; and they can help to create anaerobic zones by adding to the soil respiratory oxygen demand in conditions of reduced diffusion (high soil water content). On the other hand, water demand by the plants dries the soil and improves aeration; plant uptake of nitrate removes it from the danger of loss by denitrification; and aquatic plants with aerenchyma can increase oxygen availability near their roots by diffusion from the atmosphere.

The net effect of these influences may be positive, negative or neutral. The outcome will depend on the status of a number of other soil parameters, including availability of native soil carbon, physical factors controlling oxygen diffusion, and the rate of nitrate supply to the root zone. Haider, Mosier and Heinemeyer (1987), for example, found in experiments with maize that denitrification was not stimulated by the presence of plants in the soil during the active phase of growth, although it was greater in the planted soil when the root biomass began to decrease, presumably because of the mineralisation of carbon from the dying biomass.

Many studies have shown that denitrification activity in soils is correlated with water content, and rainfall/irrigation events. Egginton and Smith (1986) found that 50 mm applications of water to grassland that had received 100 kg of nitrogen as nitrate resulted in a maximum denitrification rate exceeding 2 kg N ha^{-1} d^{-1}. This dependence on water content is a direct consequence of the fact that the diffusion rate of oxygen through a water-filled pore is only one ten-thousandth of that through an air-filled pore, and thus the potential for the development of anaerobic zones is dependent to a greater degree on water content than on any other variable.

Reported values for oxygen concentrations in the soil atmosphere at which denitrification has been observed range from 4% to 17%. Parkin and Tiedje (1984) measured the effect of oxygen concentration on denitrification in soils of different textures in the laboratory. They found that at concentrations above 3% the rates were less than 2% of those in completely anaerobic conditions. Studies with soil cores in the laboratory (Arah et al., 1991) have shown qualitatively similar relationships, but have also indicated that there is no single concentration of oxygen which is a critical threshold above which denitrification does not occur (Figure 3.6). Rather, the critical value depends on a complex relationship between soil aggregate size, oxygen diffusion rate (and therefore water content) and soil respiration rate (which depends on available carbon). When aeration is expressed in terms of the different, though related parameter — air-filled porosity — it has been found that the critical value ranges from 11% to 14% over several soils.

Reports of minimum temperatures for the occurrence of denitrification in soil collated by Firestone (1982) ranged from 2.7°C to 10°C. The optimum temperature is 25°C and above. Where there is an abundant source of available carbon, e.g. where cattle slurry is applied to grassland, the rate of loss can be substantial at temperatures as low as 8°C. The Arrhenius relationship between reaction rate and temperature (with a Q_{10} of 2–3) is often used to describe biological reactions. However, it is not always applicable to denitrification in soils because the denitrification process often occurs within anaerobic zones in a matrix of aerated soil; the increase in aerobic respiration rate with temperature can result in a rapid growth

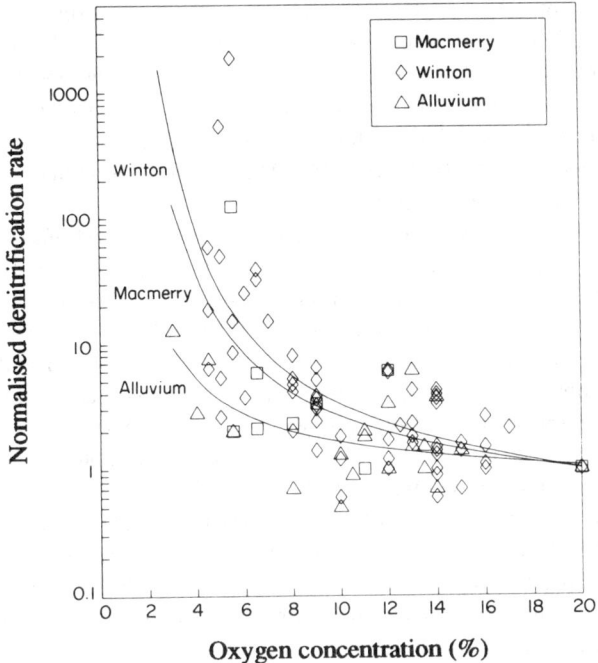

Figure 3.6 Relationship between oxygen concentration in incubation vessel and measured denitrification rate (from Arah *et al.*, 1991; reproduced by permission of Blackwell Scientific Publications Ltd)

in the size of these anaerobic zones, while the denitrification rate *per unit volume of anaerobic soil* itself increases according to the Arrhenius relationship. Thus the observed rate may increase as the *product* of these two functions.

Several factors point to the existence of chemodenitrification, i.e. gaseous losses of nitrogen from soil caused by chemical processes, as opposed to those resulting from microbial action. The evidence includes:

(1) Substantial gaseous losses observed in soils where conditions were not conducive to microbial denitrification or volatilisation of ammonia;
(2) The fact that such losses also occur under conditions that result in the accumulation of nitrite (e.g. after application of high rates of ammonium or ammonium-forming fertilisers);
(3) The rapid decomposition of nitrite, when added to acidic soils — sterilised or unsterilised — with formation of gaseous forms of nitrogen.

The mechanism appears to be as follows: nitrite accumulates in soil at the periphery of bands of ammoniacal fertiliser and around urea prills applied to acidic soils. Nitrous acid, HNO_2, is formed in these conditions, and this reacts with soil organic matter to release N_2. One approach to minimise these losses is to manage soils in such a way that rapid nitrification occurs, so conditions in which nitrite may accumulate temporarily are avoided. There is also evidence for chemodenitrification processes in the subsoil, involving ferrous iron as reducing agent, and resulting in very variable ratios of N_2 to N_2O (Lind, 1985).

Agricultural land is a significant source of emissions of N_2O. Eichner (1990) has recently reviewed the results of 104 experiments measuring N_2O emissions from fertilised soils (fallow, grassland and cropped) at nine locations in five countries, all in temperate regions. Of these experiments, 65 involved comparisons between fertilised and unfertilised control sites, and in all but one of these emissions from the fertiliser sites were higher than from the respective control. Eichner's data (Figure 3.7) indicate that increasing the fertiliser rate generally, but not always, corresponds with greater emissions. This relationship is also shown by data compiled by Bouwman (1990).

Land-use changes can contribute to N_2O release. It has been estimated that, worldwide, the total area converted into cultivated land over the last 100 years is of the order of 8.5 $\times$ 10^8 ha, which has resulted in the release from soil organic matter of 5–20 Tg per year, of which perhaps 0.2–0.6 Tg may be lost as N_2O. Several studies have shown that very high rates of N_2O emission are possible when peat soils are drained and cultivated.

A major factor contributing to the variability of N_2O emissions due to denitrification is the wide variation in the fraction of the total gaseous flux emerging as N_2O (the remainder being N_2) (Figure 3.8). It appears that soil structure and water content, affecting the balance between diffusive escape of N_2O and its further reduction to N_2, are important among the factors determining the proportions of the two gases (Arah and Smith, 1990).

Soil pH is another factor affecting the ratio of N_2O to N_2 in the gaseous products of denitrification. Inhibition of N_2O reduction to N_2 occurs at all concentrations of nitrate at low pH, thus resulting in an increased proportion of the emissions occurring as N_2O. Tracer studies with the short-lived radioactive isotope of nitrogen, ^{13}N, have shown that the effect

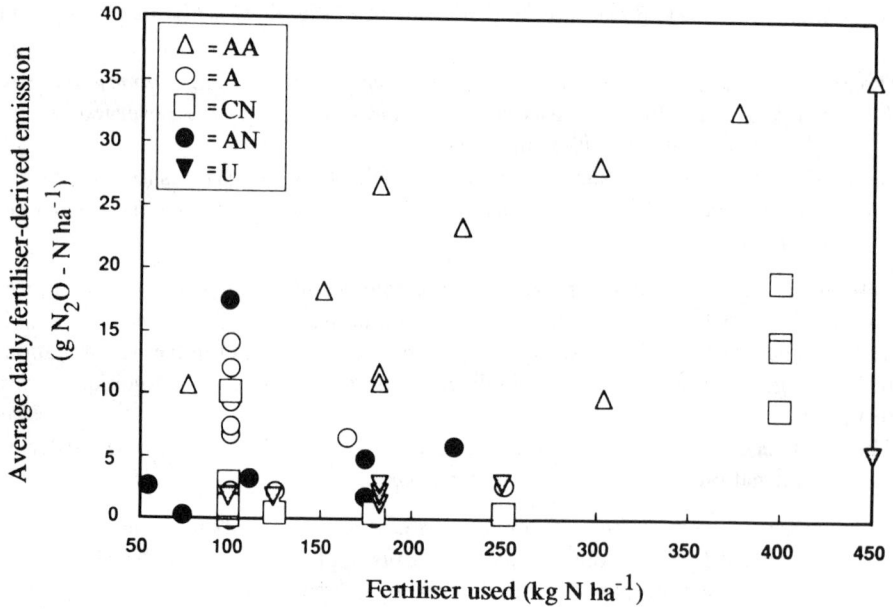

Figure 3.7 Relationship between average daily fertiliser-derived N_2O emissions and the quantity of fertiliser applied, regardless of the length of the sampling periods for controlled studies only. Excludes three points for AA-derived emissions: 72.7, 95.7 and 123 g N_2O-N ha^{-1} (from Eichner, 1990; reproduced by permission of the American Society of Agronomy, Inc., the Crop Science Society of America, Inc. and the Soil Science Society of America, Inc.)

of acidity on N_2O production is an immediate one, and thus not due to a change in the balance of the microbial population (Tiedje *et al.*, 1981).

3.3.2 AMMONIA VOLATILISATION

Nitrogen can be lost from agricultural soils by the release of gaseous ammonia, NH_3, into the atmosphere. The predominant source of the ammonia, in the farming systems of Western Europe, is the urea in the faeces and urine of livestock, either voided directly onto the land by grazing animals or spread as slurry or farmyard manure. Ammoniacal fertilisers also contribute to the release, when applied to calcareous soils. Recently, it has been estimated that of the 2.3 million tonnes of nitrogen entering the UK agricultural system each year, 0.4 million tonnes is lost as ammonia (Table 3.1; Johnston and Jenkinson, 1989).

The ammonia lost to the atmosphere is a major contributor to acid deposition (Table 3.2). From the data in Table 3.2 it can be seen that in 1975 slightly more than half of the nitrogenous component of this deposition was NO_x of industrial origin, but that most of the remainder was NH_3, and came overwhelmingly from animal sources. Some of the NH_3 deposition is very local, within a few hundred metres of the source; but at the other extreme, some is dispersed over areas as large as Western Europe. The amounts deposited on land surfaces, including farmland, in populated and industrial areas may exceed 50 kg N ha^{-1} a^{-1}. Such

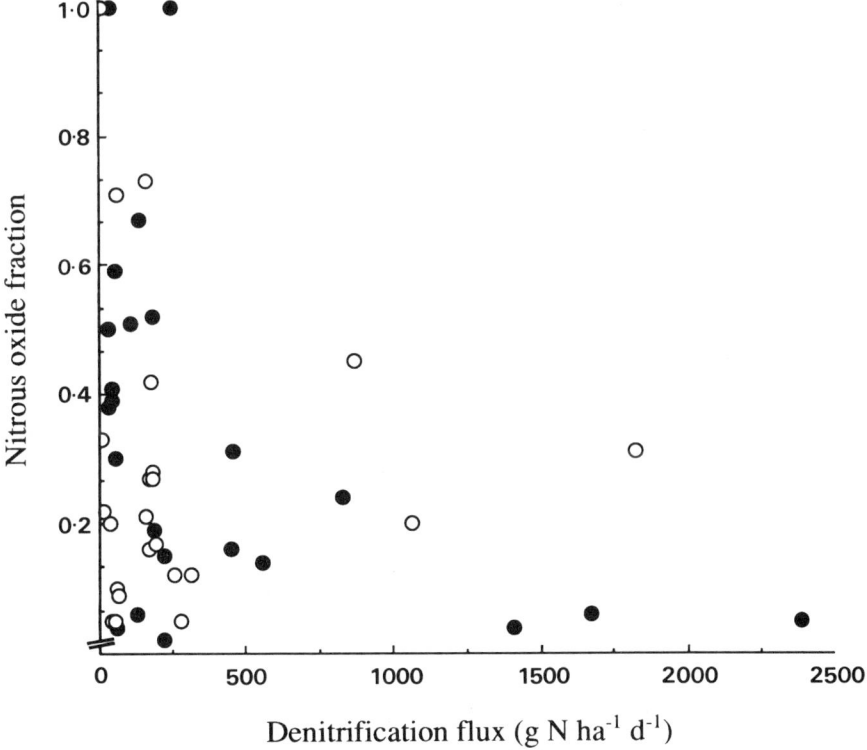

Figure 3.8 Relationship between the fraction of the denitrification flux ($N_2 + N_2O$) emitted as N_2O and the denitrification flux (from Smith and Arah, 1990; reproduced by permission of The Fertiliser Society)

Table 3.1 Nitrogen balance for the United Kingdom in 1988 (from Johnston and Jenkinson, 1989, reproduced by permission of The Fertiliser Society)

		Million tonnes N
Inputs		
Imported human food		0.1
Imported animal feed		0.1
Rainfall		0.3
Fertiliser		1.5
Biological fixation		0.3
	Total	2.3
Outputs		
To sea		0.3
Exported food		0.2
Loss of ammonia		0.4
Unaccounted for		1.4
	Total	2.3

quantities make a significant contribution to the nitrogen budgets even of relatively intensive agricultural systems. In natural ecosystems they may raise the nitrogen inputs by an order of magnitude compared with those in pristine unpolluted environments, with serious consequences for acidification of catchments and the viability of natural vegetation.

When urea is added to a soil, it is rapidly hydrolysed by the urease enzyme to ammonium and bicarbonate ions. The latter tend to raise the soil pH near the surface, and promote the loss of NH_3 by volatilisation (Rachhpal-Singh and Nye, 1988). The amounts of ammonia lost are influenced by a number of factors. Rachhpal-Singh and Nye (1988) include among these aerodynamic factors affecting the transfer of NH_3 from the soil surface to the atmosphere, the amount of urea applied, the rate of urea hydrolysis, the initial pH and the buffer capacity of the soil, the soil moisture level, and the depth of application.

The rate of hydrolysis increases with increasing temperature, as does the dissociation of NH_4^+ to NH_3, and the rate of NH_3 diffusion to the solution−air interface. Not surprisingly,

Table 3.2 NO_x and NH_x anthropogenic sources to the atmosphere and emission rates in Europe *ca* 1975; units are Tg N per year (from Melillo *et al.*, 1989, reproduced by permission of Dr Silke Bernhard)

	Sources	Emission rates
NO_x	Fossil fuels	2.3−6.2
	Automobiles	1.1−1.5
	Nitric acid production	<0.1−0.1
	Fertiliser production	0.1
	Aviation	<0.1
	Burning of waste materials	<0.1
	Total	3.5−7.9
NH_x	Domestic animals	3.0−4.5
	Fertiliser production	0.5−0.9
	Coal combustion	<0.1
	Automobiles	<0.1
	Total	3.5−5.4

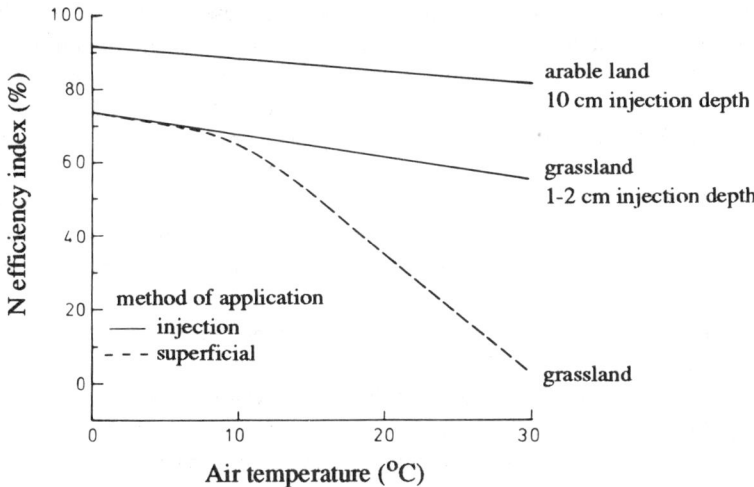

Figure 3.9 Effect of method of liquid manure application to arable land and grassland and of air temperature on the efficiency of the nitrogen component in liquid manure (from de la lande Cremer, 1986; reproduced by permission of Elsevier Science Publishers)

therefore, there is a marked dependence of rate of ammonia loss on ambient temperature. For example, marked diurnal cycles in the rate of emission, mirroring temperature cycles, have been reported by a number of workers (Haynes and Sherlock, 1986). The loss of ammonia from surface-applied slurry can be expressed in terms of the change in the efficiency of the applied material as a nitrogen fertiliser, compared with a conventional mineral nitrogen fertiliser. Figure 3.9 illustrates the decline in efficiency with increasing ambient temperature at the time of spreading. It also shows the gain in efficiency (as a result of drastically reducing the NH_3 emission), by injecting the slurry below the soil surface.

In grazed grassland ecosystems, the effective rate of application of nitrogen to the soil within urine patches may exceed 1000 kg N ha^{-1} (Kolenbrander, 1981; Ball and Ryden, 1984). This rate greatly exceeds the capacity of the herbage to utilise the nitrogen (Ball and Keeney, 1983), and thus the patches represent significant sources of ammonia volatilisation and nitrate leaching.

3.4 NITRATE LEACHING

3.4.1 TRANSPORT PROCESSES

Of the various combined forms of nitrogen present in soils, or added as fertiliser, only the nitrate ion is leached out in appreciable amounts by water passing through the soil profile. This is because, apart from some acid soils in the tropics, there is no significant adsorption of nitrate on to soil surfaces, and there are no common insoluble nitrates (Wild, 1988). Thus nitrate in the soil solution is displaced downwards by rainfall or irrigation water and if sufficient water is added it can be carried beyond the root zone and eventually to groundwater.

The rate of downward movement of nitrate is affected by the water content of the soil during leaching. The depth of displacement by a given quantity of rainfall is generally greater for sandy soils than for clays. However, nitrate movement in the field is a complex process, and the effect of soil structure increases as clay content increases (Addiscott and Bland, 1988). Variations in pore size, in the spatial distribution of pores and in their continuity all contribute to irregular movement of water down the soil profile. The effect of this is to spread out the front between the resident soil solution and the displacing rainfall: a phenomenon known as *hydrodynamic dispersion* (Wild, 1988). Superimposed on this effect is *diffusive dispersion* of nitrate in the soil solution, due to differences in concentration within the profile. Recognition of the high hydrodynamic dispersion in structured soils has led to the concept of *mobile* and *immobile water*. The immobile water is that retained in the peds, from which nitrate can only be transferred to the mobile water phase by diffusive transfer across the mobile−immobile water interface.

This concept has been used to good effect in improving simulation of solute transport in structured soils (e.g. Passioura, 1971; Addiscott, Rose and Bolton, 1978). However, under intensive rainfall, water and solute may completely bypass the mobile pore system and move via large macropores (structural cracks, worm channels, etc.). The description of water movement under these conditions is being developed (e.g, Germann and Beven, 1985; Bronswijk, 1991), but a detailed analysis of solute transport under these conditions is still lacking. Some authors (e.g. Barraclough, 1989; Addiscott and Whitmore, 1991) have developed approximate bypass routines as part of more conventional mobile−immobile water models of solute transport. One problem with improving the description of bypass transport is the highly transient nature of this type of transport. Time steps during simulation need to be of the same order as rainfall events (i.e. hours rather than days), and such data are often lacking.

3.4.2 FACTORS INFLUENCING LEACHING OF NITRATE FROM FARMING SYSTEMS

Many factors — fertilisation, soil texture, land use and crop rotation, cultivations — can have a profound effect on the quantity of nitrate leached from a soil.

3.4.2.1 Amount of fertiliser applied

An illustration of the fairly general relationship between nitrogen fertiliser application and leaching is provided by Bergstrom and Brink (1986), who made a 10-year study on an arable clay soil in Sweden. Leaching of nitrate was moderate up to a rate of application of 100 kg N ha^{-1} a^{-1}, but increased rapidly thereafter, reaching 91 kg NO$_3$-N ha^{-1} for the highest application rate of 200 kg N ha^{-1} in the wet year 1980−81 (Figure 3.10). There was also a build-up of inorganic nitrogen in the soil (increasing the potential for future leaching) when excessive nitrogen was applied (Figure 3.11). Vinten, Howard and Redman (1991) found little influence of fertiliser application on leaching of nitrogen from spring and winter barley crops on a clay loam soil, at the recommended fertiliser nitrogen rate. As the application increased above the optimum, losses increased only slightly, on both sandy loam and clay loam soils (Vinten, Vivian and Howard, 1992). Barraclough, Hyden and Davies (1983) found that the nitrate nitrogen leached from cut grassland plots, as an equivalent percentage of the nitrogen applied, increased from 1.5% to 5.4% to 16.7%, as the fertiliser application increased from 250 to 500 to 900 kg ha^{-1} a^{-1}.

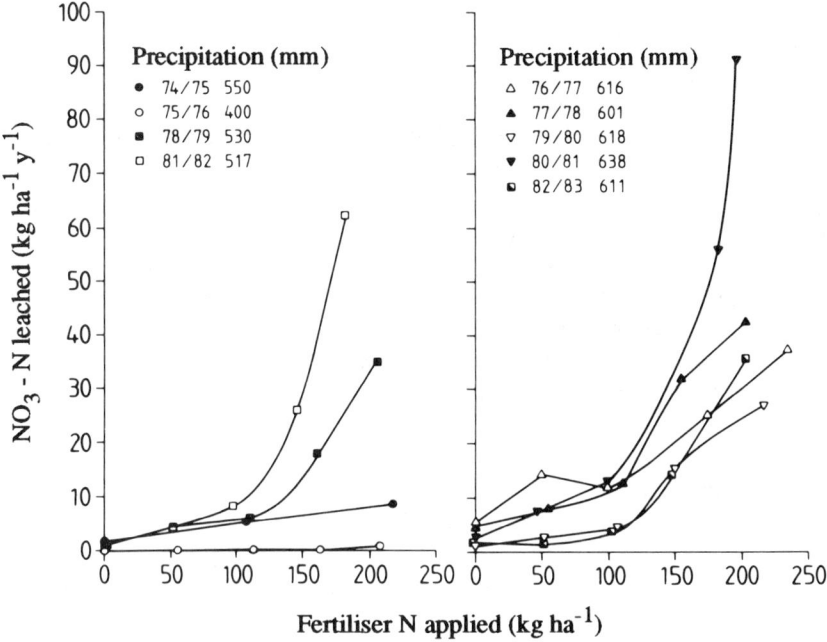

Figure 3.10 Nitrate leaching as a function of the N-fertilisation intensity during years with precipitation over 600 mm per year (*right graph*) and during years with precipitations less than 600 mm per year (*left graph*) (from Bergstrom and Brink, 1986; reprinted by permission of Kluwer Academic Publishers)

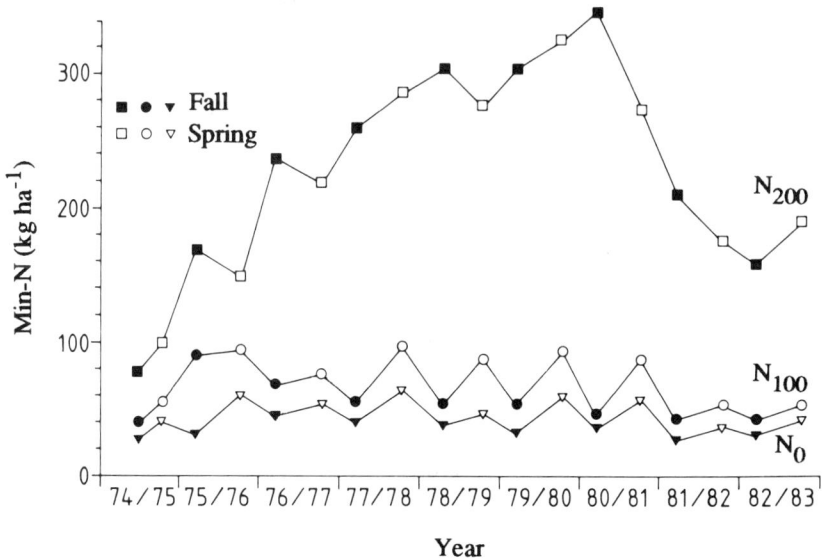

Figure 3.11 The seasonal variation of the mineral nitrogen content of the N_0, N_{100} and N_{200} profiles, down to a depth of 2 m (from Bergstrom and Brink, 1986; reprinted by permission of Kluwer Academic Publishers)

3.4.2.2 Effect of soil texture

On soils which are freely drained, nitrate leaching has to be estimated either by extracting a soil monolith for use as a lysimeter and imposing an artificial lower boundary condition on the soil, or by sequential sampling of the bulk soil or the soil solution. In this latter case, nitrate leaching may be estimated by difference or by independent estimation of water flux associated with the soil solution concentrations measured. Webster, Belford and Cannell (1986) compared the leaching of nitrogen from sandy loam and clay soils using lysimeters. Averages of 74 (range 34−102) and 41 (range 15−73) kg N ha^{-1} were leached from the sandy loam and clay soils, respectively, over six years. Kolenbrander (1981) has tried to remove some of the confounding effects which prevent comparison of results between soils of differing textures. He assumed that for a given site nitrate leaching was proportional to drainflow, and normalised data from many sites on this basis to a drainflow of 300 mm per annum.

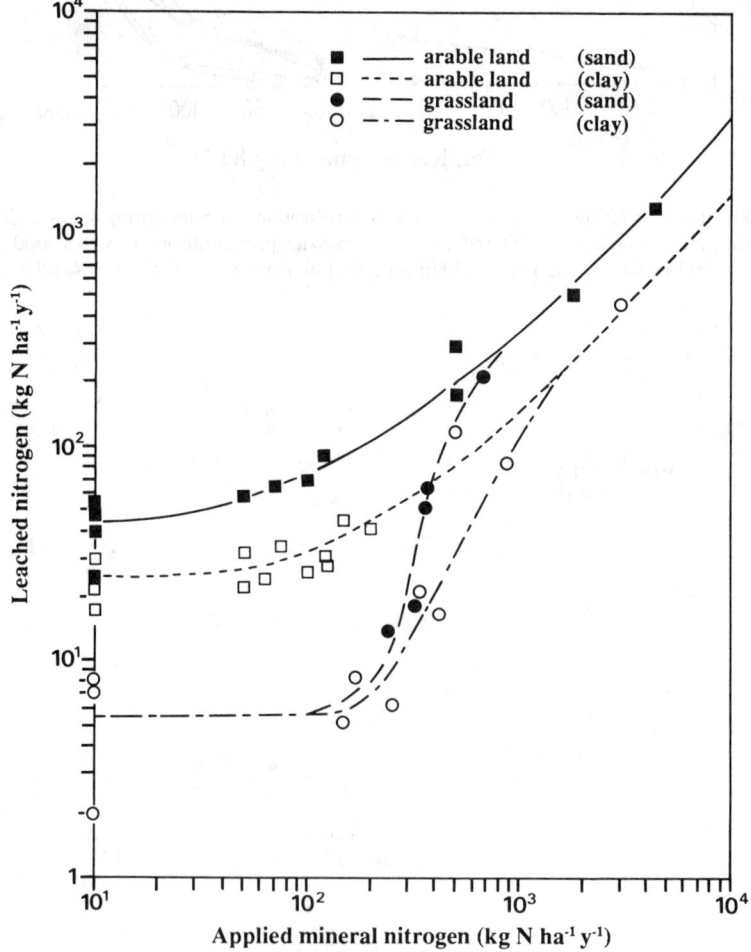

Figure 3.12 Leaching of nitrate from arable land and grassland (from Kolenbrander, 1981; reprinted by permission of Kluwer Academic Publishers)

He found that for arable soils nitrate leaching depended on soil texture (Figure 3.12), with clay soils losing about half the nitrate lost from sandy soils as long as applications did not exceed $100-200$ kg ha^{-1}. Once the application exceeded this range, leaching losses increased rapidly, and became less dependent on soil texture.

3.4.2.3 Effect of land use

The study by Kolenbrander (1981) also showed that arable land was more prone to leaching than cut grassland. This is borne out by the work of Barraclough, Hyden and Davies (1983), quoted above, but the situation is quite different if the grass is grazed *in situ*.

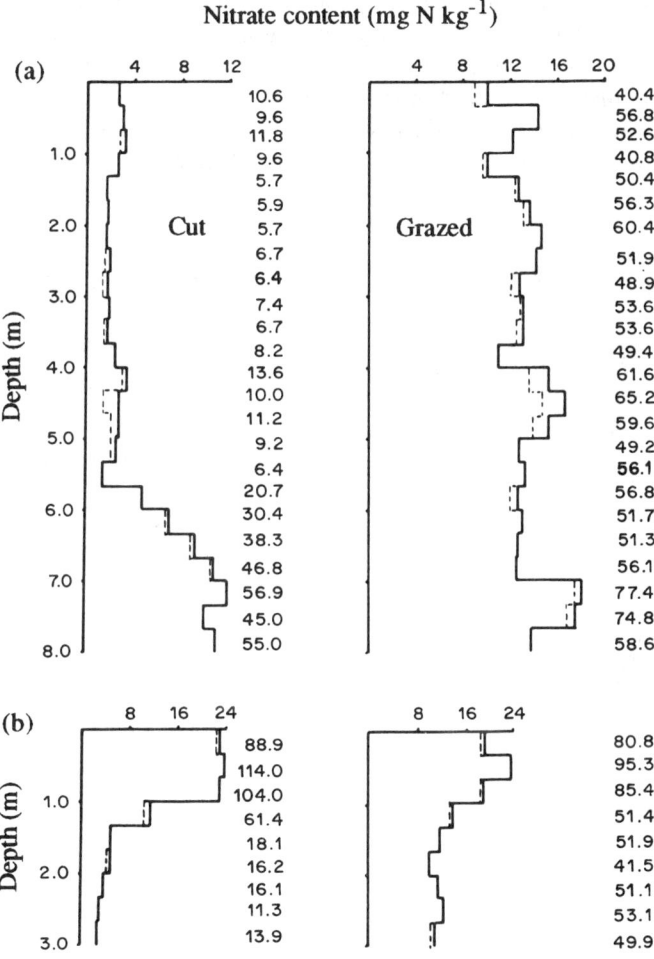

Figure 3.13 Nitrate content (mg N kg^{-1}) of samples from successive depths below cut and grazed ryegrass swards receiving 420 kg N ha^{-1} a^{-1} (a) before the swards were ploughed in December 1982 and (b) 12 months after ploughing and one crop of spring barley. The numbers alongside each profile are the nitrate concentrations in solution (mg N l^{-1}) at each depth (after Ryden, Ball and Garwood, 1984; reprinted by permission from *Nature*, **311**, 50–53; copyright © 1984 Macmillan Magazines Limited)

When nitrogen is returned in faeces and urine in grazed systems the leaching loss is generally much higher. In one experiment at the Institute for Grassland and Animal Production, Hurley, the loss was 5.6 times higher than under a cutting regime, and higher than from arable land (Ryden, Ball and Garwood, 1984; Figure 3.13). The reason for this increased leaching under grazing is twofold: first, the amount of nitrogen being removed in liveweight gain/milk production is much smaller than that removed in cut grass. Second, the very high concentrations of nitrate which can develop under urine patches (referred to in a previous section) greatly exceed the capacity of the grass for uptake.

Other data from the same institute (Garwood and Ryden, 1986) suggested leaching losses of about 160 kg N ha^{-1} a^{-1} from grazed ryegrass swards receiving 420 kg N ha^{-1} a^{-1}, whereas those from grazed grass-clover swards relying on N-fixation for nitrogen input were only one-seventh of this amount. However, in contrast, data from Northern Ireland from drained plots on which stock were grazed (Stevens et al., unpublished) suggested that nitrate leaching from grazed grass-clover swards receiving no nitrogen fertiliser (ca. 40 kg N ha^{-1}) was equivalent to that from grazed grass-only plots receiving 300 kg N ha^{-1} as fertiliser. There is other evidence that the reduction in leaching losses from grass-clover swards only applies to those with a modest proportion of clover; when the proportion is high the release

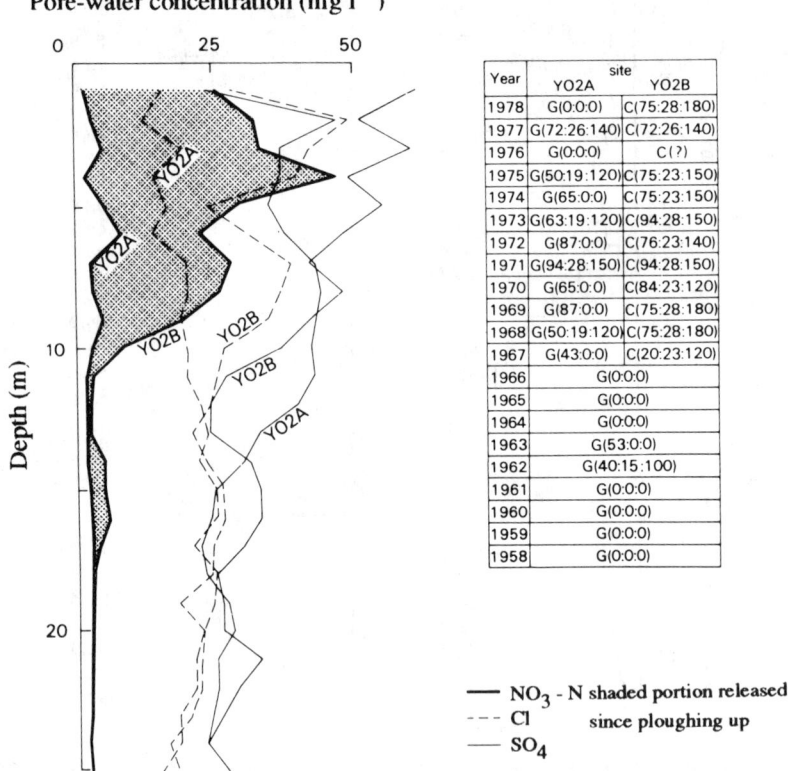

Year	site	
	YO2A	YO2B
1978	G(0:0:0)	C(75:28:180)
1977	G(72:26:140)	C(72:26:140)
1976	G(0:0:0)	C(?)
1975	G(50:19:120)	C(75:23:150)
1974	G(65:0:0)	C(75:23:150)
1973	G(63:19:120)	C(94:28:150)
1972	G(87:0:0)	C(76:23:140)
1971	G(94:28:150)	C(94:28:150)
1970	G(65:0:0)	C(84:23:120)
1969	G(87:0:0)	C(75:28:180)
1968	G(50:19:120)	C(75:28:180)
1967	G(43:0:0)	C(20:23:120)
1966	G(0:0:0)	
1965	G(0:0:0)	
1964	G(0:0:0)	
1963	G(53:0:0)	
1962	G(40:15:100)	
1961	G(0:0:0)	
1960	G(0:0:0)	
1959	G(0:0:0)	
1958	G(0:0:0)	

——— NO$_3$ - N   shaded portion released
- - - Cl   since ploughing up
——— SO$_4$

Figure 3.14 Pore-water profiles for chalk unsaturated zone in east Yorkshire showing the effects of partial conversion from pasture to arable land in autumn 1966 (G, permanent grassland; C, winter or spring cereals; (N:Cl:soil) ratio (kilograms per hectare) applied annually as fertilisers) (from Foster, Cripps and Smith-Carington, 1982; reproduced by permission of The Royal Society)

of nitrogen to the environment may be as high as from heavily fertilised grass (Macduff, Jarvis and Roberts, 1990).

Changes in land use have involved much ploughing of permanent grassland over the last few decades, with consequent acceleration in the rate of mineralisation of soil organic matter and increased nitrate leaching. In localities where the groundwater table is deep, the nitrate resulting from this change can still be on its way down to the water table, as demonstrated by analysis of deep core samples (Foster, Cripps and Smith-Carington, 1982; Figure 3.14; see also Chapter 8). The ploughing up of temporary grassland is also a major cause of nitrate leaching. The release of nitrate following mineralisation is shown in part (b) of Figure 3.13 and much of this is vulnerable to leaching. Cameron and Wild (1984) measured a loss of 100 kg of nitrogen over two years following autumn ploughing of a 3-year grass ley, and concluded that a considerable part of the loss was due to leaching in the first winter, prior to sowing with a spring crop.

The nature of the crop has a major bearing on the extent of leaching losses from British arable farming systems. Winter wheat is the crop least likely to cause nitrate leaching; potatoes, oilseed rape and some other crops have a higher risk (Tinker, 1991). The effectiveness of winter wheat is principally because of its early establishment and root development in the autumn, and thus its capacity to 'mop up' nitrate remaining from the previous crop or mineralised from crop residues after harvest. The importance of early sowing, in this connection, is illustrated by Figure 3.15.

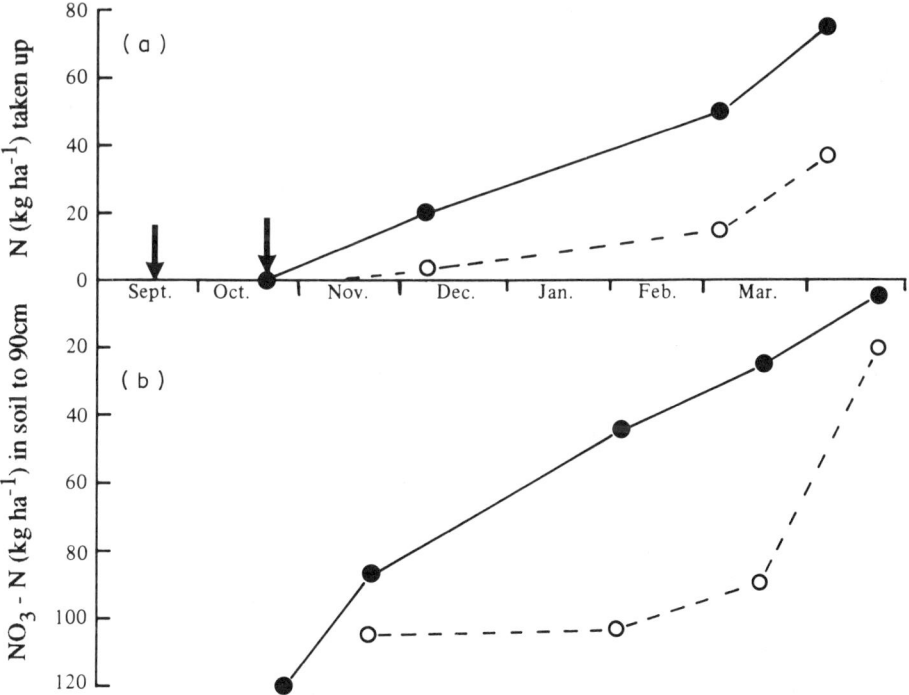

Figure 3.15 The relationship between the mean amount of NO_3-N in soil in autumn and winter and the mean uptake of N by wheat sown, either in September (- ● -) or October (- o -) in two contrasting cropping situations in Rothamsted. (a) Wheat following potatoes, 1981−3 and (b) wheat following oats, 1982−4 (from Widdowson et al., 1987; reproduced by permission of Cambridge University Press)

3.4.2.4 Effects of cultivation

Dowdell, Colbourn and Cannell (1987) reported that on direct drilled plots, over 4 years, leaching losses of nitrogen were only 48−89% of losses from ploughed plots. Vinten, Howard and Redman (1991) also found higher leaching losses from plots that had been cultivated (chisel ploughed and subsoiled) than from plots left in stubble over the winter (Figure 3.16). The probable reason is that a larger proportion of water drains through the undisturbed soils, through fairly continuous macropores, by 'bypass flow'. However, the effect of the cultivation in promoting aeration and consequently higher mineralisation and lower denitrification losses is no doubt also important.

3.4.3 METHODS OF INVESTIGATION OF NITRATE LEACHING

There is an obvious need to be able to extrapolate results of experimental work on nitrate leaching obtained in one environment to other situations, in order to form an overall picture of the extent of the problem and formulate public policy accordingly. Two different approaches to this objective are considered below.

3.4.3.1 ^{15}N Tracer experiments

The use of ^{15}N tracer studies as an aid to interpreting the results of nitrate leaching studies is widespread. For example, Powlson (1988) reported ^{15}N recoveries in soil and crop at harvest, following spring fertiliser application to 14 winter wheat crops. His data indicate average losses of 30 kg N ha^{-1}. Partitioning between denitrification and leaching was not possible. Webster, Belford and Cannell (1986) used ^{15}N-labelled ammonium nitrate fertiliser

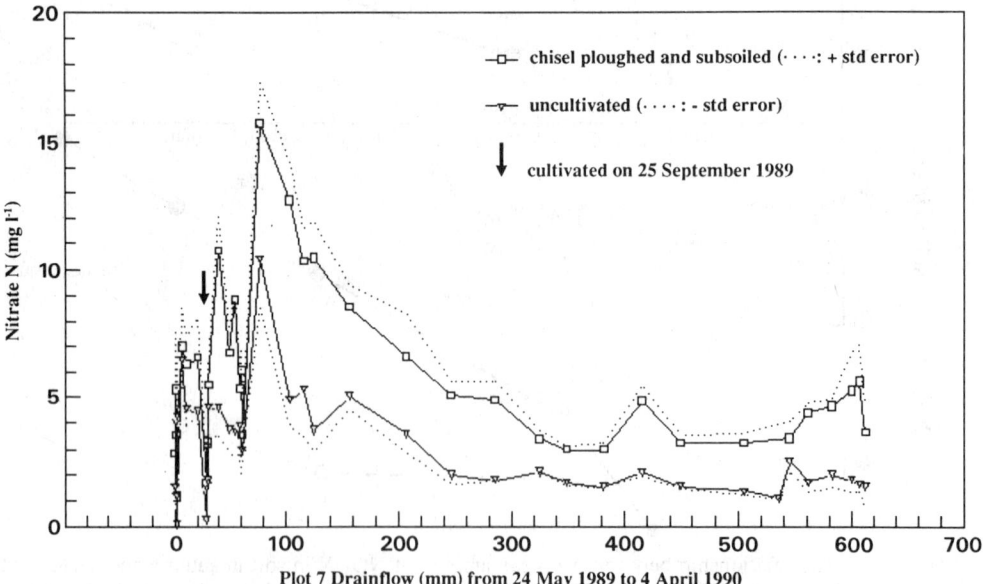

Figure 3.16 Effect of cultivation on nitrate content of water in field drains, Glencorse Mains, Midlothian, Scotland. Soil cultivated (chisel plough and subsoiler) on 25 September 1989 (Vinten, unpublished)

in the first year of the lysimeter experiments on sandy loam and clay soils referred to above. The fate of the labelled nitrogen was followed over the following 5−6 years. Over the period, only 3.8% of the labelled nitrogen appeared in the leachate in the sandy loam lysimeters and 5% in that of the clay lysimeters. On the sandy soil 22.5% (23 kg ha^{-1}) of the labelled nitrogen remained in the soil and 11.3% (12 kg ha^{-1}) was unaccounted for. Data such as these (and the results of Dowdell, Colbourn and Cannell, 1987) illustrate that much of the nitrogen found in drainage water does not derive directly from the fertiliser application of that particular year.

Other studies done with ^{15}N-labelled fertiliser have shown that, commonly, only a few kg ha^{-1} of the labelled material is left in the soil as nitrate after harvest, and is thus vulnerable to leaching (Macdonald et al., 1989; Tinker, 1991). However, this does not mean that fertiliser N has no influence on the leaching of nitrogen. As Tinker (1991) points out, there is continual interchange between labelled and non-labelled nitrogen in the soil via microbial metabolism, so the real question is whether the application of nitrogen fertiliser has led to any extra nitrate being left — labelled or not. The evidence from work such as that of Bergstrom and Brink (1986) and Addiscott, Whitmore and Powlson (1991), is that significant extra nitrate is left only when the nitrogen supply exceeds the crop's demands, but that seasonal factors cause substantial variations in this demand and thus make prediction extremely difficult. The position of the point on the yield response to nitrogen curve at which nitrate leakage begins to increase may not coincide with the economic optimum application. It also appears, from the Broadbalk experiment at Rothamsted, that the proportion of easily available nitrogen in the soil does increase after prolonged cropping, even when fertilisation does not exceed crop demand (Addiscott, Whitmore and Powlson, 1991).

3.4.3.2 Information from nitrogen balances

Long-term studies of the nitrogen balance in agricultural soils, under different climatic conditions, can provide invaluable information on the effects of farming practice on the leakage of nitrate from agricultural land to ground and surface waters. Information from such studies is essential for thorough testing of the long-term predictive power of models of the agricultural soil−plant nitrogen cycle, which include calculations of mineralisation, immobilisation, denitrification and crop uptake, as well as nitrate leaching. Such models are becoming increasingly important in making policy decisions. However, although they may be successful in short-term tests (see, for example, *Fertiliser Research*, Vol. 27 (1991) for a review and comparison), they are not necessarily successful in the medium or long term, because initial conditions of the soil organic matter, mineral nitrogen burden, etc. can have an overriding effect on their performance. Also, uncertainties about the systematic error in measurements of denitrification, even in studies such as that of Paustian et al., (1990a,b) which attempt to measure every component of the nitrogen cycle, will have a disproportionately large influence on estimates of change in organic matter content. This is important, because the rate of release of nitrogen from humified organic matter is crucial to the whole nitrogen cycle and to nitrate leaching in particular.

The early Rothamsted and Craibstone drain gauge experiments, and the long-term crop experiments at Rothamsted, provide data for model evaluation that could not be obtained in other ways. The Rothamsted drain gauges were constructed by isolating blocks of soil with a surface area of 0.001 acre (4.047 m^2), 0.5, 1.0 and 1.5 m deep, which were under-drained with minimum soil disturbance (Lawes, Gilbert and Warington, 1881). The soil was kept fallow and nitrate leakage was measured over many years. Addiscott (1988) cited the average annual loss from these blocks (or lysimeters) for the period 1887−1915 as 28−29 kg N ha^{-1} a^{-1}, and calculated a first-order rate constant for the decrease in soil nitrogen leakage of 0.0171 a^{-1}. Other work at Rothamsted gives a value for the organic nitrogen decay constant of 0.031 a^{-1} for the Hoosfield continuous barley experiment in which farmyard manure was applied every year from 1852, and one of 0.028 a^{-1} for the constant for grassland (Jenkinson, 1991). These latter constants correspond to turnover times of 32 and 36 years, respectively, whereas the smaller constant for the fallow soil in the drain gauges corresponds to a turnover time that is considerably longer (57 years). This is evidence that some at least of the organic nitrogen in cropped soils (which in normal practice have annual inputs of nitrogen as fertiliser or manure) is more labile than that in a long-term unfertilised soil.

In 1914, another drain gauge experiment was set up in a very contrasting environment from that at Rothamsted, the results of which are perhaps less well known, but which add substantially to the overall picture of long-term nitrogen transformations in British soils. A set of three 1.0 m deep gauges, similar to those at Rothamsted, was constructed at Craibstone Farm near Aberdeen, where the soil (loamy sand) and climate are very different (Hendrick, 1921, 1930; Hendrick and Welsh, 1938). The soil had about 20% more organic nitrogen than at Rothamsted (0.18% in the top 23 cm). The three lysimeters were all placed under a 6-year grass−arable rotation typical of the region and received the following three fertilisation treatments:

(1) No manure or N fertiliser;
(2) 26 kg ha^{-1} a^{-1} inorganic N + 12 tonnes FYM once in each rotation;
(3) N input as (2) above + lime (5.2 tonne ha^{-1} once in each rotation).

When the nitrogen balance is calculated it appears that the unfertilised block (1) released nitrogen (to crop + leachate) at an average rate of 71 kg ha^{-1} a^{-1} (Figure 3.17), a much larger quantity than the 28−29 kg ha^{-1} a^{-1} over the period 1877−1915 for the fallow drain gauges at Rothamsted, despite a rather similar organic nitrogen content of the top 23 cm of the soil. The overall effect of addition of farmyard manure or mineral nitrogen, in the modest quantities applied at Craibstone, on nitrate leaching losses was very small (Table 3.3). These figures from Craibstone, even though they were produced by work more than half a century ago, in a vastly different technical and economic agricultural environment to that of today, provide a useful pointer to the levels of nitrate we may expect to see leaking from some types of low-input system.

The leaching of nitrogen from artifically drained plots much larger in area than drain gauges has been measured by several workers. Such experiments have the advantage that the soil is largely undisturbed and that normal agronomic operations can occur. They have the disadvantage that only partial recovery of drainage water takes place and lateral movement of water below the depth of hydrological isolation may influence results. Drainage volumes measured from closed systems such as lysimeters may, in contrast, represent overestimations of actual drainage volumes in a field situation (Bergstrom, 1987). The results of three such studies are compared here, as they give a useful insight into the influence of climatic influence (and possibly researchers' predilections) on the nitrogen cycle and nitrate leaching.

The experiment at Brimstone, Berkshire, England (Harris et al., 1984; Dowdell, Colbourn and Cannell, 1987) has already been mentioned. Over four seasons, leaching losses ranged from 3 to 75 kg ha^{-1} a^{-1}, with an average of 34 kg ha^{-1} a^{-1}. Most of the losses took place

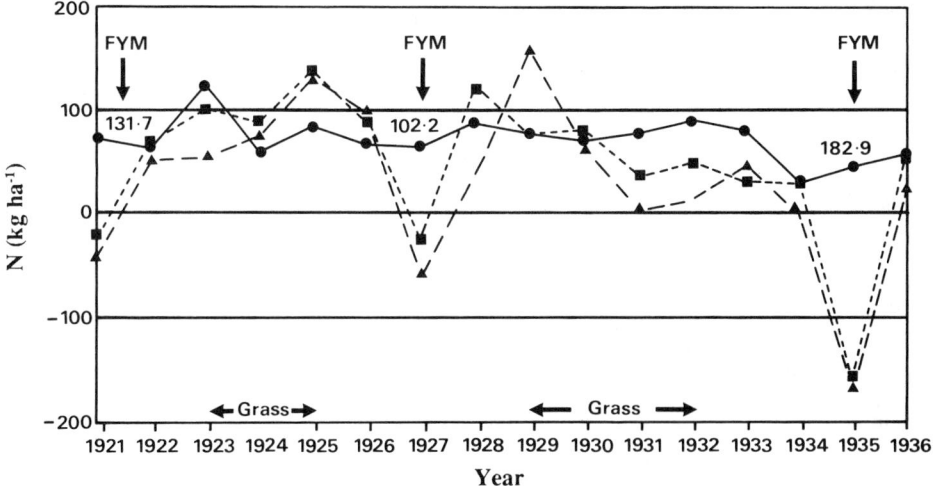

Figure 3.17 Estimated net output (offtake plus leaching minus inputs) from Craibstone drain gauges over 16-year period. Values in the figure are N inputs as FYM to gauges 2 and 3. ● Gauge 1 (no manure or lime); ■ Gauge 2 (manure, no lime); ▲ Gauge 3 (manure and lime) (based on data from Hendrick, 1930; Hendrick and Welsh, 1934; Hendrick and Welsh, 1938)

through the mole drainage system. Rough calculations show that about 50% of autumn-applied nitrogen and up to 15% of spring applications was lost by leaching.

Paustian *et al.*, (1990a,b) studied the nitrogen budget in four agro-ecosystems as part of the 'Ecology of Arable Land' project in Uppsala in Sweden. The balance sheet for a single year which they quote indicates a system close to equilibrium for a fully fertilised barley crop (Table 3.4). However, this may be deceptive; they obtained the very low figure of 5 kg ha^{-1} a^{-1} for loss by denitrification, using the acetylene-block method, and this may have been a considerable underestimate. Leaching losses were of the order of

Table 3.3 Mean annual nitrogen balance sheet for Craibstone lysimeter 1921−36

	No fertiliser, manure/lime	Fertiser, manure only	Fertiliser, manure and lime
Rainfall	40	4	4
Fertiliser	20	27	27
Manure	0	26	36
Fixation[a]	11	9	9
Total inputs	15	66	66
Harvested crop	81	90	107
Leaching	5	7	7
Total outputs	86	97	114
Annual net balance	−71	−31	−48

[a] Using data from Hendrick and Welsh (1934) to estimate clover N content of sward and the data of Heichel *et al.* (1985) to estimate fixed N contribution to clover N (54% of harvested N from clover is fixed; 6% of clover N is in roots). 1923−5 and 1929−32 were years in which a grass or grass/clover ley was present.

All figures are in kg N ha^{-1}.

Table 3.4 Typical N balance sheet from four treatments in 'Ecology of arable land' project, Uppsala, Sweden (adapted from Paustian *et al.*, 1990b; reproduced by permission of Blackwell Scientific Publications Ltd)

	Spring barley	Spring barley	Grassland ley	Lucerne ley
			kg N ha^{-1}	
Inputs				
Fertiliser	0	120	200	0
Manure	—	—	—	—
Rain	5	5	5	5
Fixation	—	—	—	380
Seed	5	5	negl	negl
Outputs				
Crop offtake	36	127	241	246
Leaching	5	10	1	1
Dentrification	4	5	10	20
Plant N	0	0	50	90
Balance	−35	−12	−47	+118

10 kg ha^{-1} a^{-1} under spring barley receiving recommended nitrogen fertiliser dressings. Losses were still lower on the zero nitrogen treatment and under grass or lucerne leys. This study is invaluable as a source of information about the internal cycling of carbon and nitrogen in the system.

Vinten, Howard and Redman (1991) have studied nitrogen leaching from drained plots on a clay loam soil in south-east Scotland. The land is class 3.2 and at 200 m altitude, and has typically much larger annual drainflow amounts than the other sites mentioned (*ca.* 400 mm). The balance sheet for the five years 1987–92 has been estimated for the arable-only plots, after measuring the fertiliser-derived and soil-derived nitrogen in the crop by ^{15}N tracer methods, and assuming no change in the mineral nitrogen content of the soil from year to year (April to March). Nitrogen return in roots and stubble was assumed to be 40% of total uptake. Any effect of added nitrogen on uptake of soil N is assumed to be real (i.e. pool substitution is not considered).

Leaching losses averaged 26 kg ha^{-1} a^{-1} over all plots, with a range of 14–47 kg ha^{-1}, over the five years (Figure 3.18). Maximum losses were not usually associated with the highest fertiliser application. The fertiliser nitrogen unaccounted for (after accounting for that in roots and stubble) averaged 52 kg ha^{-1} a^{-1}. This loss is much larger than that estimated at the site by the acetylene block method (Redman, 1991), but the results obtained with this method are questionable on all heavy soils because of poor diffusion of acetylene into the soil (Arah *et al.*, 1991). Other data on nitrogen recovery by spring and winter cereals on a similar soil to that studied by Vinten, Howard and Redman (1991) also point to large denitrification losses (Smith, Howard and Crichton, 1988). Figure 3.18 summarises the nitrogen balances of the three sites described.

Leaching losses to drainage water appear to be related to climatic conditions influencing post-harvest mineralisation of soil nitrogen. Of the three environments considered, losses were larger in southern England, despite lower drainflow, than those in southern Scotland, which in turn were larger than in southern Sweden. However, the recommended fertiliser application to winter cereals in the English work (175 kg ha^{-1}) was also larger than those to the spring cereals in Scotland or Sweden (120 kg ha^{-1}). The figures for atmospheric

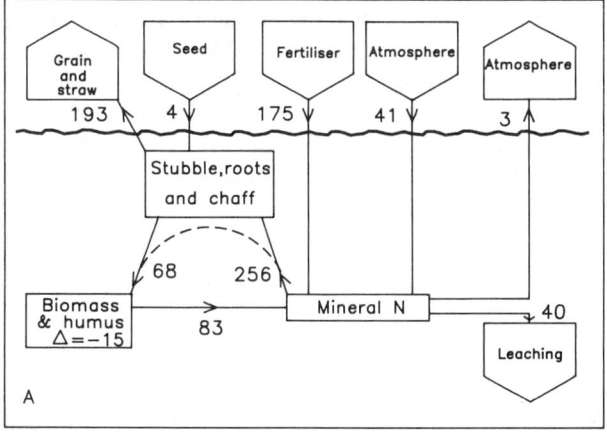

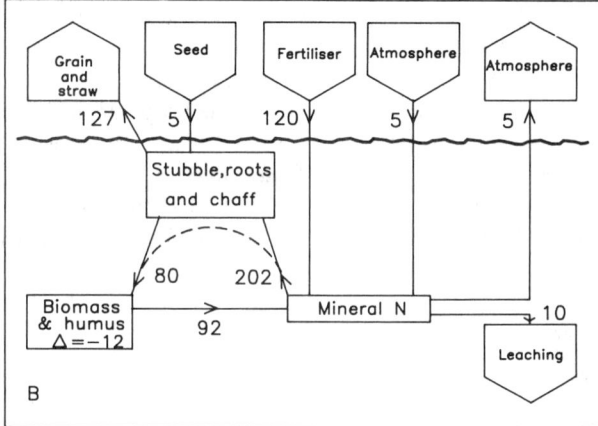

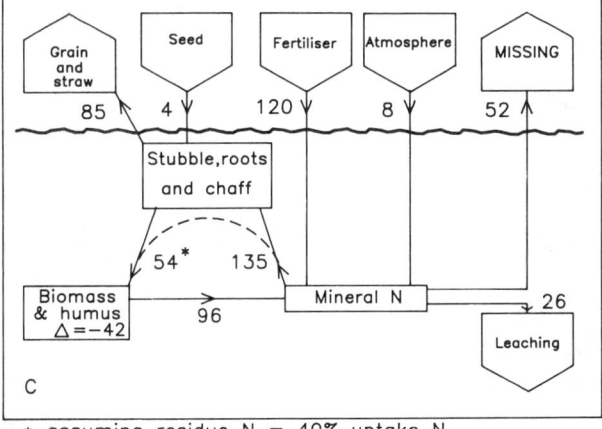

* assuming residue N = 40% uptake N

Figure 3.18 A comparison of estimated nitrogen balances on drained plots at three sites. (a) Brimstone, England (Goss, pers. comm.), winter wheat, clay soil. (b) Uppsala, Sweden (Paustian *et al.*, 1990a, b), spring barley, loam soil. (c) Glencorse Mains, Midlothian, Scotland (Vinten, Howard and Redman, 1991 and unpublished), spring barley, clay loam soil

inputs at Uppsala and Glencorse were lower than at Brimstone, but may be underestimates because only at Brimstone was dry deposition considered. Some uncertainty also exists about the denitrification figures for Uppsala and Brimstone for reasons mentioned above; the corresponding figure at Glencorse is an upper bound, because pool substitution of ^{15}N may have influenced the estimate of uptake of fertiliser N. There is also considerable uncertainty about root returns and the exudate contribution to mineralisation/immobilisation of nitrogen at Brimstone and Glencorse.

3.4.4 MODELLING NITROGEN TURNOVER AND LEACHING LOSSES IN ARABLE SOILS

3.4.4.1 Mechanistic models

There has been a tremendous growth in the number and complexity of models dealing with the dynamics of the nitrogen cycle in agricultural soils. Figure 3.19 gives a schematic view of one such model — that of Johnsson *et al.* (1987) used at Uppsala, Sweden. It illustrates the importance of the carbon cycle, which controls the mineralisation, immobilisation and denitrification of nitrogen and hence must have a central role in any successful nitrogen model. A full review of nitrogen cycle models is outside the scope of this chapter, and the reader is referred to two recent journal issues which have been devoted to this subject: *Fertiliser Research*, Vol. 27 (1991) and *Soil Use and Management*, Vol. 7, No. 2 (1991). In his analysis (in *Fertiliser Research*) of the degree of success of 14 simulation models tested against the same data sets, de Willigen (1991) concluded that: predictions of above-ground variables

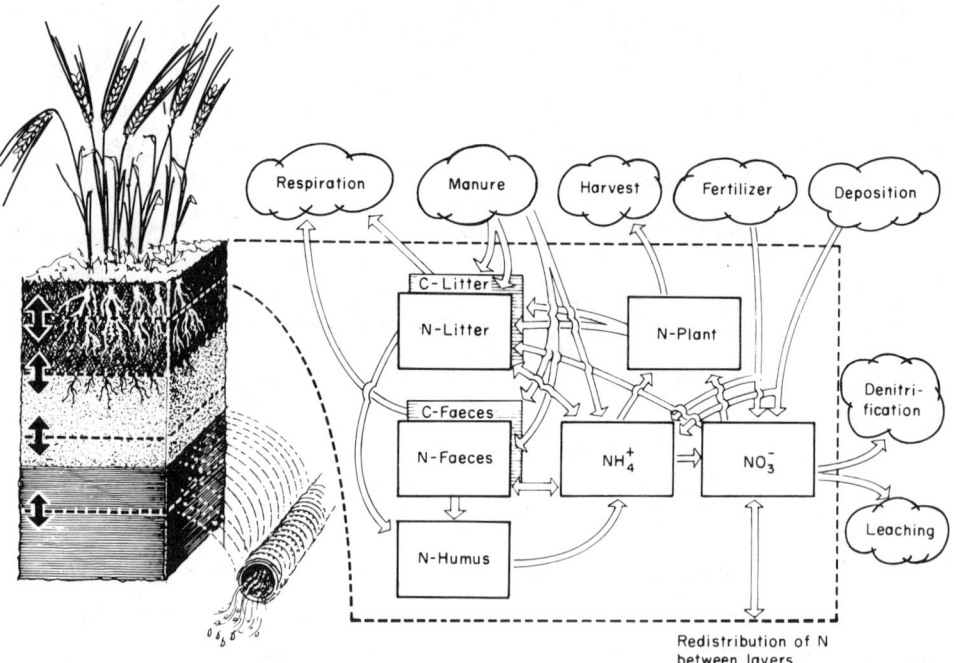

Figure 3.19 Structure of the nitrogen model of Johnsson *et al.*, (1987) (reproduced by permission of Elsevier Science Publishers)

yielded better results than those of below-ground ones; the simulation of phenomena such as the loss of mineral nitrogen immediately after fertiliser application was not good; and simple, functional water models predicted soil water content as well as (and often better than) mechanistic models.

However, on soils where structure influences nitrate leaching, accurate prediction of nitrate leaching requires the incorporation of macropore flow into the model. Bergstrom and Jarvis (1991) compared SOIL-SOILN model predictions (Johnsson *et al.*, 1987) with measured leaching from the experiments of Bergstrom and Brink (1986). The substantial errors that occurred were attributed partly to the failure of the model to predict macropore flow (Figure 3.20).

Moreover, if denitrification is an important loss mechanism a simple water-balance model may be inadequate. This can be illustrated by the results of simulations (Vinten, 1991, unpublished) of annual leaching and denitrification losses from the Glencorse site in Scotland made using the ANIMO model (Berghuis-van Dijk, Rijtema and Roest, 1985; Rijtema, Roest and Kroest, 1990) from the Netherlands, linked to a simple water-balance type model,

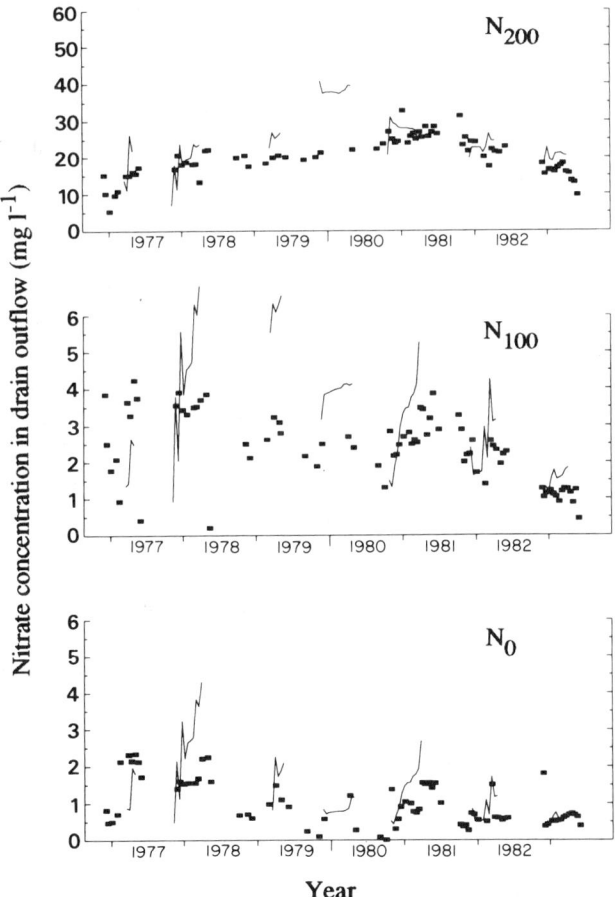

Figure 3.20 Measured (■) and predicted (solid line) nitrate concentrations in the drain outflow from the N_0, N_{100} and N_{200} plots (note different scales on the Y-axis) (from Bergstrom and Jarvis, 1991; reproduced by permission of Blackwell Scientific Publications Ltd)

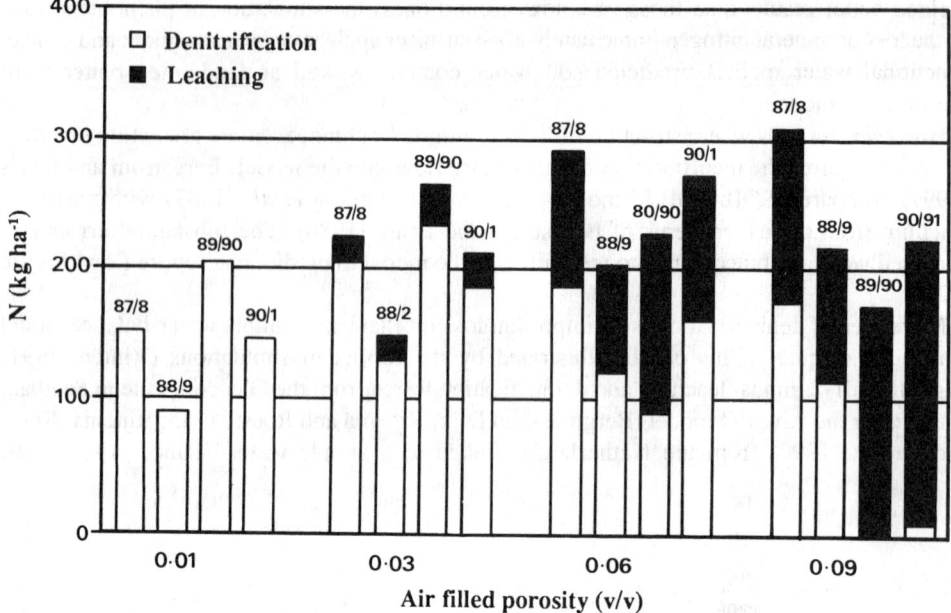

Figure 3.21 Prediction of N loss by denitrification and leaching from drained plot on clay loam soil, Glencorse Mains, Midlothian, Scotland, using ANIMO model. Effect of varying air filled porosity at −10 kPa water potential (Vinten, unpublished)

WATBAL (Berghuis-van Dijk, 1985). The simulations (Figure 3.21) were made using a daily time step. The moisture release function of the topsoil at Glencorse gives a value of 9% for airfilled porosity at −10 kPa tension (approximately field capacity). However, sensible predictions of the leaching and denitrification losses are obtained from the model only when the air-filled porosity at field capacity is reduced to about 3%. If, instead, the humus decomposition constant (0.02 a^{-1}) is decreased by about an order of magnitude, prediction of leaching losses is also reasonable, but then no denitrification is predicted, because oxygen demand is very low.

Soil aeration also influences carbon and nitrogen mineralisation and it appears therefore that modelling the influence of soil aeration on mineralisation and denitrification is essential for the improvement of nitrogen leaching predictions, at least for soils prone to poor aeration. A physically accurate water- and air-transport model must be linked to biologically accurate descriptions of the processes of organic matter breakdown and electron transfer to oxidants in the soil.

In sandy, well-aerated soils, the prediction of the release of nitrogen from soil organic matter and of leaching losses, in the absence of significant gaseous losses, is simpler. Vereecken *et al.* (1991) reported a comparison of five models, using four data sets from Belgium, Denmark and the Netherlands, on light-textured soils to which slurry and fertiliser nitrogen had been applied. Nitrate leaching was estimated by means of suction samplers or soil samplers, combined with an estimation of downward water flux. Table 3.5 gives statistical criteria for the performance of the different models on simulation of nitrate leaching. The authors concluded that additional information was needed on the measurement and prediction of evapotranspiration, mineralisation and denitrification for these models to be successfully applied.

Table 3.5 Statistical criteria for the performance of different models on simulation of N leaching (kg ha^{-1} a^{-1}) for the different sites. From Vereecken *et al.* (1991); reproduced by permission of the Commission of the European Communities.

Site	Model	n	ME	RMSE	CD	EF	CRM
Askov,	ANIMO[b]	6	13.4	20.2	1.29	0.60	−0.15
Denmark	DAISY[c]	5	24.5	30.0	0.95	0.28	0.10
(sandy loam)	EPIC[d]	6	20.4	24.4	0.69	0.42	−0.13
	RENLEM[e]	6	55.4	64.9	0.20	−3.09	−0.18
	SWATNIT[f]	6	16.0	32.9	6.38	−0.05	−0.07
Jyndevad,	ANIMO	12	41.3	23.0	1.02	0.82	−0.10
Denmark	DAISY	12	26.5	19.4	1.25	0.87	0.01
(sand)	EPIC	12	49.6	28.0	0.72	0.74	0.17
	RENLEM	12	45.2	35.5	0.68	0.58	0.17
	SWATNIT	12	54.1	28.9	0.82	0.72	−0.07
Ruurlo,	ANIMO	24	61.8	56.5	1.87	0.44	0.17
Netherlands	DAISY	12	39.7	47.1	1.29	0.68	0.15
(loamy sand)	EPIC	24	53.7	52.0	1.46	0.53	−0.13
	RENLEM	24	55.4	76.1	1.15	−−0.01	0.52
	SWATNIT	19	85.7	86.6	0.33	−0.31	−0.45
All[a]	ANIMO	43	61.8	36.2	0.96	0.78	0.00
	DAISY	29	39.7	27.2	1.09	0.89	0.05
	EPIC	43	53.7	37.1	1.17	0.77	0.01
	RENLEM	43	55.4	53.4	0.83	0.52	0.27
	SWATNIT	38	85.7	52.6	0.63	0.53	−0.27

[a] Including Pittem, Belgium.
[b] Rijtema, Roest and Kroes (1990).
[c] Hansen *et al.* (1990).
[d] Williams, Renard and Dyke (1983).
[e] Kragt and de Vries (1987).
[f] Vereecken, Van Clooster and Swerts (1990).

n	: number of observations.
ME	: maximum error (optimum = 0).
RMSE	: root mean squared error (optimum = 0).
CD	: coefficient of determination (optimum = 1).
EF	: modelling efficiency (optimum = 1).
CRM	: coefficient of residual mass (optimum = 0) (see Loague and Green, 1991, for definitions).

3.4.4.2 Regression models

An alternative approach to modelling nitrate leaching losses from land is to use a multiple-regression model to evaluate the important parameters influencing nitrate leaching. Smith and Stewart (1989) have developed a model based on multiple regression which explains 95% of the variation in nitrate load of the major rivers in the 4453 km^2 Lough Neagh catchment in Northern Ireland. The study covered the years 1971−87. The regression equation was:

$$Y = 39.2 + 0.13 \, x_1 - 5.239 \, x_2 - 2.20 \, x_3 + 0.03 \, x_4 \qquad r^2 = 0.95$$

where

x_1 = fertiliser useage (kg ha^{-1} a^{-1}),
x_2 = previous summer's rainfall (mm d^{-1}),
x_3 = mean summer temperature of the current year (°C),
x_4 = river flow, December−May (10^6 m^3 month^{-1}).

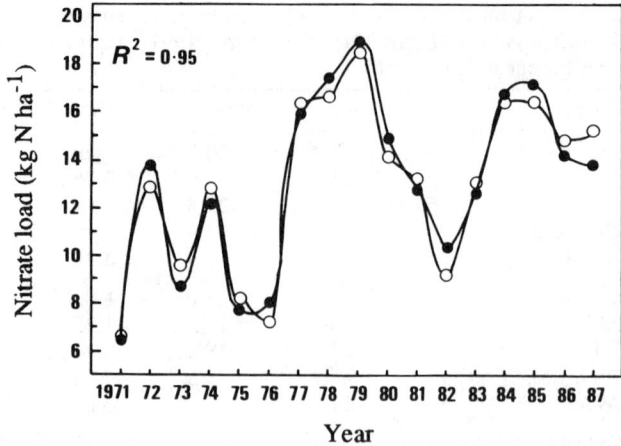

Figure 3.22 Predicted and measured nitrate load of the rivers in the Lough Neagh catchment, Northern Ireland. –o– predicted; –•– measured (from Smith and Stewart, 1989; reproduced by permission of Blackwell Scientific Publications Ltd)

Figure 3.22 shows the measured nitrate load to Lough Neagh and the load predicted by this equation.

Edwards *et al.* (1990) have estimated mean nitrate concentrations at various gauging stations along the rivers Don and Dee in north-east Scotland. A very simple relationship exists, for both catchments, between the percentage of land area under arable farming above the gauging stations and the mean NO_3^--N concentrations in the rivers between 1980 and 1986 (Figure 3.23). Such models may prove useful for predicting the future behaviour of the catchment areas for which they were calibrated, but they are likely to be of limited use when applied to other regions.

3.5 CONCLUSION

The main types of intensive agricultural ecosystem existing in Britain and the rest of Western Europe at present depend heavily on inputs of nitrogen — from mineral fertilisers, animal manures or biological N-fixation by legumes — to maintain productivity. Furthermore, the cycle is leaky, in that a considerable fraction of the annual input of nitrogen is lost to the environment. The losses are only partly as nitrate; the remainder is as ammonia or the gaseous products of denitrification.

The amounts of nitrate leached are relatively little affected by fertiliser application rate to arable crops, below the limit of the plant's capability to take up nitrogen, but thereafter the leaching rate rises. Annual losses are minimised if harvests are followed quickly by sowing of catch crops or winter wheat, rather than leaving the ground fallow until the following spring. Losses from intensively fertilised grazed grassland can be high, because of the high localised inputs where dung and urine are deposited.

The development of process-based models of nitrate leaching, which take into account the other important processes in the nitrogen cycle, is gradually improving our capability to predict

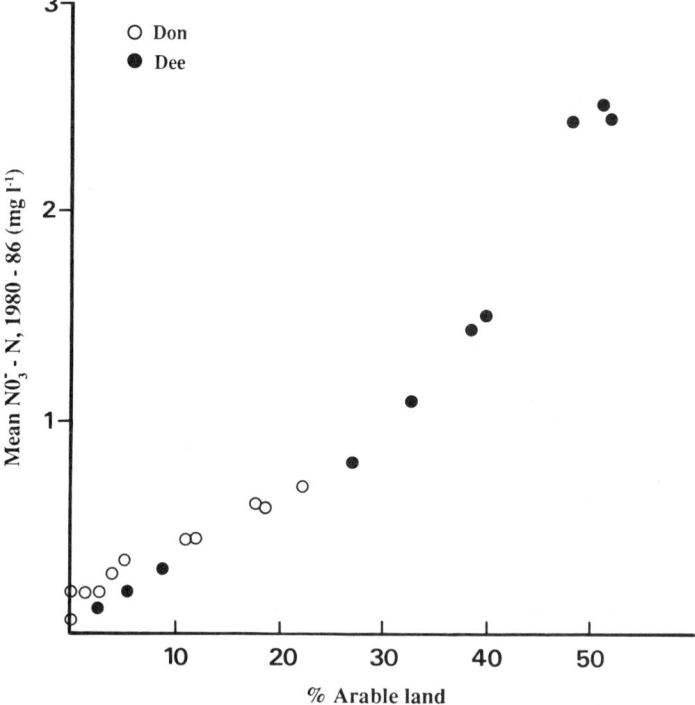

Figure 3.23   Average nitrate-N concentrations (1980−86) in Dee and Don rivers (Grampian Region), at various gauging stations, as a function of the percentage of the catchment area in arable farming (from Edwards *et al.*, 1990; reproduced by permission of Gordon and Breach Science Publishers)

the likely leaching losses in a range of important farming systems. Progress is slow, however, and more work still has to be done to decrease the uncertainties of some of the model inputs.

Present farming practice generally takes too little account of the quantities of nitrogen released in the soil in an available form from decaying crop residues, organic manures and fixation by pasture legumes. More careful adjustment of the amount of fertiliser used, to ensure that the *total* supply does not exceed the crop's needs, could bring about a significant reduction in nitrate leaching from agriculture. Given the vagaries of the weather, the influence of disease on uptake, inaccuracies in fertiliser application and the uncertainties about release of N from organic manures, precise control is almost certainly *un*achievable. However, it is to be expected that reductions in nitrate leaching (in concert with reductions in yields of commodities in surplus) will continue to be sought, in spite of the technical difficulties involved. It may well be that the methods adopted will be increasingly via legally enforced reductions in nitrogen application rates, on a national and/or EC basis (see Chapter 14).

REFERENCES

Addiscott, T.M. (1988) Long-term leakage of nitrate from bare, unmanured soil. *Soil Use and Management*, **4**, 91−5.

Addiscott, T.M. and Bland, G.J. (1988) Nitrate leaching and soil heterogeneity. In Smith, K.A. and Jenkinson, D.S. (eds), *Nitrogen Efficiency in Agricultural Soils*, Elsevier, Barking, Essex, pp. 394–408.

Addiscott, T.M. and Whitmore, A.P. (1991) Simulation of solute leaching in soils of differing permeabilities. *Soil Use and Management*, 7, 94–102.

Addiscott, T.M., Rose, D.A. and Bolton, J. (1978) Chloride leaching in the Rothamsted drain gauges: influence of rainfall pattern and soil structure. *Journal of Soil Science*, 29, 305–14.

Addiscott, T.M., Whitmore, A.P. and Powlson, D.S. (1991) *Farming, Fertilizers and the Nitrate Problem*, CAB International, Wallingford, Oxon.

Allison, F.E. and Klein, C.J. (1962) Rates of immobilization and release of nitrogen following additions of carbonaceous materials and nitrogen to soils. *Soil Science*, 93, 383–6.

Arah, J.R.M. and Smith, K.A. (1990) Factors affecting the fraction of the gaseous products of soil denitrification evolved to the atmosphere as nitrous oxide. In Bouwman, A.F. (ed.), *Soils and the Greenhouse Effect*, Wiley, Chichester, pp. 475–80.

Arah, J.R.M., Smith, K.A., Crichton, I.J. and Li, H.S. (1991) Nitrous oxide production and denitrification in Scottish arable soils. *Journal of Soil Science*, 42, 351–67.

Ball, B.C. (1990) Straw incorporation and tillage methods; straw decomposition, denitrification and growth and yield of winter barley. *Journal of Agricultural Engineering Research*, 46, 223–43.

Ball, P.R. and Keeney, D.R. (1983) Nitrogen losses from urine-affected areas of New Zealand pasture, under contrasting seasonal conditions. *Proceedings of the 14th International Grassland Congress*, 1981, pp. 342–4.

Ball, P.R. and Ryden, J.C. (1984) Nitrogen relationships in intensively managed temperate grassland. *Plant and Soil*, 76, 23–33.

Barraclough, D. (1989) A useable mechanistic model of nitrate leaching. I. The model. *Journal of Soil Science*, 40, 543–54.

Barraclough, D., Hyden, M.J. and Davies, G.P. (1983) Fate of fertiliser nitrogen applied to grassland. I. Field leaching results. *Journal of Soil Science*, 34, 483–97.

Belmans, C., Wesseling, J.G. and Feddes, R.A. (1983) Simulation model of the water balance of a cropped soil: SWATRE. *Journal of Hydrology*, 63, 271–86.

Berghuis-van Dijk, J.T. (1985) WATBAL: A simple water balance model for an unsaturated–saturated soil profile. Institute for Land and Water Management Research, Wageningen, Note No. 1670.

Berghuis-van Dijk, J.T., Rijtema, P.E. and Roest, C.W.J. (1985) ANIMO-Agricultural Nitrogen Model. Institute for Land and Water Management Research, Wageningen, Note No. 1671.

Bergstrom, L. (1987) Nitrate leaching and drainage from annual and perennial crops in the drained plots and lysimeters. *Journal of Environmental Quality*, 16, 11–18.

Bergstrom, L. and Brink, N. (1986) Effects of differentiated applications of fertilizer N on leaching losses and distribution of inorganic N in the soil. *Plant and Soil*, 93, 333–45.

Bergstrom, L. and Jarvis, N.J. (1991) Prediction of nitrate leaching losses from arable land under different fertilization intensities, using the SOIL-SOILN models. *Soil Use and Management*, 7, 79–85.

Biederbeck, V.O. and Campbell, C.A. (1973) Soil microbial activity as influenced by temperature trends and fluctuations. *Canadian Journal of Soil Science*, 53, 363–76.

Bijay-Singh, Ryden, J.C. and Whitehead, D.C. (1988) Some relationships between denitrification potential and fractions of organic carbon in air-dried and field-moist soils. *Soil Biology and Biochemistry*, 20, 737–41.

Birch, H.F. (1960) Nitrification in soils after different periods of dryness. *Plant and Soil*, 12, 81–96.

Birch, H.F. (1964) Mineralization of plant nitrogen following alternate wet and dry conditions. *Plant and Soil*, 20, 43–9.

Bronswijk, J.J.B. (1991) *Magnitude, modelling and significance of swelling and shrinkage processes in clay soils*, Doctoral thesis, Wageningen Agricultural University, Wageningen, The Netherlands.

Bouwman, A.F. (1990) *Soils and the Greenhouse Effect*, Wiley, Chichester.

Cameron, K.C. and Wild, A. (1984) Potential acquifer pollution from nitrate leaching following the plowing of temporary grassland. *Journal of Environmental Quality*, 13, 274–8.

Campbell, C.A. and Biederbeck, V.O. (1982) Changes in mineral N and numbers of bacteria and actinomycetes during two years under wheat-fallow in Southwestern Saskatchewan. *Canadian Journal of Soil Science*, 62, 125–37.

Colbourn, P. and Dowdell, R.J. (1984) Denitrification in field soils. *Plant and Soil*, **76**, 213−26.

de la lande Cremer, L.C.N. (1986) Dutch experience with slurry injection. In Dam Kofoed, A., Williams, J.H. and L'Hermite, P. (eds), *Efficient Land Use of Slurry and Manure*, Elsevier, London, pp. 99−105.

de Willigen, P. (1991) Nitrogen turnover in the soil crop system; comparisons of fourteen simulation models. *Fertiliser Research*, **27**, 141−9.

Dowdell, R.J., Colbourn, P. and Cannell, R.Q. (1987) A study of mole drainage with simplified cultivation for autumn-sown crops on a clay soil. 5. Losses of nitrate-N in surface run-off and drain water. *Soil and Tillage Research*, **9**, 317−31.

Edwards, A.C. and Cresser, M. (1992) Freezing and its effects on soil chemical and biological properties. *Advances in Soil Science* (in press).

Edwards, A.C., Pugh, K., Wright, G., Sinclair, A.H. and Reaves, G.A. (1990) Nitrate status of two major rivers in N.E. Scotland with respect to land use and fertiliser additions. *Chemistry and Ecology*, **4**, 97−107.

Egginton, G.M. and Smith, K.A. (1986) Nitrous oxide emission from a grassland soil fertilised with slurry and calcium nitrate. *Journal of Soil Science*, **37**, 59−67.

Eichner, M.J. (1990) Nitrous oxide emissions from fertilized soils: summary of available data. *Journal of Environmental Quality*, **19**, 272−80.

Firestone, M.K. (1982) Biological denitrification. In Stevenson, F.J. (ed.), *Nitrogen in Agricultural Soils*, American Society of Agronomy, Madison, Wisconsin, pp. 289−326.

Foster, S.S.D., Cripps, A.C. and Smith-Carington, A. (1982) Nitrate leaching to groundwater. *Transactions of the Royal Society of London*, **296**, 477−89.

Garwood, E.A. and Ryden, J.C. (1986) Nitrate loss through leaching and surface runoff from grassland; effects of water supply, soil type and management. In van der Meer, H.G., Ryden, J.C. and Ennik, G.C. (eds), *Nitrogen Fluxes in Intensive Grassland Systems*, Martinus Nijhoff, Dordrecht, The Netherlands, pp. 99−113.

Germann, P.F. and Beven, K. (1985) Kinematic wave approximation to infiltration into soils with sorbing macropores. *Water Resources Research*, **21**, 990−96.

Haider, K., Mosier, A.R. and Heinemeyer, O. (1987) The effect of growing plants on denitrification at high soil nitrate concentrations. *Soil Science Society of American Journal*, **51**, 97−102.

Hansen, S., Jensen, H.E., Nielsen, N.E. and Svendsen, H. (1990) DAISY: A soil plant system model. Danish simulation model for transformation of energy and matter in the soil plant atmosphere system. The National Agency for Environmental Protection, Copenhagen.

Harris, G.L., Goss, M.J., Dowdell, R.J., Howse, K.R. and Morgan, P. (1984) A study of mole drainage with simplified cultivation for autumn sown crops on a clay soil. 2. Soil water regimes, water balances and nutrient loss in drain waters. *Journal of Agricultural Science*, **102**, 561−81.

Hauck, R.D. (1986) Field measurement of denitrification — an overview. In Hauck, R.J. and Weaver, R.W. (eds), *Field Measurement of Dinitrogen Fixation and Denitrification*, Special Publication No. 18, Soil Science Society of America, Madison, Wisconsin, pp. 59−72.

Haynes, R.J. (1986) The decomposition process: mineralization, immobilization, humus formation, and degradation. In Haynes, R.J. (ed.), *Mineral Nitrogen in the Soil−Plant System*, Academic Press, Orlando, Florida, pp. 52−126.

Haynes, R.J. and Sherlock, R.R. (1986) Gaseous losses of nitrogen. In Haynes, R.J. (ed.), *Mineral Nitrogen in the Soil−Plant System*, Academic Press, Orlando, Florida, pp. 242−302.

Heichel, G.H., Vance, C.P., Barnes, D.K. and Henjum, K.I. (1985) Dinitrogen fixation, and N and dry matter distribution during 4-year stands of birds foot trefoil and red clover. *Crop Science*, **25**, 101−5.

Hendrick, J. (1921) The measurement of soil drainage with an account of the Craibstone Drain Gauges. *Transactions of the Royal Highland and Agricultural Society of Scotland*, **33**, 56−79.

Hendrick, J. (1930) A soil balance sheet for a rotation. *Transactions of the Royal Highland and Agricultural Society of Scotland*, **42**, 1−27.

Hendrick, J. and Welsh, H.D. (1934) The effect of treatment on the composition of grass. *Transactions of the Royal Highland and Agricultural Society of Scotland*, **46**, 202−23.

Hendrick, J. and Welsh, H.D. (1938) Further results from the Craibstone drain gauges. *Transactions of the Royal Highland and Agricultural Society of Scotland*, **50**, 2−20.

Jansson, P.-E., Borg, Ch.G., Lundin, L.-C. and Linden, B. (1987) *Simulation of soil nitrogen storage and leaching — applications to different Swedish agricultural systems.* National Swedish Environmental Protection Board Report No. 3356.

Jenkinson, D.S. (1988) Soil organic matter and its dynamics. In Wild. A. (ed.), *Russell's Soil Conditions and Plant Growth*, 11th edn, Longman, Harlow, pp. 564—607.

Jenkinson, D.S. (1991) The turnover of organic carbon and nitrogen in the soil. In *Soil Productivity and Pollution. Philosophical Transactions of the Royal Society of London Series B*, **329**, 361—8.

Johnsson, H., Bergstrom, L., Jansson, P.E. and Paustian, K. (1987) Simulated nitrogen dynamics and losses in a layered agricultural soil. *Agricultural Ecosystems and the Environment*, **18**, 333—6.

Johnston, A.E. and Jenkinson, D.S. (1989) The nitrogen cycle in UK arable agriculture. *Proceedings of the Fertiliser Society*, No. 286.

Kolenbrander, G.J. (1981) Leaching of nitrogen in agriculture. In Brogan, J.C. (ed.), *Nitrogen Losses and Surface Runoff from Land Spreading of Manures*, Martinus Nijhoff/Junk, Dordrecht, The Netherlands.

Kragt, J.F. and de Vries, W. (1987) Ondercoek mar de effecten van mestbe perking op de nitraatuitspoeling in water wingebieden in Ovirijssel; 1. RENLEM: een nitraatintspoelingsmodel voor toepassing op regionale schaal. Wageningen, Stichting voor Bodem Rartering, Rapport 1935.

Lawes, J.B., Gilbert, J.H. and Warington, R. (1881) On the amount and composition of the rain and drainage waters at Rothamsted. Part II. *Journal of the Royal Agricultural Society of England*, **42**, 269.

Lind, A.M. (1985) Nitrate reduction in the subsoil. In Golterman, H.L. (ed.), *Denitrification in the Nitrogen Cycle*, Plenum Press, New York, pp. 145—56.

Loague, K. and Green, P.E. (1991) Statistical and graphical methods for evaluating solute transport models: overview and application. *Journal of Contaminant Hydrology*, **7**, 51—73.

Macdonald, A.J., Powlson, D.S., Poulton, P.R. and Jenkinson, D.S. (1989) Unused fertiliser nitrogen in arable soils — its contribution to nitrate leaching. *Journal of the Science of Food and Agriculture*, **46**, 407—19.

Macduff, J.H., Jarvis, S.C. and Roberts, D.H. (1990) Nitrates: leaching from grazed grassland systems, In Calvet, R. (ed.), *Nitrates — Agriculture — Eau*, INRA, Paris, pp. 405—410.

Melillo, J.M., Steudler, P.A., Aber, J.D. and Bowden, R.D. (1989) Atmospheric deposition and nutrient cycling. In Andreae, M.O. and Schimel, D.S. (eds), *Exchange of Trace Gases between Terrestrial Ecosystems and the Atmosphere*, Wiley, Chichester, pp. 263—80.

Nieder, R., Schollmayer, G. and Richter, J. (1989) Denitrification in the rooting zone of cropped soils with regard to methodology and climate: A review. *Biology and Fertility in Soils*, **8**, 219—26.

Nyborg, M. and Hoyt, P.B. (1978) Effects of soil acidity and liming on mineralization of soil nitrogen. *Canadian Journal of Soil Science*, **58**, 331—8.

Parkin, T.B. and Tiedje, J.M. (1984) Application of a soil core method to investigate the effect of oxygen concentration on denitrification. *Soil Biology and Biochemistry*, **16**, 331—4.

Passioura, J.B. (1971) Hydrodynamic dispersion in aggregated media. I. Theory. *Soil Science*, **111**, 339—44.

Paustian, K., Bergstrom, L., Jansson, P.-E. and Johnsson, H. (1990a) Ecosystem dynamics. *Ecological Bulletins*, **40**, 153—80.

Paustian, K., Andren, O., Clarholm, M., Hansson, A.-C., Johansson, G., Lagerlof, J., Lindberg, T., Pettersson, R. and Sohlenius, B. (1990b) Carbon and nitrogen budgets of four agro-ecosystems with annual and perennial crops, with and without N fertilization. *Journal of Applied Ecology*, **27**, 60—84.

Payne, W. (1981) *Denitrification*, Wiley-Interscience, New York.

Powlson, D.S. (1988) Measuring and minimising losses of fertiliser nitrogen in arable agriculture, In Smith, K.A. and Jenkinson, D.S. (eds), *Nitrogen Efficiency in Agricultural Soils*, Elsevier, Barking, Essex, pp. 231—45.

Powlson, D.S., Brookes, P.C. and Christensen, B.T. (1987) Measurement of soil microbial biomass provides an early indication of changes in total soil organic matter due to straw incorporation. *Soil Biology and Biochemistry*, **19**, 159—64.

Rachhpal-Singh and Nye, P.H. (1988) Processes controlling ammonia losses from fertilizer urea. In Smith, K.A. and Jenkinson, D.S. (eds), *Nitrogen Efficiency in Agricultural Soils*, Elsevier, Barking, Essex, pp. 246—55.

Redman, M.H. (1991) *Nitrogen balance of an arable soil*, PhD thesis, University of Edinburgh.

Rijtema, P.E. and Kroes, J.G. (1991) Some results of nitrogen simulations with the model ANIMO. *Fertiliser Research*, **27**, 189–98.

Rijtema, P.E., Roest, C.W.J. and Kroes, J.G. (1990) Formulation of the nitrogen and phosphate behaviour in agricultural soils, the ANIMO model. Report 30 (in press), Wageningen, The Netherlands: The Winand Staring Centre.

Ryden, J.C., Ball, P.R. and Garwood, A.E. (1984) Nitrate leaching from grassland. *Nature*, **311**, 50–3.

Shields, J.A., Paul, E.A. and Low, W.E. (1974) Factors affecting the stability of labelled microbial materials in soils. *Soil Biology and Biochemistry*, **6**, 31–7.

Smith, K.A. and Arah, J.R.M. (1990) Losses of nitrogen by denitrification and emissions of nitrogen oxides by soils. *Proceedings of the Fertiliser Society*, No. 299.

Smith, K.A., Elmes, A.E., Howard, R.S. and Franklin, M.F. (1984) The uptake of soil- and fertilizer-nitrogen by barley grown under Scottish climatic conditions. *Plant and Soil*, **76**, 49–57.

Smith, K.A., Howard, R.S. and Crichton, I.J. (1988) Efficiency of recovery of nitrogen fertilizer by winter barley. In Smith, K.A. and Jenkinson, D.S. (eds), *Nitrogen Efficiency in Agricultural Soils*, Elsevier, Barking, Essex, pp. 73–84.

Smith, R.M. and Stewart, D.A. (1989) A regression model for nitrate leaching in Northern Ireland. *Soil Use and Management*, **5**, 71–6.

Sørensen, L.H. (1982) Mineralisation of organically bound nitrogen in soil as influenced by plant growth and fertilization. *Plant and Soil*, **65**, 51–61.

Stanford, G. and Epstein, E. (1974) Nitrogen mineralization–water relations in soils. *Soil Science Society of America Proceedings*, **38**, 103–7.

Stevenson, F.J. (1982) Organic forms of soil nitrogen. In Stevenson, F.J. (ed.), *Nitrogen in Agricultural Soils*, American Society of Agronomy, Madison, Wisconsin, pp. 67–122.

Tiedje, J.M., Firestone, R.B., Firestone, M.K., Betlach, M.R., Kaspar, H.F. and Sørensen, J. (1981) Use of nitrogen-13 in studies of denitrification. In Krohn, R.A. and Root, J.W. (eds), *Recent Developments in Biological and Chemical Research with Short-Lived Isotopes*, American Chemical Society, Washington, DC.

Tinker, P.B. (1991) Fertilisers in the environment. *Proceedings of the Fertiliser Society*, No. 302.

Vereecken, H., Van Clooster, M. and Swerts, M. (1990) A model for the estimation of nitrogen leaching with regional applicability. In Merckx, P., Vereecken, H. and Vlassak, K. (eds), *Fertilization and the Environment*, Leuven University Press, Leuven, Belgium, pp: 250–63.

Vereecken, H., Jansen, E.J., Hack-ten Broeke, M.J.D., Swerts, M., Engelke, R., Fabrewitz, S. and Hansen, S. (1991) Comparison of simulation results of five nitrogen models using different data sets. In *Soil and Groundwater Research Report II (CEC, 1991)*.

Vinten, A.J.A., Howard, R.S. and Redman, M.H. (1991) Measurement of nitrate leaching losses from arable plots under different nitrogen input regimes. *Soil Use and Management*, **7**, 3–14.

Vinten, A.J.A., Vivian, B.J. and Howard, R.S. (1992) The effect of nitrogen fertiliser on the nitrogen cycle of two upland arable soils of contrasting textures. *Proceedings of the Fertiliser Society* (in press).

Webster, C.P. and Goulding, K.W.T. (1989) Influence of soil carbon content on denitrification from fallow land during autumn. *Journal of the Science of Food and Agriculture*, **49**, 131–42.

Webster, C.P., Belford, R.K. and Cannell, R.P. (1986) Crop uptake and leaching losses of ^{15}N-labelled fertilizer nitrogen in relation to waterlogging of clay and sandy loam soils. *Plant and Soil*, **92**, 89–101.

Westerman, R.L. and Kurtz, L.T. (1973) Priming effect of ^{15}N-labelled fertilizers on soil nitrogen in field experiments. *Soil Science Society of America Proceedings*, **37**, 725–7.

Widdowson, A.V., Penny, A., Darby, R.J., Bird, E. and Hewitt, M.V. (1987) Amounts of NO_3-N and NH_4-N in soil, from autumn to spring, under winter wheat and their relation to soil type, sowing date, previous crop and N uptake at Rothamsted, Woburn and Saxmundham, 1979–85. *Journal of Agricultural Science, Cambridge*, **108**, 73–95.

Wild, A. (1988) Plant nutrients in soil: nitrogen. In Wild, A. (ed.), *Russell's Soil Conditions and Plant Growth*, 11th edn, Longman, Harlow, pp. 652–94.

Willliams, J.R., Renard, K.G. and Dyke, P.T. (1983) A new method for assessing the effect of erosion on productivity. The EPIC Model. *Journal of Soil and Water Conservation*, **38**, 381–3.

4 Nitrate Dynamics in Forest Soils

H. VAN MIEGROET

Oak Ridge National Laboratory, USA

and

D.W. JOHNSON

Desert Research Institute, USA

4.1 INTRODUCTION

Nitrate dynamics in forest soils are an intricate part of, and directly depend upon, the complex interactions that exist between the forest organisms and ecosystem components. The interactions which characterise forest ecosystems generally cause the nature and the rate of N dynamics (including NO_3^- fluxes) in forest soils to differ considerably from those typically encountered in soils under agricultural use (see Chapter 3). This is especially so because forest soils experience a more or less closed forest cover for a period of several decades to several centuries, depending on the forest type and management regime.

This allows the development of various strata in the vegetation, composed of different tree and understory species and leads to (1) differential rooting patterns and depths, (2) variations in the annual return of organic matter and nutrients from the canopy, (3) periodic tree mortality, (4) the accumulation of a forest floor layer of variable thickness, and (5) periodicities in the above- and below-ground processes, all of which cause great spatial and temporal heterogeneity in forest soil properties. Furthermore, forests are generally found on less fertile soils, in steeper terrain, or under climatic conditions that are less conducive for agriculture (e.g. due to cold or drought stress). In addition, the typical trend in forest soil development is one of progressive acidification, resulting from the accumulation and decomposition of the organic matter, nutrient leaching patterns and cation uptake by tree roots. Finally, if logging is part of forest management, such site disturbance does not occur annually as is typically the case with harvesting and ploughing of agricultural areas, and nutrient amendments are also far less frequent and widespread. These ecosystem characteristics create a unique set of conditions which result in N generally being less available in forest soils and nitrification being far less prevalent than in agricultural soils, particularly in temperate and cold climates.

The N in forest ecosystems is mostly in organic form and represents the cumulative retention over many decades, centuries or even millennia of N entering from the atmosphere (see Chapter 2) through (1) biological fixation of N_2; (2) wet deposition of NO_3^-, NH_4^+ and organic N; or (3) dry deposition of these forms plus nitric acid vapour. The N inputs are either immobilised by heterotrophic decomposers, or taken up by the vegetation, and only after microbial and plant demands for N are met, can excess NO_3^- begin to move through the soil profile (Figure 4.1). In many areas of the world, devoid of N-fixing tree species, current atmospheric

Nitrate: Processes, Patterns and Management. Edited by T.P. Burt, A.L. Heathwaite and S.T. Trudgill
© 1993 John Wiley & Sons Ltd

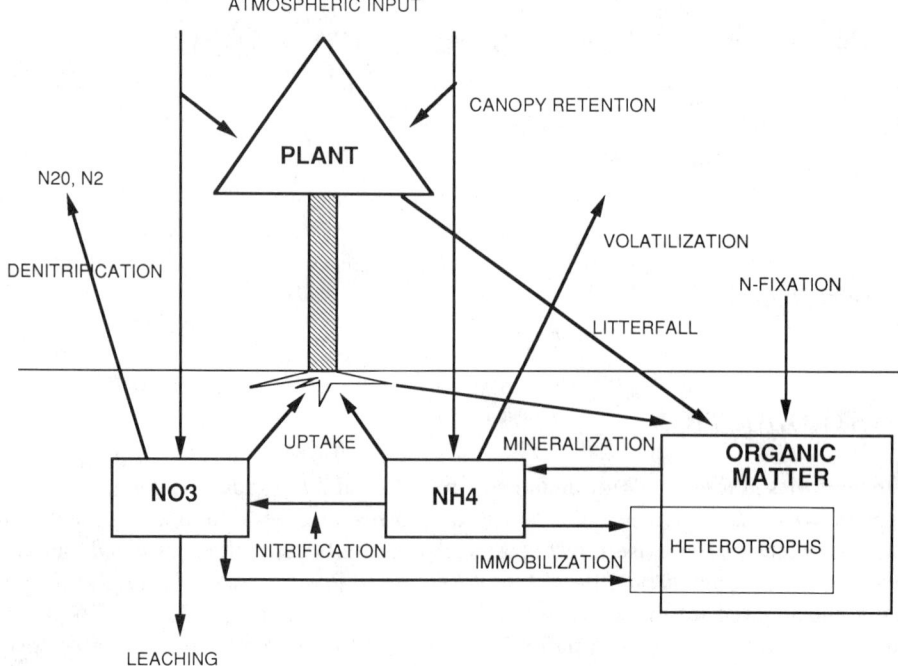

Figure 4.1 Schematic representation of nitrogen cycling in forest ecosystems

N inputs do not meet the annual demands for N of the trees, especially in younger or vigorously growing forests (Cole and Rapp, 1981), and tree growth directly depends on the amount of N supplied via mineralisation of organic N in the mineral soil and forest floor. This process is, in many forest ecosystems, restricted by suboptimal climatic conditions (such as low temperature or water stress), nutrient deficiencies, or the chemical composition of the litter (e.g. Berg and Staaf, 1981; Berg and McClaugherty, 1987). Consequently, N availability is a growth-limiting factor in many forest ecosystems, and N fertiliser is sometimes applied in commercial forest stands to alleviate this N stress. Only when the combined input of N from the atmosphere, N amendments and internal mineralisation exceeds the combined needs of decomposer organisms and higher plants at any given point in time or space will NO_3^- leaching occur. The corollary to this observation is that the occurrence of measurable nitrification and NO_3^- leaching can be considered an indication that N is no longer limited in supply, at least not at that particular point in time.

4.2 NITRIFICATION

4.2.1 NITRIFYING ORGANISMS

Since nitrification is biologically mediated, process rates strongly depend on the size, nature and activity of the nitrifying population. It is only under unusual circumstances that forest soil nitrification is inhibited for lack of nitrifiers (Nakos, 1975). In most forest soils, however, some nitrifying populations seem to be present, even though their size and activity may be curtailed as a result of specific site conditions. Contrary to observations in agricultural soils

(Chapter 3) and pure cultures of autotrophic nitrifiers (Pang, Cho and Hedlin, 1975; Laudelout, Lambert and Pham, 1976; Focht and Verstraete, 1977), acid conditions do not seem to preclude nitrification in forest soils (e.g. Smith, Bormann and Likens, 1968; Van Praag and Weissen, 1973; Klein, Kreitinger and Alexander, 1983; Mai and Fiedler, 1983; Van Miegroet and Cole, 1984; Foster and Nicolson, 1988). Nitrification is thought to be largely carried out by autotrophic bacteria, although heterotrophic pathways have also been identified (Haynes, 1986). The different pathways exhibit a different acid tolerance, utilise different N substrates (and may subsequently respond differently to N amendments) and nitrify at vastly different rates. Autotrophic nitrifiers generally nitrify more rapidly, depend entirely on the oxidation of NH_4^+ to derive their energy, and are thus more critically affected by NH_4^+ availability (Focht and Verstraete, 1977; Haynes, 1986). The biochemical pathway of heterotrophic nitrification, on the other hand, is more complex (Haynes, 1986): Heterotrophic nitrifiers mainly oxidise organic N compounds (Focht and Verstraete, 1977; Schimel, Firestone and Killham, 1984) and produce NO_3^- at lower rates than the autotrophs. They may also be involved in inorganic NH_4^+ oxidation (Focht and Verstraete, 1977; Johnsrud, 1978), although some strains fail to respond to NH_4^+ additions (Klein, Kreitinger and Alexander, 1983; Schimel, Firestone and Killham, 1984).

Some controversy exists as to the relative role of these two pathways in forest soils, the effect of forest soil acidity thereupon and the effect of atmospheric acid inputs on NO_3^- production rates. It has been suggested that heterotrophs, which are more acid tolerant, dominate the nitrification process under acid conditions (Focht and Verstraete, 1977; Verstraete, 1981). Some studies have indeed substantiated the existence of nitrifying fungi and their ability to nitrify in different forest types (Remacle, 1977; Johnsrud, 1978; Schimel, Firestone and Killham, 1984; Strayer, Lin and Alexander, 1981) but it is unlikely that heterotrophs account for all NO_3^- release in forest soils, considering their typically low oxidation rates compared to the autotrophs (Focht and Verstraete, 1977). Furthermore, several studies have identified or isolated autotrophs in acid forest soils: Smith, Bormann and Likens, (1968), for example, demonstrated their presence in hardwood forest soils of pH < 4 at Hubbard Brook Watershed, and showed that an increase in NO_3^- leaching following clearcutting was associated with a significant increase in autotroph numbers. Mai and Fiedler (1983) later isolated autotrophs from the acid humus in German fir forests. In addition, Hankinson and Schmidt (1988) substantiated the existence of acidophilic strains of autotrophic nitrifiers in acid forest soils. It is more likely, as suggested by Alexander and coworkers (Alexander, 1980; Kreitinger et al., 1985), that NO_3^- production is mediated by both groups of organisms (heterotrophs and autotrophs), with specific site conditions or external influences (e.g. acid deposition, N amendments) causing a shift in their relative importance and ultimately changing nitrification rates (e.g. Strayer, Lin and Alexander, 1981; Like and Klein, 1985; Stroo and Alexander, 1986). It is clear from the evidence in the literature that while soil acidity and external acidification may regulate nitrification rates in forest soils to some extent, they are not pivotal in determining whether or not nitrification will occur at all.

4.2.2 CONTROLLING FACTORS

It has been widely assumed that nitrifying organisms are weak competitors for available NH_4^+ compared to heterotrophic decomposers and plant uptake (Vitousek et al., 1979; Riha, Campbell and Wolfe, 1986), causing nitrification and NO_3^- leaching to be generally low in many forest soils, particularly those that are N-limited (Figure 4.2(a)). It has further been

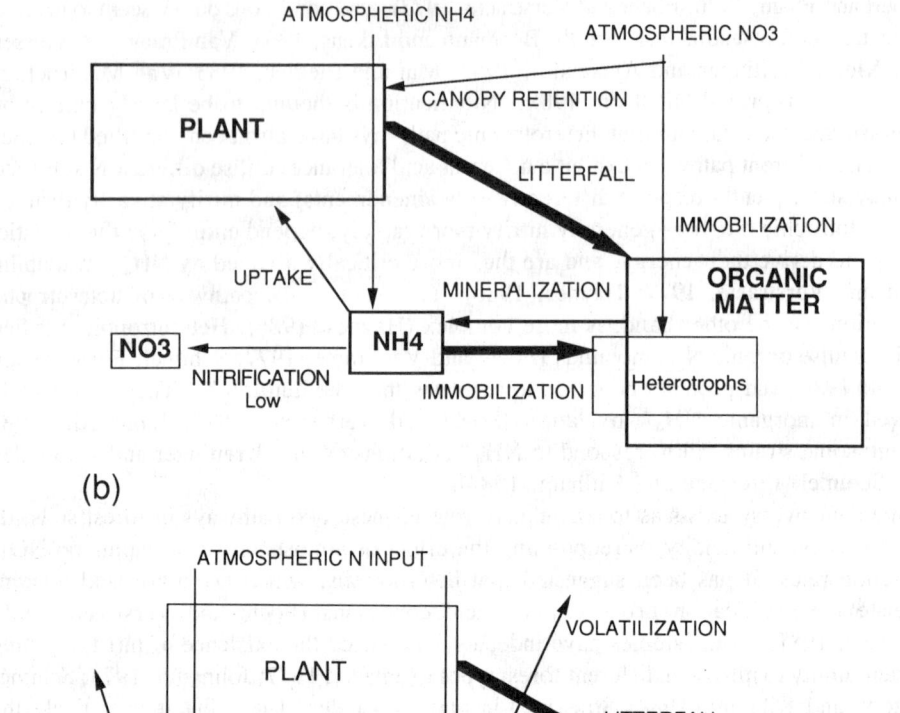

Figure 4.2 Nitrate production and leaching potential based on the NH_4-N competition model. (a) Substrate with high C/N ratio (low N availability); (b) substrate with low C/N ratio (high N availability)

the assumption that microbial uptake of NO_3^- is minimal, and that only when microbial and plant demand for NH_4^+ are satisfied will nitrification begin to occur to any significant extent. These assumptions have recently been challenged by Davidson, Stark and Firestone (1990). Using ^{15}N techniques, these authors found significant nitrification (12−46% of N mineralisation rates) and microbial NO_3^- uptake in open grassland soils and in grassland soils under oak canopy, even though soil NO_3^- pools and NO_3^- leaching rates were very low. In other words, nitrification and subsequent microbial uptake caused very rapid turnover of a small soil NO_3^- pool in this site. The authors suggested that spatial heterogeneity in NH_4^+ production and availability plays a crucial role in NO_3^- dynamics and concluded that

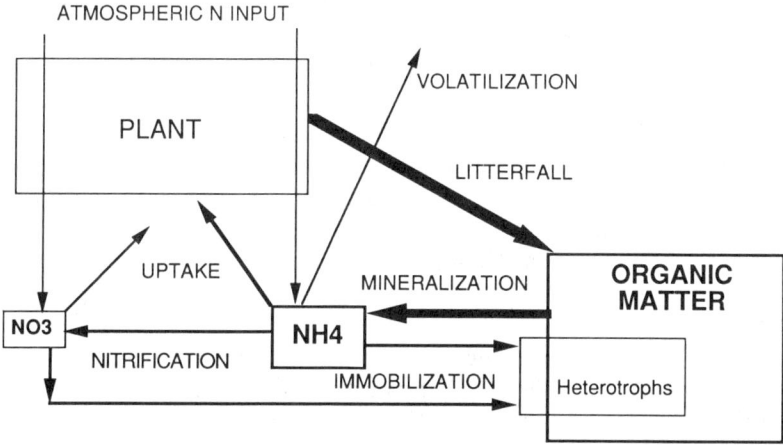

Figure 4.3 Below-ground nitrogen dynamics according to the NH_4-N diffusion model

'rather than invoking a poorly defined concept of substrate competition among soil organisms, the fate of NH_4^+ may be best described as a stochastic process of diffusion of substrate among microsites colonised by a variety of organisms'. What this means is that those organisms that are closest to the NH_4^+ production 'hot spots', irrespective of whether they are heterotrophs or nitrifiers, have the highest probability of obtaining the substrate as it diffuses away from the site of production. Similarly, microbial NO_3^- immobilisation will occur in those microsites where there is insufficient NH_4^+ to meet microbial demands (Figure 4.3).

The authors further imply that nitrification rates obtained by incubation and reported in the literature are, in fact, net nitrification rates, which they define as the difference between NO_3^- production and microbial NO_3^- uptake. Because one of their study sites was under oak cover, the results may be relevant to our present discussion of forest soil NO_3^- dynamics. However, it is yet unknown to what extent these results might apply to forest ecosystems at large. If this pattern proves to be true in general, it will require a substantial redesign of the conceptual model currently used to explain and predict nitrification and NO_3^- leaching.

Regardless of which of these conceptual models is used, it is evident that any system manipulation or disturbance that changes the relative N source−N sink strengths in forest soils is likely to affect the production and movement of NO_3^-. Additions of labile organic C have been shown to reduce net nitrification (as now defined by Davidson, Stark and Firestone, 1990) and NO_3^- leaching (Figure 4.2(a)). The explanation usually offered for this phenomenon is that the increased supply of labile C as an energy source enhances heterotrophic NH_4^+ demands, thereby reducing NH_4^+ substrate availability to the poorly competitive nitrifying bacteria (Alexander, 1977; Figure 4.2(a)). Johnson and Edwards (1979) and Van Miegroet, Johnson and Cole (1990b) demonstrated significant reductions in nitrification potential of the upper soil (measured during laboratory incubations) following C additions to a variety of forest ecosystems. While heterotrophic NH_4^+ demand is undoubtedly increased by the addition of labile organic C, we must now consider the alternative possibility that such additions also stimulate microbial NO_3^- uptake.

Conversely, an increase in soil N content, be it through natural processes or anthropogenic influences, will generally increase N availability and nitrification potential (Figure 4.2(b)).

Nitrogen enrichment of the soil via symbiotic N-fixation (Coats, Leonard and Goldman, 1976; Binkley, Kimmins and Feller, 1982; Van Miegroet and Cole, 1984), N fertilisation (e.g. Heilman, 1974; Johnson, Edwards and Todd, 1980; Popovic, 1985; Van Miegroet, Johnson and Cole, 1990b), waste-water application (Urie, 1973; Breuer, Cole and Schiess, 1979; Johnson, Breuer and Cole, 1979), or the addition of municipal sewage sludge (Wells *et al.*, 1986) has been shown to enhance nitrification in forest soils. Chronic N inputs or repeated N additions, in particular, seem to have a prolonged effect on nitrification, by allowing nitrifier populations to expand significantly (Johnson and Todd, 1988). This response was illustrated for both forest floor (Roberge and Knowles, 1966) and forest soil samples (Heilman, 1974) which showed a greater nitrification pulse with refertilisation than after the first N fertiliser application.

Removing the forest cover or reducing tree N uptake may also alleviate the competition for N and temporarily stimulate NO_3^- production and leaching (Likens, Bormann and Johnson, 1969; Johnson and Edwards, 1979; Vitousek *et al.*, 1979; Waide *et al.*, 1988), but the process tends to be reversed again with regrowth of the vegetation (Bormann *et al.*, 1974). In contrast, removing the forest cover in N-fixing systems causes reduction in NO_3^- leaching, presumably due to the cessation of N fixation (Van Miegroet, Cole and Homann, 1990a). In forest systems that are very N-limited, harvesting may not have any stimulatory effect on nitrification rates (e.g. Cole and Gessel, 1965).

Even though the conceptual model of what controls nitrification (i.e. substrate availability) is fairly well understood (e.g. Davidson, Stark and Firestone, 1990; Vitousek *et al.*, 1979; Riha, Campbell and Wolfe, 1986) and has been used successfully to explain nitrification responses to site manipulations or other forest disturbances, it is far more difficult to identify those forest ecosystem parameters that accurately predict the soil nitrification potential under relatively undisturbed conditions. The highest NO_3^- levels and the largest differences between high- and low-N soils generally occur in the upper soil horizons, which typically have the highest biological activity, and tend to decrease with depth (Federer, 1983; Van Miegroet, Johnson and Cole, 1990b). Weber and Gainey (1962) observed a close association between soil NO_3^- levels and the total N content of some acid soils. Heilman (1974) later found a similar relationship for unfertilised forest soils and noted that soils with a C/N > 27 showed little net NO_3^- production. The latter observation underscores the importance not only of total N content but also of the intrinsic quality of the substrate (expressed by its C/N ratio) on the relative rate of N mineralisation versus N immobilisation, and ultimately on the amount of NH_4^+ that becomes available for net nitrification (Figure 4.2).

As a general rule, net N mineralisation (and thus nitrification) seldom occurs until the C/N of the organic substrate is lower than 30, indicating a relative abundance of N compared to the amount of metabolisable C available to the heterotrophs (Alexander, 1977). Berg and Staaf (1981), however, noted that the critical N content or C/N ratio at which net N release from litter is observed may vary with forest ecosystem type and ranges from 0.3% N (C/N = 170) to 1.4% (C/N = 36). They indicated that the threshold C/N value is lower for systems with high decomposition rates, implying that for a given C content, the N content of the organic substrate has to be considerably higher in temperate forests than in colder forest types in order for excess N to become available for nitrification. This agrees with the model of Agren and Bosatta (1988) which also indicates a higher critical C/N ratio (i.e. lower critical N content) in low-quality litter. Vitousek *et al.* (1982) demonstrated that the C/N ratio of fresh litter was a good indicator of N-cycling properties and N availability at a forested site and concluded that it may constitute a useful predictor of soil nitrification capacity.

Robertson (1982), on the other hand, found no direct correlation between nitrification rates during soil incubation and either N content or C/N ratio of soils over a wide range of forested sites.

A recent study in the United States of the effects of atmospheric deposition (including N inputs) on nutrient flux patterns in a variety of forests across a wide range of forest types, elevations and climatic conditions, indicated that forests with small soil N pool (either young soils with a limited N accumulation history or soils whose N content had been frequently reduced through disturbances such as fire or agricultural cropping) had low nitrification potentials and insignificant NO_3^- leaching rates, irrespective of age or vigor of the forest (Van Miegroet, Cole and Foster, 1992). Those sites that had high soil N contents coupled with low C/N ratios showed higher nitrification potentials, and the annual NO_3^- leaching below the rooting zone strongly depended on a combination of atmospheric N inputs, forest age and tree N uptake rates.

Stand age is an important factor determining N uptake and annual N accumulation rates in tree biomass generally reach a maximum at canopy closure (Turner, 1977; Cole and Rapp, 1981). Nitrogen increment rates in woody tissues may remain stable, albeit lower, for a prolonged period after crown closure, but will eventually decline as stands become overmature. This declining trend in N immobilisation in standing biomass may explain why NO_3^- leaching losses are typically larger in mature than in young vigorously growing forests (Vitousek and Reiners, 1975; Van Miegroet, Cole and Foster, 1992).

None of the conceptual models described above allow for the possibility of nitrification inhibitors. An early study by Rice and Pancholy (1974) indicated that nitrification rates decrease during forest succession due to the presence of chemical nitrification inhibitors (soluble allelopathic compounds produced by plant litter). This somewhat controversial finding stimulated several follow-up studies in various ecosystems. Some of these investigations supported the contention that nitrification inhibitors were a factor in controlling NO_3^- losses from forest ecosystems (Lodhi, 1978; Olson and Reiners, 1983), but several others found no such evidence, and concluded that either competition for NH_4^+ or other nutrient limitations controlled nitrification rates (Purchase, 1974; Lamb, 1980; Robertson and Vitousek, 1981; Cooper, 1986).

4.2.3 SPATIAL AND TEMPORAL VARIABILITY

Nitrification rates may fluctuate greatly at the landscape (regional) scale, on the ecosystem level and within individual forest stands (e.g. Zak, Pregitzer and Host, 1986; Robertson et al., 1988; Strader, Binkley and Wells, 1989). This spatial variability is related to changes in environmental conditions such as temperature and soil moisture content, the composition and degree of heterogeneity of the forest cover, the influence of topography on these ecosystems characteristics and the resulting differences in litter quality and N availability (e.g. Pastor and Post, 1986). Robertson et al. (1988), for example, found differences in nitrification rates at distances of 1 m in an old field site that had previously been under hardwood cover. Such observations underscore the complex interactions that exist between rate-controlling factors and point at the logistical problems in characterising a process at the stand level or regional scale on the basis of a limited number of point measurements.

Within a particular region or forest stand, the nitrification rates further vary considerably throughout the year in response to seasonal changes in temperature and moisture regime. Spring peaks in the NO_3^- concentration in soil solutions and streams in forested watersheds are commonly observed with snowmelt (e.g. Nodvin, Driscoll and Likens, 1988; Foster, Nicolson and Hazlett, 1989; Shepard *et al.*, 1990). They result in part from the sudden release from the snowpack of previously accumulated atmospheric NO_3^- inputs, but also from an increase in microbial activity due to soil warming prior to the onset of root activity. Mineralisation assays in high-elevation spruce forests in the south-eastern United States indicated that the N mineralisation and nitrification rates were highest in spring and fall, and were associated with increased NO_3^- leaching through the soil (Strader, Binkley and Wells, 1989; Johnson *et al.*, 1991). Soil and soil water NO_3^- concentrations generally decrease in the course of the growing season and NO_3^- leaching tends to be lowest during the winter months in most forest ecosystems, especially in the upper soil horizons (e.g. Nadelhoffer, Aber and Melillo, 1983; Foster, Nicolson and Hazlett, 1989; Strader, Binkley and Wells, 1989; Shepard *et al.*, 1990).

Nitrogen dynamics in forest soils are particularly sensitive to wetting and drying cycles (Haynes, 1986), which may occur either seasonally or from one year to the next. Rewetting of dry soils is usually accompanied by a N mineralisation flush and a concomitant increase in nitrification, and is responsible for what Ulrich (1983) calls seasonal or climatic 'acidification pushes', i.e. the temporary and spatially localised acidification pulse association with NO_3^- production and leaching (see following section). Both seasonal and year-to-year variability in nitrification and NO_3^- leaching, including the effect of soil rewetting after an unusually

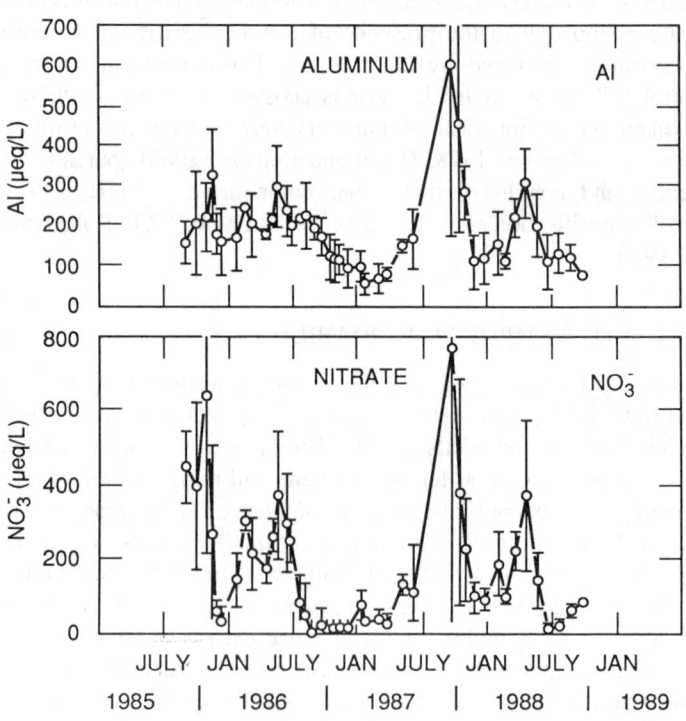

Figure 4.4 Aluminium mobilisation induced by NO_3^- leaching in the A-horizon of an acid soil of a high-elevation Spruce–Fir forest in the Great Smoky Mountains National Park (reproduced by permission from Johnson *et al.*, 1991)

dry year, was illustrated by Johnson *et al.* (1991) for a high-elevation Spruce—Fir forest in the Southern Appalachian Mountains (Figure 4.4). The study showed that this ecosystem was characterised by high nitrification potentials in the forest floor and upper mineral soil, resulting from a combination of high atmospheric N deposition, large soil N pools with low C/N ratio and low tree N uptake rates. The normal seasonal variations in the soil water NO_3^- concentrations were superimposed by the effects of a two-year drought period (1986–7) that drastically reduced NO_3^- levels. The onset of the rainfall in the fall of 1987 and the movement of the wetting front through the soil profile caused a very large NO_3^- concentration peak and restored previously disrupted seasonal patterns. Such large fluctuation in the NO_3^- concentration of soil water has important repercussions on forest soil properties and water chemistry, as will be discussed in the following section.

4.3 EFFECTS OF NITRIFICATION ON FOREST SOILS

4.3.1 NITRIFICATION AS A SOIL-ACIDIFICATION PROCESS

Nitrification has a twofold acidifying impact on the soil and the soil solution, first, due to the net H^+ release during this oxidation process (Figure 4.5), but, more importantly, because of the mobile nature of NO_3^-. Nitrate is a strong acid anion, i.e. it has a low affinity for H^+ and remains in dissociated (ionic) form even at the very low solution pH's that are sometimes observed in forest soils. Furthermore, NO_3^- is not strongly adsorbed to the positively charged soils particles (Kinjo and Pratt, 1971; Wiklander, 1976), and unless NO_3^- is removed from the solution via plant uptake, denitrification or microbial uptake it will readily move through the soil profile in this form. Because the solution has to remain electrically neutral at all times, any increase in total NO_3^- leaching must be balanced by an equivalent increase in total cation load (Nye and Greenland, 1960), with the relative abundance of the different cations determined by the exchange reactions between the solution and the cation exchange complex (Gapon, 1933; Gaines and Thomas, 1953; Reuss, 1983; Reuss and Johnson, 1986).

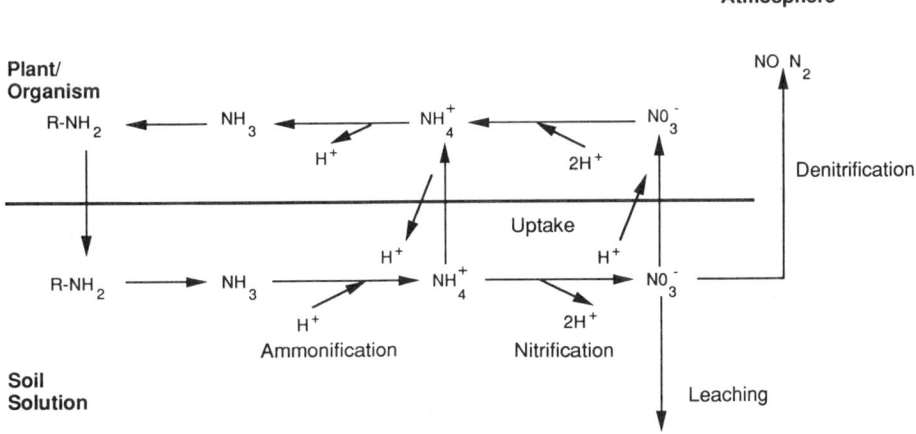

Figure 4.5 Hydrogen sources and sinks associated with the below-ground N dynamics (adapted from Reuss, 1977)

According to these theoretical exchange relationships, nitrification and NO_3^- leaching will displace primarily multivalent cations such as Mg^{2+} and Ca^{2+} in soils with medium to high base saturation, i.e. soils with relatively high nutrient cation reserves on the exchange complex. Over a prolonged period of time, this process may cause a progressive depletion of the exchangeable bases and an increase in soil acidity, if nutrients are insufficiently replaced by mineral weathering, organic matter decomposition, or nutrient amendments, and under extreme conditions nutrient deficiencies may develop. In more acid soils with low base saturation, Al^{3+} will be preferentially mobilised by NO_3^- leaching. The latter phenomenon raises concerns about potential Al toxicity and/or nutrient imbalances, which, in turn, may negatively affect forest productivity (e.g. Kelly *et al.*, 1990; Raynal *et al.*, 1990; Thornton, Schaedler and Raynal, 1987).

4.3.2 CASE STUDIES

Different examples are available in the literature that illustrate the above processes, as well as their potential impact of excessive NO_3^- leaching on forest site fertility. It needs to be pointed out that N saturation, i.e. the point where N inputs exceed the system's ability to retain N and considerable NO_3^- leaching occurs, can be the result of natural processes as well as of anthropogenic influences. The first set of examples, summarised in Van Miegroet *et al.* (1989b), involve the effect N enrichment through symbiotic N-fixation by tree species such as Alder (*Alnus* spp.) on soil N status, nitrification rates and, ultimately, on changes in soil and solution chemistry. In some studies a chronosequence was used in a space-for-time substitution to investigate the progressive effect of N enrichment on soil chemistry. In other instances, nearby conifer stands that contained no N-fixers and had been established on the same or similar soils around the same time as the pure Alder or mixed Alder–Conifer forests served as a reference condition against which changes in soil and solution chemistry

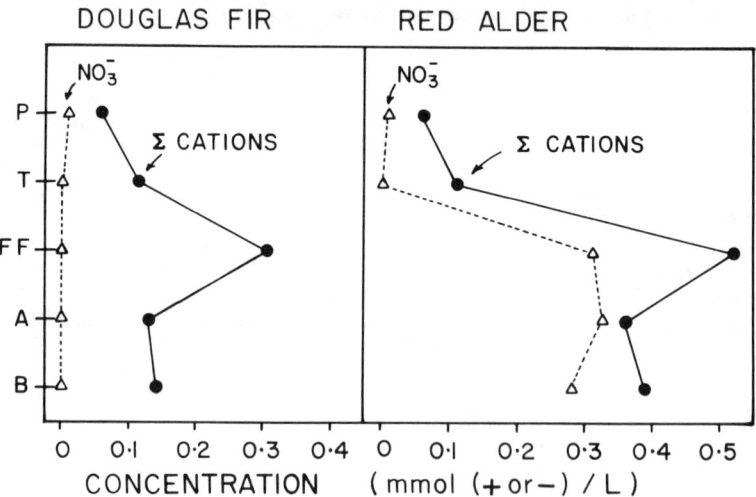

Figure 4.6 Influence of N enrichment by N-fixation on soil solution NO_3^- concentrations and base cation leaching in an Alder forest compared to an adjacent conifer forest (reproduced from Van Miegroet and Cole, 1984, by permission of the American Society of Agronomy, Inc., the Crop Science Society of America, Inc., and the Soil Science Society of America, Inc.)

Table 4.1 Difference in soil acidity expressed by pH in water, pH in $CaCl_2$, exchangeable base concentration, exchangeable Al and percentage base saturation between adjacent conifer and Alder forests at three study sites in the Pacific north-western United States (reproduced from Van Miegroet *et al.*, 1989, by permission of the Air and Waste Management Association)

Location stand	Depth (cm)	pH_{water}	pH_{CaCl_2}	Exch. bases	Exch. Al (mmol kg^{-1})	BS (%)
THOMPSON SITE						
Douglas Fir	0–15	5.3(0.2)[c,a]	4.5(0.2)[c]	13	89(15)[c]	12.5(5)[c]
	15–30	5.3(0.2)[c]	4.8(0.2)[c]	11	81(16)[c]	11.5(6)[c]
	30–45	5.3(0.1)[c]	4.7(0.1)[c]	7	72(15)[c]	9(5)
Red Alder	0–15	4.5(0.3)[c]	4.0(0.2)[c]	11	136(25)[c]	7(3)[c]
	15–30	4.8(0.2)[c]	4.4(0.1)[c]	9	119(25)[c]	7(3)[c]
	30–45	4.9(0.2)[c]	4.4(0.2)[c]	11	107(29)[c]	8.5(5)
WIND RIVER						
Conifer	0–15	5.4(0.1)[b]	4.3(0.1)	16	32(7)	32[b]
	15–40	5.3(0.1)[b]	4.3(0.1)	8	28(7)[b]	22[b]
	40–65	5.2(0.1)[b]	4.1(0.1)	26	83(30)	24[b]
Alder/conifer	0–15	5.1(0.3)[b]	4.3(0.2)	34	42(5)	45[b]
	15–40	5.1(0.2)[b]	4.2(0.1)	16	46(3)[b]	26[b]
	40–65	5.0(0.2)[b]	4.1(0.1)	15	81(10)	16[b]
CASCADE HEAD						
Conifer	0–15	5.4(0.4)[b]	4.4(0.4)[b]	37	89(6)[b]	29[b]
	15–30	5.4(0.2)[b]	4.4(0.2)[b]	21	51(16)	29[b]
	30–60	5.2(0.2)[b]	4.3(0.2)	11	53(12)	17[b]
Alder/conifer	0–15	4.3(0.4)[b]	3.7(0.3)[b]	14	138(4)[b]	9[b]
	15–30	4.8(0.4)[b]	4.1(0.1)[b]	16	76(17)	18[b]
	30–60	4.8(0.3)[b]	4.2(0.1)	5	49(9)	9[b]

[a] Mean with standard deviation in parentheses, followed by significance of difference between Alder and Douglas Fir soils at a given depth.   [b] $p = 0.1$.   [c] $p = 0.05$.

were evaluated. Nitrogen-enrichment invariably resulted in an increase in nitrification and, to the extent that it was measured, to an acceleration of NO_3^--mediated cation leaching loss (Figure 4.6) and an increase in soil solution acidity, particularly in the upper part of the soil profile (Table 4.1).

In some case studies this leaching merely resulted in a downward redistribution of exchangeable Ca^{2+} and Mg^{2+} with no change in the total exchangeable pool (Van Miegroet and Cole, 1984). Other studies have reported a distinct depletion of the exchangeable base pool over the entire soil rooting zone (e.g. Franklin *et al.*, 1968). Overall, the upper horizons of N-enriched soils had significantly lower soil pH and percentage base saturation. Reuss (1989) further noted that the progressive acidification of the upper soil horizon (i.e. lowering of the base saturation) may drive the system to the point where NO_3^- pulses start to mobilise Al^{3+} in addition to the base cations. Such drastic soil changes are likely to influence nutrient availability to the trees, as illustrated by Binkley, Lousier and Cromack (1984), who found that foliar levels for various nutrients were significantly lower in Douglas Fir trees (*Pseudostuga menziesii*) in the mixed Alder–conifer stands compared to those in pure conifer stands. Along those same lines, Wang *et al.* (1990) demonstrated that excessive NO_3^-

leaching induced Mg deficiencies in Grand Fir (*Abies grandis*) and Douglas Fir trees growing on glacial outwash soils that had been N-amended with municipal sewage sludge. Interestingly, subsequent studies in the conifer/Alder sites have shown that the removal of Alder can lead to a drastic reduction in NO_3^- leaching for several years after harvesting (Van Miegroet, Cole and Homann, 1990a). It is not clear, however, if and how quickly soil chemical properties will return to pre-leaching conditions.

High-elevation Spruce—Fir forests in the south-eastern United States offer another example of the extent to which soil nitrification and NO_3^- leaching may impact rhizosphere solution chemistry and site fertility (Johnson *et al.*, 1991; Joslin *et al.*, 1987; Van Miegroet, Cole and Foster, 1992). In this instance the 'leakiness' of the system was the result of a combination of natural and anthropogenic factors. Compared to similar forest types in the north-eastern United States, the Spruce—Fir zone in the southern Appalachians has remained relatively undisturbed (e.g. no glaciation effects, wildfires, logging history, etc.), allowing large N pools with low C/N to accumulate in the forest floor and mineral soil of these ecosystems. These soil conditions alone may have contributed to a large degree to the high N mineralisation and nitrification rates typically observed in these systems (Strader, Binkley and Wells, 1989; Johnson *et al.*, 1991; Van Miegroet, Cole and Foster, 1992). In addition, the Red Spruce (*Picea rubens*) overstory in these forests has reached a stage of overmaturity. Slow growth and low tree uptake rates coupled with increased tree mortality, especially to the Fraser Fir (*Abies fraseri*), caused by balsam woody adelgid (*Adelges piceae*), account for low N immobilisation in woody biomass. These high-elevation forests further receive large N inputs via atmospheric deposition, and the progressive break-up of the forest canopy may have exposed the soil more directly to the impact of external influences. This combination of inherent ecosystem characteristics and atmospheric N deposition regime are responsible for the generally low N retention capacity and the high NO_3^- leaching rates observed (Johnson *et al.*, 1991; Joslin *et al.*, 1987; Van Miegroet, Cole and Foster, 1992). The soils in these systems are extremely acid (base saturation < 10%) with the exchange complex largely dominated by exchangeable acidity rather than by base nutrients. Consequently, seasonal NO_3^- pulses, associated both with atmospheric deposition and internal nitrification patterns, preferentially mobilise Al^{3+} into the rhizosphere solution, and peak Al^{3+} values in the A-horizon solutions (Figure 4.4) occasionally approach threshold toxicity levels reported for Red Spruce seedlings (Joslin and Wolfe, 1988; Thornton, Schaedle and Raynal, 1987). It has been hypothesised by some that increases in Al concentrations may cause an imbalance in the Ca and Mg nutrition of the trees, which may ultimately result in a reduction in forest productivity (e.g. Shortle and Smith, 1988). Joslin *et al.* (1992) have reported significantly lower Ca concentrations in Red Spruce needles in the Southern Appalachians compared to those in the north-east, and results from a field-fertilisation trial indicate a positive foliar response to Ca amendments (Van Miegroet, unpublished data). It is not unlikely that NO_3^- leaching, irrespective of its causes, contributed to some extent to this reduction in nutrient availability.

4.4 DENITRIFICATION

Nitrate may be lost from forest soils through denitrification and volatilisation of N gases (including NO, N_2O and N_2), but, as stated by Pluth and Nommik (1981), information on the occurrence and actual rates of denitrification in forested ecosystems are scarce. Furthermore, because the soil conditions most commonly found under forest cover are generally unfavourable to denitrifier activity (high acidity, low N availability and NO_3^- levels) denitrification rates that have been reported are, for the most part, low. The process requires the presence of NO_3^- in the soil as N substrate, and because most forest soils are N-limited resulting in low nitrification rates, measurable denitrification is frequently observed following N amendments (Pluth and Nommik, 1981; Mahendrappa, 1974), or site disturbance that cause an increase in N availability (Robertson et al., 1987).

Denitrification increases with increasing soil pH (Waring and Giliam, 1983), reaching optimum conditions around pH 7 to 8 (Knowles, 1981), and is somewhat reduced by acid treatments (Davidson and Swank, 1987). Nevertheless, denitrification has been measured in acid forest soils in forested watersheds containing hardwoods in the Southern Appalachian Mountains (Davidson and Swank, 1986, 1987), under various mixed hardwood stands in Michigan (Robertson and Tiedje, 1984), and in cultivated and virgin soil with low pH in the lower Coastal Plains of North Carolina (Waring and Giliam, 1983).

Because there is an inverse relationship between O_2 partial pressure and soil moisture content, wetter soils in coves and along streamsides will generally show greater denitrification capacity than well-drained soils on midslope and toeslope positions in the landscape (Davidson and Swank, 1986). Being heterotrophs, denitrifiers require organic C for their energy supply (Knowles, 1981), and Davidson and Swank (1987) found that glucose amendments increased denitrification in hardwood forest soils at Coweeta Watershed in North Carolina, in mineral soil samples at greater depth but not in the upper soil. They interpreted the differential response with soil depth as an indication that organic C was not a limiting factor to denitrification in the surface soils but that C limitations become progressively larger with increasing soil depth.

Robertson calculated net N_2O release rates of <1 kg ha^{-1} a^{-1} in undisturbed (low-nitrification) soils in Loblolly Pine forests (Pinus taeda) in the south-eastern United States and rates of $3-6$ kg ha^{-1} a^{-1} following site disturbance and increased nitrification caused by harvesting and site preparation (Robertson et al., 1987). He also estimated denitrification rates as high as $2-12$ kg ha^{-1} month^{-1} in some of the mixed hardwood stands in Michigan; rates which are comparable to or even exceed the atmospheric N inputs to most forests in the United States (Robertson and Tiedje, 1984). It is questionable whether these rates can be extrapolated to an annual timescale, however, because of large temporal variations. Indeed, Groffman and Tiedje (1989) showed in a separate study that over 80% of the annual denitrification N loss from hardwood forest soils in Michigan occurred during a very brief period ($3-6$ weeks) of intensive activity in spring and autumn with annual denitrification estimates ranging from 1 kg ha^{-1} a^{-1} in a well-drained sandy soil to over 40 kg ha^{-1} a^{-1} in a poorly drained clay loam soil. The study indicated that lack of available NO_3^- was the primary factor limiting the process in summer. Insignificant denitrification rates (<0.5 kg ha^{-1} a^{-1}) were found in a variety of high-elevation Spruce−Fir forests in the Southern Appalachians (Wells, Jones and Craig, 1988), despite the seemingly favourable site conditions (e.g. rainfall pattern, soil C content, high mineralisation and nitrification rates (Johnson et al., 1991; Sasser and Binkley, 1989; Strader, Binkley and Wells, 1989)).

4.5 NITRATE UPTAKE

It is generally accepted that tree roots can take up both inorganic NH_4^+ and NO_3^-, and will likely do so when N is added to N-deficient systems (e.g. Cole, 1981; Kramer and Kozlowski, 1979). There is, however, still some controversy regarding actual rates of NO_3^- uptake, especially when NH_4^+ is also present in the soil, and on the cost to the plant that is associated with the production of nitrate reductase needed to metabolise NO_3-N. Soils data and the previous discussions on N availability suggest that a large portion of forest ecosystems typically grow on NH_4^+-dominated soils, where NO_3^- uptake will be low by default because little nitrification is actually taking place. However, with the increased use of N fertilisers in forest management and the increased N input to forest ecosystem via atmospheric deposition, soil NO_3^- levels are likely to increase in some forests, rendering the question of NO_3^- uptake more pertinent.

A few studies, mostly using conifer seedlings, have investigated the N uptake preference of different tree species, yielding divergent results depending on the species (e.g. McFee and Stone, 1968; Ingestad, 1971; Krajina, Madoc-Jones and Mellor, 1973; Ingestad, 1979; Bledsoe, 1976; Bledsoe and Zasoski, 1983). Even for the same tree species, results often differed between studies. A study by Bledsoe (1976), for example, demonstrated a clear preference of NH_4^+ uptake over NO_3^- by Douglas Fir, while Krajina, Madoc-Jones and Mellor (1973) came to the opposite conclusion. It has been suggested that the observed variability in N uptake preference between tree species may be a reflection of the specific soil conditions under which a particular tree species has typically evolved (Cole, 1981; Krajina, Madoc-Jones and Mellor, 1973).

Mycorrhizal infections of the root system were found to increase the uptake of NO_3^- by Douglas Fir seedlings (Bledsoe and Zasoski, 1983; Rygiewicz, Bledsoe and Zasoski, 1984b). More acid nutrient solutions caused a decline in NO_3^- uptake, but less so than in the case of NH_4^+ uptake (Rygiewicz, Bledsoe and Zasoski, 1984a,b). Earlier studies with excised barley roots had shown that NO_3^- uptake was lower in solutions at pH 4 compared to solutions at pH 6, but that the presence of Al^{3+} in the solution at pH 4 increased NO_3^- uptake. The latter effect was ascribed to a relatively lower proportion of H^+ uptake associated with NO_3^- uptake and the prevention of H^+-induced root injury (Bassioni, 1971). Comparable information is not available for tree species.

The fact that NH_4^+ is sometimes taken up in excess of NO_3^- is not conclusive evidence that the presence of NH_4^+ inhibits the uptake of NO_3^-, nor that the form of N supply will ultimately affect the total N uptake by the plant. Ingestad (1979) found that N supplied at various NH_4^+ to NO_3^- ratios to Scots Pine (*Pinus sylvestris*) and Norway Spruce (*Picea abies*) did not cause differences in plant growth. Ammonium, however, was more readily taken up than NO_3^-, suggesting that NH_4^+ uptake was more easily adjusted to N demands. Indeed, NH_4^+ can be metabolised directly, whereas NO_3^- has to be reduced first before it can be incorporated into organic plant constituents (Haynes, 1986). It has been suggested that energy (carbon) costs associated with the formation of nitrate reductase in the roots of high-elevation Spruce trees in forest soils that have become increasing dominated by NO_3^- may have caused an additional stress on the system, ultimately contributing to a growth decline (Aber *et al.*, 1989; Schulze, 1989). However, a causal relationship between NO_3^- uptake and a reduction in tree vigor has not yet been conclusively established. It is further unclear to what extent the results from seedlings studies conducted under controlled conditions are related to the behaviour of mature forest trees under field conditions.

Nitrate uptake is accompanied by a compensatory extrusion of OH^- or HCO_3^- (Figure 4.5), which tends to increase the pH of the soil and the soil solution, particularly in the rhizosphere (Bledsoe and Zasoski, 1983; Rollwagen and Zasoski, 1988). Rygiewicz and Bledsoe (1984b) noted that the OH^- release was not always equivalent to the NO_3^- uptake, and that the pH effect was generally less pronounced for mycorrhizal roots. It has also frequently been noted that the uptake of anionic NO_3^- is accompanied by an increased influx of cations such as Ca^{2+} into the roots (e.g. Haynes, 1986; Rygiewicz and Bledsoe, 1984b) and an improvement in base cation nutrition of tree seedlings (McFee and Stone, 1968), a process which would tend to further diminish the pH effect of NO_3^- uptake.

There is recent evidence that NO_3^- assimilation by trees may not be limited to the roots, but may also be occurring at the level of the canopy. Some small-scale seedling exposure studies, for example, have shown that wet-deposited NO_3^- can be absorbed by conifer needles and by some hardwood leaves (Bowden, Geball and Bowden, 1989; Eberhardt and Pritchett, 1971; Garten and Hanson, 1990). Comparisons of annual atmospheric NO_3^- input and throughfall fluxes in a variety of forest ecosystems have also indicated that forest canopies can act as net NO_3^- sinks (Lovett, 1992). Garten and Hanson (1990) were unable, however, to demonstrate substantial incorporation of the captured NO_3-N in White Oak (*Quercus alba*) and Red Maple (*Acer rubrum*) leaves, but rather showed the NO_3^- deposited on the leaves to remain water soluble even several days after exposure. Norby, Weerasuriya and Hanson (1989) on the other hand, showed that exposure of the needles of Red Spruce seedlings to NO_2 gas or HNO_3 vapour stimulated NO_3^- reductase activity in the foliage, which is an indication of the NO_3^- assimilation capacity. Although canopy uptake of NO_3-N seems plausible, it is not clear at this point what portion of annual N requirements by the tree is derived via this pathway. Calculations by Garten and Hanson (1990) and Bowden, Geball and Bowden (1989) indicate that the role of foliar uptake of NH_4^+ is small ($<20\%$), suggesting foliar NO_3^- should be even less important.

4.6 EFFECT OF DISTURBANCE ON NITRATE FLUXES

From the previous discussions it follows that any perturbation, be it natural or anthropogenic, that modifies the relative strength of N sources (external from the atmosphere, via fertilisation; internal via mineralisation) versus N sinks (plant uptake, microbial immobilisation, denitrification) has the potential to affect NO_3^- cycling rates.

Numerous examples can be found in the literature that illustrate an increase in nitrification and NO_3^- leaching losses following harvesting or clearcutting of forests. Such increased NO_3^- flux can be ascribed to one or a combination of the following: (1) a reduction or elimination of plant N uptake, (2) an increase in root mortality and turnover, and (3) an increase in N mineralisation rates in the exposed soil and forest-floor material (e.g. Bormann *et al.*, 1974; Mann *et al.*, 1988; Vitousek *et al.*, 1982; Waide *et al.*, 1988). The magnitude and longevity of the response is related to the severity of the manipulation and the rate of vegetation regrowth (e.g. harvesting intensity, use of herbicides) as well as to the inherent properties of the forest ecosystem. Indeed, some systems are so strongly N-limited that N immobilisation continues to dominates the N cycle even after harvesting, causing no net increase in nitrification with vegetation removal (e.g. Cole and Gessel, 1965; Miller and Newton, 1983). In other instances, the removal of N-fixing tree species from the site may actually reduce nitrification and NO_3^- leaching (Van Miegroet, Cole and Homann, 1990a).

Natural tree mortality and forest gap formation affects the soil N dynamics very much as forest harvesting does, generally leading to increased N availability and accelerated nitrification rates (Matson and Boone, 1974; Mladenoff, 1987; Sasser and Binkley, 1989). Soil disturbance in and of itself also appears to increase nitrification, but the actual causes for such stimulation are not entirely clear. Disturbance-induced nitrification is illustrated by the NO_3^- solution peaks commonly noted following lysimeter installation (e.g. Johnson $et\ al.$, 1991), and may likewise be responsible for the elevated soil NO_3^- levels that were observed following rooting activity of wild boars in mixed hardwood stands in the Smoky Mountains in North Carolina (Singer, Swank and Lebsch, 1984). Periodic outbreaks of defoliating insects in the canopy of mixed hardwood forests at Hubbard Brook and Coweeta watersheds were also observed to stimulate nitrification, resulting in elevated soil and streamwater NO_3^- levels (Bormann and Likens, 1979; Swank $et\ al.$, 1981; Swank, 1986).

Wildfires, burning of slash following harvesting, or prescribed burning in existing forests volatilise large portions of the N previously incorporated into the forest-floor material (in addition to carbon and sulphur), particularly during high-intensity fires (Grier, 1975; Raison, 1979). Although some increases in NH_4^+ availability and leaching have been reported following fire (Khanna and Raison, 1986; various citations in Raison, 1979), there is little evidence that this type of site disturbance causes a significant increase in NO_3^- movement through the forest soil profile.

In recent years, the effect on soil N dynamics of anthropogenic changes in atmospheric deposition and global climatic conditions has received increasing attention and has led to the introduction of the terms 'N-saturation', 'critical N loads', and 'excess N deposition'. They are generally used in the context of atmospheric N deposition and refer to the capacity of forest systems to retain N (particularly atmospherically deposited N) and curtail NO_3^- leaching losses (Dempster and Manning, 1988; Nilsson and Grennfelt, 1988). A soil is considered N saturated when N inputs (from atmospheric deposition and internal N mineralisation) exceed N retention through microbial immobilisation and plant N uptake, and measurable amounts of NO_3^- leach below the rooting zone (Agren and Bosatta, 1988; Dempster and Manning, 1988; Nilsson and Grennfelt, 1988; Van Miegroet, Cole and Foster, 1992). This soil condition is the integrated result of vegetation type, age and vigour; past N accumulation history; climatic conditions; current and past N input regime; and soil characteristics, factors that were discussed earlier in the context of nitrification and NO_3^- leaching controls.

It is sometimes difficult to totally divorce N saturation from anthropogenic N inputs, as illustrated by the high NO_3^- leaching rates in southern Appalachian Spruce–Fir forests discussed earlier in Section 4.3.2 (e.g. Johnson $et\ al.$, 1991; Van Miegroet, Cole and Foster, 1992). This condition can also be attained naturally, for example, through N-fixation as discussed in Section 4.3.2 (e.g. Van Miegroet and Cole, 1984; Van Miegroet, Cole and Homann, 1990a), or may develop in response to changing environmental conditions. It may even occur within particular parts of the ecosystem, seasonally, from year to year or during stand/ecosystem development as a result of changes in the spatial and temporal balance between N sources and sinks (e.g. Vitousek and Reiners, 1975; Foster, Nicolson and Hazlett, 1989; Shepard $et\ al.$, 1990). Forest ecosystems characterised by a high nitrification potential and a low N retention capacity are likely to show increased NO_3^- leaching with increasing N deposition inputs. Likewise, in systems that are approaching N saturation (due to a combination of soil N content, C/N ratio and forest characteristics) an increase in N input or warming trend may stimulate mineralisation and nitrification rates. It is difficult to predict at what point nitrification will be induced by either prolonged N inputs or temperature changes in systems that are currently N-limited and still exhibit strong N retention.

4.7 CONCLUSIONS

Until now, the conceptual model for the control of NO_3^- production and movement in forest soils has been based upon the assumption that nitrification rates are largely controlled by the competition for NH_4^+ substrate between nitrifiers, decomposers and plants, and that microbial NO_3^- uptake is minimal. It has been assumed that nitrification and NO_3^- cycling are not prevalent in temperate and boreal forest ecosystems compared to soils under agricultural use on the basis of the observation that many forest ecosystems grow under N-deficient conditions, and that N amendments are not a widespread forest management practice. Recent evidence of high rates of nitrification and microbial NO_3^- uptake in grassland soils (with or without tree cover) that otherwise show very low soil NO_3^- pools and low NO_3^- leaching rates challenge this model and its assumptions. Research is needed to determine the extent to which nitrification and microbial NO_3^- uptake occur in forest ecosystems. In any event, it is clear that nitrate leaching (and perhaps nitrification itself) can be stimulated by any disturbance, natural or anthropogenic, that causes a shift in the relative relationship between N sources and N sinks in the ecosystem in such a way that excess N becomes available to the nitrifiers or that NO_3^- production exceeds microbial NO_3^- uptake.

Nitrate that enters from the atmosphere or is formed within the upper soil layer freely moves through the soil profile, carrying with it an equivalent amount of cations, unless NO_3^- is taken up again by the vegetation or is denitrified in anoxic zones in the soil. Nitrification and NO_3^- leaching have a strongly acidifying impact on the soil in terms of (1) the direct release of H^+ to the system during nitrification, (2) the stripping of base nutrients from the cation exchange complex in soil with medium to high base saturation and the leaching export caused by the mobile NO_3^- anion, and (3) the displacement of Al^{3+} from the exchange complex in very acid soils. Each of these aspects of soil and solution acidification can have important negative repercussions on nutrient availability and forest-site productivity.

Nitrate dynamics in forests, their effect on forest soil properties and ultimately on site fertility and ecosystem stability have recently received increased attention in the context of atmospheric N deposition and global warming trends, because both phenomena can cause forest soils to become N saturated. The sensitivity of forests to such disturbance-induced nitrification and an increase in cation leaching loss or Al mobilisation that is triggered by the movement of NO_3^- through the soil profile differs between ecosystems. It is particularly high in forest soils that currently have a marginally low N retention capacity (because of a combination of high N content and low C/N ratio of the forest soil and low or declining forest growth rates) and already show signs, at least periodically, of elevated NO_3^- levels in the soil and soil solution. Even forest ecosystems that are currently still N-deficient and have soils with low nitrification potentials may eventually reach the saturation point if large amounts of N continue to be added to the soil (via atmospheric deposition, N-fixation or N fertilisation), or if gradual soil warming stimulates the N mineralisation process enough to increase N availability. However, it is not known at this time how long and what kind of N input rates or climatic changes are necessary before NO_3^- starts to leach out of such forest soils.

ACKNOWLEDGEMENTS

The senior author's contribution was supported by Oak Ridge National Laboratory, which is managed by Martin Marietta Energy Systems, Inc., under contract DE-AC05084OR21400 with the US Department of Energy. Dale Johnson's contribution was sponsored by the Electric Power Research Institute under contract RP-2621 and the Spruce–fir Cooperative, under the US Forest Service/Environmental Protection Agency Forest Response Program. Publication No. 3661, Environmental Sciences Division, Oak Ridge National Laboratory.

REFERENCES

Aber, J.D., Nadelhoffer, K.J., Steudler, P. and Melillo, J.M. (1989) Nitrogen saturation in northern forest ecosystems. *Bioscience,* **39,** 378–87.

Agren, G.I. and Bosatta, E. (1988) Nitrogen saturation in terrestrial ecosystems. *Environ. Pollut.,* **54,** 185–97.

Alexander, M. (1977) *Introduction to Soil Microbiology,* 2nd edition, Wiley, New York.

Alexander, M. (1980) Effects of acidity on microorganisms and microbial processes in soil. In Hutchinson, T.C. and Havas, M. (eds), *Effects of Acid Precipitation on Terrestrial Ecosystems,* Plenum Press, New York, pp. 363–74.

Bassioni, N.H. (1971) Temperature and pH interaction in NO_3^- uptake. *Plant and Soil,* **35,** 445–8.

Berg, B. and Staaf, H. (1981) Leaching, accumulation and release of nitrogen in decomposing leaf litter. *Ecol. Bull.,* **33,** 163–78.

Berg, B. and McClaugherty, C. (1987) Nitrogen release from litter in relation to disappearance of lignin. *Biogeochemistry,* **4,** 219–24.

Binkley, D., Kimmins, J.P. and Feller, M.C. (1982) Water chemistry profiles in early- and mid-successional forest of British Columbia. *Can. J. For. Res.,* **12,** 240–48.

Binkley, D., Lousier, J.D. and Cromack, K. Jr (1984) Ecosystem effects of sitka alder in Douglas Fir plantation. *Forest Sci.,* **30,** 26–35.

Bledsoe, C.S. (1976) Effect of nitrate and ammonium supply on nutrient uptake by solution-grown Douglas Fir seedlings in solution culture. *Plant Physiol.,* **57,** (Suppl.), 49.

Bledsoe, C.S. and Zasoski, R.J. (1983) Effects of ammonium and nitrate on growth and nitrogen uptake by mycorrhizal Douglas Fir seedlings. *Plant and Soil.,* **7,** 445–54.

Bormann, F.H. and Likens, G.E. (1979) *Pattern and Process in Forest Ecosystems,* Springer-Verlag, New York.

Bormann, F.H., Likens, G.E., Siccama, T.G., Pierce, R.S. and Eaton, J.S. (1974) The export of nutrients and recovery of stable conditions following deforestation at Hubbard Brook. *Ecol. Monogr.,* **44,** 255–77.

Bowden, R.D., Geball, G.T. and Bowden, W.B. (1989) Foliar uptake of ^{15}N from simulated cloud water by Red Spruce (*Picea rubens*) seedlings. *Can J. For. Res.,* **19,** 382–6.

Breuer, D.W., Cole, D.W. and Schiess, P. (1979) Nitrogen transformation and leaching associated with wastewater irrigation in Douglas Fir, poplar, grass and unvegetated systems. In Sopper, W.E. and Kerr, S.N. (eds), *Utilization of Municipal Sewage Effluent and Sludge on Forest and Disturbed Land,* Pennsylvania State University Press, University Park, PA, pp. 19–33.

Coats, R.N., Leonard, R.L. and Goldman, C.B. (1976) Nitrogen uptake in a forested watershed, Lake Tahoe, California. *Ecology,* **57,** 995–1004.

Cole, D.W. (1982) Nitrogen uptake and translocation by forest ecosystem. *Ecol. Bull. (Stockholm),* **33,** 219–32.

Cole, D.W. and Gessel, S.P. (1965) Movement of elements through a forest soil as influenced by tree removal and fertilizer addition. In Youngberg, C.T. (ed.), *Forest–Soil Relationships in North America,* Oregon State University Press, Corvallis, OR, pp. 95–104.

Cole, D.W. and Rapp, M. (1981) Elemental cycling in forest ecosystems. In Reichle, D.E. (ed.), *Dynamic Properties of Forest Ecosystems,* Cambridge University Press, Cambridge, pp. 341–409.

Cooper, A.B. (1986) Supression of nitrate formation within an exotic conifer plantation. *Plant Soil,* **93,** 383–94.

Davidson, E.A. and Swank, W.T. (1986) Environmental parameters regulating gaseous nitrogen losses from two forested ecosystems via nitrification and denitrification. *Appl. Environ. Microbiol.,* December, 1287–92.

Davidson, E.A. and Swank, W.T. (1987) Factors limiting denitrification in soils from mature and disturbed southeastern hardwood forests. *Forest Sci.,* **33.,** 135–44.

Davidson, E.A., Stark, J.M. and Firestone, M.K. (1990) Microbial production and consumption of nitrate in an annual grassland. *Ecology*, **71**, 1968−75.

Dempster, J.P. and Manning, W.J. (eds) (1988) Special Issue: Excess nitrogen deposition. *Environ. Pollut.*, **54**, 159−298.

Eberhardt, P.J. and Pritchett, W.L. (1971) Foliar application of nitrogen to slash pine seedlings. *Plant and Soil*, **34**, 731−40.

Federer, C.A. (1983) Nitrogen mineralization and nitrification: depth variation in four New England soils. *Soil. Sci. Soc. Am. J.*, **47**, 1008−14.

Focht, D.D. and Verstraete, W. (1977) Biochemical ecology of nitrification and denitrification. *Adv. Microb. Ecol.*, **1**, 135−214.

Foster, N.W. and Nicolson, J.A. (1988) Acid deposition and nutrient leaching from deciduous vegetation and podzolic soils at the Turkey Lakes Watershed. *Can. J. Fish. Aquat. Sci.*, **45**, (Suppl. 1), 96−100.

Foster, N.W., Nicolson, J.A. and Hazlett, P.W. (1989) Temporal vatiation in nitrate and nutrient cations in drainage waters from a deciduous forest. *J. Environ. Qual.*, **18**, 238−44.

Franklin, J.F., Dyrness, C.T., Moore, D.G. and Tarrant, R.F. (1968) Chemical soil properties under coastal Oregon stands of Alder and conifers. In Trappe, J.M. *et al.* (eds), *Biology of Alder*, USDA Pacific Northwest Forest Range Exp. Stn, Portland, OR, pp. 157−72.

Gaines, G.L. and Thomas, H.C. (1953) Adsorption studies on clay minerals. II. A formulation of the thermodynamics of exchange adsorption. *J. Chem. Phys.*, **21**, 714−18.

Gapon, E.N. (1933) On the theory of exchange adsorption in soils (in Russian). *J. Gen. Chem. USSR*, **3**, 144.

Garten, C.T., Jr and Hanson, P.J. (1990) Foliar retention of ^{15}N-nitrate and ^{15}N-ammonium by Red Maple (*Acer rubrum*) and White Oak (*Quercus alba*) leaves from simulated rain. *Environ. Exp. Botany*, **30**, 333−42.

Grier, C.C. (1975) Wildfire effects on nutrient distribution and leaching in a coniferous ecosystem. *Can J. For. Res.*, **5**, 599−607.

Groffman, P.M. and Tiedje, J.M. (1989) Dentrification in north temperate forest soils I. Spatial and temporal patterns at the landscape and seasonal scales. *Soil Biol. Biochem.*, **21**, 613−20.

Hankinson, T.R. and Schmidt, E.L. (1988) An acidophilic and neutrophilic *Nitrobacter* strain isolated from the numerically predominant nitrite-oxidizing population of an acid forest soil. *Appl. Environ. Microbiol.*, 1536−40.

Haynes, R.J. (1986) *Mineral Nitrogen in the Plant−Soil System*, Academic Press, Orlando, FL.

Heilman, P. (1974) Effect of urea fertilization on nitrification in forest soils of the Pacific Northwest. *Soil Sci. Soc. Am. Proc.*, **38**, 664−7.

Ingestad, T. (1971) A definition of optimum nutrient requirements in birch seedlings. II. *Physiol. Plant.*, **24**, 118−25.

Ingestad, T. (1979) Mineral nutrient requirements in *Pinus silvestris* and *Picea abies* seedlings. *Physiol Plant.*, **45**, 373−80.

Johnson, D.W. and Edwards, N.T. (1979) Effect of stem girdling on biogeochemical cycles in a mixed deciduous forest of eastern Tennessee. II. Soil mineralization and nitrification rates. *Oecologia*, **40**, 259−71.

Johnson, D.W. and Todd, D.E. (1988) Nitrogen fertilization of young Poplar and Loblolly Pine plantations at differing frequencies. *Soil Sci. Soc. Am. J.*, **52**, 1468−77.

Johnson, D.W., Breuer, D.W. and Cole, D.W. (1979) The influence of anion mobility on ionic retention in waste-water-irrigated soils. *J. Environ. Qual.*, **8**, 246−51.

Johnson, D.W., Edwards, N.T. and Todd, D.E. (1980) Nitrogen mineralization, immobilization and nitrification following fertilization of a forest soil under field and laboratory conditions. *Soil Sci. Soc. Am. J.*, **44**, 610−16.

Johnson, D.W., Van Miegroet, H., Lindberg, S.E., Harrison, R.B. and Todd, D.E. (1991) Nutrient cycling in Red Spruce forests of the Great Smoky Mountains. *Can. J. For. Res.*, **21**, 769−87.

Johnsrud, S.C. (1978) Heterotrophic nitrification in acid forest soils. *Holarct. Ecol.*, **1**, 27−30.

Joslin, J.D. and Wolfe, M.H. (1988) Response of Red Spruce seedlings to change in soil aluminium in six amended forest soil horizons. *Can. J. For. Res.*, **18**, 1614−23.

Joslin, J.D., Kelly, J.M. and Van Miegroet, H. (1992) Soil chemistry and nutrition of North American Spruce−Fir stands: Evidence of recent change. *J. Environ. Qual.*, **21**, 12−30.

Joslin, J.D., Mays, P.A., Wolfe, M.H., Kelly, J.M., Garber, R.W. and Brewer, P.F. (1987) Chemistry of tension lysimeter water and lateral flow in spruce and hardwood stands. *J. Environ. Qual.*, **16**, 152–60.

Kelly, J.M., Schaedle, M., Thornton, F.C. and Joslin, J.D. (1990) Sensitivity of tree seedlings to aluminium: II. Red Oak, Sugar Maple, and European Beech. *J. Environ. Qual.*, **19**, 172–9.

Khanna, P.K. and Raison, R.J. (1986) Effect of fire intensity on solution chemistry of surface soil under *Eucalyptus pauciflora* forest. *Austr. J. Soil. Res.*, **24**, 423–34.

Kinjo, T. and Pratt, P.F. (1971) Nitrate adsorption: II. In competition with chloride, sulfate and phosphate. *Soil Sci. Soc. Am. Proc.*, **35**, 725–8.

Klein, T.M., Kreitinger, J.P. and Alexander, M. (1983) Nitrate formation in acid soils of the Adirondacks. *Soil Sci. Soc. Am. J.*, **47**, 506–8.

Knowles, R. (1981) Denitrification. *Ecol. Bull. (Stockholm)*, **33**, 315–29.

Krajina, V.J., Madoc-Jones, S. and Mellor, G. (1973) Ammonium and nitrate in the nitrogen economy of some conifers growing in Douglas–Fir communities of the Pacific North-West of America. *Soil Biol. Biochem.*, **5**, 143–7.

Kramer, P.J. and Kozlowski, T.T. (1979) *Physiology of Woody Plants*, Academic Press, New York.

Kreitinger, J.P., Klein, T.M., Novick, N.J. and Alexander, M. (1985) Nitrification and nitrifying microorganisms in an acid forest soil. *Soil Sci. Soc. Am. J.*, **49**, 1407–10.

Lamb, D. (1980) Soil nitrogen mineralisation in a secondary rainforest succession. *Oecologia*, **47**, 257–63.

Laudelout, H., Lambert, R. and Pham, M.L. (1976) Influence du pH at de la pression partielle d'oxygène sur la nitrification. *Ann. Microbiol. (Inst. Pasteur)*, **127a**, 367–82.

Like, D.E. and Klein, R.M. (1985) The effect of simulated acid rain on nitrate and ammonium production in soils from three ecosystems of Camels Hump Mountain, Vermont. *Soil Sci.*, **140**, 352–5.

Likens, G.E., Bormann, F.H. and Johnson, N.M. (1969) Nitrification: Importance of nutrient losses from cutover forested ecosystem. *Science*, **163**, 1205–6.

Lodhi, M.A.K. (1978) Comparative inhibition of nitrifiers and nitrification in a forest community as a result of the allelopathic nature of various tree species. *Amer. J. Bot.*, **65**, 1135–7.

Lovett, G.M. (1992) Atmospheric deposition and canopy interaction of nitrogen. In Johnson, D.W. and Lindberg, S.E. (eds), *Atmospheric Deposition and Nutrient Cycling in Forest Ecosystems*, Springer-Verlag, New York, pp. 152–66.

Mahendrappa, M.K. (1974) Volatilization of oxides of nitrogen from nitrate-treated black spruce raw humus. *Soil Sci. Soc. Am. Proc.*, **38**, 522–4.

Mai, H. and Fiedler, H.J. (1983) Nitrifikation in sauren Waldböden unter Fichte. *Arch. Acker- u. Pflanzenbau u. Bodenk. (Berlin)*, **27**, 499–507.

Mann, L.K., Johnson, D.W., West, D.C., Cole, D.W., Hornbeck, J.W., Martin, C.W., Riekerk, H., Smith, C.T., Swank, W.T., Tritton, L.M. and Van Lear, D.H. (1988) Effects of whole-tree and stem-only clearcutting on post-harvest hydrologic losses, nutrient capital, and regrowth. *Forest Science*, **34**, 412–28.

Matson, P.A. and Boone, R.D. (1984) Natural disturbance and nitrogen mineralization. wave-form dieback of Mountain Hemlock in the Oregon Cascades. *Ecology*, **65**, 1511–16.

McFee, W.W. and Stone, E.L. Jr (1968) Ammonium and nitrate as nitrogen sources for *Pinus radiata* and *Picea glauca*. *Soil Sci. Soc. Am. Proc.*, **31**, 879–84.

Miller, J.H. and Newton, M. (1983) Nutrient losses from disturbed forest watersheds in Oregon's Coast Range. *Agro-Ecosystems*, **8**, 153–67.

Mladenoff, D.J. (1987) Dynamics of nitrogen mineralization and nitrification in Hemlock and hardwood treefall gaps. *Ecology*, **68**, 1171–80.

Nadelhoffer, K.J., Aber, J.D. and Melillo, J.M. (1983) Leaf-litter production and soil organic matter dynamics along a nitrogen-availability gradient in southern Wisconsin. *Can. J. For. Res.*, **13**, 12–21.

Nakos, G. (1975) Absence of nitrifying microorganisms from a Greek forest soil. *Soil Biol. Biochem.*, **7**, 335–6.

Nilsson, J. and Grennfelt, P. (1988) Critical loads of sulphur and nitrogen — Report from a Nordic working group. *Nord 1986*, 11, Nordic Council of Ministers.

Nodvin, S.C., Driscoll, C.T. and Likens, G.E. (1988) Soil processes and sulfate loss at the Hubbard Brook Experimental Forest. *Biogeochemistry*, **5**, 185–99.

Norby, R.J., Weerasuriya, Y. and Hanson, P.J. (1989) Induction of nitrate reductase activity in Red Spruce needles by NO_2 and HNO_3 vapor. *Can. J. For. Res.*, **19**, 889–96.

Nye, P.H. and Greenland, D.J. (1960) The soil under shifting cultivation Commonwealth Bureau Soils Tech. Comm # 51, Commonwealth Agric. Bureaux, Farnam Royal, Bucks, UK.

Olson, R.K. and Reiners, W.A. (1983) Nitrification in subalpine Balsam Fir soils: Tests for inhibitory factors. *Soil Biol. Biochem.*, **15**, 413—18.

Pang, P.C., Cho, C.M. and Hedlin, R.A. (1975) Effects of pH and nitrifier population on nitrification of band-applied and homogeneously mixed urea nitrogen in soils. *Can. J. Soil. Sci.*, **55**, 560—61.

Pastor, J. and Post, W.M. (1986) Influence of climate, soil moisture, and succession on forest carbon and nitrogen cycles. *Biogeochemistry*, **2**, 22—7.

Pluth, D.J. and Nommik, H. (1981) Potential denitrification affected by nitrogen source of a previous fertilization of an acid forest soil from Central Sweden. *Acta Agric. Scand.*, **31**, 235—41.

Popovic, B. (1985) The effect of nitorgenous fertilizer on nitrification of forest soils. *Fertiliser Research*, **6**, 139—47.

Purchase, B.S. (1974) Evaluation of the claim that grass root exudates inhibit nitrification. *Plant and Soil*, **41**, 527—39.

Raison, R.J. (1979) Modification of the soil environment by vegetation fires, with particular reference to nitrogen transformations: A review. *Plant and Soil*, **51**, 73—108.

Raynal, D.J., Joslin, J.D., Kelly, J.M., Thornton, F.C., Schaedle, M. and Henderson, G.S. (1990) Sensitivity of tree seedlings to aluminum: III. Red Spruce and Loblolly pine. *J. Environ. Qual.*, **19**, 180—87.

Remacle, J. (1977) The role of heterotrophic nitrification in acid forest soils — Preliminary results. *Ecol. Bull. (Stockholm)*, **25**, 560—61.

Reuss, J.O. (1977) Chemical and biological relationships relevant to the effect of acid rainfall on the soil—plant system. *Water Air Soil Pollut.*, **7**, 461—78.

Reuss, J.O. (1983) Implications of the calcium—aluminum exchange system for the effects of acid precipitation on soil. *J. Environ. Qual.*, **12**, 591—5.

Reuss, J.O. (1989) Soil solution equilibria in lysimeter solutions under red alder. In Olson, R.K. and LeFohn, A.S. (eds), *Effects of Air Pollution on Western Forests*, APCA Transaction Series No. 16, Air and Waste Manage. Assoc., Pittsburgh, PA, pp. 547—59.

Reuss, J.O. and Johnson, D.W. (1986) Acid deposition and acidification of soils and waters. *Ecol. Series*, Vol. 59, Springer-Verlag, New York.

Rice, E.L. and Pancholy, S.K. (1974) Inhibition of nitrification by climax ecosystems. *Am. J. Bot.*, **59**, 1033—40.

Riha, S.J., Campbell, G.S. and Wolfe, J. (1986) A model of competition for ammonium among heterotrophs, nitrifiers, and roots. *Soil Sci. Soc. Am. J.*, **50**, 1463—6.

Roberge, M.R. and Knowles, R. (1966) Ureolysis, immobilization, and nitrification in Black Spruce (*Picea mariana Mill.*) humus. *Soil Sci. Soc. Am. Proc.*, **30**, 201—4.

Robertson, G.P. (1982) Nitrification in forested ecosystems. *Philos. Trans. R. Soc. London Biol. Sci.*, **296**, 445—57.

Robertson, G.P., Huston, M.A., Evans, F.C. and Tiedje, J.M. (1988) Spatial variability in a successional plant community: patterns of nitrogen availability. *Ecology*, **69**, 1517—24.

Robertson, P.G. and Tiedje, J.M. (1984) Denitrification and nitrous oxide production in successional and old-growth Michigan forests. *Soil Sci. Soc. Am. J.*, **48**, 383—9.

Robertson, P.G. and Vitousek, P.M. (1981) Nitrification potentials in primary and secondary succession. *Ecology*, **62**, 376—86.

Robertson, P.G., Vitousek, P.M., Matson, P.A. and Tiedje, J.M. (1987) Denitrification in a clearcut Loblolly Pine (*Pinus taeda*) plantation in the southern US. *Plant Soil*, **97**, 119—29.

Rollwagen, B.A. and Zasoski, R.J. (1988) Nitrogen source effects on rhizosphere pH and nutrient accumulation by Pacific Northwest conifers. *Plant and Soil*, **105**, 79—86.

Rygiewicz, P.T., Bledsoe, C.S. and Zasoski, R.J. (1984a) Effects of ectomycorrhizae and solution pH on [^{15}N] ammonium uptake by conifer seedlings. *Can. J. For. Res.*, **14**, 885—92.

Rygiewicz, P.T., Bledsoe, C.S. and Zasoski, R.J. (1984b) Effects of ectomycorrhizae and solution pH on [^{15}N] nitrate uptake by conifer seedlings. *Can. J. For. Res.*, **14**, 893—9.

Sasser, C.L. and Binkley, D. (1989) Nitrogen mineralization in high-elevation forests of the Appalachians. II. Patterns with stand development in Fir waves. *Biogeochemistry*, **7**, 147—56.

Schimel, J.P., Firestone, M.K. and Killham, K.S. (1984) Identification of heterotrophic nitrification in a Sierran forest soil. *Appl. Environ. Microbiol.*, **48**, 802—6.

Schulze, E.D. (1989) Air pollution and forest decline in a Spruce (*Picea abies*) forest. *Science*, **244**, 766–83.

Shepard, J.P., Mitchell, M.J., Scott, T.J. and Driscoll, C.T. (1990) Soil solution chemistry of an Adirondack spodosol: Lysimetry and N dynamics. *Can. J. For. Res.*, **20**, 818–24.

Shortle, W.C. and Smith, K.T. (1988) Aluminum-induced calcium deficiency syndrome in declining Red Spruce. *Science*, **220**, 1017–18.

Singer, F.J., Swank, W.T. and Lebsch, E.C. (1984) Effects of wild pig rooting in a deciduous forest. *J. Wildl. Manage.*, **48**, 464–73.

Smith, W.T., Bormann, F.H. and Likens, G.E. (1968) Response of chemoautotrophic nitrifiers to forest cutting. *Science*, **106**, 471–3.

Strader, R.H., Binkley, D. and Wells, C.G. (1989) Nitrogen mineralization in high elevation forests of the Appalachians. I. Regional patterns in southern Spruce–Fir forests. *Biogeochemistry*, **7**, 131–45.

Strayer, R.F., Lin, C.J. and Alexander, M. (1981) Effect of simulated acid rain on nitrification and nitrogen mineralization in forest soils. *J. Environ. Qual.*, **10**, 547–51.

Stroo, H.F. and Alexander, M. (1986) Role of organic matter in the effect of acid rain on nitrogen mineralization. *Soil Sci. Soc. Am. J.*, **50**, 1218–23.

Swank, W.T. (1986) Biological control of solute losses from forest ecosystems. In Trudgill, S.T. (ed.), *Solute Processes*, Wiley, New York, pp. 85–136.

Swank, W.T., Waide, J.B., Crossley, D.A. Jr and Todd, R.L. (1981) Insect defoliation enhanced nitrate export from forest ecosystems. *Oecologia (Berlin)*, **51**, 297–9.

Thornton, F.C., Schaedle, M. and Raynal, D.J. (1987) Effect of Al on Red Spruce seedlings in solution culture. *Environ. Exp. Bot.*, **27**, 489–98.

Turner, J. (1977) The effect of nitrogen availability on nitrogen cycling in a Douglas–Fir stand. *Forest Sci.*, **23**, 307–26.

Ulrich, B. (1983) A concept of forest ecosystem stability and of acid deposition as a driving force for destabilization. In Ulrich, B. and Pankrath, J. (eds), *Effects of Accumulation of Air Pollutants in Forest Ecosystems*, D. Reidel, Dordrecht, pp. 1–31.

Urie, D.H. (1973) Phosphorus and nitrate levels in groundwater related to irrigation of Jack Pine with sewage effluent. In Kardos, L.T. and Sopper, W.E. (eds), *Recycling treated municipal Wastewater and Sludge through Forest and Cropland*, Pennsylvania State University Press, University Park, PA, pp. 176–83.

Van Miegroet, H. and Cole, D.W. (1984) The impact of nitrification on soil acidification and cation leaching in a red alder forest. *J. Environ. Qual.*, **13**, 586–90.

Van Miegroet, H., Cole, D.W., Binkley, D. and Sollins, P. (1989) The effect of nitrogen accumulation and nitrification on soil chemical properties in Alder forests. In Olson, R.K. and LeFohn, A.S. (eds), *Effects of Air Pollution on Western Forests*, APCA Transaction Series No. 16, Air and Waste Manage. Assoc., Pittsburgh, PA, pp. 515–28.

Van Miegroet, H., Cole, D.W. and Homann, P.S. (1990a) The effect of alder forest cover and alder forest conversion on site fertility and productivity. In Gessel, S.P., Lacate, D.S., Weetman, G.F. and Powers, R.F. (eds), *Sustained Productivity of Forest Soils*, Proceedings of the 7th North American Forest Soils Conference. University of British Columbia, Faculty of Forestry Publications, Vancouver, BC, pp. 333–54.

Van Miegroet, H., Johnson, D.W. and Cole, D.W. (1990b) Soil nitrification as affected by N fertility and changes in forest floor C/N ratio in four forest soils. *Can. J. For. Res.*, **20**, 1012–19.

Van Miegroet, H., Cole, D.W. and Foster, N.W. (1992) Nitrogen distribution and cycling. In Johnson, D.W. and Lindberg, S.E. (eds), *Atmospheric Deposition and Nutrient Cycling in Forest Ecosystems*, Springer-Verlag, New York, pp. 178–99.

Van Praag, H.J. and Weissen, F. (1973) Elements of functional definition of oligotroph humus based on nitrogen nutrition of forest stands. *J. Appl. Ecol.*, **10**, 569.

Verstraete, W. (1981) Nitrification. *Ecol. Bull. (Stockholm)*, **33**, 303–14.

Vitousek, P.M. and Reiners, W.A. (1975) Ecosystem succession and nutrient retention: A hypothesis. *BioScience*, **25**, 155–77.

Vitousek, P.M., Gosz, J.R., Grier, C.C., Melillo, J.M., Reiners, W.A. and Todd, R.L. (1979) Nitrate losses from disturbed ecosystems. *Science*, **204**, 469–74.

Vitousek, P.M., Gosz, J.R., Grier, C.C., Melillo, J.M. and Reiners, W.A. (1982) A comparative analysis of potential nitrification and nitrate mobility in forest ecosystems. *Ecol. Monogr.*, **52**, 155–77.

Waide, J.B., Caskey, W.H., Todd, R.L. and Boring, L.R. (1988) Changes in soil nitrogen pools and transformations following clearcutting. In Swank, W.T. and Crossley, D.A. Jr (eds), *Forest Hydrology and Ecology at Coweeta*, Springer-Verlag, New York, pp. 221−43.

Wang, L.J., Harrison, R.B., Dongsen, X. and Henry, C.L. (1990) Nitrate leaching induced Mg deficiency in Douglas−Fir and Grand Fir growing on a coarse outwash soil amended with sewage sludge. *Agronomy Abstracts*, **50**.

Waring, S.A. and Giliam, J.W. (1983) The effect of acidity on nitrate reduction and denitrification in lower coastal plain soils. *Soil Sci. Soc. Am. J.*, **47**, 246−51.

Weber, D.F. and Gainey, P.L. (1962) Relative sensitivity of nitrifying organisms to hydrogen ions in soils and in solutions. *Soil Sci.*, **94**, 138−45.

Wells, C.G., Jones, A. and Craig, J. (1988) Dentrification in southern Appalachian Spruce−Fir forests. In Hertel, G.D. (Tech. Coord.) *Effects of Atmospheric Pollutants on the Spruce−Fir Forests of the Eastern United States and the Federal Republic of Germany*, US Forest Service, General Technical Report NE-120, USDA-Forest Service, Northeastern Forest Experiment Station, Broomall, PA, pp. 117−22.

Wells, C.G., Murphy, C.E., Davis, C., Stone, D.M. and Hollord, G.J. (1986) Effect of sewage sludge from two sources on elemental fluxes in soil solution of Loblolly Pine plantations. In Cole, D.W., Henry, C.L. and Nutter, W.L. (eds), *The Forest Alternative for Treatment and Utilization of Municipal and Industrial Wastes*, University of Washington Press, Seattle, WA, pp. 154−67.

Wiklander, L. (1976) The influence of anions on adsorption and leaching of cations in soils. *Gründforbättring*, **27**, 125−35.

Zak, D.R., Pregitzer, K.S. and Host, G.E. (1986) Landscape variation in nitrogen mineralization and nitrification. *Can. J. For. Res.*, **16**, 1258−63.

5 Nitrogen Cycling in Surface Waters and Lakes

A.L. HEATHWAITE
Department of Geography, University of Sheffield

5.1 AN OVERVIEW: THE AQUATIC NITROGEN CYCLE

The rising nitrogen concentrations in streams and rivers and accelerated eutrophication of lakes and reservoirs have focused attention on the pathways and cycling of nitrogen in the aquatic system. Nitrogen is important because it controls the productivity of freshwater ecosystems, and together with phosphorus, is a rate-limiting nutrient in freshwater eutrophication.

Two factors are often taken to suggest that nitrogen is unlikely to be the controlling element in the eutrophication of freshwaters. First, nitrogen concentrations in most freshwaters are usually higher than phosphorus. Average nutrient concentrations are commonly of the order 16:1 N:P (Ryding and Rast, 1989). Second, cyanobacteria are able to fix nitrogen from the atmosphere and are not, therefore, responsive to limitations in the nitrogen supply from, for example, catchment sources. Such assumptions are an oversimplification of the processes and pathways of nitrogen cycling in aquatic systems, and do not account for factors such as the seasonal fluctuation in nutrient supply to any freshwater body from its catchment, and the ability of freshwater biota to utilise and release nitrogen and phosphorus. In this chapter, the sources, pathways and transformations of nitrogen in freshwater ecosystems will be examined, and the mechanisms by which nitrogen is involved in the eutrophication of lake systems will be identified.

Nitrogen (N) in freshwaters takes several forms, and their dynamic interrelationships can be described as the aquatic nitrogen cycle. The dominant combined N species in waters (excluding molecular N_2) are dissolved inorganic N (NH_4^+, NO_2^-, NO_3^-), dissolved organic N, and particulate N, which is usually organic but can contain inorganic N. Organic nitrogen usually exists either as an integral part of protein molecules or in the partial breakdown of these molecules (for example, peptides, urea and amino acids). Ammoniacal N is usually present in freshwaters as a result of the biological decomposition of organic N.

Figure 5.1 illustrates the magnitude of the flow pathways of each nitrogen fraction within the terrestrial and aquatic systems. The importance of the terrestrial nitrogen fraction as a source of N for the aquatic system is clear. Up to 40% of the total nitrogen flux reaches the aquatic system through direct surface runoff or subsurface flow from the catchment (see Chapter 3). Consequently, the patterns of use, particularly land use, and processes of nitrogen release in the terrestrial system must, to a large extent, determine the magnitude of input to the aquatic system and the nature of the response.

Nitrate: Processes, Patterns and Management. Edited by T.P. Burt, A.L. Heathwaite and S.T. Trudgill
© 1993 John Wiley & Sons Ltd

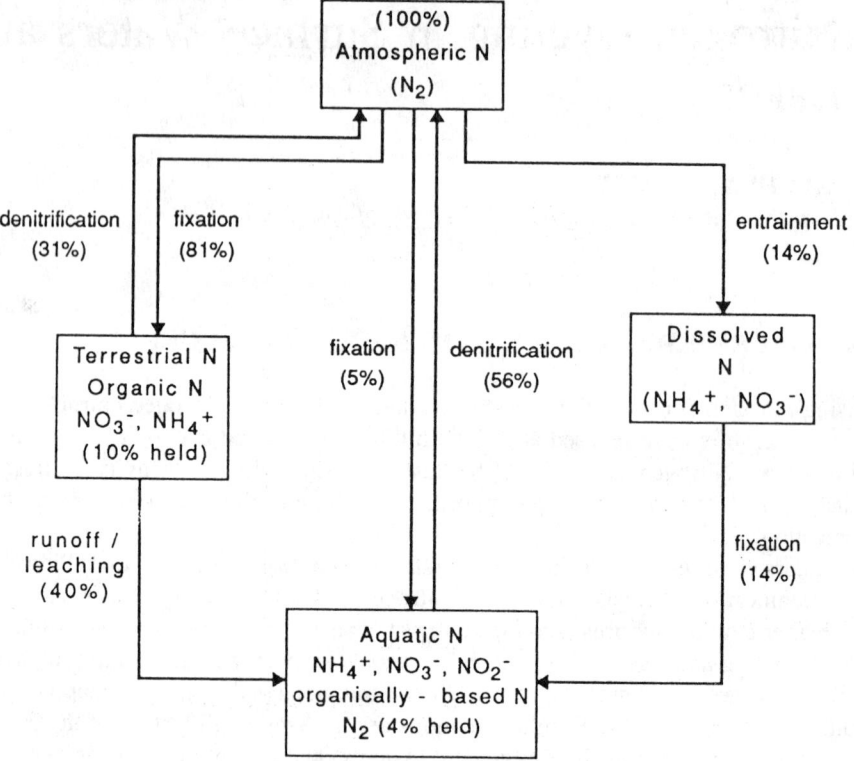

Figure 5.1 The nitrogen cycle (redrawn from Henderson-Sellers and Markland, 1987)

Figure 5.2 illustrates the nitrogen speciation in aquatic systems. The aquatic N cycle is similar in form to the terrestrial cycle, the key difference being the size of the biomass pool. In the aquatic nitrogen cycle, the biomass pool may be only one-thirtieth of the terrestrial pool (Henderson-Sellers and Markland, 1987). However, energy utilisation is more efficient so that net primary productivity may be double that of terrestrial ecosystems. Furthermore, the utilisation and turnover of nitrogen in aquatic systems in more rapid than that of terrestrial systems.

Five possible nitrogen-cycling reactions are shown in Figure 5.2. These are: fixation, nitrification, assimilation, denitrification and mineralisation or ammonification. Immobilisation may occur where there is limited O_2 but this process is usually seasonal and depends on biomass availability. Whether these reactions occur or not is largely determined by the energy required or lost during the reaction process (see Table 5.1). Actual energetics in the aquatic system must consider:

(1) The rate of reaction;
(2) Organic matter availability;
(3) Partial reactions which, when taken together may constitute a thermodynamically viable process;
(4) The energy source; and
(5) The oxidation state of the environment.

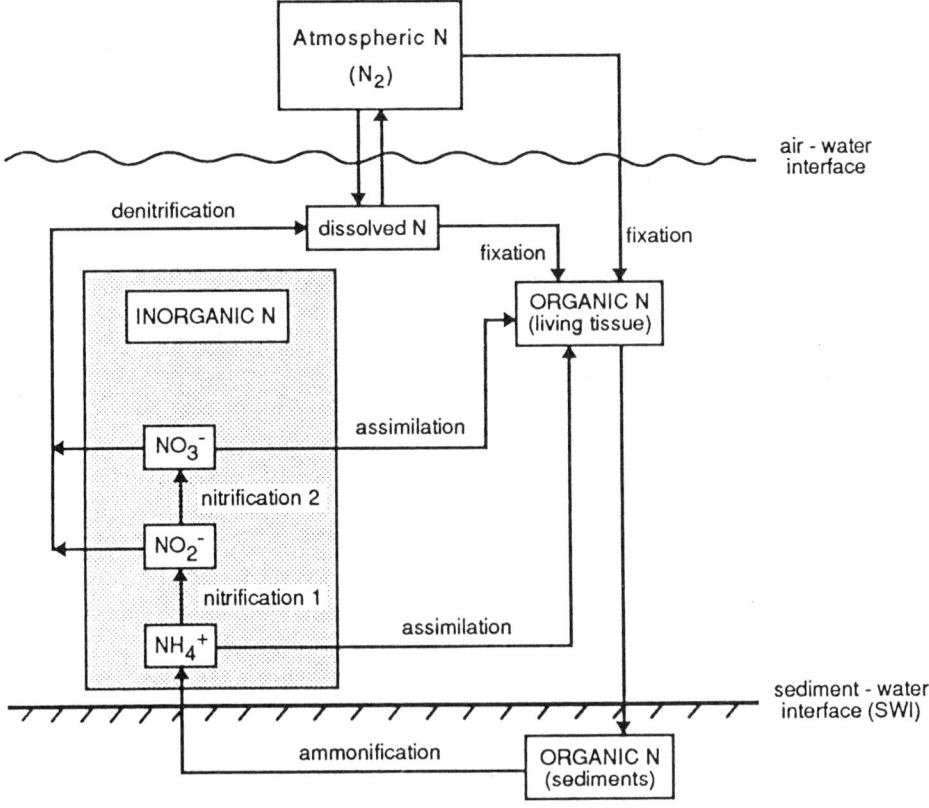

Figure 5.2 The aquatic nitrogen cycle

A further constraint on the likelihood of nitrogen transformation in aquatic systems is the supply of nitrogen. The total N in a freshwater system is less important than the rate of supply when N is a limiting factor. This can often occur in lakes in summer months (Stewart *et al.*, 1982; Van Vlymen, 1980; Heathwaite, 1989). In general, all aquatic systems have some constraints on nitrogen cycling, primarily as a result of some degree of anaerobiosis. To some extent this can be balanced by, for example, ammonium which, in the absence of cation binding sites, is more available in aquatic than in terrestrial systems. Light penetration is important in aquatic systems and may differentially affect micro-organisms and photosynthetic organisms. For example, nitrite-oxidising species are more light sensitive than ammonium-oxidising species; these differences will affect activity profiles in the water column and cause diurnal nitrogen cycling fluctuations. Table 5.2 summarises some of the factors affecting N cycling in freshwaters and these processes are examined in more detail below. The focus in primarily on lake systems, although the processes described do occur in streams and rivers but are usually at different rates.

5.1.1 NITROGEN FIXATION

Nitrogen fixation is a bacterially mediated, exergonic reduction process which converts molecular N to ammonia:

$$8H^+ + N_2 + 8e^- \rightarrow 2NH_3 + H_2$$

On an annual basis, total nitrogen fixation in aquatic systems rarely exceeds 20 kg N ha^{-1} (Royal Society, 1983). In general, N fixation requires adenosin triphosphate (ATP) which is generated by photosynthesis; so this process is inefficient at night. However, cyanobacteria (primarily *Anabaena, Aphanizomenon, Gloeotrichia*) can fix nitrogen directly, so do not have this diurnal limitation (Sprent, 1987). These species can form dense planktonic mats and their vertical distribution in the water column can be controlled by means of a gas vacuole. *Microcystis aeruginosa*, although an abundant bloom-forming bacterium, does not fix nitrogen.

There is still much uncertainty over the relative importance of nitrogen fixation in comparison with the total nitrogen input to lake systems from catchment sources. There is evidence to suggest that N fixation is primarily important in eutrophic lakes with large populations of cyanobacteria. Torrey and Lee (1976), for example, suggested that N fixation accounted for almost 10% of the total nitrogen load in a eutrophic North American lake.

5.1.2 ASSIMILATION OF NITROGEN IN FRESHWATERS

The importance of plankton assimilation of nitrate in freshwaters was demonstrated in Blelham

Table 5.1 Nitrogen cycle reactions: standard free energy changes

Reaction	Free energy change (kJ mol^{-1})	Process
$NO_3^- \rightarrow NO_2^-$	-142 to -161	Anaerobic NO_3 respiration
$NO_2^- \rightarrow NH_4^+$	-347 to -433	Anaerobic NO_2 reduction
$NO_3^- \rightarrow NH_4^+$	$+348$	$NO_2 + NO_3$ reduction
$\frac{1}{2}N_2 \rightarrow NO_2^-$	$+480$	N fixation
$NH_4^+ \rightarrow NO_2^-$	-229 to -338	*Nitrosomonas*
$NO_2^- \rightarrow NO_3^-$	-65 to -88	*Nitrobacter*

$+$ Net energy gain; $-$ net energy loss (after Sprent, 1987).

Table 5.2 Factors affecting the nitrogen cycle in freshwater lakes

Process	Factors
Nitrogen fixation	Cyanobacteria at lake surface
	Photosynthetic bacteria at anoxic zone
	N Fixation when soluble N concentration low
Mineralisation	More important in lake sediments
	Rapid mineralisation when plankton biomass dominates lake (low C:N)
Nitrification	Autotrophic so NH_4^+ and O_2 dependent
	Seasonal variation in site: spring and autumn: at SWI;[a] summer: open water (O_2 depletion at interface)
Dentrification	Seasonal, affected by NO_3^- supply
	Generally occurs at SWI
Assimilation	Phytoplankton; varies with NO_3^- concentration
	NH_4^+ can be assimilated if available
Immobilisation	Seasonal, depends on biomass, N rapidly mineralised if sufficient O_2

[a]SWI = sediment–water interface (modified from Sprent, 1987).

Tarn in the English Lake District (Preston, Stewart and Reynolds, 1980; Stewart *et al.*, 1982). The authors found that added ^{15}N-enriched sodium nitrate was removed from the water column within 14 days. The added N accumulated in the resulting *Microcystis* bloom. These results suggest that nitrogen stripping by primary producers may be an important mechanism for the removal of nitrate from freshwaters, although this depends on the ultimate fate of the nitrogen once it reaches the lake sediments because N may be available for re-release. Heathwaite and O'Sullivan (1991), for example, showed that lake sediments can store a large authigenic nitrogen fraction. The authors were able to correlate a significant peak in authigenic nitrogen in the sediments of Slapton Ley in south-west Devon, with the high nitrate load from the catchment linked to autumn rainfall following the 1976 drought.

5.1.3 AMMONIFICATION OR MINERALIZATION

Ammonium production occurs both in the water column of rivers and lakes and in their sediments. Microbial decomposition converts organic nitrogen to ammoniacal form. This process is oxygen-demanding and regenerates available nitrogen for re-assimilation by primary producers. Ammonification can result in rapid nitrogen cycling between the sediment and the water column. The rate of release of N from decomposing organic matter can be an important factor in determining nutrient limitation in freshwaters. Where N release is relatively slow, as in the example given in Table 5.3, the process of assimilation can become N-limited.

Ammonia in freshwaters can exist as the ammonium cation (NH_4^+) or as the un-ionised ammonia molecule (NH_3). High temperatures and high pH ($>$ pH 8) encourage the conversion of ammonium to ammonia. Ammonia (NH_3) is more toxic than ammonium, and acute toxicity can occur at low concentrations. Fortunately, high concentrations of ammonia are usually only associated with wastewater discharges where biological treatment is minimum. Freshwater plants and invertebrates are more resistant to ammonia than fish. Acute toxicity data for fish suggest that lethal concentrations (96-h LC_{50}) of ammonia range from 0.3 mg l$^-$ for rainbow trout to 3.2 mg l^{-1} for mosquito fish (Ellis, 1989).

5.1.4 NITRIFICATION

Nitrification is a two-stage oxidation process mediated by the chemoautotrophic genera

Table 5.3 Inorganic N and P release in solution from decomposing plankton (after Vaccaro, 1965)

Days	Inorganic N	PO_4-P	Atomic N:P
0	0	24.8	0
7	12.3[a]	24.5	1.1
17	39.3[a]	26.0	3.3
30	43.1[a]	32.6	2.8
48	54.0[b]	30.4	3.9
72	57.6[c]	30.6	4.2
87	58.0[c]	32.2	4.1

All results in μg l^{-1}.
[a]NH_4-N.
[b]NO_2-N.
[c]NO_3-N.

Nitrosomonas (NH_3 to NO_2^-) and *Nitrobacter* (NO_2^- to NO_3^-). In this exothermic reaction, more energy for biosynthesis is obtained from the oxidation of NH_4^+ to NO_2^- (-84 kcal $mole^{-1}$) than the subsequent oxidation to NO_3^- (-18 kcal $mole^{-1}$). The net reaction is:

$$NH_4^+ + 2O_2 \rightarrow NO_3^- + H_2O + 2H^+$$

The oxidation of ammonia to nitrite by *Nitrosomonas* is usually rate-limiting, so nitrite is rarely present in appreciable concentrations in freshwaters. Nitrate, the end-product, is highly oxidised, soluble and biologically available.

Nitrification is oxygen-demanding and can, in some aquatic systems, create anoxic conditions. This is because *Nitrosomonas* and *Nitrobacter* are strict aerobes, requiring minimum oxygen concentrations around 2 mg l^{-1} to function efficiently. The conversion of 1 mg NH_3 to NO_3^- theoretically requires 4.56 mg O_2. However, as this reaction is autotrophic, CO_2 may be used as the carbon source and the amount of free oxygen finally used may be less than the stoichiometric requirement at around 4.33 mg oxygen per mg ammonia.

The nitrifying bacteria are also pH and temperature susceptible, with an optimum pH of 8.4–8.6 (Wild, Sawyer and McMahon, 1971) and requiring a temperature above 15°C. *Nitrosomonas* has a wider temperature tolerance than *Nitrobacter*, and the growth rate constant for these bacteria increases by approximately 10% per degree Celsius up to about 25°C. This temperature limitation suggests that optimum rates of nitrification in British rivers and lakes will be rare (Ellis, 1989). There is also evidence to suggest that the action of nitrifying bacteria at pH 7 may be reduced by as much as 50% (Krenkel and Novotny, 1980). For these reasons, nitrification in surface waters and lakes has traditionally received little attention because the population of nitrifying organisms will often be low and the reaction rate is also low compared with carbonaceous oxidation by heterotrophs. However, nitification is important where retention times are long and temperatures high (for example, in shallow lakes in summer months). Furthermore, the discharge of nitrifying wastewater effluent to rivers and lakes will shift the balance in favour of increased nitrification.

A high rate of nitrification is essential for efficient N cycling in freshwaters, particularly as nitrate is an important substrate for denitrification (see Figure 5.2). Chemoautrophic nitrifying bacteria are usually dominant in freshwaters and their activity is generally highest at the sediment−water interface where ammonium-N generation is maximum (Stewart *et al.*, 1982). However, in eutrophic waters in particular, nitrate generated internally through nitrification is often relatively unimportant in comparison with the nitrate load received from the drainage basin. An example for Balgavies Loch, Scotland (Stewart *et al.*, 1982), suggested that nitrification accounted for only 22% of the nitrate load entering the loch in inflowing waters. However, N production *in situ* cannot be entirely ignored as an important component of the aquatic nitrogen cycle because nitrification is often high during summer when water temperatures are high. During this period, catchment inputs are often minimum and algal utilisation of nitrogen is maximum. Nitrification during this period could be critical to the efficient cycling of nitrogen within the aquatic system.

5.1.5 DENITRIFICATION

Loss of nitrate from river and lake systems can occur through denitrification or dissimilatory nitrate reduction. Denitrification is quantitatively more important, particularly in lake sediments, and is high in summer months (Royal Society, 1983). The rate and extent of denitrification is controlled by the oxygen supply and available energy provided by organic matter. It is seen as an important mechanism in the reduction of nitrate concentrations in reservoirs, but is limited by the requirement for anaerobic conditions and a fixed bacterial carbon supply. Stewart *et al.* (1982) concluded that although denitrification rates in UK eutrophic waters were substantial, they were not sufficient to cope adequately with NO_3^- inputs from agriculture. They suggested that NO_3^- concentration in lakes are likely to continue to increase, unless inputs to the aquatic system from the terrestrial system waters are reduced, or techniques for manipulating the nitrogen cycling processes *in situ* are more effectively exploited. These mechanisms, which are examined in more detail later, may include algal-stripping or encouraging denitrification by placing nitrate-rich waters near the anoxic zone at the sediment–water interface in lakes.

5.1.6 NITROGEN CYCLING BETWEEN FRESHWATERS AND LAKE SEDIMENTS

Lake sediments and sediment deposition zones in slow-moving rivers are both important sinks and sources of nitrogen. Nitrogen in lake sediments is present mainly as organic N. Sedimentation of particulate algae is an important route whereby fixed N reaches the lake sediment. Once there, it is rapidly mineralised. Nitrate transformations in lake sediments may be the result of both preferential microbial utilisation of nitrate over ammonium and N_2 fixation by anaerobes. The overall reaction may have the following form (after Keeney, 1974):

---

overlying water

↑

NO_3-N → NO_2-N → (NO) → N_2O → N_2

↑

NO_2-N → NH_4-N   organic N

The net result of these transformations is that a proportion of the nitrogen reaching the lake sediments will be converted to biologically available forms.

   Stewart *et al.* (1982) describe the sediment–water interface as the 'engine-room' of the aquatic nitrogen cycle, and as lake sediments are generally anoxic, except in highly turbulent lakes, the processes of denitrification and nitrogen fixation dominate at the sediment–water interface. This means that several of the nitrogen pathways in lake sediments are controlled by the reduction-oxidation status at the sediment–water interface. Other factors include pH, available carbonate, available calcium and phosphorus (Keeney, 1974). Within the lake sediments, ammonification is more important, particularly in summer months when temperatures are high. Nitrogen immobilisation can be important in lake sediments in winter months.

   Liao and Lean (1978) evaluated the nitrogen dynamics in the Bay of Quinte, Ontario, using limnocorrals by measuring seasonal changes in the concentrations of nitrate, ammonia and dissolved organic and particulate nitrogen in the lake waters and pore-waters of the lake sediments. They found that the seasonal changes in the nitrogen status of the lake began with a decline in particulate N in summer, which was followed by an increase in NH_3, and finally resulting in an increase in NO_3 throughout the winter. Their results suggest that nitrification is an essential process in nitrogen cycling in lakes. However, although the importance of lake sediments as a net source of biologically available N was small, Liao and Lean (1978) concluded that the flux of nitrogen to and from the lake sediments was a significant component of the nitrogen cycle. This conclusion was based on nitrogen budgets for the lake, which showed that the catchment nitrate loading together with N_2 fixation were approximately equal to the net change in the total combined N in the lake water column. This suggests that denitrification at the sediment—water interface is not important. However, most of the nitrogen entering the lake system from the catchment was lost over the year, which implies that denitrification *must* be important at least for some periods of the year. The results suggest that an accurate interpretation of nitrogen cycling in lakes must be seasonally based and must incorporate the role of the lake-sediment N in the nitrogen budget. The authors also suggest that lake ecosystems are capable of compensating for low- or high-catchment nitrogen loads.

Contrary to the finding of Liao and Lean (1978), for Alderfen Broad in the Norfolk Broads, Phillips (1977) suggested that sediment was an important nitrogen source in summer because changes in the nutrient budget of the Broad could not be accounted for by changes in the nutrient load of the inflowing streams. Reductions of NO_3-N in the Broad in spring are thought to be the result of removal by benthic algae and planktonic material.

5.2 SOURCES OF NITROGEN IN FRESHWATERS

Nitrogen in freshwaters is derived from three key sources:

(1) Rainfall and dry deposition;
(2) Domestic sewage and industrial effluents;
(3) Agricultural land.

While geochemical and topographical catchment factors will establish a regional *potential* for the nitrogen content of receiving waters, human impact can result in greater nutrient export than can natural factors. Agricultural and industrial activities usually result in *additions* to the nitrogen cycle either as inorganic and organic fertilisers or as atmospheric pollutants. A further source is domestic sewage discharge.

5.2.1 THE EVIDENCE FOR NITROGEN INCREASES IN SURFACE WATERS AND LAKES

Any examination of nitrogen sources in freshwaters, which invariably include increased inputs as a result of human activity, must start by examining the evidence for increased nitrogen concentrations in surface waters and lakes. The WHO guidelines for nitrate in drinking water are:

(1) Recommended concentration: < 50 mg l^{-1} NO_3^-
                                                             $(< 11.3$ mg l^{-1} NO_3-N)
(2) Acceptable concentration: $50-100$ mg l^{-1} NO_3^-
(3) Not-recommended: > 100 mg l^{-1} NO_3^-

The 1980 EC Directive is more stringent, with a guide level of 25 mg l^{-1} NO_3^- and a maximum acceptable concentration of 50 mg l^{-1} NO_3^- for water intended for human consumption. Current EC legislation will extend these human health considerations to environmental limits. There is no current human health legislation for ammonium in freshwaters. Ammonium concentrations in rivers rarely exceed 0.5 mg l^{-1} NH_4-N except where the sewage or industrial load is high; here concentrations may reach 2 mg l^{-1} NH_4-N. Instream nitrogen transformations prevent high ammonium concentration being maintained in surface waters except in polluted environments.

5.2.2 RIVERS: CONCENTRATIONS AND TRENDS

While national monitoring of nitrogen in rivers covers nitrate, and in some cases ammonium concentrations, there is a paucity of information on organic or total nitrogen concentrations. Any evaluation of the long-term trends in nitrogen concentrations in rivers is therefore only covered in part here; for further details see Chapter 10.

There are few rivers in the UK for which long-term water quality records exist. Casey and Clarke (1979) examined an 11-year nitrate record for the River Frome, UK, based on weekly nitrate samples. Their results suggest an average increase in the nitrate concentration in the Frome of 0.11 mg l^{-1} NO_3-N per year. Figure 5.3 shows nitrate concentration data for five UK rivers with long-term records. The progressive increase in river nitrate concentration over the period of the records is clearly shown. All the records pinpoint the high nitrate concentration recorded in rivers following the 1976 drought.

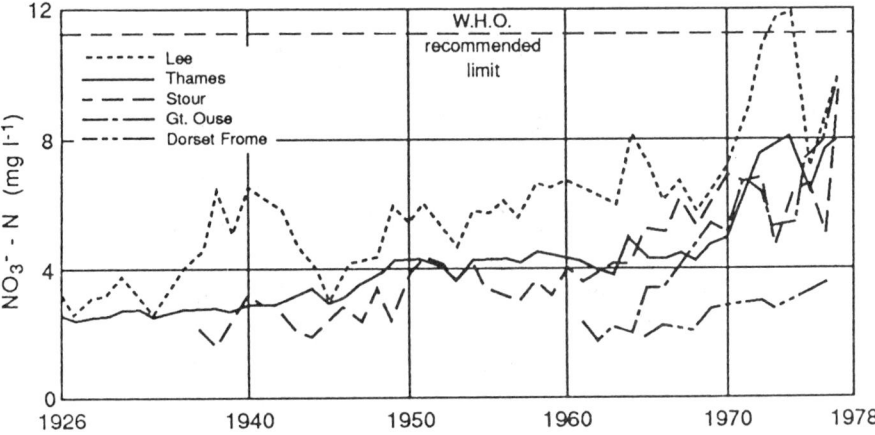

Figure 5.3 Trends in mean and annual nitrate concentration for five UK rivers (modified from Royal Society, 1983)

Figure 5.4 illustrates the 20-year nitrate record (1970–90) for a small grassland and arable catchment in south-west Devon (Heathwaite and Burt, 1991). The record was derived largely from weekly sampling with two periods of more frequent sampling (1983–5 and 1987–9), where 15-minute intervals were used during storm events and daily samples at times of low flow. The impact of the 1976 and 1984 droughts are clearly reflected in the nitrate concentration increase recorded in the autumn of those years. Over the period 1970 to 1990 a significant increase in the average stream nitrate concentration was recorded from 5 mg l^{-1} in 1970 to 8 mg l^{-1} NO_3-N in 1990 (Heathwaite and Burt, 1991). Wilkinson and Green (1982) report similar nitrate concentration increases between 50% and 400% for 12 UK rivers over a 20-year period up to 1982.

108

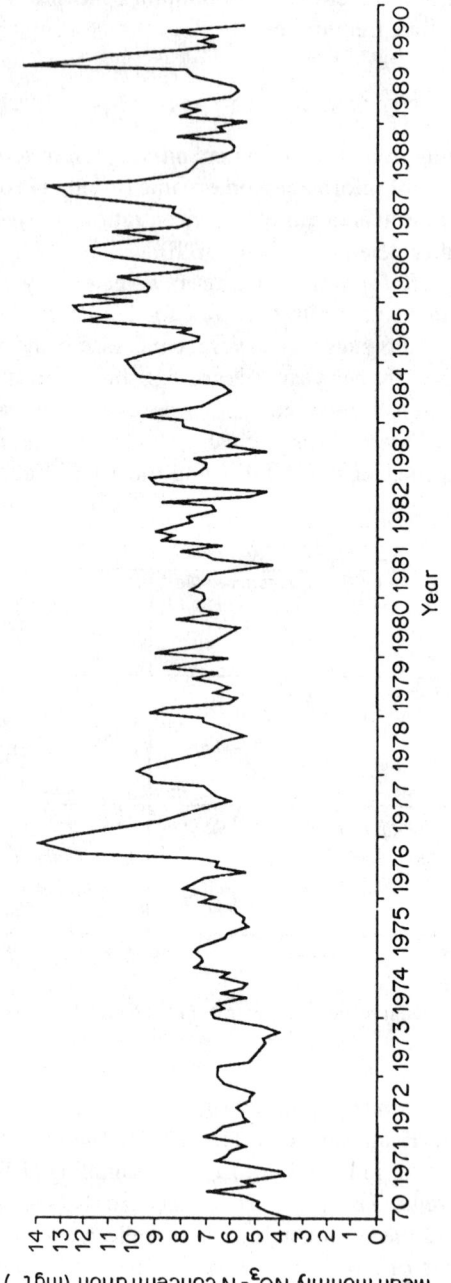

Figure 5.4 The 20-year nitrate record for the Slapton Wood catchment, Devon (after Heathwaite and Burt, 1991)

Figure 5.5 illustrates the seasonal fluctuation in stream nitrate concentration through the year for four subcatchments of the Slapton Ley catchment in south-west Devon, UK. The nitrate load is closely linked to the rainfall:runoff relationship for each subcatchment. The variation in the stream nitrate load for the different subcatchments may be accounted for by variations in catchment topography and land use (Heathwaite, Burt and Trudgill, 1989). Similar seasonal trends in stream and river nitrate concentrations have been reported by Casey and Clarke (1979) for the river Frome. Dermine and Lamberts (1987) also report seasonal variation in nitrate concentration in the Belgian Meuse which they attribute to variations in river discharge. At low discharge (less than 150 $m^3 s^{-1}$) nitrate leaching from catchment soils increased river nitrate concentration. At higher discharges, nitrate dilution by surface runoff from the catchment lowered river nitrate concentrations. Such patterns are obviously dependent on the runoff characteristics of the catchment and antecedent soil moisture conditions (see also Chapter 10).

Models of river nutrient loads are available for the prediction of future nitrate trends in rivers — for example, the Thames Catchment Model (Onstad and Blake, 1980). This simple model has three components:

(1) The annual mean nitrate concentration in the Thames based on the long-term nitrate record (1928+);
(2) The catchment input based on precipitation, land use, nitrogen mineralisation rates and fertiliser application rates, denitrification and volatilisation estimates, and removal of nitrogen in crops;
(3) A transfer component, relating (1) to (2).

Similar models are being developed for other catchments with long-term records (e.g. Johnes and O'Sullivan, 1989; Heathwaite and Burt, 1991).

5.2.3 LAKES: CONCENTRATIONS AND TRENDS

Upland lakes and reservoirs in the UK are often naturally oligotrophic; here concentrations of nitrate seldom exceed 2 mg l^{-1} (Royal Society, 1983). The dissolved organic nitrogen fraction in these lakes is normally greater than the dissolved inorganic nitrogen fraction, and primary productivity is low. The main source of organic nitrogen is catchment runoff.

In eutrophic lakes, the proportion of inorganic to organic nitrogen dissolved in the water column usually increases, largely as a result of high nitrogen export from the catchment. High inorganic nitrogen inputs to lakes have been recorded from heavily fertilised agricultural catchments (Stewart et al., 1982; Foy, Smith and Stevens, 1982; Heathwaite and Burt, 1991), and where urban effluent is discharged into the lake (Smith, 1977; Foy, Smith and Stevens, 1982; Stewart et al., 1982).

The fractionation of the total N pool in lakes generally follows the pattern recorded by Stewart et al. (1982) in Balgavies Loch, Scotland. They found that the major N pools were dissolved inorganic N, dissolved organic N and particulate organic N. Keeney (1973) suggests the following nitrogen concentration ranges for lake waters: nitrate 0−4 mg l^{-1}, ammonium 0−5 mg l^{-1} (higher concentrations are recorded in anoxic bottom waters and sediment interstitial waters); nitrite 0−0.01 mg l^{-1} (higher in sediment interstitial waters); dissolved organic N 0−2.5 mg l^{-1}. Organic N often accounts for over 50% of total dissolved nitrogen in lake waters, although this is dependent on the nutrient status of the lake.

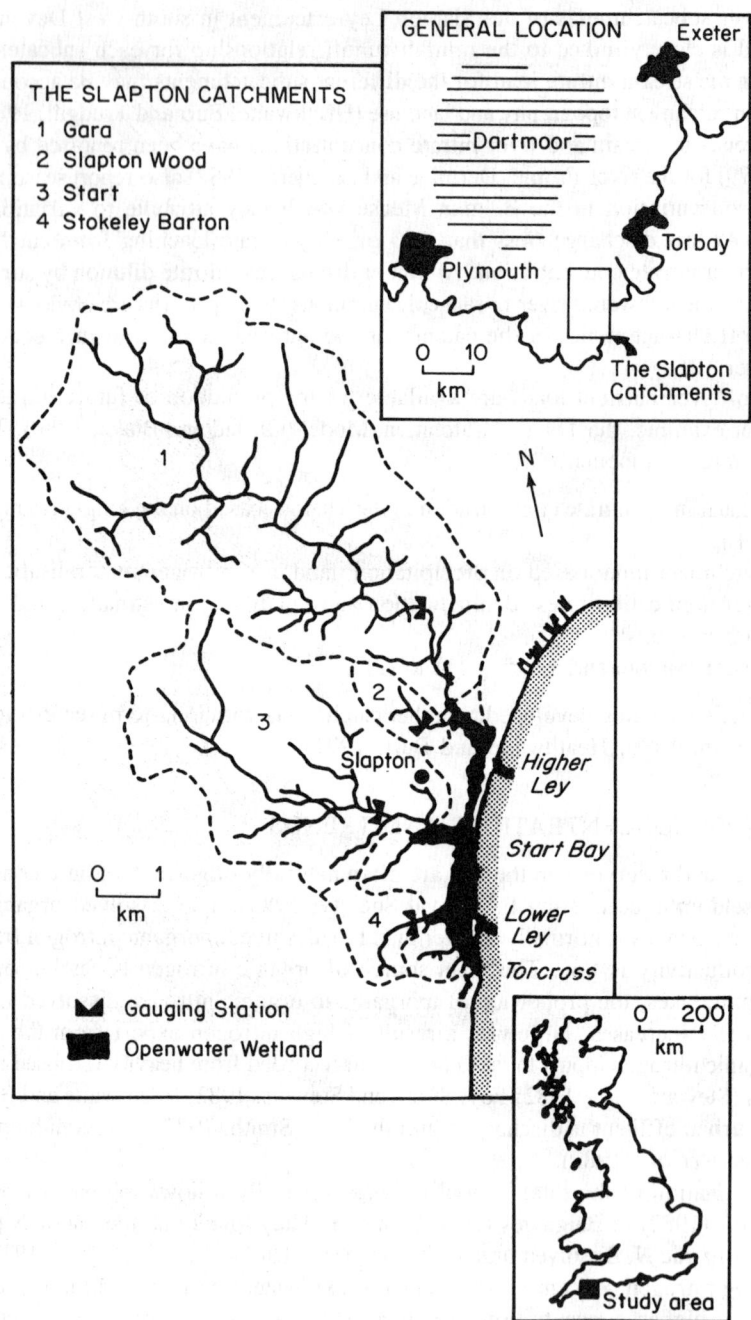

Figure 5.5 The seasonal relationship between nitrate load (kg km^{-2}) and catchment runoff for the Slapton catchments (after Heathwaite *et al.*, 1989)

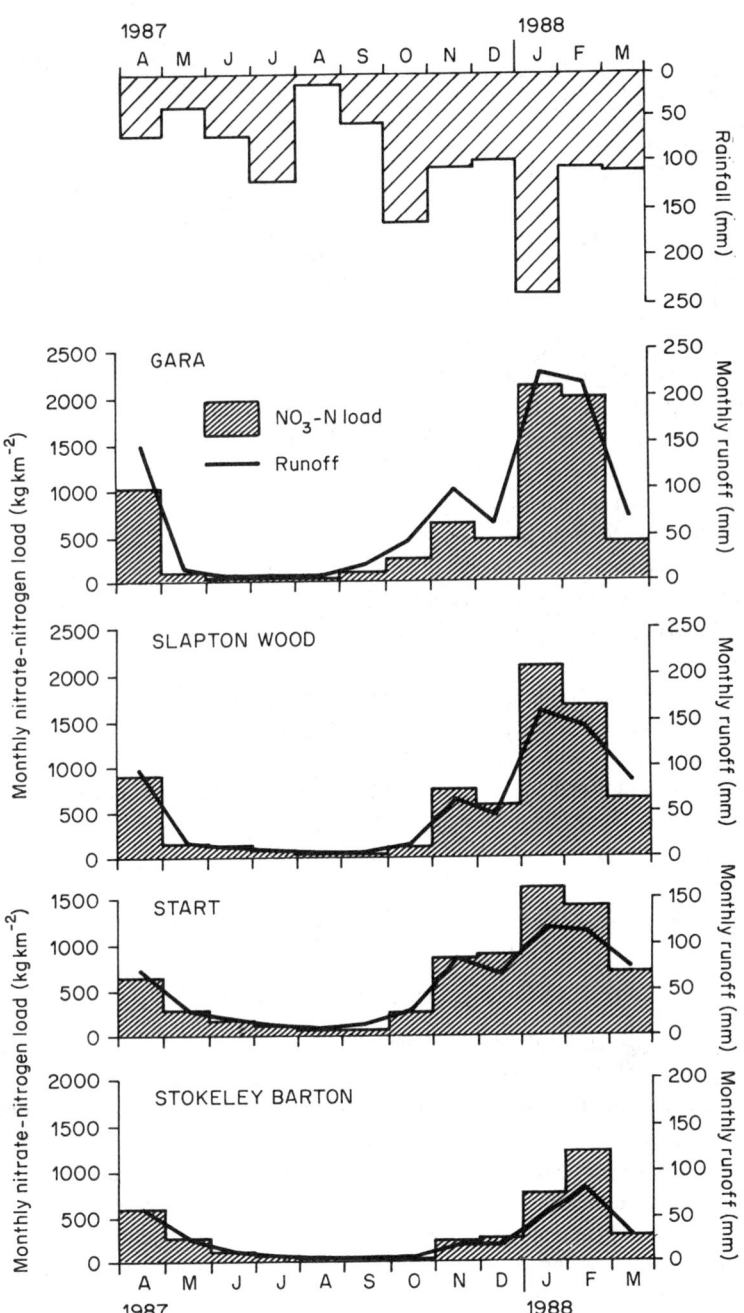

Figure 5.5 (*continued*)

When evaluating nitrogen cycling in lakes (and surface waters), it is important to consider seasonal fluctuations in the catchment nitrogen load and the rate of nitrogen cycling within the lake. Stewart *et al.* (1982), for example, found that the dissolved organic N pool showed little seasonal variation, whereas the dissolved inorganic pool, dominated by nitrate, shows the strong seasonal variation that has also been recorded in river systems (see, for example, Burt *et al.*, 1988; Heathwaite and Burt, 1991).

The processes and patterns of N cycling in lakes is, to a large extent, determined by the morphology of the lake and the influence of its catchment. The characteristics of lakes vary markedly from deep, oligotrophic lakes such as Loch Morar in Scotland (Maitland, 1981) to relatively shallow and commonly more eutrophic lakes such as Lough Neagh in Northern Ireland (Foy, Smith and Stevens, 1982) and Slapton Ley (O'Sullivan *et al.*, 1991) in south-west Devon. Peterson and Stewart (in Royal Society, 1983) found that the total nitrogen concentration in lakes of varying nutrient enrichment ranged from 0.51 mg l^{-1} for Craiglush, an oligotrophic Scottish lake, to 6.86 mg l^{-1} for hypertrophic Forfar loch. The dissolved organic nitrogen fraction formed approximately 30% of total N in the oligotrophic lake and nearly 90% of total N in the hypertrophic lake, demonstrating the importance of catchment sources in nutrient-rich lakes.

Nitrogen cycling in reservoirs has also been shown to vary with catchment characteristics and upland or lowland location. In Welsh upland reservoirs, for example, nitrate concentrations are low enough to make regular monitoring unnecessary (Scott, 1975). However, in lowland reservoirs, the increase in nitrate concentration in rivers has also led to a rise in the nitrate concentration of a number of reservoirs (for example, Grafham Water Reservoir, a water supply for Anglian Water, UK (Wilkinson and Greene, 1982)). Similarly, the average annual concentration of nitrate in Farmoor Reservoir which draws water from the river Thames near Oxford has risen from 2 mg N l^{-1} in 1961 to 6 mg N l^{-1} in 1980. Denitrification in reservoirs is an important process by which this increased catchment nitrogen load entering reservoirs is reduced (Wilkinson in Royal Society, 1983).

5.3 MODIFICATION OF THE NITROGEN CYCLE BY HUMAN ACTIVITY

5.3.1 THE EFFECT OF LAND USE ON NITROGEN IN FRESHWATERS

Nutrient export from agricultural land, and the ultimate expression of this export in the trophic status of a waterbody, is determined both by physical parameters (for example, precipitation intensity and duration, relief, geology) and land-use modification, particularly where inorganic or organic fertilisers are added or the land is grazed (see Chapters 10 and 12). The catchment factors affecting the trophic status of a lake system are summarised in Figure 5.6, which demonstrates how the terrestrial and aquatic components of the drainage basin are linked by the transfer of sediment and solutes from land to water. Note that neither external catchment controls nor in-lake nutrient concentrations exclusively control lake productivity. The physical and biotic structure of the waterbody (for example, lake depth and flushing rate) are also important. Nitrogen input to a waterbody is also affected by the degree of biological activity in the catchment, particularly microbial and plant activity: biological nitrification and denitrification affect the nitrogen flux in the catchment.

Land use, together with the efficiency of chemical cycling and hydrological processes in the catchment, are key parameters determining the quantity of nitrogen exported from the

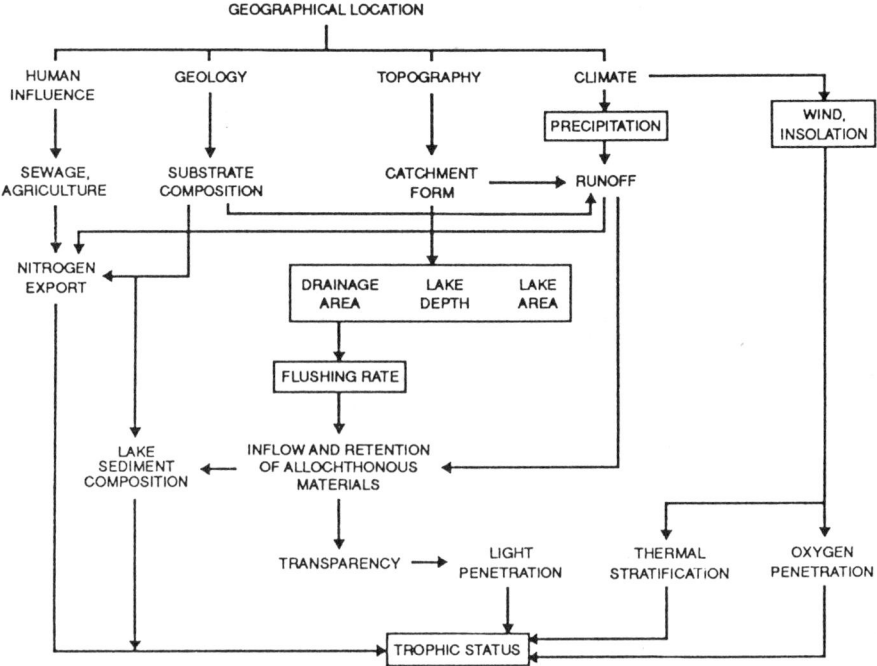

Figure 5.6 Interrelationships affecting lake trophic status (modified from Ryding and Rast, 1989)

catchment. For example, a negative correlation between nitrate input to the aquatic system and the proportion of riparian marshland or floodplains in the catchment might be anticipated because anaerobic decomposition and hence denitrification are likely to be high in waterlogged zones (see, for example, Haycock, 1991). The land-use controls given below are likely to be important in determining the magnitude of nitrogen inputs to the aquatic system:

(1) Presence or absence of grazing animals (organic N);
(2) Timing of ploughing and related operations;
(3) Amount and timing of fertiliser application;
(4) Type of manure or inorganic fertiliser used;
(5) Extent of crop cover;
(6) Soil type and drainage;
(7) Quantity and frequency of fertiliser application.

Figure 5.7 indicates the range of total nitrogen loads for a number of non-point catchment sources including rural, urban and atmospheric contributions. Table 5.4 provides an estimate of the nitrogen pollution load from different non-point sources in the USA.

Forest ecosystems, for example, have a relatively low nitrogen load (Table 5.4) because most mineralised nitrogen is recycled to vegetation. However, removal of this vegetation can accelerate the export of nitrogen to surface waters. Furthermore, changes in soil temperature and moisture as a result of the removal of forest cover can accelerate nitrogen mineralisation and lead to further nitrogen losses as a result of the low potential for vegetation uptake (Likens et al., 1977). Smolen (1981) examined arable non-point sources of nitrogen in a paired catchment study in Virginia. He recorded a doubling of the discharge-weighted

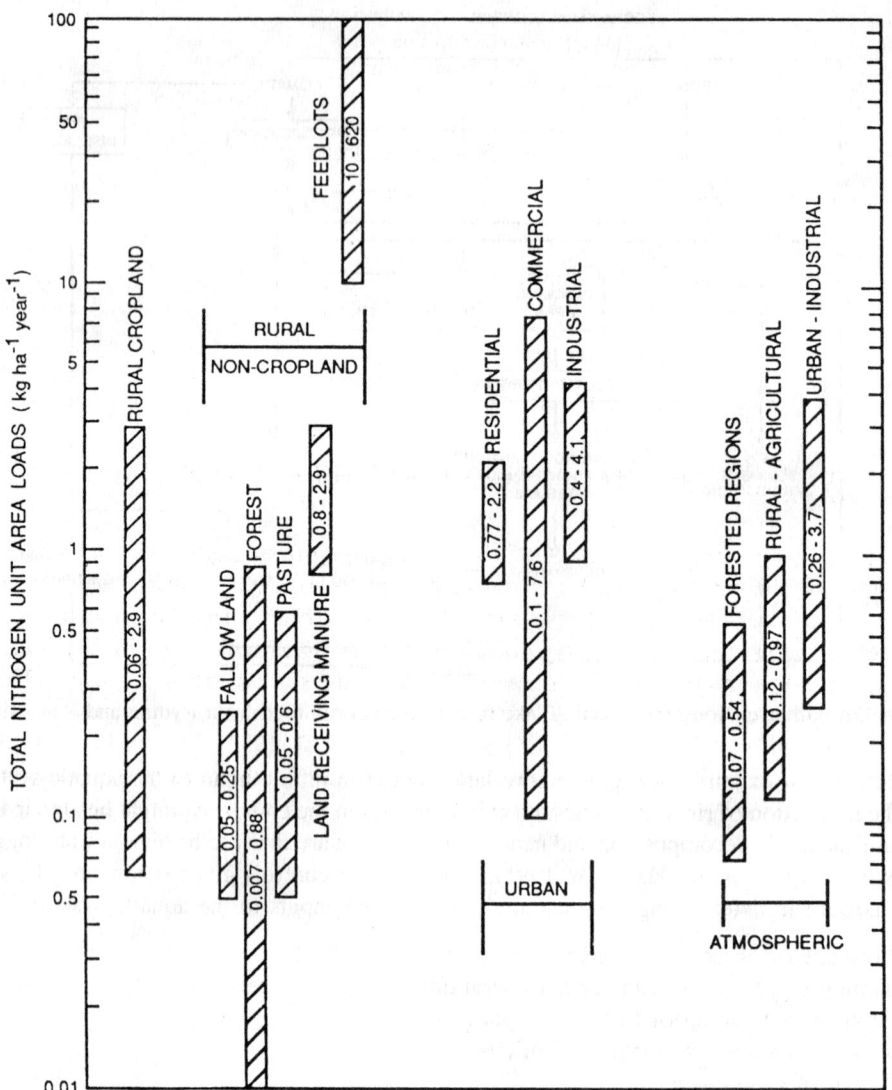

Figure 5.7 Total nitrogen loads for a range of non-point sources (reproduced by permission from Loehr, 1974)

nitrogen concentration of stream waters as a result of agricultural activity. Most of the increase in stream nitrogen was found in the soluble inorganic fraction. In comparison with arable land, nitrogen export from grassland is usually lower because grass roots rapidly exploit available soil nitrogen. Wild and Cameron (1980), for example, found that less than 10% of fertiliser applied to grassland was leached. The figure for arable land was around 15% N leached. The efficiency of fertiliser N use will depend on climatic conditions during the growing season and the timing of fertiliser application in relation to rainfall. Enhanced nitrate leaching may occur when grassland soils are ploughed.

Table 5.4 Nitrogen contribution to surface waters from non-point sources (modified from Loehr, 1974)

Non-point source	Nitrogen load[a]
Arable	4.3
Pasture	2.5
Forest	0.30
Urban runoff	0.15
Rural roadways	0.0005
Small feedlots	0.17
Landfill	0.026

[a]N load in MT a^{-1}.

5.3.2 NITROGEN EXPORT FROM URBAN DOMESTIC AND INDUSTRIAL DISCHARGES

Although values for nitrogen export from urban land ranging from 1.9 to 14 kg N ha^{-1} a^{-1}, are reported in the literature (see, for example, Loehr, 1974), detailed studies on nitrogen loads from urban areas are scarce. Urban surface water runoff can contribute a high nitrogen load to the aquatic system. The nitrogen load from urban areas will vary with the annual precipitation and the proportion of impermeable surface areas in the urban catchment. This relationship is illustrated in Figure 5.8. Surface water runoff from urban areas may contain over four times the nitrogen load of wooded ecosystems (OECD, 1982). The two key controls on non-point source nitrogen losses from urban areas are rainfall (as in rural areas) and the impermeable surface area. The relationship between these two controls is demonstrated in Figure 5.8.

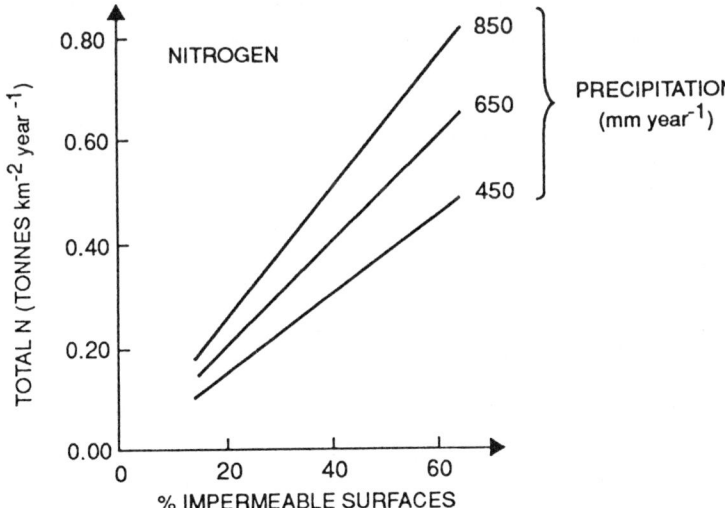

Figure 5.8 The estimated total nitrogen load from urban areas as a function of precipitation and the proportion of impermeable surfaces (redrawn from Ryding and Rast, 1989)

Point sources of nitrogen are largely associated with urban areas and, in particular, domestic sewage because nutrients in domestic and industrial wastes are rarely removed. However, important agricultural point sources do exist (for example, manure and silage waste from intensive livestock units).

The nutrient load from human waste is estimated at 10.8 g nitrogen per head per day. Nitrogen in wastewaters exists in four forms: NH_4-N (50–60%), NO-N (up to 5%), NO_3-N (up to 5%) and organic nitrogen (40–60%). Untreated sewage contains roughly 55 g m^{-3} total nitrogen and 25 g m^{-3} NH_3-N. After secondary sewage treatment, this load is reduced to between 20 g m^{-3} to 50 g m^{-3} total nitrogen, and negligible concentrations of NH_3-N. Moss et al. (unpublished) report ammonium concentrations in sewage works effluent in the Norfolk Broads ranging from 1.0–6.2 g m^{-3} NH_4-N. The variable degree of nitrification during sewage treatment results in a wide concentration range for ammonium. Nitrate concentrations were found to fluctuate between 40–50 g m^{-3} NO_3-N. Annual nitrogen loads to the river Brue, Norfolk, were calculated as 15 tonnes NO_3-N and 0.7 tonnes NH_4-N (Moss et al., unpublished). Nitrate concentrations ranges for the Slapton sewage works, which discharges into Slapton Ley, a large, eutrophic freshwater lake in south-west England, are of the order 15–25 g m^{-3} NO_3-N. Here a natural reedbed system was found to be very effective at reducing nitrate concentrations in sewage effluent to around 0.5 mg NO_3-N within 10 m of the sewage works outflow (Trudgill, Heathwaite and Burt, 1990).

Table 5.5 indicates the annual nitrogen, phosphorus and BOD loads from human and animal sources, together with an estimate of the load from cropped agricultural land. Nitrogen loads from human populations are high where there is limited biological treatment of the waste. However, these nitrogen loads are small in comparison with the potential nitrogen export from livestock. This form of N-loading is primarily diffuse; the contribution to the surface water nitrogen load will depend on the transport mechanisms from the land to the stream. The diffuse nitrogen load from livestock is potentially higher than that from fertiliser-amended land (Ryding and Rast, 1989).

Table 5.5 Potential catchment nutrient loads (modified from Ryding and Rast, 1989)

Type of load	Annual population[a] equivalent		
	BOD	TN	TP
One person; no treatment	1	1	1
One person; septic tank	0.7	0.8	0.8
One person; primary treatment	0.7	0.8	0.7
One person; secondary treatment	0.2	0.4	0.4
One person; tertiary treatment	0.1	0.3	0.1
One livestock unit;[b] diffuse loss, manure amendment	2.0	12	5.0
One livestock unit; 6 months permanent grass, 6 months dry feed	0.5	3	0.6
Cropped agricultural land;[c] no organic manure, inorganic fertiliser applied	—	9	0.5

[a]One population equivalent — 54 g BOD, 13 g N, 2 g P.
[b]One livestock unit = 1 cattle/horse, 6 pigs, 14 sheep, 150 poultry.
[c]Assumes N fertiliser applied at rate: 30 kg ha^{-1} cereal, 40 kg ha^{-1} roots, 80 kg ha^{-1} vegetables.

5.3.3 NITROGEN EXPORT FROM NON-POINT AGRICULTURAL SOURCES

Although point sources of nitrogen can be important, on a national scale, non-point or diffuse nitrogen sources are thought to dominate nitrogen transport pathways form terrestrial to aquatic systems (Royal Society, 1983).

5.3.3.1 Livestock and animal wastes

Livestock waste can be a valuable soil amendment because its high organic content gives it slow nutrient release characteristics and enhanced soil moisture-retention capacity. The main limitation on its value as a nitrogen fertiliser is its uneven spatial distribution together with the relatively high losses through ammonia volatilisation, particularly in warm weather. A further problem relates to trampling ('poaching') of the land surface by livestock. This can significantly reduce the infiltration capacity of the upper soil layers which can lead to high ammonium losses in surface runoff (Heathwaite, Burt and Trudgill, 1990).

The intensification of livestock production, achieved by increasing the input of allochthonous nutrients, such as artificial feed, increases the potential nitrogen load from livestock wastes. For example, the number of sheep and poultry on agricultural holdings in England and Wales doubled from 1950 to 1980. Over the same period the number of pigs trebled. Only cattle show relatively small increases (18% rise) (MAFF Agricultural statistics). Greater allochthonous inputs to the system are likely to mean greater nutrient export from the system. This is either in the form of livestock waste from intensive animal rearing units or as slurry and farmyard manure (see Table 5.5). Loehr (1974) suggests that the nitrogen load for land receiving manure is around $4-13$ kg N ha^{-1} a^{-1}, whereas an intensive animal rearing unit may have a potential nitrogen load as high as 1600 kg N ha^{-1} a^{-1}.

Nitrogen waste from animal-rearing units is at least a *point* source of nutrient input to the aquatic system, and so *theoretically* is relatively easily controlled. Application of organic manures to land creates a *non-point* or diffuse source of nitrogen enrichment, the effect of which depends largely on timing of application and antecedent soil characteristics. Slurry application requires at least a threefold dilution for spray application to the land, which in turn creates a high runoff and nitrogen leaching potential. Unfortunately the availability of livestock waste is high in early spring following winter housing. Surface application in this period is likely to result in high runoff because the soil moisture content will be high (for further discussion see Chapter 11).

Cattle manures and slurries typically contain 1.5% N, whereas poultry manure contains up to 10% N (Royal Society, 1983). UK studies suggest that approximately 50% of the nitrogen load from cattle and 100% of the nitrogen load from sheep was directly voided onto the land (Royal Society, 1983). In England and Wales, the total nitrogen load voided in faeces and urine (739 Kt; 1978 figures) approaches that of inorganic fertilisers (924 Kt). It has been calculated that on the east coast of Scotland approximately 75% of the nitrogen present in organic manures applied to the land between November and December is lost to the aquatic system by the following March (Gostick, 1982). Organic wastes exert a high biological, chemical and nitrifying oxygen demand on the aquatic system. Furthermore, a high concentration of ammonia is toxic to many organisms (Ellis, 1989).

5.3.3.2 Inorganic fertilisers

One of the key problems with the application of both inorganic and organic nitrogen fertilisers to land is the 'balancing act' required to match crop demand to fertiliser supply. While rarely practicable on economic grounds, repeat applications also cause physical damage to the soil from farm machinery. Fertiliser applications, therefore, typically occur in large batches. This nutrient stockpile is potentially a major non-point source of nitrogen for aquatic systems.

In 1981 an average of 130 kg N ha^{-1} was added to UK agricultural crops, with average dressings for 2−7-year grass leys of 170 kg N ha^{-1} and permanent grassland of 97 kg N ha^{-1} (Royal Society, 1983). At these rates, the annual application of nitrogen to agricultural land is approximately 3 Mt. The annual output of nitrogen from agricultural land in the UK can be subdivided into: crops and grass (1.4 Mt), leaching 0.3 Mt and ammonia volatilisation from livestock waste (0.5 Mt) (Royal Society, 1983). These figures suggest that an equivalent of roughly one third of the annual application of fertiliser nitrogen is lost through leaching (Foster, Cripps and Smith-Carington, 1982; Burns and Greenwood, 1982).

Nitrate leaching from agricultural land in autumn is one of the key non-point sources of stream nitrogen (see Figure 5.5 and Chapter 9). Nitrogen may be derived from unused fertiliser residues applied earlier in the year or from mineralisation of soil organic N. The relative importance of each fraction depends on factors such as crop cover, soil texture, topography and proximity to the stream, fertiliser treatment and rainfall intensity and duration. For a range of different UK soils, Macdonald et al. (1989) suggest that mineralisation of organic N, rather than unused fertiliser-N, is the major source of potentially leachable inorganic nitrogen in autumn.

5.4 NITROGEN PATHWAYS

The main outputs of nitrogen from the biosphere occur through the processes of denitrification and leaching. Denitrification, although representing a loss form the system, is a potentially useful process in nitrogen-enriched ecosystems because a reduction in nitrate entering the aquatic system is possible. Leaching is primarily a nitrate transfer process via subsurface water movement. Nitrogen loss through leaching from the terrestrial system is particularly important in arable soils where most soil nitrogen is present as nitrate. In forest and grassland soils, inorganic nitrogen concentrations are generally lower (Royal Society, 1983), and a larger soil ammonium fraction may exist. Furthermore, as the positively charged ammonium ion tends to be adsorbed onto soil particles, its key hydrological pathway is in association with surface runoff and erosion, rather than in subsurface flow.

5.4.1 NITROGEN TRANSFER FROM HILLSLOPE TO STREAM

In most surface waters and lakes, the catchment nitrate load dominates the total N pool for most of the year with the exception of the summer period. This means that the terrestrial system exerts an important control on N cycling in freshwaters (Stewart et al., 1982; Heathwaite, Burt and Trudgill, 1989). Nitrate loading is usually highest in winter and spring and lowest in summer and early autumn. High nitrogen concentrations in early winter are a result of soil water recharge in autumn and the increase in nitrogen mineralisation that occurs when soil drying is followed by rewetting. In Slapton Ley in south-west Devon, for example,

Table 5.6 The effect of increased nitrate load on freshwater lakes (after Stewart *et al.*, 1982)

Factor	Blelham	Balgavies
Nitrate load	46	507
Primary productivity	146	136
Denitrification	38	151
Dissimilation to ammonium	7	151

All figures in kg N ha^{-1}.

the total annual nitrate load is 260 tonnes from a 46 km^2 arable and grassland catchment. Over 85% of this annual nitrate load is delivered between November and February. Other nitrogen sources for this eutrophic lake are negligible in comparison with the nitrate load of the inflowing waters (Heathwaite and Burt, 1991). Hill (1974) similarly found that up to 80% of the annual nitrate load of rivers in the Lee drainage basin, UK, was carried during the winter period.

In surface water and lakes, provided other nutrient and environmental factors are not limiting, primary productivity may increase as a result of increased nitrate load (Table 5.6). If other nutrients are limiting (which is commonly the case with, for example, phosphorus), denitrification and dissimilatory nitrate reduction may increase. Control of excess nitrate usually focuses, therefore, on optimising denitrification (Stewart *et al.*, 1982), often by adding a suitable carbon substrate for the denitrifying organisms. Dissimilatory nitrate reduction does not reduce the nitrate load.

The main routes for nitrogen transfer from hillslope to steam are:

(1) Nitrate leaching to surface waters or groundwater;
(2) Input of animal excreta at watering sites;
(3) Surface runoff of excreta from farmyards or grazed land;
(4) Leakage of silage effluent;
(5) Soil erosion.

Leaching is the dominant process of nitrogen transfer from hillslope to stream owing to the high solubility of nitrate; subsurface flow is therefore the major hydrological pathway. The greatest leaching losses of nitrate from soils occur in winter months when residual nitrogen is leached below the root zone (Wild and Cameron, 1980). Depending on local stratigraphy, mobilised nitrate will enter either the surface water or the groundwater zone. A downwards migration rate in the porous sandstones and Chalk limestones of southern England has been estimated at 1 m a^{-1}. At this rate it would take 20−30 years for surface-applied nitrate to reach the water table. Therefore the doubling of nitrogen fertiliser applications in the UK in the 1960s has yet to manifest itself in some surface waters.

5.4.2 INSTREAM NITROGEN TRANSFORMATIONS

Once nitrogen enters the aquatic system, instream nitrogen transformations, as a result of both physical and chemical discontinuities in the stream, will occur. Instream nitrogen transformations are primarily concerned with the metabolic processes of the stream flora and can involve aquatic macrophytes, stream bank macrophytes, epilithic algae, micro-heterotrophs and denitrifying bacteria. Cooper and Cooke (1984), for example, found that aquatic macrophytes accounted for 73–100% of stream biotic nitrate uptake; Howard-Williams and Downes (1984) suggest that uptake by stream bank macrophytes may be of the order of $0.9–1.5$ g N m^{-2} day^{-1}.

Stream biota are involved not only in nitrogen uptake but also in nitrogen-transformation processes such as nitrification and denitrification, which remove N from the stream water by transferring it to the biota, atmosphere or stream sediments. Denitrification in wetlands is thought to be particularly important (Swank and Caskey, 1982; Hill and Sanmugadas, 1985). Sloey, Spamgler and Fetter (1978) recorded denitrification rates up to 3.5 kg N ha^{-1} day^{-1} for US wetlands during the growing season. Denitrification rates in New Zealand pastoral streams were less than 1 kg N ha^{-1} day^{-1} (Howard-Williams and Downes, 1984) but were found, in some streams, to match maximum macrophyte uptake rates. Swank and Caskey (1982) suggest that rates of denitrification are positively correlated with total nitrogen and organic matter and negatively correlated with stream sediment nitrate concentrations. However, in order for denitrification to be a significant factor in removing N from wetlands, N$_2$ must be able to escape to the atmosphere before it is fixed as nitrate. Aerated water overlying anaerobic sediments will restrict N$_2$ escape, so flow in wetland systems is critical in the denitrification process.

Howard-Williams et al. (1982, 1988) attribute nitrate losses in New Zealand streams to macrophyte uptake. They found that in macrophyte-dominated streams, nitrate was reduced to concentrations below 5 mg m^{-3} and increases in dissolved organic nitrogen were negligible (Howard-Williams, Pickmere and Davies, 1988). Modification of stream nitrate concentrations varied seasonally, with high retention of dissolved nitrogen by macrophytes in summer. Similar seasonal trends were recorded by Hill (1979, 1988) and by Haycock (1991). However, Cooper and Cooke (1984) concluded that instream nitrogen removal occurred throughout the year.

Richey, McDowell and Likens (1985), working in the Hubbard Brook Experimental Forest, New Hampshire, did not detect nitrogen depletion where ammonium and urea were added to a stream. This, they suggest, was due to nitrogen cycling by detritus and bryophytes on the streambed, with uptake of ammonium being followed by release of nitrate, so there was no net loss of nitrogen. Wyer and Hill (1984) also found that nitrification and ammonification occurred simultaneously with nitrate reduction in laboratory experiments. In streams receiving NH$_4$-N, nitrifying bacteria are important in the conversion to NO$_3$-N and subsequent potential losses through denitrification.

There are a number of important controls on the rate of denitrification: these include temperature, the availability of organic matter as a metabolisable energy source, the nitrate concentration in stream water overlying stream sediments and the stream flow regime and water residence time (Terry and Nelson, 1975). Hill (1988) found that nitrate removal efficiency in streams was low where the stream nitrate load was high. This was thought to be the result of shorter residence times and a higher ratio of water volume to stream sediment areas, although increased flow should increase the oxygen diffusion rate and exchange of nitrogen between organisms and water. Casey and Farr (1982) studied the effect of artificially increased flow on a chalk stream in Dorset. They suggest that, regardless of allochthonous inputs, nitrate concentrations in rivers may increase at high discharge because of the reduction in benthic denitrification at high flow. Pinay et al. (1990) use the concept of nutrient spiralling to explain the strong interactions between a stream and its riparian zone, where they suggest that major exchanges of matter and energy take place. Nutrient spiralling is related to the mean distance travelled by a nutrient atom during its cycle and the number of times that nutrient atom is utilised in a section of stream or river. Interactions in the terrestrial–aquatic zone are thought to lead to changes in the recyling of nitrogen and other nutrients and, in particular, the retention of nutrients during displacement downstream. On a larger scale, Admiraal and Botermans (1989) investigated nitrogen transformations in the lower river Rhine which receives high ammonium inputs from its industrialised catchment. They found that differences in river discharge, sediment grain-size composition and the aeration and turbulence caused by shipping influenced 'in-river' nitrogen transformations and, in particular, nitrification rates.

5.4.3 NITROGEN PATHWAYS IN LAKES AND RESERVOIRS

Once nutrients enter a lake they may be recycled many times between the sediments, aquatic plants and water column. On a longer timescale, lake sediments generally form the ultimate nutrient sink, although this depends on the lake flushing rate and the nature of catchment inputs. For nitrogen, biological fixation can form a potentially significant diffuse source, particularly for eutrophic lakes.

5.4.3.1 Seasonal cycling of nitrogen compounds in lakes

In temperate lake ecosystems, nitrate concentrations show winter maxima which are independent of lake trophic status. This winter maxima arises because (1) surface and subsurface catchment inputs are high as a result of high rainfall and (2) microbial activity is low as a result of low temperatures. Lake trophic status then affects the way in which nitrogen is utilised through the spring−autumn growing season. In eutrophic lakes, and particularly in shallow lakes, the nitrate concentration drops rapidly from winter through to spring as biotic demand exceeds supply once temperatures rise (Figure 5.9). In oligotrophic lake systems, from an initially lower nitrate concentration in the water column, nitrate availability also decreases in summer. The relative decrease is less than that in eutrophic lake systems because biotic demand is lower.

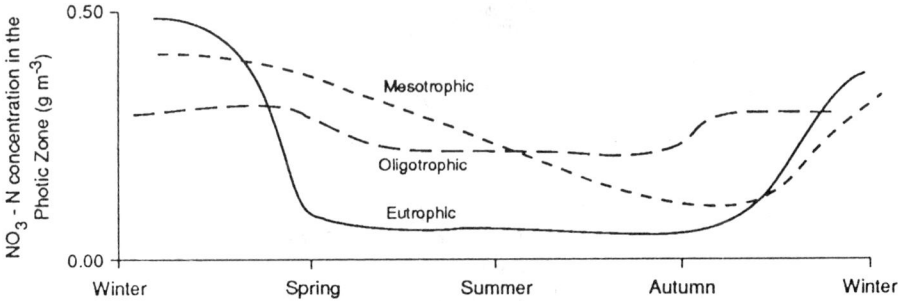

Figure 5.9 The variation in nitrate concentration as a function of lake trophic status (modified from Henderson-Sellers and Markland, 1987)

The hypothetical pattern of ammonium concentration variation in lakes is shown in Figure 5.10. Oligotrophic lakes are expected to show little variation in ammonium concentration throughout the year because lake biomass is small and the organic nitrogen store is therefore small. Mesotrophic lakes generally show an increase in ammonium concentrations at the autumn overturn, although this pattern is only found in deep, stratified lakes. Eutrophic lakes show irregular patterns of ammonium concentration fluctuation throughout the spring−autumn

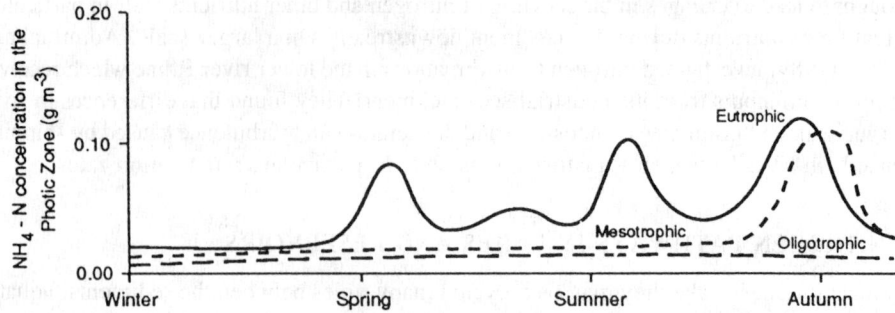

Figure 5.10 Hypothetical seasonal variation in NH$_4$-N concentration in lakes of different trophic status (modified from Henderson-Sellers and Markland, 1987)

(a) Seasonal variation in the contribution of different N fractions to the total N concentration of a eutrophic lake.

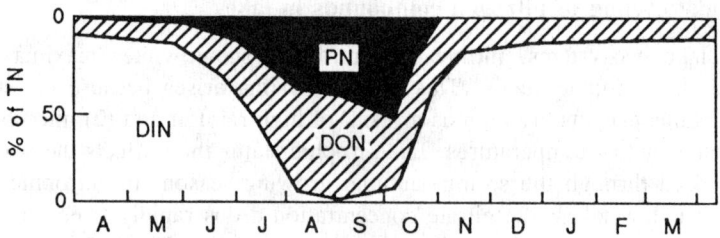

(b) Seasonal variation in the inorganic N fraction of a eutrophic lake.

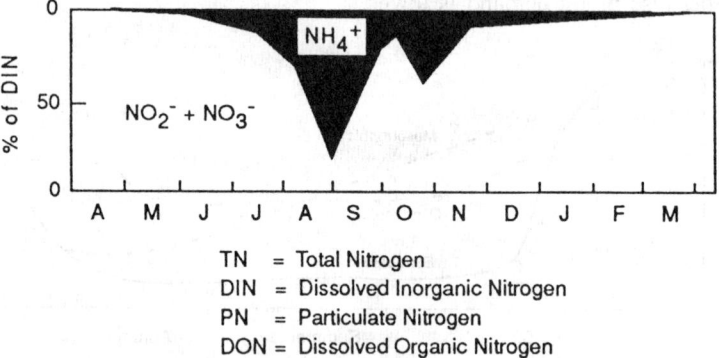

TN = Total Nitrogen
DIN = Dissolved Inorganic Nitrogen
PN = Particulate Nitrogen
DON = Dissolved Organic Nitrogen

Figure 5.11 Seasonal variation in the contribution of nitrogen fractions to eutrophic Rescobie Loch, Scotland (after Peterson and Stewart, unpublished)

period. This is a consequence of the pattern of algal bloom and algal decay which is characteristic of eutrophic waters.

Figure 5.11(a) indicates the contribution of different nitrogen species to nitrogen cycling in eutrophic Rescobie Loch (Peterson and Stewart in Royal Society, 1983). It clearly illustrates the reduced contribution from dissolved inorganic nitrogen in summer which is partially balanced by the larger contribution from dissolved organic and particulate nitrogen derived

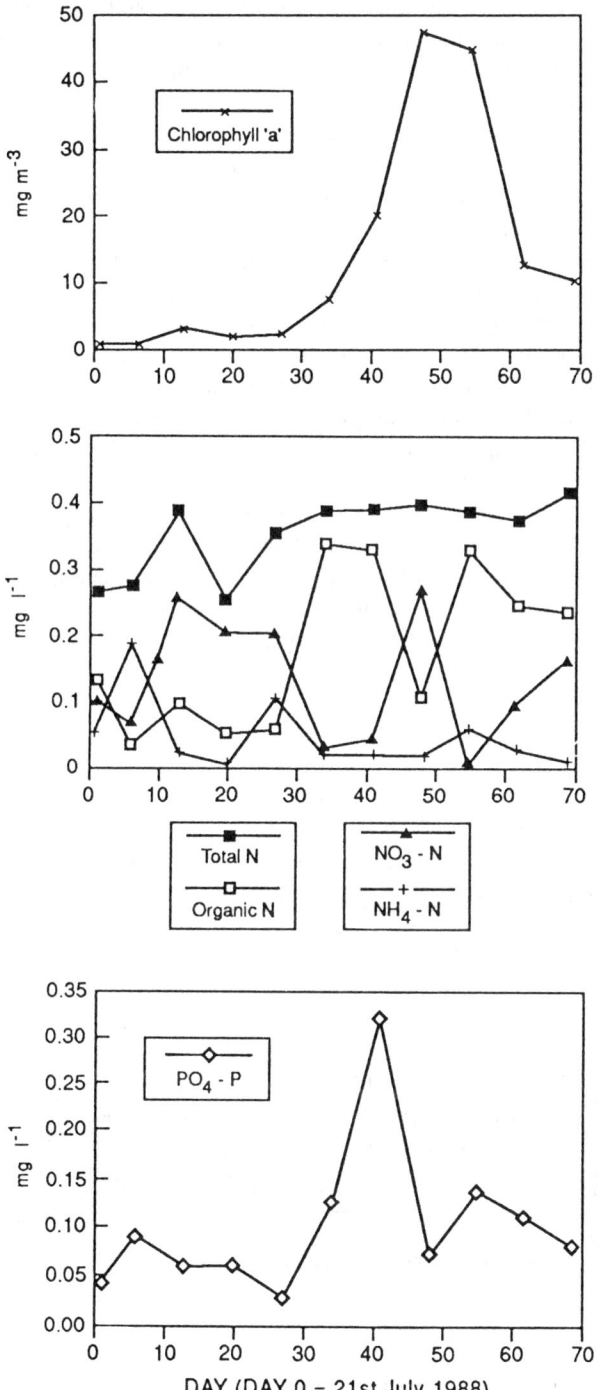

Figure 5.12 Chlorophyll 'a', nitrogen and soluble reactive phosphorus variation in Slapton Ley during an algal bloom, July to September, 1989 (after Heathwaite, 1989)

from lake biota. Figure 5.11(b) illustrates the dissolved inorganic nitrogen fractions for the same period. Nitrate dominates this fraction for most of the year, except when biotic activity is high in summer, when the concentration of ammonium in the lake waters increases.

The pattern of chemical transformations in eutrophic Slapton Ley, south-west Devon, between mid-July and mid-September are shown in Figure 5.12. Blue-green algae begin to appear in Slapton Ley in early May and remain in the water column until December. *Anabaena* are dominant in early summer and there is a shift to *Microcystis aeruginosa, Coelosphaerium naegelianum* and *Gloeotrichia echinulata* in autumn (Benson-Evans *et al.*, 1967; Van Vlymen, 1980). High primary productivity indicated by chlorophyll '*a*' concentrations above 50 mg m^{-3} and cyanophyte blooms are evidence of eutrophication in Slapton Ley. Chlorophyll '*a*' concentrations up to 330 mg m^{-3} in Slapton Ley are reported by Van Vlymen (1980).

The algal bloom represented by peak chlorophyll '*a*' concentrations in September (Figure 5.12) appears to have been nitrogen-funded because it coincides with the maximum total nitrogen concentrations in the lake. Biologically available phosphorus peaks before the algal blooms, but concentrations are low during the bloom itself. The concentrations of dissolved inorganic nitrogen, which fall below 0.01 mg TON l^{-1} in summer, suggest that Slapton Ley, for some periods, is a nitrogen-limited lake. This observation is supported by Van Vlymen (1980). More recent data for the same lake showed that, although inorganic nitrogen concentrations fell in summer, the total nitrogen concentration remained relatively constant and appeared to be maintained by a supply of organic nitrogen. However, because this form of nitrogen is less biologically available, nitrogen may still be limiting in this lake.

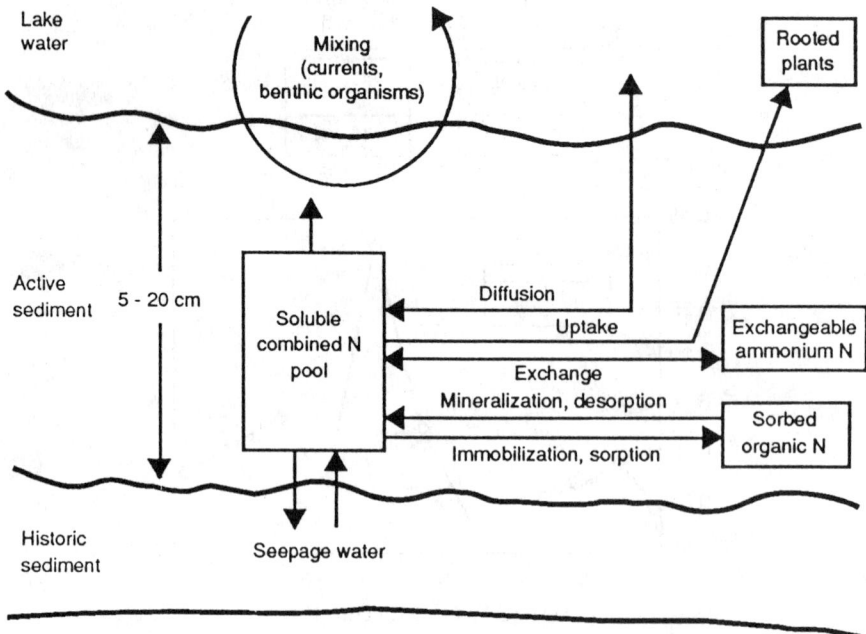

Figure 5.13 Nitrogen cycling at the sediment–water interface (after Keeney, 1973)

5.4.3.2 Nitrogen in lake sediments

Nitrogen reaches lake sediments by sedimentation, nitrogen fixation and nitrate immobilisation (Figure 5.13). Lake sediments typically contain 50−200 kg N per 10 cm sediment depth per hectare (Keeney, 1973), and are thought to act both as a sink and a source of nitrogen, depending on the N status of interstitial or overlying waters.

Readily available nitrogen in lake sediments is held in the ammonium and organic nitrogen forms present in interstitial water, and the ammonium nitrogen on the exchange complex. Diffusion and sediment mixing can release nitrogen in these forms to the water column, although, in general, denitrification is the key process. Chen and Keeney (1974), for example, showed that denitrification was a significant N sink in Wisconsin lakes. In sediment core experiments they found that 15−26% added ^{15}N-labelled NO_3-N was immobilised to organic N and NH_4-N, the remainder being accounted for by denitrification. Chen, Keeney and McIntosh (1983) studying Lake Michigan, concluded that only 1.2% of the annual catchment nitrogen load was present in the active sediment layer of the lake. They suggested that nitrification and subsequent denitrification at the sediment−water interface was a major nitrogen sink. These nitrogen-cycling processes were, however, thought to be minor relative to the high catchment nitrogen load reaching the lake. Their study concluded that non-point nitrogen sources would be sufficient to maintain eutrophic conditions in the lake even if point sources of nitrogen were reduced.

5.5 THE ROLE OF NITROGEN IN FRESHWATER EUTROPHICATION

One of the key products of eutrophication is excess algal growth. Hayes and Greene (1984), for UK reservoirs, suggested that algal growth becomes a problem where inorganic N concentrations greater than 1.5 mg l^{-1} and inorganic P in excess of 0.02 mg l^{-1} are recorded. Although nitrogen is acknowledged as one of the key contributory nutrients in the eutrophication of freshwaters (see Table 5.7), most of the focus is on phosphorus inputs and phosphorus control. There are a number of reasons for this. First, nitrogen is usually derived from non-point sources, so control is difficult. Second, cyanobacteria can fix atmospheric nitrogen rendering control ineffective in some lakes, and third, phosphorus is generally viewed as the 'limiting' (and therefore controling) nutrient. For British lakes and reservoirs, however, there is good correlation between chlorophyll 'a' concentration and nitrate (Figure 5.14), and hence a good link between lake nitrogen concentration and trophic status.

Table 5.7 Lake loading level (g m^{-2} a^{-1}) for total N and total P (after Vollenweider, 1975)

Mean depth	Permissible loading		Dangerous loading	
(m)	Nitrogen	Phosphorus	Nitrogen	Phosphorus
5	1.0	0.07	2.0	0.13
10	1.5	0.10	3.0	0.20
50	4.0	0.25	8.0	0.15
100	6.0	0.40	12.0	0.80
150	7.5	0.50	15.0	1.00
200	9.0	0.60	18.0	2.00

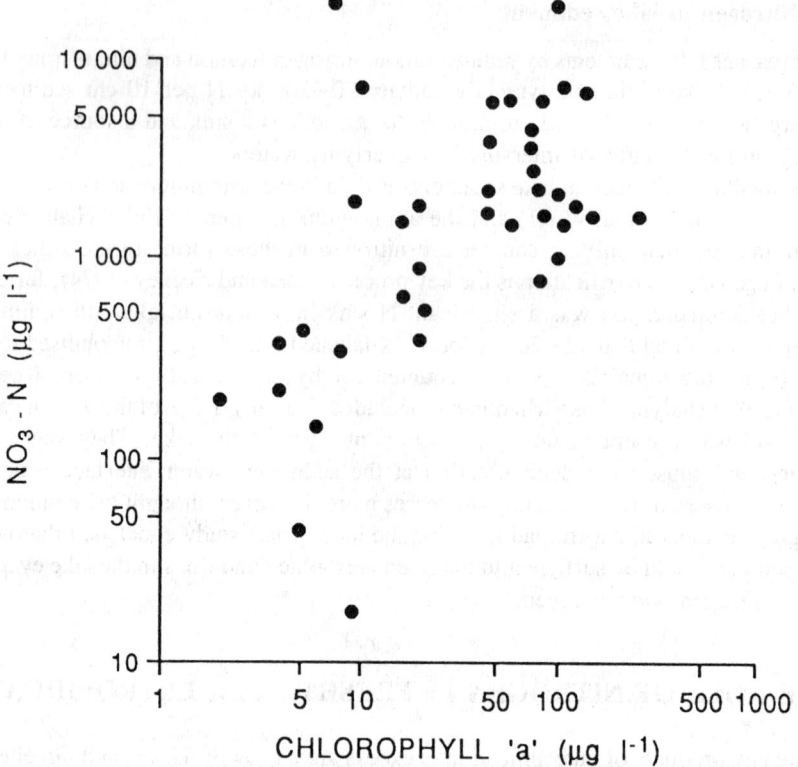

Figure 5.14   The relationship between maximum annual chlorophyll 'a' concentration and maximum NO$_3$-N concentration in British lakes and reservoirs (after Collingwood, 1977)

Nitrogen and phosphorus concentrations tend to increase in parallel as the trophic status increases from oligotrophic through to eutrophic. The N:P ratio decreases from roughly 100N:1P in oligotrophic lakes to less than 10N:1P in eutrophic ones. This suggests a tendency for lakes to shift from phosphorus dependency towards nitrogen dependency with increasing trophic status.

5.5.1 THE TROPHIC STATUS OF FRESHWATER SYSTEMS

Lakes and reservoirs can be broadly described as ranging from oligotrophic (nutrient-poor) to eutrophic (nutrient-rich). The terms have no absolute meaning, but are valuable in reflecting the nutrient status of a waterbody. Eutrophication generally refers to standing waterbodies rather than streams or rivers, where the rate of throughput of nutrients and the rate of water movement are assumed to be too rapid for conditions characteristic of eutrophication to develop.

The identification of trophic status in freshwaters is usually based on establishing boundary values for a number of water quality criteria. Total phosphorus, chlorophyll 'a' and Secchi depth are commonly used (Table 5.8). Nitrogen, although an important determinant of the productivity of a waterbody, is normally excluded owing to the difficulty in establishing effective control strategies based on nitrogen.

Table 5.8 Water quality parameters and trophic status

Trophic category	Total phosphorus ($\mu g\ l^{-1}$)	Chlorophyll 'a' ($\mu g\ l^{-1}$)	Secchi depth (m)
Oligotrophic	<10	<2.5	>6.0
Mesotrophic	10–35	2.5–8.0	6.0–3.0
Eutrophic	35–100	8.0–25.0	3.0–1.5

Using the OECD boundary values given in Table 5.8 (OECD, 1982), a waterbody falls into one of the three main trophic categories where no more than one of the parameters deviates from its geometric mean value by ±2 standard deviations. Collingwood (1977) concluded that 15, 11 and 4 lakes in England, Wales and Scotland, respectively, fall into the eutrophic category on the basis of their maximum chlorophyll 'a' concentrations alone.

5.5.2 THE CONCEPT OF LIMITING NUTRIENTS: HOW IMPORTANT IS NITROGEN?

The whole-lake experiments in Canada (Schindler and Fee 1974; Schindler, 1977) confirmed the importance of phosphorus and nitrogen in lake primary productivity, previously only investigated in laboratory studies. Nitrogen is often seen as the nutrient determining lake productivity, whereas phosphorus is the nutrient limiting growth.

Phytoplankton biomass (indicated by chlorophyll concentration) in a waterbody appears to be proportional to its nutrient load up to a point, beyond which other factors such as carbon or dissolved oxygen concentration exert primary control (Sridharan and Lee, 1977; Lee, Jones and Rast, 1980). The concept of limiting nutrients in waterbodies relies on the fact that both absolute and relative quantities of essential nutrients regulate primary productivity (mainly algal biomass) in the waterbody. Therefore it is assumed that the ratio at which nutrients are taken up and used by algae reflect the relative composition of these nutrients in their cellular material. On this basis, the ratio 106C:16N:1P has become widely cited as the standard reference value for assessing the limiting nutrient in waterbodies (Ryding and Rast, 1989). Carbon is rarely limiting in freshwaters. The assumption is that control of the nitrogen or phosphorus load entering the waterbody from its catchment, or manipulation of the ratio of N:P within the waterbody, should control algal growth.

The limiting nutrient concept is based on the photosynthesis reaction. Conceptually, the reaction can be expressed as follows:

$$CO_2 + NO_3^- + PO_4^{2-} + H_2O + H^+\ (+\ \text{trace elements: sunlight})$$

photosynthesis
$$\rightarrow$$
algal protoplasm $+\ O_2$
$$\leftarrow$$
respiration

This equation suggests that several variables are involved in the reaction, so theoretically control of any one variable may offer control of eutrophication in waterbodies. Most attention has been focused on control of the external nutrient supply, partly because of the relative complexity of algal processes and partly because some macro-nutrients, particularly phosphorus, offer a *relatively* simple means of limiting nutrient control.

Table 5.9 Approximate levels of nitrogen and phosphorus limitation using N:P ratios (modified from OECD, 1982)

Total N by wt	Total P moles	Inorganic N by wt	Inorganic P moles	Limiting nutrient
< 10	22	< 5	10	N
10–17		5–12		N/P
> 17	37.6	> 12	26.5	P

There is no sharp boundary between phosphorus and nitrogen limitation (Table 5.9). Low N:P ratios favour nitrogen-fixing blue-green algae which are undesirable from a water quality perspective (Schindler, 1977). Sewage effluents, which contain a low N-P ratio because of the high concentration of soluble reactive phosphorus (SRP) released (4N:1P by weight), can shift a waterbody from phosphorus to nitrogen limitation. The N:P ratio in sewage is much lower than that of agricultural runoff, which is commonly between 30:1 and 50:1. However, seasonal variation in both the supply and uptake of these nutrients in aquatic systems complicates this simplistic pattern. Furthermore, the ratio of N:P in surface waters and lakes is not always reflected in the biota. Moss *et al.* (unpublished), for example, recorded a ratio of N:P of 26:1 by weight (64:1 by atoms) for the river Brue in the Norfolk Broads. These values are much higher than the optimum quoted for algal growth (Golteman and Kouwe, 1980). Despite these results, nitrate was often undetectable in the Brue in summer, and severe nitrogen depletion was reported in phytoplankton cells.

The 'biologically' available fraction of nitrogen and phosphorus in the lake epilimnion, rather than total nitrogen and phosphorus concentrations, are critical in funding algal blooms (Table 5.9). The biologically available fractions are: dissolved or soluble reactive phosphorus (SRP) and ammonium-N plus nitrate-N. A biologically available phosphorus concentration less than 5 $\mu g\ l^{-1}$ and a biologically available nitrogen concentration less than 20 $\mu g\ l^{-1}$ are usually taken to suggest phosphorus and nitrogen limitation, respectively (Ryding and Rast, 1989).

As the trophic status of a waterbody increases, the mineral nitrogen component becomes relatively more dominant. The inorganic N fraction forms approximately 60% of total N at total N concentrations less than 500 mg m^{-3} and increases to 70% at total N concentrations greater than 5000 mg m^{-3} (OECD, 1982). This trend effectively means that increasing trophic status corresponds with an increase in the 'biologically available' nitrogen fraction in a lake.

Calculation of the atomic ratio of nitrogen to phosphorus in the waterbody and comparison with the algal uptake ratio of 16N:1P is an indication of the potential growth-limiting nutrient in the waterbody. If biologically available N and P are measured in concentration units of mg l^{-1} the 16N:1P atomic reference value then corresponds to a mass ratio of 7.2N:1P. Therefore, if the N:P ratio is less than 7N:1P, nitrogen is potentially limiting because the more rapid removal of nitrogen would limit biotic growth; if the ratio is greater than 7N:1P, phosphorus is limiting. In Swedish waterbodies, Fosberg and Ryding (1980) used a ratio of less than 5N:1P to denote nitrogen limitation and greater than 12N:1P to indicate phosphorus limitation.

A major restriction on applying the concept of limiting nutrients too liberally to freshwaters is that although it is robust under steady-state conditions, it fails where the waterbody is under transient conditions (for example, where a waterbody receives intermittent inputs of phosphorus or nitrogen). Limiting nutrients will also vary seasonally but practical use of the limiting factor concept tends to assume that a single nutrient, usually phosphorus, is limiting *throughout the year*.

Finally, different uptake and assimilation rates for different primary producers will affect the response of the waterbody to nutrient limitation. Algal populations will compete for the available nutrient resources. Smith (1983) suggested that low N:P ratios encourage dominance by blue-green algae in temperate lakes. Blue-green algae were usually absent above a total N to total P ratio greater than 29N:1P. Some species of blue-green algae are capable of 'luxury' uptake of limiting nutrients, which are stored for later use. Diatoms, for example, are superior competitiors for phosphorus, but inferior competitors for nitrogen (Tilman, Kilham and Kilham, 1982). A further complication is that N_2-fixing blue-green algae can be phosphorus-limited, while non-N_2 fixing algae can be nitrogen-limited. This can occur simultaneously in the same lake.

5.5.3 MODELLING EUTROPHICATION

5.5.3.1 Catchment-driven (nutrient load) models

The key objective of this form of model is control of eutrophication by estimating the nutrient load, particularly from non-point sources, reaching a lake from its catchment. The basis for measurement is normally the subcatchment scale. Simple models are based on steady-state nutrient loadings, whereas more realistic models attempt some subdivision of biologically available and unavailable nutrient forms. Seasonal variations in dissolved and solid-phase nitrogen and phosphorus inputs to the aquatic system may also be included.

The simplest empirical models for estimating the nutrient load to a waterbody are usually of the form:

$$A = a_0 + a_1X_1 + a_2X_2 + \ldots + am^Xm$$

where A = average annual nutrient load to waterbody (kg a^{-1}),
  X = catchment area (ha) for land use 1, 2…m,
  a = export coefficients (kg ha^{-1}).

The export coefficients (a) are ideally based on intensive water quality sampling in small catchments with a single predominant land use! In practice, export coefficients are usually obtained by collating published information on nutrient export from land uses covering wide geographic ranges (see, for example, Rast and Lee, 1983; Johnes and O'Sullivan, 1989; Johnes, 1990). Ryding and Rast (1989) suggest that nutrient export coefficients should be viewed primarily as a method of obtaining rough estimates of annual nutrient loads.

Coote, MacDonald and De Haan (1979) applied the simple model given above to two years of water quality data for dissolved and total nitrogen from ten Canadian agricultural watersheds. Catchment parameters included land use, fertiliser and manure application, soil erosion, slope, and soil texture. The effect of control mechanisms such as fertiliser reduction or erosion prevention can then be estimated in terms of the potential decrease in the average annual nitrogen load. The main limitation of nutrient export models is that they are usually site-specific as well as being sampling-specific. That is, the sampling interval and sampling season will be reflected in the reliability of the results. Finer resolution of nutrient loading to waterbodies requires evaluation of water and associated nutrient *fluxes* from the land surface and the soil profile. The complexity of these models depends on the required time resolution and the extent to which biochemical processes are incorporated. The general relationships are:

$$D_{kt} = d_{kt}Q_{kt}TD_{kt}$$

where D_{kt} = dissolved nutrient loss (kg) in surface runoff or baseflow from area k in time t,
    d_{kt} = dissolved nutrient concentration in surface runoff or baseflow (kg m^{-3}),
    Q_{kt} = surface runoff or baseflow (m^3),
    TD_{kt} = fraction of dissolved nutrients reaching the waterbody from area k.

The total nitrogen load to a waterbody in time t is computed from the sum of nitrogen losses from each catchment source. Non-point nutrient sources can be evaluated on the basis of runoff (Q_{kt}) or by including a term for erosion (X_{kt}). Baseflow nitrogen will be primarily in the dissolved inorganic nitrogen fraction. Assuming the groundwater nitrogen concentration is relatively stable, the dissolved nitrogen load can be expressed as:

$$D_k = d_kQ_{kt}$$

where d_k = dissolved nutrient concentration in groundwater from area k (kg m^{-3}),
    Q_{kt} = baseflow (m^3).

   Haith and Tubbs (1981) and Haith (1982) describe a number of mechanisms for estimating non-point nutrient loads from agricultural land in the north-east United States. They found that the main limitation of the form of model given above is the transport factor (TD_{kt}) which is not easily quantified. For dissolved nutrients it is often assumed that $TD_{kt} = 1$ and that the delivery of solid-phase nutrients is equal to the sediment delivery ratio of the catchment (Haith and Tubbs, 1981).

5.5.3.2 Waterbody-driven (mass-balance) models

Based on nutrient mass balance calculations, these models are used to estimate the 'biological' response of a lake to nutrient load reductions normally using simple statistical (usually correlation or regression) parameters. Empirical waterbody models such as those of Vollenweider (1968, 1975, 1976; Rast and Lee, 1978; Ryding, 1980) are valuable in lake-eutrophication control because they are simple and are not data-demanding. In Vollenweider's 1976 model, for example, lake total phosphorus concentrations is computed from the annual areal phosphorus loading (mg m^{-2} a^{-1}), the annual areal water loading (m^2 a^{-1}), hydraulic residence time and mean lake depth. Such models are limited, however, because they do not give any indication of the *processes* involved or the magnitude of the various transport mechanisms responsible for delivering nutrients from land to water. Errors in predicting phosphorus concentrations, for example, can be at least $\pm30\%$ (Ryding and Rast, 1989).
   Waterbody-driven models are usually biased towards prediction of lake phosphorus concentrations. Nitrogen cycling, owing to the complexities of accounting for diffuse inputs and the possibility that nitrogen is not limiting, is not usually considered. Extensions of the basic model exist for other water quality parameters such as, chlorophyll 'a' (Dillon and Rigler, 1974; Smith and Shapiro, 1981), Secchi depth (Rast and Lee, 1978) and dissolved oxygen (Ryding, 1980; Vollenweider and Janus, 1982).
   Baker, Brezonik and Kratzer (1985) developed a simple *nitrogen*-loading model based on the OECD-Vollenweider model (OECD, 1982) for Florida lakes which are commonly nitrogen — rather than phosphorus — limited, owing to excess P supply from phosphate-rich catchment soils. Canfield et al. (1983, 1984) again for Florida lake systems suggested the following

model for algal biomass prediction:

$$\log \text{chlorophyll } a = -2.49 + 0.269 \log \text{TP} + 1.06 \log \text{TN}$$
$$(\mu g \, l^{-1})$$

The model is best applied to hypertrophic lake systems ($r^2 = 0.82$ for the Florida Lake systems) and can be extended to include aquatic macrophyte infestation which also affects chlorophyll 'a' concentration.

5.5.3.3 Dynamic waterbody models

Dynamic models include some assessment of physical, chemical and biological processes within the lake system. The resolution of spatial and temporal detail is therefore greater and they are less 'site-specific'. These models usually contain:

(1) Hydrologic and hydrodynamic characteristics;
(2) Chemical and biological transformations;
(3) 'Interface' terms such as input, output or exchange of material across boundaries (usually driven by (1)).

Interface terms in dynamic models are primarily concerned with the link between lake and catchment and exchange processes across both the air−water and the sediment−water interfaces, although these complex processes remain highly simplified.

The forcing variables driving dynamic models are usually external (for example, temperature or solar radiation). Hydrological components are integrated in the models and include lake flushing rate and nutrient mixing within a thermally stratified lake. The complexity of such processes usually results in an emphasis on transport rather than nutrient exchange (Shanahan and Harlemann, 1984).

Figure 5.15 illustrates the basic carbon, nitrogen and phosphorus cycles important in the

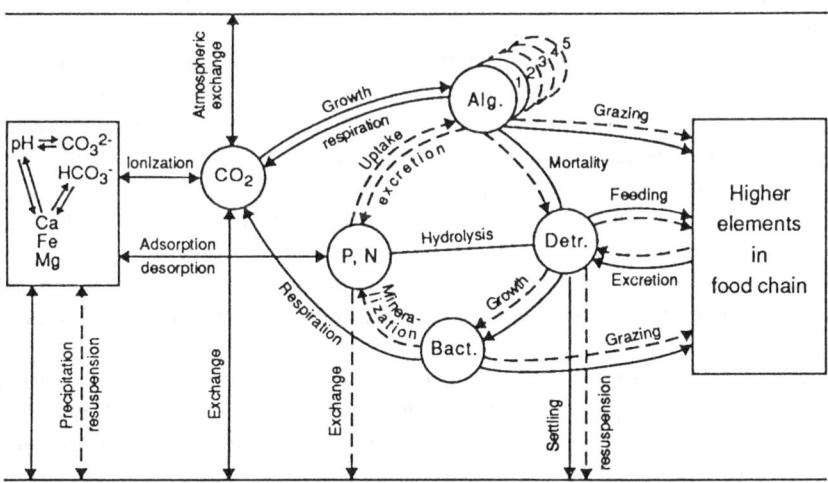

Figure 5.15 The carbon, nitrogen and phosphorus cycles involved in lake eutrophication (modified from Ryding and Rast, 1989) ——— = carbon; - - - - = nutrients; Alg. = algae; Detr. = detritus; Bact. = bacteria; P = phosphorus; N = nitrogen; CO_2 = carbon dioxide; CO_3^{2-} = carbonate; HCO_3^- = biocarbonate; Ca = calcium; Fe = iron; Mg = magnesium

eutrophication process. To be fully effective, dynamic models need to incorporate at least some of the accepted linkages outlined in Figure 5.15. The emphasis is usually on the biological rather than the chemical component of the cycle, although DeRooij (1980) describes a dynamic model based on chemical processes in lakes.

5.5.4 NITROGEN AND THE CONTROL OF EUTROPHICATION

As discussed earlier, the main non-point sources of nitrogen are urban and agricultural runoff. Urban control mechanisms include creation of permeable surfaces in urban areas and construction of storage and detention ponds. Neither mechanism is particularly useful for nitrogen control, primarily because nitrate is highly soluble and readily leached from such devices. Agricultural control mechanisms include construction and control of manure storage facilities, attention to timing of manure and inorganic fertiliser application and the use of slow-release fertilisers. An auxiliary measure for the reduction of agricultural nitrogen transfer is the establishment of 'protection zones' or vegetative buffer strips which restrict fertiliser application close to waterbodies (see also Chapter 12). Other mechanisms of nitrogen control include the use of nitrification inhibitors such as 'N-serve' to reduce nitrate leaching losses. N-serve inhibits *Nitrosomonas* but its volatile nature means that it is difficult to incorporate into the soil.

Riparian forest 'ecotones' are acknowledged as forming natural filters against diffuse nitrogen input from agricultural land (Peterjohn and Correll, 1984; Jacobs and Gilliam, 1985). Lowrance, Todd and Asmussen (1984) found that a riparian forest 'ecotone' in Georgia, USA, retained 68% of the diffuse nitrogen load as a result of denitrification. Chauvet and Décamp (1989) found that the nitrate concentration in groundwater varied with land use in the Garonne river catchment, France. In order of increasing nitrate concentration the authors found: woodland < pasture < orchard < fallow < cereals < market gardening < urban. However, nitrate appeared to be eliminated from groundwater *before* it reached the river *after* it had passed through the floodplain. Pinay (1986) and Pinay and Décamp (1988) examined the process of nitrate elimination from groundwater with flow towards the river system. They concluded that denitrification was responsible for allochthonous nitrate removal (Figure 5.16). Howard-Williams, Pickmere and Davies (1988) concluded that the riparian vegetation of stream channels and steam banks, together with the associated bacterial communities, were able to remove dissolved inorganic nitrogen from the stream water. The rate of removal depends on factors such as stream morphometry, discharge and season. The authors state that headwater streams are valuable for the removal of diffuse nitrogen inflow. They suggest that the grazing and watering of livestock in riparian zones should be minimised in order to maximise the nitrogen removal potential of these zones. Further discussion of riparian buffer zones is given in Chapter 12.

A simplified approach for selecting the appropriate eutrophication-control measure where nitrogen enrichment is a possible source of eutrophication is shown in Figure 5.17. The approach relies largely on external nutrient control and focuses on the nutrient status of the waterbody. Note that the starting point for control is *nitrogen* (rather than phosphorus), and some mechanisms of nitrogen control are shown in Table 5.10. Note also that it is implicit in the diagram that nitrogen will not be the limiting nutrient and that phosphorus will be the key nutrient in the *control* of eutrophication. In many eutrophic freshwaters it is not so important that phosphorus *is* the specific algal growth-limiting factor at a given time but rather

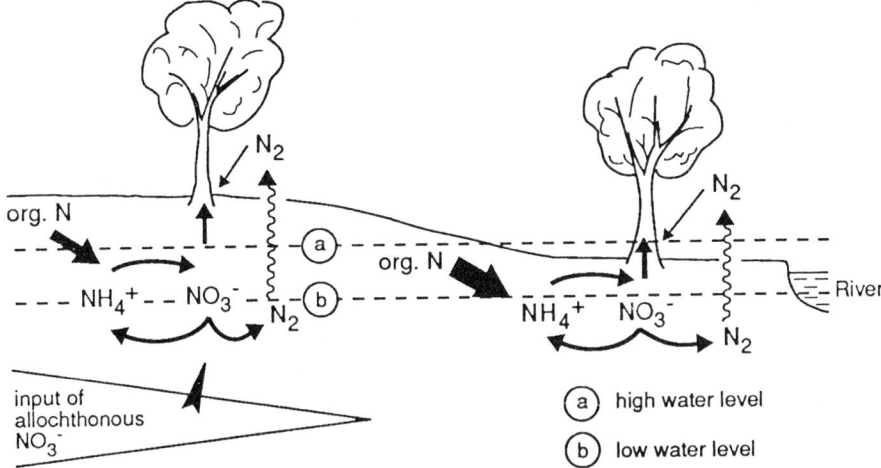

Figure 5.16 The nitrogen cycle in riparian zones illustrating the importance of denitrification in N removal (after Pinay and Décamps, 1988)

that it is *made* limiting by reducing external catchments loads or by in-lake control measures (PLUARG, 1987; Ryding, 1980; OECD, 1982).

The control of point sources of nitrogen usually focuses on sewage works discharges. One approach is to incorporate selected stages of the nitrogen cycle into the treatment process (for example, adding lime to raise the pH and volatise ammonia). However, much nitrogen still remains in the resulting sludge which must be disposed of. Optimisation of denitrification in sewage treatment requires addition of a carbon source (for example, methanol and the creation of an anoxic environment).

Table 5.10 Methods for reducing nitrogen losses from agricultural lands (modified from United States Environmental Protection Agency, 1980)

Control practice	Description
Eliminate excess use of fertiliser	Cut in N leaching; cut in fertiliser cost, no effect on yield
Leaching control	
(1) Timing of N application	Reduction in N leaching through efficient use; inconvenient
(2) Crop rotations	Reduction in N input and erosion; not compatible with many farm enterprises
(3) Animal wastes as fertilisers	Creates non-point (slow-spreading) N source from point source
(4) Ploughing green legume crops	Reduces N fertiliser input; not always feasible
(5) Use winter cover crops	Uses nitrate, reduces percolation, reduces erosion
Control of nutrients in runoff	
(1) Incorporate surface applications	Decreases N in runoff; no yield effects; not always possible
(2) Control surface applications	Use when incorporation not feasible
(3) Use legumes in grasslands	Replaces N fertiliser; limited applicability; difficult to manage

134

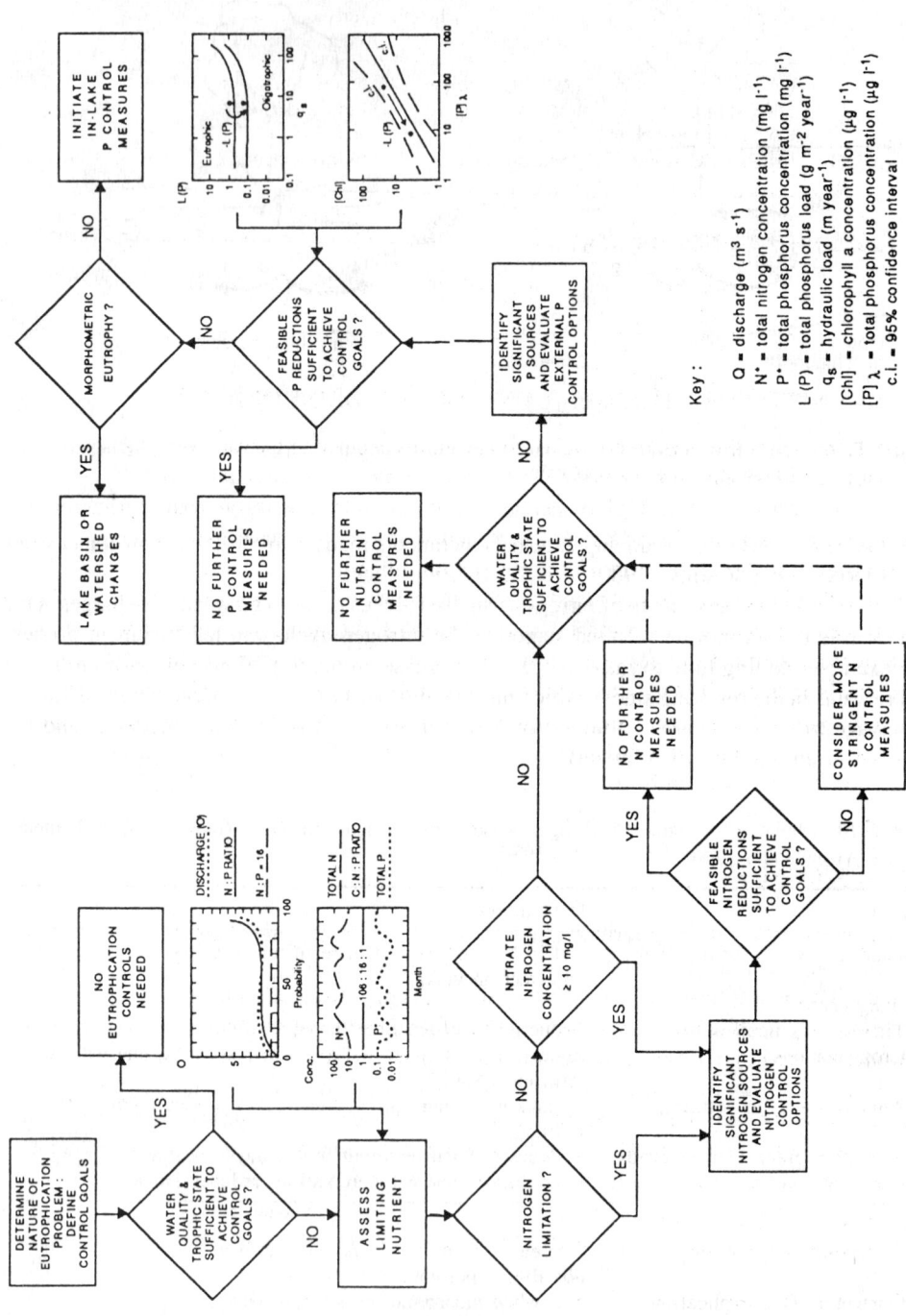

Figure 5.17 A general approach for the control of eutrophication based on nitrogen and phosphorus criteria (modified from Ryding and Rast, 1989)

Table 5.11 Water use and trophic status

Use	Trophic state	
	Required	Acceptable
(1) Drinking water	Oligotropohic	Mesotrophic
(2) Bathing	Mesotrophic	Slightly eutrophic
(3) Water sports	Mesotrophic	eutrophic
(4) Process water	Mesotrophic	Slightly eutrophic
(5) Cooling water	—	Eutrophic
(6) Irrigation	—	Strongly eutrophic
(7) Energy production	—	Strongly eutrophic
(8) Fish culture:		
salmonid	Oligotrophic	Mesotrophic
cyprinid	—	Eutrophic

5.5.5 THE NEED FOR NITROGEN CONTROL FOR DRINKING WATER

Table 5.11 indicates the variation in lake trophic status according to water use. Oligotrophic lake waters are desirable for drinking water because nitrate is potentially toxic to humans either in the form of methaemoglobinaemia in young children or through the formation of carcinogenic nitrosamines in the human gut. Control of nitrogen for drinking-water supplies may be necessary, therefore, even if it is not the limiting nutrient in the aquatic system. The problem is that land-use changes together with treatment of inflowing lake waters is much more efficient for phosphorus control than for dissolved nitrogen compounds. For example, even though measures to increase the infiltration rate of surface soils would reduce excess nitrate levels in the soil, this excess nitrate is simply passed on to the groundwater and eventually to the lake system. Nitrogen-control measures such as runoff or leaching control (see Table 5.10) are much less efficient at low temperatures. The effect of nitrogen on human health is examined more fully in Chapter 13.

5.6 CONCLUSIONS

This chapter has sought to underline some of the fundamental criteria involved in nitrogen cycling in surface waters and lakes. The sources of nitrogen for freshwater systems are examined and the implications for lake trophic status discussed. Control strategies and the implications for both natural ecosystems and human health are outlined.

Nitrogen is the key nutrient determining the primary productivity of surface waters and lakes. There is strong evidence that the concentration of nitrogen, particularly nitrate, has increased in rivers and lakes in the past 20 years. This increase is primarily the result of land-use change and agricultural intensification in the catchments of these aquatic systems. The contribution of nitrogen to freshwaters from sewage has also increased. The pathways of nitrogen transfer from the land to the stream have been discussed in this chapter and the importance of the nitrogen cycle in transforming the nitrogen export has been examined. Riparian 'ecotones' are felt to be important zones for modifying nitrogen export in the land-to-stream pathway. Instream nitrogen transformations are also viewed as critical in modifying the nitrogen load once it enters the aquatic system. In both these pathways, denitrification is the critical process governing the extent of nitrogen 'loss' from the system.

The role of nitrogen in the eutrophication of freshwaters has also been examined. It is noted that the control of nutrients to limit eutrophication normally focuses on phosphorus. However, it is not always the case that phosphorus is the limiting nutrient, and hence the controlling factor, in freshwater eutrophication. The relative importance of nitrogen and phosphorus as limiting nutrients will vary seasonally. This is particularly true of nitrogen, which is primarily dependent on catchment sources. Phosphorus is less dependent on this source and, furthermore, has a secondary source of phosphorus in lake sediments. This difference in N and P sources means that there is a high potential for nitrogen limitation in eutrophic waters when the catchment supply of nitrogen is low, that is, in summer months. The extent to which nitrogen actually becomes limiting also depends on the algal composition of lake waters. It is possible that nitrogen-fixation by blue-green algae may reduce nitrogen limitation. However, it is just as feasible that the initial dominance of blue-green algal blooms in many eutrophic waters is itself triggered by a limited nitrogen supply, which blue-greens can adapt to and exploit. More research on the role of nitrogen in freshwater eutrophication is needed if control measures are to be effective.

REFERENCES

Admiraal, W. and Botermans, Y.J.H. (1989) Comparison of nitrification rates in three branches on the lower river Rhine. *Biogeochemistry*, **8**, 135–51.

Baker, L.A., Brezonik, P.L. and Kratzer, C.-R. (1985) Nutrient loading models for Florida lakes. In *Lake and Reservoir Management — Practical Applications*, Proceedings 4th Annual Conference and International Symposium, New Jersey, USA, 16–19 October 1984, pp. 253–8.

Benson-Evans, K., Fish, P., Pickup, G. and Davies, P. (1967) The Natural History of Slapton Ley Nature Reserve II: Preliminary studies on the freshwater algae. *Field Studies*, **2**, 493–519.

Burns, I.G. and Greenwood, D.J. (1982) Estimation of the year-to-year variations in nitrate leaching in different soils and regions of England and Wales. *Agriculture and Environment*, **7**, 35–45.

Burt, T.P., Arkell, B.P., Trudgill, S.T. and Walling, D.E. (1988) Stream nitrate levels in a small catchment in southwest England over a period of 15 years (1970–1985). *Hydrological Processes*, **2**, 567–84.

Cameron, K.C. and Wild, A. (1984) Potential aquifer pollution from nitrate leaching following ploughing of temporary grassland. *J. Env. Qual.*, **13**, 274–8.

Canfield, D.E., Langeland, S.B., Maceina, M.J., Haller, W.T., Shireman, J.V. and Jones, J.R. (1983) Trophic state classification of lakes with aquatic macrophytes. *Can. J. Fish Aquatic Sci.*, **40**, 1713–18.

Canfield, D.E., Shireman, J.V., Coole, D.E., Haller, W.T., Watkins, C.E. and Maceina, M.J. (1984) Prediction of chlorophyll *a* concentrations in Florida Lakes: Importance of aquatic macrophytes. *Can. J. Fish. Aquatic Sci.*, **41**, 497–501.

Casey, H. and Clarke, R.T. (1979) Statistical analysis of nitrate concentrations from the River Frome (Dorest) for the period 1965–1976. *Freshwater Biol.*, **9**, 91–7.

Casey, H. and Farr, I.S. (1982) The influence of within-stream disturbance on dissolved nutrient levels during spates. *Hydrobiologia*, **92**, 447–62.

Chauvet, E. and Décamp, H. (1989) Lateral interactions in a fluvial landscape: the River Garonne, France. *J. North American Benthological Society*, **8**, 9–17.

Chen, R.L. and Keeney, D.R. (1974) The fate of nitrate in lake sediment columns. *Water Resources Bulletin*, **10**, 1162–72.

Chen, R.L., Keeney, D.R. and McIntosh, T.H. (1983) The role of sediments in the nitrogen budget of Lower Green Bay, Lake Michigan. *J. Great Lakes Res.*, **9**, 23–31.

Collingwood, R.W. (1977) A survey of eutrophication in Britain and its effects on water supplies. Water Research Centre Technical Report No. TR40. WRc, Stevenage, UK.

Cooper, A.B. and Cooke, J.G. (1984) Nitrate loss and transformation in two vegetated headwater streams. *New Zealand Journal of Marine and Freshwater Research*, **18**, 441–50.

Coote, D.R., MacDonald, E.M. and De Haan R. (1979) Relationships between agricultural land and water quality. In Loehr, R.C., Haith, D.A., Walter, M.F. and Martin, C.S. (eds), *Best Management Practices for Agriculture and Silviculture*, Ann Arbor Science Publishers, Ann Arbor, MI, pp. 79−92.

DeRooij, N.M. (1980) A chemical model to describe nutrient dynamics in lakes. In Barica, J. and Murs, L.R. (eds), *Hypertrophic Systems*, Developments in Hydrobiology, Volume 2, W. Junk, The Hague, pp. 139−49.

Dermine, B. and Lamberts, L. (1987) Nitrate nitrogen in the Belgian course of the Meuse river — fate of the concentrations and origin of the inputs. *J. Hydrology, 93*, 91−9.

Dillon, P.J. and Rigler, F.H. (1974) The chlorophyll−phosphorus relationship in lakes. *Limnol. Oceanogr., 19*, 767−73.

Ellis, K.V. (1989) *Surface Water Pollution and its Control*. Macmillan.

Forsberg, C. and Ryding, S.-O. (1980) Eutrophication parameters and trophic state indices in 30 Swedish waste-receiving lakes. *Arch. Hydrobiol., 89*, 189−207.

Foster, S.S. P., Cripps, A.C. and Smith-Carington, A. (1982). Nitrate leaching to groundwater. *Phil. Trans. Roy. Soc. Lond., B296*, 341−9.

Foy, R.H., Smith, R.V. and Stevens, R.J. (1982) Identification of factors affecting nitrogen and phosphorus loadings to Lough Neagh. *J. Env. Managt, 15*, 109−29.

Golterman, H.L. and Kouwe, F.A. (1980) Chemical budgets and nutrient pathways. In Le Cren, E.D. and Lowe-McConnell, R.H. (eds), *The Functioning of Freshwater Ecosystems*, Cambridge University Press, Cambridge, pp. 85−140.

Gostick, K.G. (1982) Agricultural Development and Advisory Service recommendations to farmers on manure disposal and recycling. *Phil. Trans. Roy. Soc. London., B296*, 329−32.

Haith, D.A. (1982) Development and testing of watershed loading functions for nonpoint sources. In Vogt, W.G. and Mickle, M.H. (eds), *Modeling and Simulation*, Volume 13, School of Engineering, University of Pittsburg, pp. 1463−7.

Haith, D.A. and Tubbs, L.J. (1981) Watershed loading functions for non-point sources. *J. Am. Soc. Civ. Engr, Env. Engr. Div., 107*, 121−37.

Haycock, N.E. (1991) *Riparian land as buffer zones in agricultural catchments*, Unpublished PhD thesis, Oxford University.

Hayes, C.R. and Greene, L.A. (1984) The evaluation of eutrophic impact in public water supply reservoirs in East Anglia. *Water Pollution Control, 83*(1), 42.

Heathwaite, A.L. (1989) An excursion guide to Slapton Ley. *Vth International Symposium on Palaeolimnology*, Excursion Guide A, September 1989, Polytechnic South West, UK.

Heathwaite, A.L. and Burt, T.P. (1991) Predicting the effect of land use on stream water quality. In Peters, N.E. (ed.), *Sediment and Stream Water Quality in a Changing Environment: trends and explanation*, IAHS, pp. 203, 209−218.

Heathwaite, A.L., Burt, T.P. and Trudgill, S. (1989) Runoff, sediment, and solute delivery in agricultural drainage basins — a scale dependent approach. In Lagone, S. (ed.), *Regional characterization of water quality*, IAHS 182, 175−91.

Heathwaite, A.L., Burt, T.P. and Trudgill, S.T. (1990) The effect of agricultural land use on nitrogen, phosphorus and suspended sediment delivery to streams in a small catchment in South West England. In Thornes, J.B. (ed.), *Vegetation and Erosion*, Wiley, Chichester, pp. 161−79.

Heathwaite, A.L. and O'Sullivan, P.E. (1991) Sequential inorganic chemical analysis of a core from Slapton Ley, Devon. *Hydrobiologia, 214*, 125−35.

Henderson-Sellers, B. and Markland, H.R. (1987) *Decaying Lakes*, Wiley, Chichester.

Hill, A.R. (1979) Denitrification in the nitrogen budget of a river ecosystem. *Nature, 281*, 291−2.

Hill, A.R. (1988) Factors influencing nitrate depletion in a rural stream. *Hydrobiologia, 160*, 111−22.

Hill, A.R. and Sanmugadas, K. (1985) Denitrification rates in relation to stream sediment characteristics. *Water Res., 19*, 1579−86.

Hill, J.M. (1974) Nitrates in surface waters: observations from some rivers in the Lee drainage basin. Water Research Centre Technical Report No. TR203, WRc, Stevenage, UK.

Howard-Williams, C., Davies, J. and Pickmere, S. (1982) The dynamics of growth and the effects of changing area and nitrate uptake by watercress *Nasturtium officinale* in a New Zealand stream. *J. Appl. Ecol., 19*, 589−601.

Howard-Williams, C. and Downes, M.T. (1984) Nutrient removal by streambank vegetation. In *Land Treatment of Wastes, Water and Soils*, Miscellaneous Publication 70, pp. 409−22.

Howard-Williams, C., Pickmere, S. and Davies J. (1988) Nutrient retention and processing in New Zealand streams: the influence of riparian vegetation. *New Zealand Agricultural Science*, **20**, 110–14.

Jacobs, T.C. and Gilliam, J.W. (1985) Riparian losses of nitrate from agricultural drainage waters. *J. Env. Qual.*, **14**, 472–8.

Johnes, P.J. (1990) *An investigation of the effects of land use upon water quality in the Windrush catchment*, Unpublished DPhil thesis, Oxford University.

Johnes, P.J. and O'Sullivan, P.E. (1989) The Natural History of Slapton Ley Nature Reserve XVIII: Nitrogen and phosphorus losses from the catchment — an export-coefficient approach. *Field Studies*, **7**, 285–309.

Keeney, D.R. (1973) The nitrogen cycle in sediment–water systems. *J. Env. Qual.*, **2**, 15–27.

Keeney, D.R. (1974) Protocol for evaluating the nitrogen status of lake sediments. US Environmental Protection Agency Report No. EPA-660/3-73-024, Washington, DC.

Krenkel, P.A. and Novotny, V. (1980) *Water Quality Management*, Academic Press, London.

Lee, G.F., Jones, R.A. and Rast, W. (1980) Availability of phosphorus to phytoplankton and its implications for phosphorus management strategies. In Loehr, R.C., Martin, C. and Rast, W. (eds), *Phosphorus Management Strategies*, Ann Arbor Publishers, Ann Arbor, MI, pp. 259–308.

Liao, C.F.-H. and Lean, D.R.S. (1978) Seasonal changes in nitrogen compartments of lakes under different loading conditions. *J. Fish Res. Bd. Can.*, **35**, 1095–1101.

Likens, G.E., Bormann, F.H., Pierce, R.J., Eaton, J.S. and Johnson, N.M. (1977) *Biogeochemistry of a Forested Ecosystem*, Springer-Verlag, New York.

Loehr, R.C. (1974) Characteristics and comparative magnitude of non-point sources. *J. Water Pollut. Cont. Fed.*, **46**, 1849–72.

Lowrance, R.R., Todd, R.L. and Asmussen, L.E. (1984) Nutrient cycling in an agricultural watershed: I. Phreatic movement. *J. Env. Qual.*, **13**, 22–7.

Macdonald, A.J. Powlson, D.S., Poulton, P.R. and Jenkinson, D.S. (1989) Unused fertiliser nitrogen in arable soils — Its contribution to nitrate leaching. *J. Sci. Food Agric.*, **46**, 407–19.

Maitland, P.S. (1981) Introduction and catchment analysis. In Maitland, P.S. (ed.), *The Ecology of Scotland's Largest Lochs*, W. Junk, The Hague, pp. 1–27.

Moss, B., Balls, H., Booker, I., Manson, K. and Timms, M. (unpublished) Problems in the construction of a nutrient budget prior to restoration of the R. Bure and Its Boards, Norfolk, from eutrophication. University of East Anglia.

Oakes, D.B., Young, C.P. and Foster, S.S.D. (1981) The effects of farming pratices on groundwater quality in the UK. *Sci. Tot. Env.*, **21**, 17–30.

OECD (1982) *Eutrophication of Waters: monitoring, assessment and control*, Paris.

Onstad, C.A. and Blake, J. (1980) Thames Basin nitrate and agricultural relations. In *Proc. Symp. Watershed Management.*, Am. Soc. Civil Engineers, pp. 961–73.

O'Sullivan, P.E., Heathwaite, A.L., Appleby, P.G., Brookfield, D., Crick, M.W., Moscrop, C., Mudler, T.B., Vernon, N.J. and Wilmhurst, J.M. (1991) Paleolimnology of Slapton Ley, Devon, UK. *Hydrobiologia, 214*, 115–24.

Peterjohn, W.T. and Correll, D.L. (1984) Nutrient dynamics in an agricultural watershed: Observations on the role of a riparian forest. *Ecology, 65*, 1466–75.

Phillips, G.L. (1977) The mineral nutrient levels in three Norfolk Broads differing in trophic status, and an annual mineral content budget for one of them. *J. Ecol.*, **65**, 447–74.

Pinay, G. (1986) *Relations sol-nappe dans les bois riverains de la Garonne. Etude de la dénitrification.* Dissertation, Université de Lyon I, Lyon, France.

Pinay, G. and Décamp, H. (1988) The role of riparian woods in regulating nitrogen fluxes between the alluvial aquifer and surface water: a conceptual model. *Regulated Rivers*, **2**, 507–16.

Pinay, G., Décamp, H., Chauvet, E. and Fustec, E. (1990) Functions of ecotones in fluvial systems. In Naiman, R.J. and Décamps, G. (eds), *The Ecology and Management of Aquatic–Terrestrial Ecotones*, Parthenon, Paris, pp. 141–70.

PLUARG (Pollution From Land Use Activities Reference Group) (1987) *Environmental management strategy for the Great lakes ecosystem*, Final Report, Pollution From Land Use Activities Reference Group to the International Joint Commission, Great Lakes Regional Office, Windsor, Ontario, Canada.

Preston, T., Stewart, W.D.P. and Reynolds, C.S. (1980) The bloom-forming cyanobacterium *Microcystis aeruginosa* overwinters on the sediment surface: evidence from a ^{15}N tracer study. *Nature, Lond.*, **288**, 365–7.

Rast, W. and Lee, F.G. (1978) *Summary Analysis of the North American OECD Eutrophication Project: Nutrient loading*—lake response relationships and trophic state indices, Ecological Research Series, No. EPA-600/3-78-008, US Environmental Protection Agency, Environmental Research Laboratory, Corvallis, Oregon.

Rast, W. and Lee, G.F. (1983) Predictive capability of US. OECD phosphorus loading—eutrophication response models. *J. Water Pollut. Cont. Fed.*, **55**, 990−1003.

Richey, J.S., McDowell, W.H. and Likens, G.E. (1985) Nitrogen transformations in a small mountain stream. *Hydrobiologia*, **124**, 129−39.

Royal Society (1983) *the Nitrogen Cycle of the United Kingdom*, The Royal Society, London.

Ryding, S.-O. (1980) *Monitoring of inland waters: OECD eutrophication programme — The Nordic Project*, Publication 1980:2, Nordic Cooperative Organisation for Applied Research, Helsinki, Finland.

Ryding, S.-O. and Rast, W. (eds) (1989) *The Control of Eutrophication in Lakes and Reservoirs*, Man and the Biosphere Series, Volume 1, Parthenon, Paris.

Schindler, D.W. (1977) Evolution of phosphorus limitation in lakes. *Science*, **195**, 260−62.

Schindler, D.W. and Fee, E.J. (1974) Experimental Lake Area: Whole-lake experiments in eutrophication. *J. Fish Res. Bd. Can.*, **31**, 937−53.

Scott, R.N. (1975) Studies on some Welsh reservoirs with special reference to Talybont Reservoir. In *Proceedings of the Water Research Centre Symposium on the Effects of Storage on Water Quality*, Marlow, Bucks, WRc.

Shanahan, P. and Harlemann, D.R.F. (1984) Transport in lake water quality modeling. *J. Am. Soc. Civil Engr., Env. Engr. Div.*, **110**, 42−57.

Sloey, W.E., Spamgler, F.L. and Fetter, C.W. (1978) Management of freshwater wetlands for nutrient assimilation. In Good, R.E. *et al.* (eds), *Freshwater Wetlands*, Academic Press, New York.

Smith, R.V. (1977) Domestic and agricultural contributions to the inputs of phosphorus and nitrogen to Lough Neagh. *Water Research*, **11**, 453−9.

Smith, V.H. (1983) Low nitrogen to phosphorus ratios favour dominance by blue-green algae in lake phytoplankton. *Science*, **221**, 669−70.

Smith, V.H. and Shapiro, J. (1981) Chlorophyll−phosphorus relations in individual lakes. Their importance to lake restoration strategies. *Environ. Sci. Technol.*, **15**, 444−51.

Smolen, M.D. (1981) Nutrient runoff from agricultural and non-agricultural watersheds. *Trans. Am. Soc. Agric. Engnr.*, **24**, 981−7.

Sprent, J.I. (1987) *The Ecology of the Nitrogen Cycle*, Cambridge University Press, Cambridge.

Sridharan, N. and Lee, G.F. (1977) Algal nutrient limitation in Lake Ontario and its tributary waters. *Water Res.*, **11**, 849−58.

Stewart, W.D.P., Preston, T., Peterson, H.G. and Christofi, N. (1982) Nitrogen Cycling in eutrophic freshwaters. *Phil. Trans. Roy. Soc. Lond.*, **B296**, 491−509.

Swank, W.T. and Caskey, W.J. (1982) Nitrate depletion in a second-order mountain stream. *J. Env. Qual.*, **11**, 581−4.

Terry, R.E. and Nelson, D.W. (1975) Factors influencing nitrate transformations in sediments. *J. Env. Qual.*, **4**, 549−54.

Tilman, D., Kilham, S.S. and Kilham, P. (1982) Phytoplankton community ecology: The role of limiting nutrients. *Ann. Rev. Ecol. Sys.*, **13**, 349−72.

Torrey, M.S. and Lee, G.F. (1976) Nitrogen fixation in Lake Mendota, Madison, Wisconsin. *Limnol. Oceanogr.*, **21**, 365−79.

Trudgill, S.T., Heathwaite, A.L. and Burt, T.P. (1990) Wetland control of sewage point sources of nitrate and phosphate, Slapton, S. Devon, UK. *Proceedings of the Use of Constructed Wetlands in Water Pollution Control*, Water Research Centre, Medmenham, pp. 575−7.

Vaccaro, R.F. (1965) Inorganic nitrogen in seawater. *Chem. Oceanogr.*, **1**, 365−408.

Van Vlymen, C.D. (1980) *The water balance, physico-chemical environment, and phytoplankton studies of Slapton Ley, Devon*, Unpublished PhD thesis, University of Exeter.

Vollenweider, R.A. (1968) *Scientific fundamentals of the eutrophication of lakes and flowing waters, with particular reference to nitrogen and phosphorus as factors in eutrophication*, Technical Report DAS/CSI/68.27, Environmental Directorate, OECD, Paris.

Vollenweider, R.A. (1975) Input−output models with special reference to phosphorus loading concept in limnology. *Schweiz. Zeit. Hydrol.*, **37**, 53−84.

Vollenweider, R.A. (1976) Advances in defining critical loading levels for phosphorus in lake eutrophication. *Mem. Ist. Ital. Idrobiol.*, **33**, 53–83.

Vollenweider, R.A. and Janus, L.L. (1982) Statistical models for predicting hypolimnetic oxygen depletion rates. *Mem. Ist. Ital. Idrobiol.*, **40**, 1–24.

Wild, A. and Cameron, K.C. (1980) Soil nitrogen and nitrate leaching. In Tinker, P.B. (ed.), *Soils and Agriculture*, Blackwell, Oxford.

Wild, H.E., Sawyer, C.N. and McMahon, J.C. (1971) Factors affecting nitrification kinetics. *J. Wat. Pollut. Control Fed.*, **43**(9), 1845–54.

Wilkinson, W.B. and Greene, L.A. (1982) The water industry and the nitrogen cycle. *Phil. Trans. Roy. Soc. Lond.*, **B296**, 459–75.

Wyer, M.D. and Hill, A.R. (1984) Nitrate transformations in Southern Ontario stream sediments. *Water Resources Bulletin*, **20**, 729–37.

6 Nitrogen Cycling in Wetlands

C. HOWARD-WILLIAMS and
M.T. DOWNES
Taupo Research Laboratory, New Zealand

6.1 INTRODUCTION

In recent years there has been considerable interest in the contributions that wetlands make to the cycling of atmospheric gases in the biosphere. Wetlands have been shown to process and transfer globally significant quantities of CH_4, CO_2, H_2S and N_2O (Armento and Menges, 1986; Aselman and Creutzen, 1989; Buresh, Casselman and Patrick, 1980). Fixation of nitrogen gas from the atmosphere also occurs on an extensive scale in wetlands, particularly artificial wetlands (rice paddies). The anoxic conditions in wetlands which contribute to atmospheric methane and H_2S flux also provide for denitrification.

Wetlands are areas which share ecosystem properties with both terrestrial and aquatic systems and may be in standing or running waters. Wetlands can be defined as areas where the water table is at or above land surface for long enough each year to promote the formation of hydric soils and to support the growth of aquatic vegetation, much of which has its photosynthetic tissue above the water surface (Cowardin *et al.*, 1979; Howard-Williams, 1985).

Large areas of the world's land surface are covered by wetland ecosystems. The global distribution of natural freshwater wetlands is documented, in some detail, by Aselman and Creutzen (1989), who added up the areas of six defined wetland types over a $2.5° \times 5.0°$ grid across the world. A total global wetland area of 5.7×10^6 km^2 was computed.

The largest areas of wetland are, not surprisingly, in cold wet boreal regions between latitudes $50°$ and $70°N$. However, extensive areas also occur to $15°$ on either side of the equator, associated with the huge swamps and floodplains of South America and tropical Africa (Figure 6.1(a)).

A very large area (1.2×10^6 km^2) of wetlands occur only on a seasonal basis. In Africa, for instance, the seasonal variability in wetland area is from 355×10^3 km^2 to perhaps 60×10^3 km^2 during the dry seasons (see e.g. Howard-Williams and Thompson, 1985). Aselman and Creutzen (1989) point out that in the former USSR the northern wetland areas decrease from 1512×10^3 km^2 to 0 when they freeze. Artificial wetlands (mainly rice paddies) may, on a global scale, be as important to nitrogen cycling as natural wetlands are, as the areas are considerable (Figure 6.1(b)).

The ecosystem structure of wetlands is such that almost all processes in the conventional nitrogen cycle can occur in close proximity, either spatially or temporally. For instance, seasonal wetting and drying cycles allow for very efficient ammonification and nitrification (McLachlan, 1970). Wetlands promote nitrogen fixation by both heterotrophic and chemotrophic bacteria (e.g. azotobacter) and autotrophic cyanobacteria (e.g. *Nostoc*). In

Nitrate: Processes, Patterns and Management. Edited by T.P. Burt, A.L. Heathwaite and S.T. Trudgill
© 1993 John Wiley & Sons Ltd

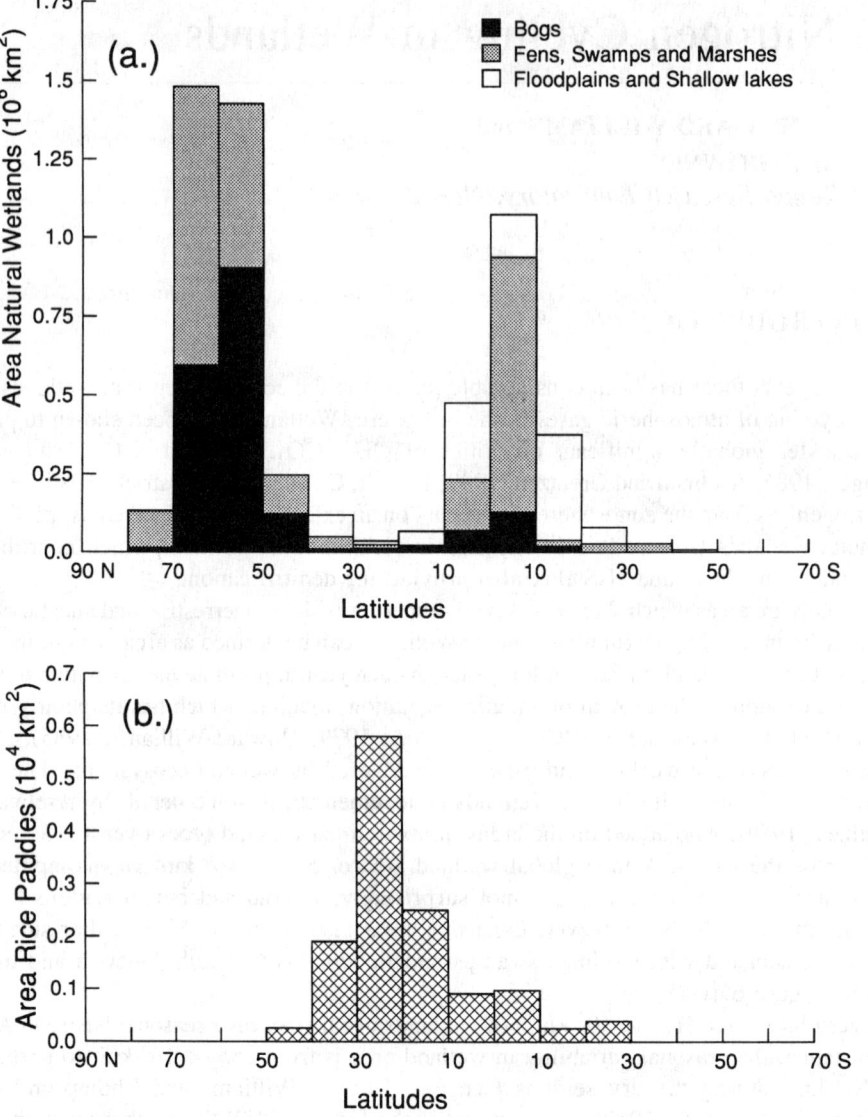

Figure 6.1 Global distribution of natural wetlands (a) and of rice paddies (b) along 10° latitudinal belts (after Aselmann and Creutzen, 1989)

addition, slow oxygen diffusion rates in hydric soils, combined with an oxygen demand generated from the high primary production, produce ideal conditions for the anaerobic pathways of the nitrogen cycle.

In this chapter we view the nitrogen cycle in wetlands with nitrate as the central focus and consider pathways of assimilation and dissimilation to and from NO_3-N. The chapter first examines the physical structures of wetlands which enhance N cycling both on a coarse (m) and fine (μm to mm) scale. We review the evidence for assimilation, ammonification, nitrification and consequent dissimilatory pathways of nitrate. Mass flows of nitrogen are discussed together with the role of wetland hydrodynamics, and finally we consider the applied issues of nitrogen in wetlands. Nitrate control through land-use regulation in wetlands is discussed further in Chapter 12. The link between nitrogen cycling in wetlands and surface waters is examined in Chapter 5.

6.2 STRUCTURE OF WETLANDS

6.2.1 PHYSICAL STRUCTURE

For the purposes of a review on nitrogen cycling we can consider three useful wetland subdivisions:

(1) Permanently inundated versus periodically inundated wetlands. The periodic (seasonal) drying of a wetland has a marked effect on nutrient cycling.
(2) Permanently non-frozen wetlands versus periodically snow-covered or frozen wetlands where nitrogen cycling is intermittent due to low temperatures, physical freezing disruptions, freeze concentration of salts, etc.
(3) Rooted versus floating wetlands (cf. Howard-Williams, 1985; Denny, 1987). In rooted wetlands the dominant wetland vegetation (e.g. *Phragmites, Typha*) is rooted in the sediment, and the sediments play the major role in nitrogen cycling (Figure 6.2). In floating wetlands the vegetation forms a dense floating mass, and recycling of nitrogen (and other

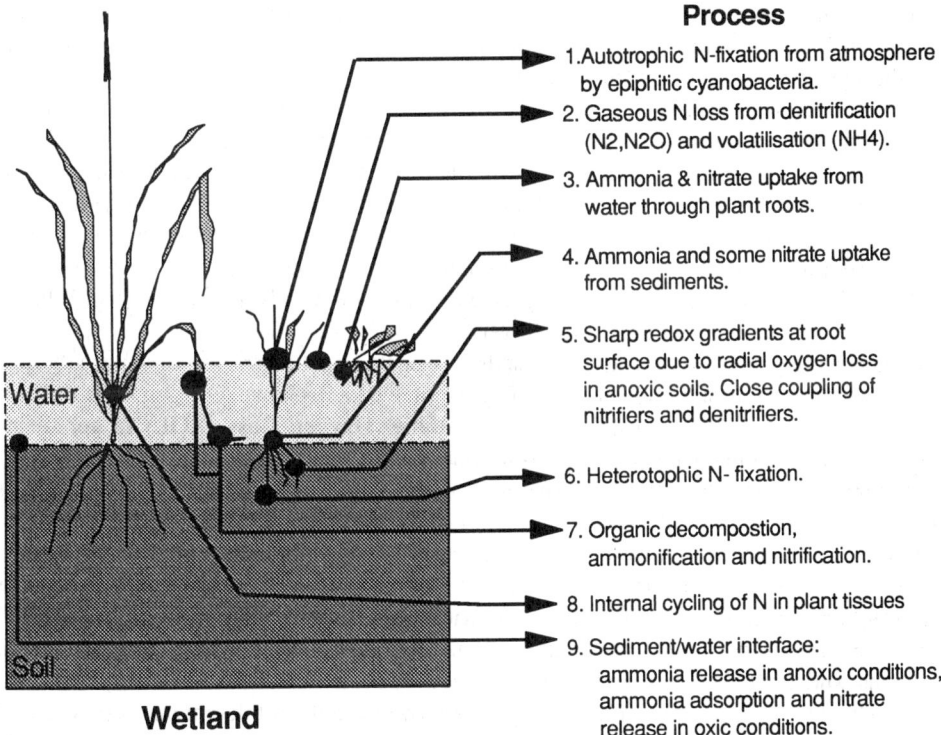

Process

1. Autotrophic N-fixation from atmosphere by epiphitic cyanobacteria.
2. Gaseous N loss from denitrification (N2,N2O) and volatilisation (NH4).
3. Ammonia & nitrate uptake from water through plant roots.
4. Ammonia and some nitrate uptake from sediments.
5. Sharp redox gradients at root surface due to radial oxygen loss in anoxic soils. Close coupling of nitrifiers and denitrifiers.
6. Heterotophic N- fixation.
7. Organic decompostion, ammonification and nitrification.
8. Internal cycling of N in plant tissues
9. Sediment/water interface: ammonia release in anoxic conditions, ammonia adsorption and nitrate release in oxic conditions.

Figure 6.2 Structure of wetlands showing the principal sites of processes in the nitrogen cycle

nutrients) is from plant mat to water and back. Such wetlands are typical of large areas of the tropics and examples include the papyrus swamps of Africa (Gaudet, 1977), the floating meadows of South America (Junk, 1983), perched bogs (Gorham, 1957) or central European 'plaur' formations (Hejny, 1971).

Wetland soils may be mineral (gleys) or organic (peat) and have markedly different patterns of N cycling. Gley soils generally dry out periodically and the resulting variability in the soil redox potentials allows for nitrate accumulation in dry periods and reduction in wet ones. Thus during inundation these soils are nitrate deficient. Peats are formed of partially decomposed organic matter and are general oligotrophic in permanently waterlogged conditions. Most are acidic (pH < 6).

Figure 6.2 is a diagrammatic representation of the coarse and fine structure of a wetland with the features of importance to nitrogen cycling. The bulk of photosynthesis occurs in the air but respiration and decomposition is in the water or the sediments, hence the tendency to anoxia. Autotrophic nitrogen fixers are usually found in the surface waters attached to larger substrates (plant stems, roots, etc.). Floating plants are common constituents of wetlands so that two layers of roots occur, one just below the water surface and one in the sediments. In true floating wetlands all the root tissue is associated with the floating mat.

6.2.2 DISSOLVED OXYGEN

Of particular importance in the wetland nitrogen cycle is the distribution of dissolved oxygen. There is considerable variability depending on latitude (e.g. higher respiration rates in the tropics), wetland flushing rates, vegetation type, etc. This variability extends to a vertical scale through the wetland, and to a range of temporal scales (hourly to seasonal).

Oxygen under-saturation in wetland waters is a feature of those systems where the water surface is heavily shaded by emergent vegetation. In such conditions, ammonium-N is likely to be the dominant inorganic N species. An extreme example is found in the floodplain wetlands of the large south American rivers (Melack and Fisher, 1983; Junk, 1983). Oxygen levels can increase in clearings or open-water areas and when flushing rates increase by either wind-induced water movements between wetland and open water or during floods. Marked diurnal variability in oxygen concentrations in natural and artificial wetlands is common (e.g. Whitton et al., 1988; Howard-Williams et al., 1989).

Nitrogen cycling in the sediments of wetlands is regulated by the transport of oxygen through the aerenchymatous tissue and lacunae of roots of wetland plants. Oxygen then diffuses outwards from the roots by a process called Radial Oxygen Loss (ROL) to aerate the rhizosphere in the immediate vicinity of the root (Dunbabin, Pokorny and Bowmer, 1988; Reddy, Patrick and Lindau, 1989). The extent of oxygenation will depend on the plant species (Armstrong, 1964) and on the oxygen demand of the sediments. However, the mosaic of oxygenated microzones (Figure 6.3) in an environment in which inorganic N is dominated by NH_4^+ has major repercussions, as it allows for nitrification, adjacent denitrification as well as N_2O formation and uptake (Reddy, Patrick and Lindau, 1989). apRees et al. (1987) suggest that the majority of flood-tolerant plants are not capable of significant growth in anoxic conditions, and that the metabolism of such species probably does not differ in any way from flood-intolerant species. Rather, it is the greater ability of the flood-tolerant plants to transport oxygen to submerged tissue which is important (apRees et al., 1987). ROL from wetland plant roots is considerably greater than non-wetland plants (Figure 6.4).

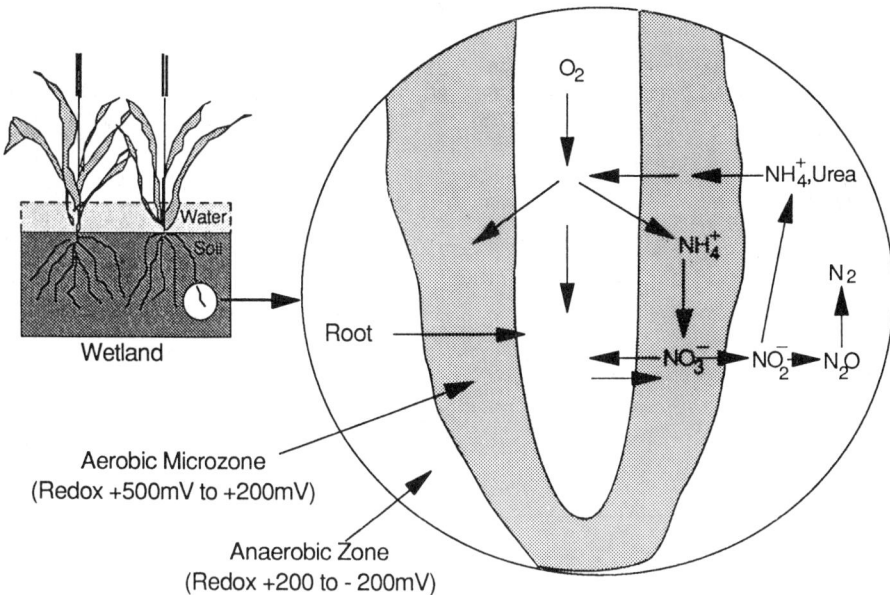

Figure 6.3 Schematic presentation of rhizosphere oxygenation and affected metabolic pathways of the nitrogen cycle (after Reddy *et al.*, 1989)

6.2.3 STANDING STOCKS OF NITROGEN

Standing stocks of nitrogen in wetlands will vary greatly depending on the seasonality of the growth cycles of the dominant vegetation. Tissue levels of nitrogen vary greatly between species of wetland plant, with some such as watercress (*Nasturtium officinale* R. Br.) having concentrations of 44 mg N g^{-1} dry weight (over 4%) in the leaves, while at the other extreme, some species in the oligotrophic 'black waters' of the Amazon had less than 10 mg N g^{-1} dry weight (Howard-Williams and Junk, 1977). A mean concentration of nitrogen in the tissues of water plants of 2.37% was derived by Riemer and Toth (1968). The composition of the total nitrogen content of wetland plants shows that 60−65% of the tissue nitrogen was protein-N (Boyd, 1969). The standing stocks of nitrogen in contrasting wetlands are shown in Table 6.1. The bulk of the total nitrogen pool (80−90%) is in the sediments with a very small amount 0.1−0.3% as interstitial water with dissolved-inorganic N. Of this, most was as NH_4-N. Similarly NH_4-N concentrations were generally much higher than those of NO_3-N in the free water compartment, although the absolute amounts of both species were very small. Nitrogen concentrations in wetland waters are generally inadequate to provide for the plant growth, and in wetlands where the throughflow of water is slow, the sediments act as the major supply source for growth (Kadlec and Kadlec, 1978; Howard-Williams, 1985).

6.3 WETLAND PROCESSES

6.3.1 ASSIMILATION PATHWAYS

There are two biological processes by which nitrogen can be assimilated in wetlands: nitrogen fixation and active nitrogen uptake.

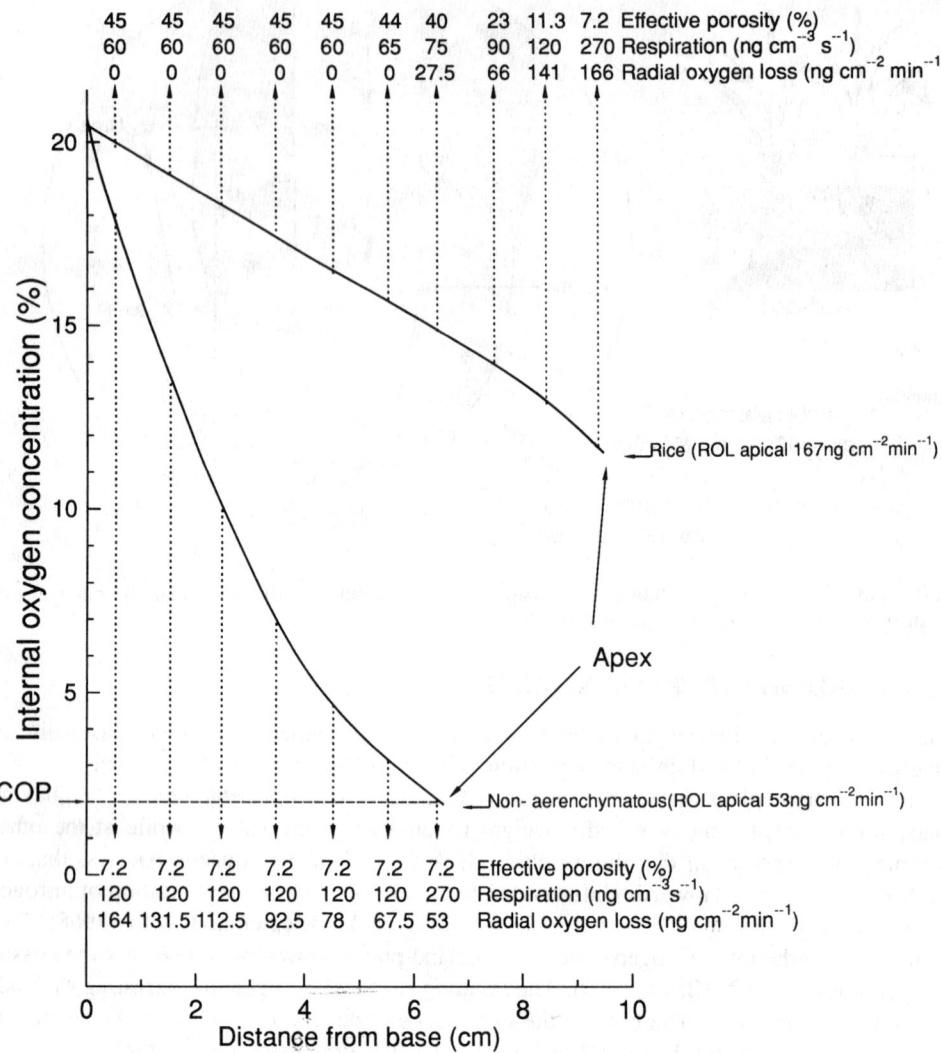

Figure 6.4 Analogue analysis showing how the various characteristics of wetland (aerenchymatous rice root) and non-wetland (non-aerenchymatous) roots contribute to the oxygen status of the root in the wetland condition. The data were compiled on the assumption that the wetland soil, where aerated, consumes oxygen at the uniform rate of 5.27×10^{-8} g cm^{-3} s^{-1}, and that oxygen diffusivity in the soil was a uniform 1×10^{-5} cm^2 s^{-1}. It was assumed also that wall permeability to oxygen of the rice root declined from a maximum of 100% at the apex, to zero at 5 cm from the apex; in the non-aerenchymatous root the minimum value (60%) was attained at 6 cm. ROL = Radial Oxygen Loss, COP = Critical Oxygen Pressure for respiration (modified from Gaynard and Armstrong, 1977)

Table 6.1 Standing stocks of nitrogen in a range of wetlands

Wetland component	Lago Infernão Brazil, 0.2 km² *Eichhornia azurea* (Howard-Williams et al., 1989)	Kawaga Swamp Kenya, 12.5 km² *Cyperus papyrus* (Gaudet, 1977)	Houghton Lake Fen USA, 7.2 km² *Carex, Salix, Betula* (Richardson et al., 1978)	Tundra Pond Alaska, <0.01 km² *Carex aquatilis Arctophila fulva* Hobbie, 1980
Water: NH_4-N	50 (<1)	1005 (0.01)	182[a] (<1)	1.6 (<1)
NO_3-N	<10 (<1)	1 (<1)	8[a]	1.8 (<1)
DON	923 (3)			98.6 (<1)
Vegetation shoots	4 872 (8)	29 596 (0.5)	6 200 (0.01)	404 (1)
Rhizomes/roots	538 (1)	92 524 (1.21)		1 616 (4)
Benthic algae				84 (<1)
Plankton				0.1 (<1)
Benthic fauna				
Detritus standing	581 (1)	435 884 (15.2)	9 800 (0.01)	240 (0.5)
Sediment (inc. detrital material)	43 029 (88)	2 351 480 (82.1)	683 000 (97.5)	38 460 (93)
Sediment interstitial H_2O NH_4-N	22 (<1)		732 (<1)	6.4 (<1)
NO_3-N			18 (<1)	
DON				223 (0.5)

[a] Assume water depth = 25 cm.

Data as mg N m⁻². DIN = Dissolved inorganic N, DON = Dissolved organic N. Numbers in parentheses are percentage of total.

Table 6.2 Nitrogen fixation rates from a contrasting range of wetland habitats

Wetland system	N-fixation rate $(g \, N \, m^{-2} \, a^{-1})$	Reference
Florida cypress dome	0.39	Dierberg and Brezonik (1981)
Sphagnum bog, USA	1.0	Hemond (1983)
Spartina salt marsh (enriched)	52	Hanson (1977)
Spartina salt marsh (natural)	6—7	DeLaune and Patrick (1990)
Papyrus swamp, Kenya	8.0	Viner (1982)
Nasturtium bed, New Zealand	2.2	Howard-Williams and Downes (unpublished)
Amazon floodplain[a] $(nm \, C_2H_2 \, g^{-1} \, h^{-1})$	2—90	Melack and Fisher (1988)
Oxbow lake, South Brazil	1.8	Howard-Williams *et al.* (1989)
Tundra pond, Alaska	0.028	Hobbie (1980)
Salt marsh, UK	0.02—3	Abd. Aziz and Nedwell (1986)
Deep-water rice, Bangladesh	1.7	Rother *et al.* (1988)

[a]Units are not convertible to $g \, N \, m^{-2} \, a^{-1}$ from available data.

6.3.1.1 Nitrogen fixation

Wetland nitrogen fixation rates have been shown to vary by three orders of magnitude with low rates in the arctic and some oligotrophic freshwater wetlands (Table 6.2), and high rates in some enriched systems (Hanson, 1977, Table 6.2). Considerable small-scale spatial variability in N-fixation occurs (Hanson, 1983) and, where N-fixation is carried out by autotrophic cyanobacteria, considerable diurnal variation in rates may be expected (Howard-Williams *et al.*, 1989).

The close interaction between wetland plants and oxygenated microzones (Figures 6.2 and 6.4) may extend to N-fixers also. Root exudates (e.g. carbohydrates) may enhance the growth of heterotrophic N-fixers (Dicker and Smith, 1980). In reverse, *Sphagnum* has been shown to make direct use of N fixed by associated *Nostoc* colonies (Basilier, 1980). Conversion of fixed N to nitrate requires only oxygen and bacterial nitrification. Nitrogen fixation generally provides only a small source of N to wetland processing (*ca* 5%) although in a *Sphagnum* bog Hemond (1983) showed that N-fixation was the single largest source of N, equivalent to 25% of the N taken up by the plants. In a recent study of nitrogen cycling in a *Spartina* marsh, DeLaune and Patrick (1990) estimated that N-fixation was the primary source of N for the marsh providing $6.7 \, g \, N \, m^{-2} \, a^{-1}$ compared to $2—4 \, g \, m^2 \, a^{-1}$ from sediment inputs.

6.3.1.2 Nitrogen uptake

Nitrogen is taken up by wetland plants (macrophytes and microphytes) as NH_4-N, NO_3-N or urea-N and, it has been recently demonstrated, as N_2O and N_2 gas (Reddy, Patrick and Lindau, 1989). Uptake as nitrate requires that the organism has the capacity to produce nitrate reductase. Preferential uptake of NH_4-N is likely in most rooted wetland plant species where, in spite of aerobic microzones (Figure 6.3), NH_4-N dominates the inorganic N pool. In well-aerated waters nitrate uptake can, however, be rapid as seen by watercress uptake rates. Uptake rates of NO_3-N by watercress in a New Zealand stream varied on a diurnal basis from 100 to 200 $\mu g \, N \, g$ dry wt. $root^{-1} \, h^{-1}$ (Vincent and Downes, 1980) with maximal nitrate reductase

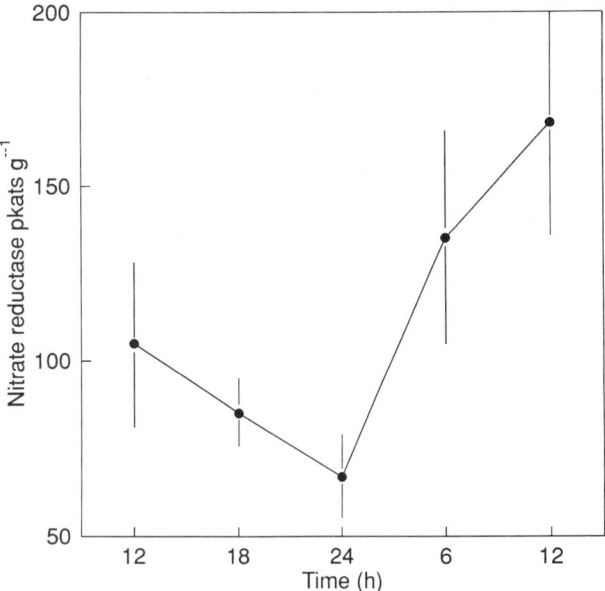

Figure 6.5 Diurnal variations in nitrate reductase in watercress growing in a macrophyte-dominated stream (after Vincent and Downes, 1980)

activity being up to ten times higher in the leaves than the roots. Nitrate reductase activity followed a diurnal pattern (Figure 6.5). The assimilation capacity for nitrate in wetland plants appears to vary with the intracellar concentration of the substrate (cf. Vincent and Downes, 1980). Mangroves, for instance, show little evidence of nitrate utilisation, as measured by nitrate reductase activity, but when grown in nitrate this enzyme increased in the roots allowing for its utilisation (Stewart and Orebamjo, 1984).

Rates of nitrogen uptake by wetland plants vary by three orders of magnitude from as low as 0.5 g m^{-2} a^{-1} in a tundra pond (Hobbie, 1980), through 3 g m^{-2} a^{-1} by *Sphagnum* in a temperate bog (Hemond, 1983), and 94 g m^{-2} a^{-1} in a *Typha* marsh (Davis and van der Valk, 1983) to as much as 178 g m^{-2} a^{-1} in a fast-growing perennial watercress community (Howard-Williams, Davies and Pickmere, 1982). A mean rate of nitrogen uptake for wetlands is *ca* 20 g N m^{-2} a^{-1}. The global coverage of wetlands is 7×10^6 km^2 (Aselman and Creutzen, 1989). Thus the annual store of nitrogen in wetland vegetation on a global scale is likely to be *ca* 140 Tg a^{-1}.

6.3.1.3 Internal cycling of nitrogen

Nitrate taken up by plants must first be reduced to ammonia prior to its conversion via glutamine and glutamate to other nitrogen compounds associated with cell metabolism. This reduction is facilitated by the enzyme nitrate reductase. High nitrate reductase activity occurs in the leaves of some wetland plants (Vincent and Downes, 1980; Figure 6.5). The close association of the enzyme with chloroplasts shows the existence of an internal upward transport of nitrogen as nitrate in those plants. In anoxic conditions nitrogen will be taken up as ammonium-N and converted to glutamine or glutamate by the enzymes glutamine synthetase (GS) and glutamate synthase. These enzymes occur in both roots and shoots and subsequent translocation as amides and amino acids has been demonstrated in mangroves (*Avicenna nitida*) (Stewart and Orebamjo, 1984). Beevers (1976) provides an excellent overview of internal nitrogen metabolism of plants, outside the scope of the present review. Internal translocations of nitrogen have been suggested as mechanisms for the conservation of this element in wetland plants and a distinct seasonality of carbon and nutrients are characteristic of plants with overwintering (or in the case of warmer climates, a dry season) strategies (Madsen, 1991). Retraction of nitrogen from dying leaves is well documented. Twenty-five to fifty per cent of the N in *Phragmites* can be recycled internally after storage in the rhizomes during winter (van der Linden, 1980). In *Typha glauca*, 45% of the nitrogen in senescent leaves is translocated to the rhizomes and stored over winter (Davis and van der Valk, 1983). These N stocks are again moved up to the young shoots in the early stages of growth the following spring (Boyd, 1970; Boyd and Hess, 1970, Mason and Bryant 1975), as photosynthetic tissues require higher nitrogen concentrations than non-photosynthetic tissue (Madsen, 1991). The distribution of nitrogen in *Spartina alterniflora* stands in midwinter showed that 63% was in the rhizomes, 17% in the roots and 20% in the aerial portions. In spring rhizome storage had decreased to 18%, with 68% being located in the aerial tissues (Hopkinson and Schubauer, 1984).

Internal translocations of a significant pool of the assimilated nitrogen are probably the reason why correlation between tissue nitrogen concentrations and those in the wetland sediments or water are seldom found.

6.3.2 AMMONIFICATION

While ammonification (mineralisation) is a well-studied process in sediments (see Chapters 3 and 4), open-water planktonic systems, and particularly in industrial waste-treatment plants (see Chapter 13), there are few data on this process in wetlands. The process may proceed more rapidly than nitrification (see Section 6.3.3 below) as a wide range of micro-organisms may be involved in the initial mineralisation of decaying wetland plant tissue. Rates of ammonification of 480 ($\pm$ SE 139) and 657 ($\pm$ SE 118) μg NH_4-N g (dry wt)$^{-1}$ day^{-1} were recorded for decaying watercress at 10°C and 20°C (Table 6.3). Jones and Simon (1981) found that the ammonification rates of 42 mg m^{-2} day^{-1} in the littoral zone of lakes were higher than in the porfundal zone (28–30 mg m^{-2} day^{-1}), and in a *Spartina* marsh mineralisation rates of 25 g m^{-2} a^{-1} (68 mg m^{-2} day^{-1}) were recorded by DeLaune and Patrick (1980).

6.3.3 NITRIFICATION

The production of nitrate and other oxidised forms of nitrogen is the result of two bacterial processes. Autotrophic nitrification is the sequential oxidation of ammonia to nitrite by bacterial groups such as *Nitrosomonas* sp. and *Nitrosolobus* sp. followed by oxidation of nitrite by groups such as *Nitrobacter* sp. The energy generated by these oxidations is used to fix carbon dioxide, the main carbon source of autotrophic nitrifiers. Heterotrophic nitrification is the oxidation of reduced organic nitrogen compounds to oxidised nitrogen species. Unlike autotrophic nitrification, which has been intensively studied, little is known about the factors which control heterotrophic nitrification or its ecological significance (Schimel, Firestone and Kilham, 1984).

Although nitrification is a strictly aerobic process ammonia oxidation can take place at oxygen concentrations as low as 0.5 Kpa in pure culture (Goreau *et al.*, 1980) and at 0.1 Kpa in marine sediment (Jørgensen, Jensen and Sørensen, 1984). However, oxygen appears to be one of the main factors limiting nitrification in wetlands. Radial oxygen loss from wetland plant roots (Section 6.2.2. above and Figures 6.2 and 6.3) provides a mosaic of oxygenated environments in wetland soils for nitrification. Nitrification rates in wetland systems range from $<0.1-56$ mg N m^{-2} h^{-1} (Table 6.4). Cooper (1984), in a study of four streams in the North Island of New Zealand, showed that nitrification was principally sediment based and that the depth of oxygen penetration into the sediments was an important factor governing the extent of nitrification. He also reported a seasonal variation in oxygen penetration with minimum penetration in the summer when sediment BOD was highest and maximum penetration in the winter giving a greater sediment volume for nitrification. Cooper (1984) estimated nitrate flux rates of up to 2.8 g N m^{-2} per day. Nitrification rates of decomposing watercress from a New Zealand stream (Table 6.3) were *ca* 600 μg dry wt^{-1} day^{-1}. This was equivalent to 238 mg m^{-2} day^{-1} during the 24-week decomposition period.

In a study of coastal Danish waters Henriksen, Hansen and Blackburn (1981) also concluded that the extent of nitrification was limited to a zone in the top 1.5−5.5 mm of sediment where oxygen could penetrate. Nitrification rates were in the range 4.2−19.6 mg N m^{-2} per day. Nitrate flux out of the sediment/water interface was 0−9.8 mg N m^{-2} per day. They found no significant difference in nitrification rate between sandy or muddy sediments or between sediments from shallow and deeper waters. In addition, there was no correlation between the gradient in nitrate concentration across the sediment/water interface and the measured flux of nitrate (see Chapter 5). Fifty per cent of the nitrate produced was released into the water column with the rest denitrified by diffusing into the deeper, anoxic sediment. In the relatively aerobic sediments of oligotrophic Lake Taupo, New Zealand, nitrifier activity was restricted to the top few surficial millimetres of sediment where Eh was well above 200 mV and nitrate and ammonia concentrations were far in excess of values in the overlying water (Vincent and Downes, 1981). No ammonia gradient was detected in the water column, suggesting that the ammonia was oxidised before it left the sediment.

Nitrifiers appear to be able to adapt to low temperatures. Dodds and Jones (1987) studied nitrification potentials in sediments cores taken from a cold (5°C) spring. The pool bottom was covered with patches of large *Nostoc pruniform* Ag. colonies which probably contributed to the high sediment Eh (>200 mV) and oxygen concentration (6 mg l^{-1} near the spring,

Table 6.3 Nitrogen loss parameters associated with decomposing watercress at two different temperatures (reproduced from Howard-Williams *et al.*, 1983, by permission of Kluwer Academic Publishers)

Parameter	10°C		20°C	
	x	SE	x	SE
Rate of ammonification: μg NH$_4$-N g^{-1} (dry wt) d^{-1}	480	139	657	118
Rate of nitrification: μg NO$_3$-N g^{-1} (dry wt) d^{-1}	640	166	571	69
% N regenerated:				
as NO$_3$-N	57	6.3	51	3.6
as DON	18	1.1	21	3.6
as NH$_4$-N	3	0.7	3	0.3
Refractory N%	22	3.7	25	6.0

x = mean, S.E. = standard error.

Table 6.4 Nitrification rates in wetland systems

Wetland type	Nitrification rate	Reference
Streams		
Unenriched stream	$0.09-0.21$ mM N m^{-2} d^{-1}	Chatarpaul *et al.* (1980)
Enriched stream	10.2 mM N m^{-2} d^{-1}	Schwert and White (1974)
Sediment	0.015 μM ng^{-1} h^{-1}	Cooke and White (1987b)
Sediment (pot)	$6.3-12.1$ nM N cm^{-3} h^{-1}	Dodds and Jones (1987)
Riparian zone (pot)	$0.019-0.025$ μg N g^{-1} h^{-1}	Hill and Warwick (1987)
Decomposing wetland plants	$571-640$ μg N g^{-1} (dry wt) d^{-1} ($\equiv 17$ mM N m^{-2} d^{-1} for 6 months)	Howard-Williams *et al.* (1983)
Freshwater sediments		
Marsh	$2.8-3.4$ nM N cm^{-3} h^{-1}	Bowden (1986)
Oligotrophic lake (0−1 cm)	$10-52$ μg N g^{-1} h^{-1}	Vincent and Downes (1981)
	$0.1-0.7$ μg N g^{-1} h^{-1}	Hall (unpublished) cited in Hall (1986)
Mesotrophic lake	$0.7-3.1$ μg N g^{-1} h^{-1}	Hall (unpublished) cited in Hall (1986)
Eutrophic lake	$0-2.3$ μg N g^{-1} h^{-1}	Hall (unpublished) cited in Hall (1986)
Marine sediments		
Estuarine (nitrification/ denitrification)	$77-89$ μM N cm^{-3} h^{-1}	Jenkins and Kemp (1984)
Nearshore	$0.01-0.015$ μM N m^{-2} h^{-1}	Horrigan and Capone (1985)
Coastal (sandy)	0.003 μM N g^{-1} h^{-1}	Koike and Hattori (1978)
Coastal (muddy)	0.004 μM N g^{-1} h^{-1}	Koike and Hattori (1978)
Offshore (0−5 cm)	$0.3-1.4$ mM N m^{-2} d^{-1}	Henriksen *et al.* (1981)
	$2.4-4.0$ mM N m^{-2} h^{-1}	Billen (1976)
Salt marsh	$0.01-0.03$ μM N g^{-1} d^{-1}	Abd. Aziz and Nedwell (1986)

pot = potential rates. Units are those published in the original papers.

2 mg l^{-1} under the *Nostoc*). Nitrification potentials were $88-169$ ng N cm^{-3} h^{-1} in the top $0-3$ cm of the sediment and $10-70$ ng N cm^{-3} h^{-1} in the bottom $12-15$ cm. High nitrification potentials correlated with pore water content (probably greater oxygen diffusion), low Eh and, with organic carbon content and were one to three orders of magnitude higher than potential denitrification rates. Collos *et al* (1988) reported tight coupling between nitrification and decreases in microalgal biomass in a shallow coastal seawater pond at zero and sub-zero temperatures. Half the phytoplankton standing crop was degraded and transformed to nitrate within $2-3$ days without accumulation of ammonia or nitrite.

6.3.4 NITROUS OXIDE PRODUCTION

The gas nitrous oxide (N_2O) is produced as an intermediate during denitrification and as a by-product of ammonia oxidation. In the field, factors which control these microbial processes (e.g. oxygen concentration, moisture content, temperature and availability of substrates) will control N_2O emissions from wetlands. It is difficult to partition N_2O emissions between nitrification and denitrification as the number of aerobic and anaerobic sites is usually unknown and complex. Even in well-aerated soils there may be reduced 'microniches' where denitrification may take place (Tiedje *et al.*, 1984; Sexstone *et al.*, 1985).

Decreasing pH increases the mole fraction of N_2O produced during denitrification (Wijler and Delwich, 1954; Nommik, 1956) which suggests that N_2O may be a major product of this process in acidic wetland soils. Although under conditions of low pH and high nitrite concentrations chemo-denitrification may produce N_2O and other oxides of nitrogen, in most wetland systems nitrite does not usually accumulate due to its oxidation to nitrate by nitrifiers.

Very low oxygen concentrations may result in increased N_2O production by both nitrification and denitrification. In a marine sediment Jørgensen, Jensen and Sørensen (1984) found that N_2O production from nitrification reached an apparent maximum at 0.1 kPa O_2 where N_2O production accounted for 25% of the ammonia oxidised. In the same sediment the proportion of N_2O to $N_2O + N_2$ produced by denitrification reached a maximum of 50% at 5 kPa O_2. Except when oxygen was between 0.1 and 0.2, kPa denitrification was the main source of N_2O.

The presence of sulphide can inhibit the reduction of N_2O to N_2 by denitrifiers (Tam and Knowles, 1979; Sørensen, Tiedje and Firestone,, 1980) and it has been suggested that this mechanism could account for reported accumulations of N_2O in coastal marine sediments (Sørensen, 1978a,b). Production of N_2O is essentially a nitrogen-loss process from wetlands as further metabolism of this gas is only to N_2. N_2O is lost directly by diffusion to the atmosphere, by uptake into plants (Reddy, Patrick and Lindau, 1989; see also Figure 63) and by conversion to N_2. Data on specific rates of wetland N_2O production, separate from those of denitrification, are not readily available in the literature.

6.3.5 DISSIMILATION PATHWAYS

6.3.5.1 Dissimilation of nitrate to ammonia

The dissimilatory nitrate and nitrite reductases responsible for dissimilation of nitrate to ammonia differ from the assimilatory enzymes in that they are membrane bound and only function under anaerobic conditions.

Nitrate dissimilation to ammonia via nitrite is favoured by low redox and high organic carbon content at the expense of denitrification and hence may be a common feature of wetland habitats. It is thought that the reasons that ammonifiers do well under highly reduced conditions are:

(1) Ammonifiers may have constitutive enzymes for nitrate reduction whereas enzymes for denitrification may have to be induced. This could favour ammonifiers in situations such as wetlands where nitrate availability is highly variable.
(2) Ammonifiers have a more versatile metabolism and may grow under low-nitrate conditions.
(3) Where electron donors (e.g. organic carbon) are abundant dissipation of electrons may be a more urgent requirement than energy production. Dissimilation to ammonia is a highly efficient mechanism for this, accepting eight electrons as against five for denitrification (Tiedje et al., 1982).

One of the important consequences of dissimilatory nitrate reduction is that, unlike denitrification, the reduced nitrogen is retained within the ecosystem rather than being lost as a gaseous product. Published measurements of nitrate dissimilation to ammonia are sparse relative to those of other nitrogen-cycling processes, particularly in fresh waters. Priscu and Downes (1987) and Downes (1991) found that less than half the nitrate added to surficial eutrophic lake sediments was denitrified and attributed the balance to nitrate dissimilation to ammonia.

In a study of coastal marine sediments, Sørensen (1978a) found that the capacity for denitrification decreased rapidly with depth, whereas the capacity for reduction of nitrate to ammonia was significant even at sediment depths of 6−9 cm.

In intertidal sediments Kaspar (1983) found that dissimilation to ammonia accounted for 5−30% of the dissimilatory nitrate reduction with the rest being denitrified. Koike and Hattori (1978) considered that nitrate dissimilation to ammonia was an important process in coastal sediment with between 7% and 52% of nitrate reduction going to ammonia. They reported rates of nitrate dissimilation to ammonia between $0.1-4.62$ μM N g^{-1} h^{-1}.

6.3.5.2 Denitrification

Denitrification is a key nitrogen sink and has been studied in many types of wetland environment (Table 6.5). In stream systems oxygenated conditions in the stream water usually inhibit denitrification and the process is restricted to the stream sediments and riparian zones. Christensen et al. (1989) studied the microzonation of denitrification in dark incubated stream sediments of fine sand and silt using a combined oxygen/nitrous oxide microsensor. They found that oxygen penetrated about 1 mm into the sediment and that denitrification was restricted to the thin, anoxic layer immediately below the oxygen zone. The thickness of this denitrification zone was dependent upon the nitrate concentration in the water above the sediment.

In stream sediments the temporal and spatial patterns of denitrification activity may vary according to the source and concentration of nitrate and the nature of the sediment. In streams with organic-rich sediments and high nitrate loads, denitrification tends to be controlled by nitrate supply from the water phase. Christensen and Sørensen (1988) found a strong relationship between sediment denitrification rates and stream nitrate concentration in a small Danish lowland stream during summer. During this time the diffusive flux of nitrate from the overlying water was a limiting factor for the decreasing denitrification activity. Cooper and Cooke (1984) also reported concentration-dependent denitrification in a New Zealand stream which was rich in organic matter. In a small New Zealand headwater stream organic riparian soils comprised 12% of soils bordering the stream but accounted for 56−100% of nitrate depletion (Cooper, 1990). This was due to a combination of high denitrification activity in the soil and a high percentage (37−81%) of the groundwater entering the stream through these soils. Warwick and Hill (1988) found little nitrate removal in a riparian zone of a small woodland stream, despite a high denitrification enzyme activity. They concluded that this was due to the short residence time (<1 h) of nitrate in contact with the riparian soil (see Chapter 12).

In general, the dentrification capacity of organic riparian soils tends to be under-utilised with in-situ rates of denitrification being limited by the diffusion rate at low nitrate concentrations. Cooper (1990) found nitrate-dependent denitrification rates when nitrate concentrations were below about 300 μg N l^{-1}. Warwick and Hill (1988) found denitrification potentials were highly correlated with the initial nitrate concentration of substrate samples, implying that low nitrate levels in the riparian substrate may have been an important factor in controlling denitrification rates. Similar conclusions were reached by Schipper et al. (1989) for a riparian zone below a sewage land-treatment system. When nitrate supply is not limiting denitrification may be limited by organic carbon substrate. Hill and Sanmugadas (1985) found that potential daily nitrate loss rates in sediments from three Ontario streams were positively correlated with sediment organic carbon, total nitrogen, sediment ammonia and percentage silt + clay. However, the strongest correlation was with water-soluble carbon, which accounted for 67−79% of the variation in nitrate loss.

In streams with organic-poor sediments denitrification rates tend to be lower. Cooke and

Table 6.5 Denifrification rates in wetland systems

Wetland type	Denitrification	Reference
Streams		
Sediment	$0.08-0.12\ \mu M\ N\ g^{-1}\ h^{-1}$	Cooke and White (1987a)
Coarse sand/gravel	$0.2-0.3\ \mu M\ g^{-1}\ h^{-1}$	Cooke and White (1987b)
Fine sand	$0.05-0.75\ \mu M\ g^{-1}\ h^{-1}$	Cooke and White (1987b)
Sandy	$4\ mM\ N\ m^{-2}\ d^{-1}$	Christensen and Sørensen (1988)
Vegetated	$11\ mM\ N\ m^{-2}\ d^{-1}$	Christensen and Sørensen (1988)
	$34\ nM\ N\ cm^{-2}\ h^{-1}$	Christensen *et al.* (1989)
Silt	$0.5-1.2\ nM\ N\ cm^{-2}\ h^{-1}$	Christensen *et al.* (1989)
Stream	$1.3\ \mu g\ N\ g^{-1}\ h^{-1}$	Cooke and Cooper (1988)
Seepage zone	$6.48\ \mu g\ N\ g^{-1}\ h^{-1}$	Cooke and Cooper (1988)
Sediments (beneath cress)	$0.001\ \mu g\ N\ g^{-1}\ h^{-1}$	Cooper (1990)
Open water (coarse sand)	$<0.0002\ \mu g\ N\ g^{-1}\ h^{-1}$	Cooper (1990)
Periphyton mats (pot)	$91.6\ \mu M\ N\ m^{-2}\ h^{-1}$	Triska and Oremland (1981)
Light	$18.6\ \mu M\ N\ m^{-2}\ h^{-1}$	Duff *et al.* (1984)
Dark	$104.2\ \mu M\ m^{-2}\ h^{-1}$	Duff *et al.* (1984)
in situ	$1302\ \mu M\ N\ m^{-2}\ d^{-1}$	Duff *et al.* (1984)
Epilithon	$0.2-1.4\ mM\ N\ m^{-2}\ d^{-1}$	Sørensen *et al.* (1988)
Riparian soils		
	$0.5-0.7\ \mu g\ N\ g^{-1}\ h^{-1}$	Schipper *et al.* (in press)
	$0.5-3.19\ \mu g\ N\ g^{-1}\ h^{-1}$	Warwick and Hill (1988)
Mineral (pot)	$0.013-0.044\ \mu g\ N\ g^{-1}\ h^{-1}$	Cooper (1990)
Organic (pot)	$2.61\ \mu g\ N\ g^{-1}\ h^{-1}$	Cooper (1990)
Organic (*in situ*)	$0.0004-1.35\ \mu g\ N\ g^{-1}\ h^{-1}$	Cooper (1990)
Littoral sediments		
Vegetated, winter	$5\ \mu M\ N\ m^{-2}\ h^{-1}$	Christensen and Sørensen (1986)
Root zone, summer	$50\ \mu M\ N\ m^{-2}\ h^{-1}$	Christensen and Sørensen (1986)
Peat and low pH		
spodsols		
(pot)	$0.12-53.8\ \mu g\ N\ cm^{-3}\ d^{-1}$	Müller *et al.* (1980)
Fresh and saline ponds	$3.7\ mg\ N\ m^{-2}\ d^{-1}$	Smith and Delaune (1983)
Rice (*Oryza Sativa*)	$113\ mg\ N\ m^{-2}\ d^{-1}$	Reddy *et al.* (1989)
Salt marsh	$0.3-1.5\ \mu g\ N\ g^{-1}\ d^{-1}$	Kaplan *et al.* (1977)
Low marsh	up to $3\ mg\ N\ m^{-2}\ h^{-1}$	Valiela and Teal (1979)
High marsh	$1\ mg\ N\ m^{-2}\ h^{-1}$	Valiela and Teal (1979)
Pannes	$5\ mg\ N\ m^{-2}\ h^{-1}$	Valiela and Teal (1979)
Intertidal sediments	$1.6-2.4\ \mu M\ N\ m^{-2}\ h^{-1}$	Oremland *et al* (1984)
Juncus maritimus marsh	$1.2-11.0\ mg\ N\ m^{-2}\ d^{-1}$	Kaspar (1983)
Euglena and Oscillatoria	$1.6-7.9\ mg\ N\ m^{-2}\ d^{-1}$	Kaspar (1983)
Zostera novazelandica		
bed	$1.2-6.0\ mg\ N\ m^{-2}\ d^{-1}$	Kaspar (1983)
Aquifer soil (14–40 cm)	$17\ nM\ N\ g^{-1}\ d^{-1}$	Slater and Capone (1987)
Sediment (40 cm)	$0.14-2.8\ nM\ N\ g^{-1}\ d^{-1}$	Slater and Capone (1987)
Estuarine sediments		
Nit. dinit coupling	$77-89\ \mu M\ N\ m^{-2}\ h^{-1}$	Jenkins and Kemp (1984)
	$0.079-0.994\ ng\ N\ g^{-1}\ d^{-1}$	Smith and Delaune (1985)
Early spring	$5.1\ mM\ N\ m^{-2}\ d^{-1}$	Anderson *et al.* (1985)
Early summer	$0.3\ mM\ N\ m^{-2}\ d^{-1}$	Anderson *et al.* (1985)
Nearshore sediments	$100\ \mu M\ N\ m^{-2}\ h^{-1}$	Seitzinger *et al.* (1980)
Marine sediment	$1.57\ \mu M\ N\ cm^{-3}\ h^{-1}$	Goyens *et al.* (1987)

pot = potential rates. Units are those published in the original papers.

White (1987a) reported that the maximum activity found in a sandy gravel sediment core was 20% of the maximum activity in a predominantly silt sediment core. On the basis of Eh and nitrate profiles in their cores they estimated that denitrification took place at a depth below 7−10 cm in sandy gravel, 2−3.5 cm in fine sand and <2 cm in silt. Highest *in-situ* denitrification activity was associated with accumulations of fine-grained sediments at meander bends. There was a correlation between denitrification rate and the mineralisable carbon content of the sediment.

The growth of wetland vegetation in stream channels can enhance denitrifier activity through the deposition of organic matter. Christensen and Sørensen (1988) found that increasing denitrification activity in Rabis Bæk spring followed the growth pattern of *Batrachium peltatum*. Several studies (see Section 6.2.2 above) report significant release of oxygen from the roots of aquatic macrophytes. This raises the possibility of closely coupled nitrification− denitrification in the root zone as shown in Figure 6.2 and 6.3 (see also Reddy, Patrick and Lindau, 1989). Christensen and Sørensen (1988) found 25−50% depletion of the nitrate pool in the top 1−5 cm of vegetated sediment cores, treated with acetylene to inhibit nitrification, compared to untreated cores.

The presence of macrophytes can also impose seasonal and diurnal cycles on denitrification activity. Christensen and Sørensen (1986) reported a tenfold increase in denitrification between summer and winter in littoral lake sediments colonised by the perennial macrophyte *Littorella uniflora* (L.) Aschers. In winter, denitrification was restricted to the top 0−1 cm of debris, but by midsummer it could be measured throughout the root zone to a depth of 10 cm. Release of organic substrates from the roots was thought to be responsible for this high summer activity. There was also a diurnal variation in denitrification activity and it was suggested that inhibition of denitrification by oxygen, produced by benthic algae and the roots of *L. uniflora* during the day, controlled this diurnal variation. In the growth season there was evidence for a significant population of denitrifiers closely associated with the root surface.

Periphyton and epilithon biofilms have also been shown to sustain denitrification activity. Triska and Oremland (1981) showed that scrapings of decomposing *Cladophora* sp. mats covering stream-bed rocks exhibited denitrifying enzyme activity when incubated in the field in the dark. Denitrification was inhibited in the light due to algal photosynthetic oxygen production. The diurnal, periphyton-associated denitrification rate was estimated to be 91.6 μM N m^{-2} d^{-1} based on a 14 h light, 10 h dark day. Sørensen, Jørgensen and Brandt (1988) found light/dark effects on denitrifiation activity in undisrupted epilithic biofilm communities. Light-exposed epilithon denitrification rates were two to three times lower than under dark conditions. They also observed a 30−100% increase in denitrification activity in the dark when the water flow across the biofilm was reduced from 15 cm s^{-1} to 5 cm s^{-1}.

The effect of temperature on bacterial denitrification in a saltmarsh was studied by Kaplan, Teal and Valiela (1977). *In-situ* incubations suggested that the resident populations of denitrifying bacteria were not optimally adapted to their thermal environment but to temperatures 5−10°C higher, suggesting controls other than temperature on denitrification rates. Valiela and Teal (1979) found that low marsh and creek bottoms were salt marsh habitats where denitrifiers were most active and contributed most to losses of N by denitrification. In all habitats there was a strong seasonal pattern associated with the annual temperature cycles. However, the seasonality of denitrification may be linked to seasonal differences in organic production as well as temperature. A salt marsh at Sepelo Island, Georgia, showed a seasonal variation that appeared to be tightly coupled to the phases of productivity and exudation of photosynthate from the roots of cord grass, *Spartina alterniflora*, that covered most of the area (Sherr and Payne, 1978). When plots were cleared of the above-ground production parts of *S. alterniflora* and of roots below, the denitrifier population dropped significantly over a period of 18 months. Addition of organic material (sewage sludge, ground root extract or glucose) to the uprooted plots decreased the denitrification even more, whereas plots with undisturbed *spartina* cover maintained unaltered denitrifier populations. Additions of nitrate along with organic substrate prevented the decline of the denitrifier population and it was discovered that the roots of *S. alterniflora* exude oxygen which supported nitrification in the root zone.

6.3.5.3 Nitrifier denitrification at intermediate oxygen levels

Under low-oxygen conditions nitrifying bacteria may also produce nitrite reductase and reduce nitrite to nitrous oxide (Hooper, 1968; Ritchie and Nicholas, 1972). This process has been demonstrated in cultures of *Nitrosomonas europaea* (Hynes and Knowles, 1984; Poth and Focht, 1985). More recently, Poth (1986) has shown that nitrifiers can also reduce nitrous oxide to dinitrogen. Downes (1988) suggested that the high concentrations of nitrous oxide which accumulated in the hypolimnion of a mesotrophic New Zealand lake were produced by nitrifier denitrification of nitrite. The process took place at dissolved oxygen concentrations between 1.0 and 0.2 g mm^{-3} where oxidation of nitrite to nitrate was inhibited and nitrite could accumulate. When oxygen levels were higher than 1.0 g m^{-3}, nitrite was completely oxidised to nitrate and no nitrous oxide accumulated. When oxygen concentrations fell below 0.2 g m^{-3} complete reduction of all oxides of nitrogen to dinitrogen took place.

6.4 HYDROLOGICAL EFFECTS ON NITRATE CONCENTRATIONS

6.4.1 DETERMINATION OF A NUTRIENT BUDGET

Determination of a nutrient budget for a wetland, or quantification of wetland nutrient fluxes requires an understanding of wetland hydrology. Novitski (1978) gives an introduction to this subject and defined four wetland hydrological classes on the basis of a Wisconsin (USA) wetland classification:

(1) Surface water depressions — where precipitation and overland flow collect in a depression, and water leaves by leakage to the ground or by evaporation.
(2) Surface water slope wetlands — occur along the margins of lakes and streams.
(3) Groundwater depression wetlands — where a depression intercepts the water table and no surface drainage away from the site occurs.
(4) Groundwater slope wetlands — where groundwater discharges as springs or seeps at the land surface. Surface water flows downslope away from the site.

Groundwater plays a negligible role in the first two types. In type (4), however, groundwater may represent 50% or more of the water budget. Very few studies of wetland nutrient cycles take groundwater flows into account, but Valiela *et al.* (1978) found that the major source of nutrient nitrogen to a US tidal wetland was in the groundwater. This entered primarily as NO$_3$-N and DON. The input of nitrogen-poor groundwater to a wetland results in a dilution of the concentrations of N in the wetland, and hence Tilton and Kadlec (1979) cautioned against the reporting of percentage reductions in nutrients passing through wetlands unless the hydrology was well understood. Kadlec (1981), for instance, reported a 91% reduction in the concentrations of dissolved inorganic N passing through a wetland. However, considerations of the wetland hydrology showed that 58% of the reduction was due to dilution alone. The influence of dilution, evapotranspiration (which has a nutrient concentrating effect), floods and rates of throughflow of water on wetland nutrient pathways are discussed in Howard-Williams (1985).

Increasing rates of throughflow of water may reduce the ability of a wetland to process nitrogen (Hill, 1988) and may also alter the type of nitrogen transformations that occur. For instance, Cooper and Cooke (1984) showed that the concentration of nitrate in a *Glyceria fluitans*-dominated stream fell from 2000 to 900 mg m^{-3} at a stream discharge of 0.95 l s^{-1} and from 2000 to 200 mg m^{-3} when the discharge was 0.35 l s^{-1}. Independent modelling of nitrate in a spring-fed watercress-dominated ecosystem showed a similar effect (Hearne and Howard-Williams, 1989). Hill (1988) showed a negative relationship of nitrate removal versus stream discharge (Figure 6.6). However, this relationship varied over different reaches of the same stream, showing that factors other than discharge were also involved.

Hill's analyses precluded differences in temperature and nitrate concentration as being important in the two reaches shown in Figure 6.6. However, he noted that the residence time of the water in reach B for a given discharge was three times that of reach A.

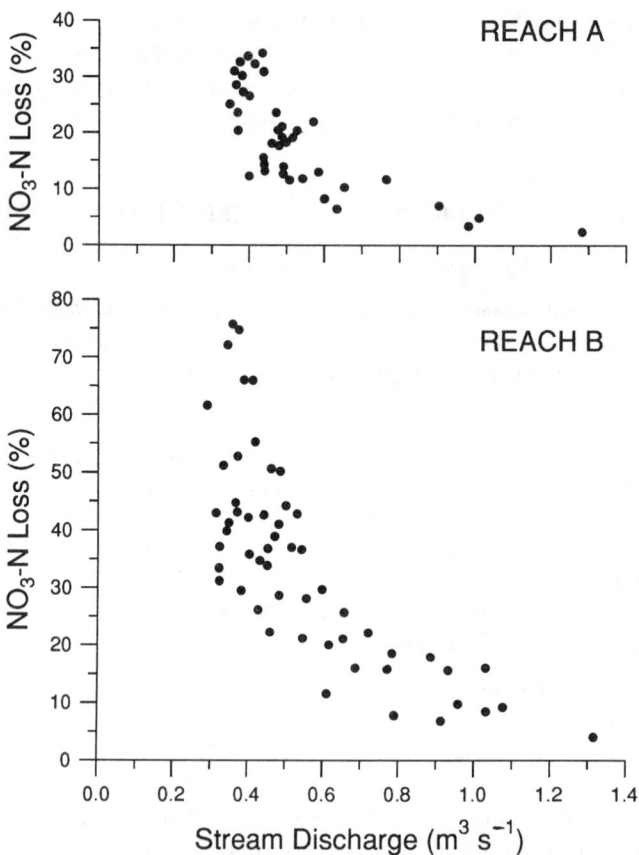

Figure 6.6 Nitrate removal from a stream as a function of discharge. Note that removal rates for a given discharge vary between stream reaches (after Hill, 1988)

The residence time of water in a stream reach will affect the rate for nutrient processing (see e.g. Elwood *et al.*, 1983; Howard-Williams, 1985) and will, in turn, directly influence the removal of nutrients such as nitrate. Long residence times lead to greater processing and removal of dissolved nutrient. In wetlands, physical barriers to water flow set up by aquatic vegetation will greatly increase the path length that a parcel of water travels per unit distance downstream, and hence increase residence time. Similarly, wetland flow will incorporate a large 'dead-zone' component. Dead zones are regions of stagnant water which occur along the line of flow such as backwaters, holes in the bed, etc. These have the effect of increasing dispersion and decreasing average velocity, both of which will facilitate biological uptake mechanisms.

Increasing average water velocities through a wetland may have two effects on nitrate: biological removal over a given downstream distance will decrease (see e.g. Figure 6.6) and, conversely, in some conditions nitrification may be favoured. thus when increasing current velocities result in increased oxygenation of wetland waters a change in the nitrogen species in wetland waters can be expected. Gaudet (1979), for instance, found that increasing dissolved oxygen levels associated with increased throughflow resulted in the reduced forms of N (NH_3, DON) declining in a tropical swamp, and a compensating increase in NO_3-N occurred.

6.5 APPLIED ISSUES

6.5.1 NITRATE POLLUTION AND EUTROPHICATION OF WETLANDS

Nitrate pollution of wetlands is often difficult to separate from the general trends in eutrophication involving the combined effects of both N and P. A major study of wetland eutrophication has been that on the Norfolk Broads (Moss, Leah and Clough, 1979; Phillips, Eminson and Moss, 1978). Increasing eutrophication in the nineteenth century resulted in elevated nutrient concentrations in the Broads and the subsequent conspicuous development of filamentous algae. A consistent feature of enriched wetlands appears to be a proliferation of filamentous green algae, usually at the expense of other more desirable aquatic plants (see review in Howard-Williams, 1985). Duckweeds (*Lemna* spp., *Spirodela* spp.) are also able to take advantage of increased wetland nutrient pollution. In tropical Waigani swamp, Papua New Guinea, nutrient enrichment caused a reduction in the wetland communities resulting in increased areas of open water (Osborne and Leach, 1983).

Nutrient enrichment of salt marshes has been shown in several studies to increase primary production of salt marsh vegetation and in one case (Patrick and DeLaune, 1976) it was found that this increase in production was due to additions of nitrogen rather than phosphorus. *Typha* spp. is one genus that appears to respond well to nutrient additions. In studies in the USA (see review in Howard-Williams, 1985) and in New Zealand (Cooke, Cooper and Clunie, 1990) *Typha orientalis* has replaced other species in similar conditions following nutrient additions to the wetlands.

Inorganic nitrogen additions to wetlands have a variable effect on the microbial communities. Nitrogen fixation may be inhibited (van Raalte *et al.*, 1974) or stimulated (see Section 6.3.1.1. above), depending on the type of wetland. Cooke and White (1988) showed that additions of nitrate to sediment in water microcosms simulating lowland English streams resulted in an enhanced loss of ammonium-N from the sediment. This was because nitrate diffusion and denitrification progressively consumed organic matter through the sediment profile. This allowed oxygen to diffuse further into the sediment, increasing the depth of nitrification in the sediment and hence' a greater transformation of ammonium-N to nitrate.

The addition of inorganic nitrogen to wetlands appears to alter the species composition of wetlands and the relative importance of different aspects of the nitrogen cycle. However, wetlands are not destroyed by the eutrophication process as they are by drainage.

6.5.2 WETLAND USE FOR NITRATE REMOVAL

The use of artificial and natural wetlands for wastewater purification has been a subject of considerable international interest over the last decade or so (e.g. Howard-Williams, 1985; Fetter, Sloey and Spangler, 1978; Tilton and Kadlec, 1979; Roser et al., 1987; Finlayson and Chick, 1983; Finlayson et al., 1986).

In spite of the extensive literature on nutrient removal by wetlands it is difficult to find studies where the following criteria were met:

(1) The hydrological balance was measured so that mass flows (loads) of nitrogen can be calculated (see Section 6.4 above);
(2) There were suitable controls to ensure that the processes observed were distinctly wetland processes and not merely a simple filter effect;
(3) Total-N budgets were calculated so that, for instance, nitrogen transformations could be examined. For example, measurement of the mass flows in a Glyceria-dominated stream by Cooper and Cooke (1984) illustrated removal rates of up to 66.5 g NO_3-N h^{-1}. Analysis of the complete nitrogen budget showed that these removals were balanced by exports of particiulate N and dissolved organic N.

Table 6.6 details four studies on artificial wetlands which met these criteria. The proportion of total N in the inflow removed by the wetlands was between 6% and 74%. In two of these

Table 6.6 Removal of nitrogen (as total N) by artificial wetland systems which meet specified criteria (see text)

Reference	Description	Wetland type or species	N removal (%)
Finlayson and Chick (1983)	Abbatoir effluent, low in oxidised N (66% as PN)	Typha Phragmites Scirpus	42 62 74
Gersberg et al. (1982)	Clarified, nitrified secondary sewage effluent (96% as NO_3-N)	Typha Scirpus	6 13
Gersberg et al. (1983)	As above	Gravel bed only Typha Phragmites Scirpus	25 27 27 27
Bavor (1986)	Secondary sewage effluent, 80% as NH_4-N, 2% as NO_3-N	Water only Gravel bed only Myrophyllum Typha Schoenoplectus in gravel Mixed Typha gravel, water	41 49 43 51 51 64

Table 6.7 Percentage removal of nitrogen from secondarily nitrified sewage effluent by artificial wetlands with and without organic additions

Type	$(NO_3 + NO_2)$	Total inorganic nitrogen	Total nitrogen
Unamended wetlands	6 (± 7.0)	8 (± 7.0)	8 (± 7.4)
Wetlands plus plant mulch	76 (± 19)	77 (± 19)	70 (± 17)
Wetlands plus methanol	99 (± 2)	97 (± 4)	94 (± 3)

Data from Gersburg *et al* (1982). Data as means ($\pm$ SD).

wetlands it was recognised that most of the N removal was due to the filtering of particulate nitrogen. These studies also showed that a very small proportion of the N removed from the effluent was absorbed by the wetland vegetation. The only real long-term loss of nitrogen was via denitrification or sedimentation of particulate matter.

Thus Gersberg *et al.* (1982, 1983) in two separate wetland experiments found that by enhancing denitrification rates with labile organics they achieved very high removal efficiencies of up to 94% of total N (Table 6.7). However, the waste water that was fed into these wetlands was secondarily treated, clarified, and nitrified so that 96% of the total N was present as NO_3-N. The main process limiting inorganic N transformations and finally gaseous loss from waste water effluent applied to wetlands appears to be nitrification (Bowmer, 1987) and nitrifier populations there are generally slow to develop (Howard-Williams, Pickmere and Davies, 1983; Roser *et al.*, 1987).

Removal of nitrate by wetland vegetation bordering streams is an effective process, particularly where the NO_3-N enters the streams as a diffuse pollutant from agricultural catchments (Hoare, 1979; Howard-Williams, Pickmere and Davies, 1983; Hill, 1988; Cooper, 1990; Schipper, Cooper and Dyck, in press). In most cases, stream bank wetlands are the only feasible mechanism for diffuse NO_3-N removal.

6.6 CONCLUSIONS

The wetland nitrogen cycle is controlled by the sharp gradients in dissolved oxygen that occur spatially, both on a horizontal and vertical scale and in time. High production rates by wetland vegetation result in an abundance of carbon providing an organic substrate for bacterial processes. The capacity of wetland plants to transport oxygen into anaerobic wetland sediments results in a mosaic of aerobic—anaerobic conditions. This mosaic has been shown to enhance denitrification in its various forms, leading to losses of nitrogen as N_2O or N_2 from wetland sediments. These losses may be facilitated by transport upwards through wetland vegetation. The nitrification step has been identified as the rate-limiting step in the N-cycle in several wetland studies. Hence, wetlands are most effective at removing dissolved nitrogen compounds from through-flowing waters when the nitrogen is in the form of nitrate rather than ammonium-N or dissolved organic N. In areas where diffuse runoff of nutrients is seen as a long-term problem (e.g. agricultural (see Chapters 3 and 11) or forestry (see Chapter 4) catchments) wetlands may have a significant role to play as nitrogen sinks.

ACKNOWLEDGEMENTS

Thanks to Dr Eddie White and Dr Tony Viner for review and to Janet Simmiss for typing the many drafts and Jan Symes for final manuscript preparation.

REFERENCES

Abd. Aziz, S.A. and Nedwell, B.D. (1986) The nitrogen cycle of an east coast, U.K. saltmarsh: II. Nitrogen fixation, Nitrification and Denitrification, Tidal exchange. *Estuarine, Coastal and Shelf Science*, **22**, 689–704.

Andersen, T.K., Jensen, M.H. and Sørensen, J. (1984) Diurnal variation of nitrogen cycling in coastal, marine sediments. I. Denitrification. *Mar. Biol.*, **83**, 171–6.

apRees, T., Jenkin, L.E.T., Smith, A.M. and Wilson, P.M. (1987) The metabolism of flood-tolerant plants. In Lees, J.A., McNiell, S. and Rorison, I.M. (eds), *Nitrogen an Ecological Factor*, Blackwell Scientific Publications, Oxford, pp. 167–88.

Armento, T.V. and Menges, E.S. (1986) Patterns of change in the carbon balance of organic soil-wetlands of the temperate zone. *J. Ecol.*, **74**, 755–74.

Armstrong, W. (1964) Oxygen diffusion from the roots of British bog plants. *Nature (Lond.).*, **204**, 801–2.

Aselman, I. and Creutzen, P.J. (1989) Global distribution of natural freshwater wetlands and rice paddies, their net primary productivity, seasonality and possible methane emissions. *J. Atmos. Chem.*, **8**, 307–58.

Basilier, K. (1980) Fixation and uptake of dinitrogen in *Sphagnum* blue-green algal association. *Oikos*, **34**, 239–42.

Bavor, H.J. (ed.) (1986) *Joint Study on Nutrient Removal Using Shallow Lagoon — Macrophyte Systems*, Interim Report, March 1984–May 1985, Water Research Laboratory, Hawkesbury Agricultural College, Australia.

Beevers, L. (1976) *Nitrogen metabolism in Plants*, Edward Arnold, London.

Billen, G. (1976) Evaluation of nitrifying activity in sediments by dark ^{14}C-bicarbonate incorporation. *Wat. Res.*, **10**, 51–7.

Bowden, W.B. (1986) Nitrification, nitrate reduction and nitrogen immobilisation in a tidal freshwater marsh sediment. *Ecology*, **67**, 88–99.

Bowmer, K.H. (1987) Nutrient removal from effluents by an artificial wetland: influence of rhizosphere aeration and preferential flow studied using bromide and dye tracers. *Wat. Res.*, **21**, 591–9.

Boyd, C.E. (1969) Production, mineral nutrition, absorption and biochemical assimilation by *Justicia americana* and *Alternanthera philoxeroides*. *Arch. Hydrobiol.*, **66**, 139–60.

Boyd, C.E. (1970) Production, mineral accumualtion and pigment concentrations in *Typha latifolia* and Scirpus americanus. Ecology, **51**, 285–90.

Boyd, C.E. and Hess, L.W. (1970) Factors influencing shoot production and mineral nutrient levels in *Typha latifolia*. *Ecology*, **51**, 296–300.

Buresch, R.J., Casselman, M.E. and Patrick, W.H. (1980) Nitrogen fixation in flooded soil systems, a review. *Adv. Agronomy*, **33**, 149–92.

Chatarpaul, L., Robinson, J.B. and Kaushik, N.K. (1980) Effects of tubificid worms on denitrification and nitrification in stream sediment. *Can. J. Fish. Aquat. Sci.*, **37**, 656–63.

Christensen, P.B. and Sørensen, J. (1986) Temporal variation in denitrification activity in plant-covered littoral sediment from Lake Hampden, Denmark. *Appl. Environ. Microbiol.*, **51**, 1174–9.

Christensen, P.B. and Sørensen, J. (1988) Denitrification in sediment of lowland streams: Regional and seasonal variation in Gelbæk and Rabis Bæk, Denmark. *FEMS*, **53**, 335–44.

Christensen, P.B., Nielsen, L.P., Revsbech, N.P. and Sørensen, J. (1989) Microzonation of denitrification activity in stream sediments as studied with a combined oxygen and nitrous oxide microsensor. *Appl. Environ. Microbiol.*, **55**, 1234–41.

Collos, Y., Linley, E.A.S., Frikha, M.G. and Ravail, B. (1988) Phytoplankton death and nitrification at low temperatures. *Est. Coast. and Shelf Sci.*, **27**, 341–7.

Cooke, J.G. and Cooper, A.B. (1988) Sources and sinks of nutrients in a New Zealand hill pasture catchment III. Nitrogen. *Hydrological Processes*, **2**, 135–49.

Cooke, J.G. and White, R.E. (1987a) Spatial distribution of denitrifying activity in a stream draining an agricultural catchment. *Freshwat. Biol.*, **18**, 509–19.

Cooke, J.G. and White, R.E. (1987b) The effect of nitrate in stream water on the relationship between denitrification and nitrification in a stream-sediment microcosm. *Freshwat. Biol.*, **18**, 213–26.

Cooke, J.B. and White, R.C. (1988) Nitrate enhancement of nitrification depth in sediment/water microcosms. *Environ. Geol. Wat. Sci.*, **11**, 85–94.

Cooke, J.G., Cooper, A.B. and Clunie, N.M.U. (1990) Changes in the water, soil and vegetation of a wetland after a decade of receiving sewage effluent, *NZ J. Ecol.*, **14**, 37–47.

Cooper, A.B. (1984) Activities of benthic nitrifiers in streams and their role in oxygen consumption. *Microbiol. Ecol.*, **10**, 317–34.

Cooper, A.B. (1990) Nitrate depletion in the riparian zone and stream channel of a small headwater catchment. *Hydrobiologia*, **202**, 13–26.

Cooper, A.B. and Cooke, J.G. (1984) Nitrate loss and transformation in two vegetated headwater streams. *NZ J. Mar. Freshwat. Res.*, **18**, 441–50.

Cowardin, L.M., Carter, V., Golet, F.C. and La Roe, E.T. (1979) *Classification of Wetlands and Deepwater Habitats in the United States*, US Dept of the Interior, Fish and Wildlife Service Report FWS/OBS — 79/31.

Davis, C.B. and van der Valk, A.G. (1983) Uptake and release of nutrients by living and decomposing *Typha glauca* Godr. tissues at Eagle Lake, Iowa. *Aquat. Bot.*, **16**, 75–89.

DeLaune, R.D. and Patrick, W.H. (1980) Nitrogen and phosphorus cycling in a Gulf Coast salt marsh. in Kennedy, V.S. (ed.), *Estuarine Perspectives*, Academic Press, New York, pp. 143–51.

DeLaune, R.D. and Patrick, W.H. (1990) Nitrogen cycling in Louisiana Gulf Coast brackish marshes. *Hydrobiologia*, **199**, 73–9.

Denny, P. (1987) Mineral cycling by wetland plants — a review. *Arch. Hydrobiol. Beihefte*, **27**, 1–25.

Dicker, H.J. and Smith, D.W. (1980) Acetylene reduction (nitrogen fixation) in a Delaware, USA, salt marsh. *Mar. Biol.*, **57**, 241–50.

Dierberg, F.E. and Brezonik, P.L. (1981) Nitrogen fixation (acetylene reduction) associated with decaying leaves of Pond Cypress (*Taxodium distinctum* var. nutans) in a natural and sewage enriched Cypress dome. *Appl. Environ. Microbiol.*, **41**, 1413–18.

Dodds, W.K. and Jones, R.D. (1987) Potential rates of nitrification and denitrification in an oligotrophic freshwater sediment system. *Microb. Ecol.*, **14**, 91–100.

Downes, M.T. (1988) Aquatic nitrogen transformations at low oxygen concentrations. *Appl. environ. Microbiol.*, **54**, 172–5.

Downes, M.T. (1991) Production and consumption of nitrate in an eutrophic lake during early stratification. *Arch. Hydrobiol.*, **122**, 257–74.

Duff, J.H., Triska, F.J. and Oremland, R.S. (1984) Denitrification associated with stream periphyton: Chamber estimates from undisrupted communities. *J. Environ. Qual.*, **13**, 514–18.

Dunbabin, J.S., Pokorny, J. and Bowmer, K.M. (1988) Rhizosphere oxygenation by *Typha domingensis* Pers. in miniature artificial wetland filters used for metal removal from wastewaters. *Aquat. Bot.*, **29**, 303–17.

Elwood, J.W., Newbold, J.D., O'Neill, R.V. and van Winkle, W. (1983) Resource spiralling: 1 an operational paradigm for analysing lotic ecosystems. In Fontaine, T.D. and Bartell, S.M. (eds), *Dynamics of Lotic Ecosystems*, Ann Arbor Science Publishers, Ann Arbor, pp. 3–27.

Fetter, C.W., Sloey, W.E. and Spangler, F.L. (1978) Use of a natural marsh for wastewater polishing. *J. Wat. Poll. Contr. Fed.*, **50**, 290–307.

Finlayson, C.M. and Chick, A.J. (1983) Testing the potential of aquatic plants to treat abbatoir effluent. *Wat. Res.*, **17**, 415–22.

Finlayson, M., Cullen, P., Mitchell, D. and Chick, A. (1986) An assessment of a natural wetland receiving sewage effluent. *Aust. J. Ecol.*, **11**, 33–47.

Gaudet, J.J. (1977) Uptake accumulation and loss of nutrients by papyrus in tropical swamps. *Ecology*, **58**, 415–22.

Gaudet, J.J. (1979) Seasonal changes in nutrients in a tropical swamp: North Swamp, Lake Naivasha, Kenya. *J. Ecol.*, **67**, 953–81.

Gaynard, T.J. and Armstrong, W. (1977) Some aspects of internal plant aeration in amphibious habitats. In Crawford, R.M.M. (ed.), *Plant Life in Aquatic and Amphibious Habitats*, Blackwell Scientific Publications, Oxford, pp. 303–20.

Gersberg, R.M. Elkins, B.V. and Goldman, C.R. (1983) Nitrogen removal in artificial wetlands. *Wat. Res.*, **18**, 1009–14.

Gersberg, R.M., Goldman, C.R. and Elkins, B.V. (1982) The use of artifical wetlands to remove nitrogen compounds from wastewater. *Aqua*, **6**, 492–6.

Goreau, T.J., Kaplan, W.A., Wofsy, S.J., McElroy, M.B., Valoise, F.W. and Watson, S.W. (1980) Production of NO_2^- and N_2O by nitrifying bacteria at reduced concentrations of oxygen. *Appl. Environ. Microbiol.*, **40**, 526–32.

Gorham, E. (1957) The development of peatlands. *Quart. Rev. Biol.*, **32**, 145–66.

Goyens, L., De Vries, R.T.P., Bakker, J.F. and Helder, W. (1987) An experiment on the relative importance of denitrification, nitrate reduction and ammonification in coastal marine sediment. *Neth. J. Sea Res.*, **21**, 171–5.

Hall, G.H. (1986) Nitrification in lakes. In Prosser, J.I. (ed.), *Nitrification*, Spec. Pub. Soc. Gen. Microbiol., **20**, 127–56, IRL Press, Oxford.

Hanson, R.B. (1977) Nitrogen fixation (acetylene reduction) in a salt marsh amended with sewage sludge and organic carbon and nitrogen compounds. *Appl. Environ. Microbiol.*, **33**, 846–52.

Hanson, R.B. (1983) Nitrogen fixation activity (acetylene reduction) in the rhizosphere of saltmarsh angiosperms, Georgia, USA. *Botanica Marina*, **26**, 49–59.

Hearne, J. and Howard-Williams, C. (1988) Modelling nitrate removal by riparian vegetation in a springfed stream: the influence of land-use practices. *Eco. Modelling*, **42**, 179–98.

Hejny, S. (1971) The dynamic characteristics of littoral vegetation with respect to changes in water level. *Hidrobiologia* (Bucuresti), **12**, 71–86.

Hemond, H.F. (1983) The nitrogen budget of Thoreaus bog. *Ecology*, **64**, 99–109.

Henriksen, K., Hansen, J.I. and Blackburn, J.H. (1981) Rates of nitrification, distribution of nitrifying bacteria, and nitrate fluxes in different types of sediment from Danish waters. *Mar. Biol.*, **61**, 299–304.

Hill, A.R. (1988) Factors influencing nitrate depletion in a rural stream. *Hydrobiologia*, **160**, 111–22.

Hill, A.R. and Sanmugadas, K. (1985) Denitrification rates in relation to stream sediment characteristics. *Wat. Res.*, **19**, 1579–86.

Hoare, R.A. (1979) Nitrate removal from streams draining experimental catchments. *Prog. Wat. Tech.*, **11**, 303–14.

Hobbie, J. (ed.) (1980) *Limnology of Tundra Ponds*, US/IBP Synthesis series 13, Dowden, Hutchinson and Ross Inc., Stroudsburg, Penn.

Hooper, A.B. (1968) A nitrite reducing enzyme from *Nitrosomonas europaea*: Preliminary characteristation with hydroxylamine as electron donor. *Biochim. Biophys. Acta*, **51**, 12–20.

Hopkinson, C.S. and Schubauer, J.P. (1984) Static and dynamic aspects of nitrogen cycling in the salt marsh graminoid *Spartina alterniflora*. *Ecology*, **65**, 961–9.

Horrigan, S.G. and Capone, D.G. (1985) Rates of nitrification and nitrate reduction in nearshore marine sediments at near ambient substrate concentrations. *Mar. Chem.*, **16**, 317–27.

Howard-Williams, C. (1985) Cycling and retention of nitrogen and phosphorus in wetlands: A theoretical and applied perspective. *Freshwat. Biol.*, **15**, 391–431.

Howard-Williams, C. and Junk, W.J. (1977) The chemical composition of central Amazonian aquatic macrophytes with special reference to their role in the ecosystem. *Arch. Hydrobiol.*, **79**, 446–64.

Howard-Williams, C., Davies, J. and Pickmere, S. (1982) The dynamics of growth, the effects of changing area and nitrate uptake by watercress *Nasturtium officinale* R.Br. in a New Zealand stream. *J. Appl. Ecol.*, **19**, 589–601.

Howard-Williams, C., Pickmere, S., and Davies, J. (1983) Decay rates and nitrogen dynamics of decomposing watercress *Nasturtium officinale* R.Br. *Hydrobiologia*, **99**, 207–14.

Howard-Williams, C. and Thompson, K. (1985) The conservation and management of African wetlands. In Denny, P. (ed.), *The Ecology and Management of African Wetland Vegetation*, W. Junk, Dordrecht, pp. 203–30.

Howard-Williams, C., de Esteves, F., Santos, J.E. and Downes, M.T. (1989) Short term nitrogen dynamics in a small Brazilian wetland (Lago Infernão, São Paulo). *J. Trop. Ecol.*, **5**, 323–35.

Hynes, R.K. and Knowles, R. (1984) Production of nitrous oxide by *Nitrosomonas europaea*: effect of acetylene, pH and oxygen. *Can. J. Microbiol.*, **30**, 1397–1404.

Jenkins, M.C. and Kemp, W.M. (1984) The coupling of nitrification and denitrification in two estuarine sediments. *Limnol. Oceanogr.*, **29**, 609–19.

Jones, G. and Simon, B.M. (1981) Differences in microbial decomposition processes in profundal and littoral lake sediments with particular reference to the nitrogen cycle. *J. Gen. Microbiol.*, **123**, 297–312.

Jørgensen, K.S., Jensen, H.B. and Sørensen, J. (1984) Nitrous oxide production from nitrification and denitrification in marine sediment at low oxygen concentration. *Can. J. Microbiol.*, **30**, 1073–8.

Junk, W.J. (1983) Ecology of swamps on the middle Amazon. In Gore, A.J.P. (ed.), *Mires: Swamp, Bog, Fen and Moor. Ecosystems of the World*, Vol. 4B, Elsevier, Amsterdam, pp. 269–94.

Kadlec, R.H. (1981) *Monitoring report on the Bellaire Wastewater Treatment Facility*, Wetlands Ecosystem Research Group, University of Michigan, Ann Arbor.

Kadlec, R.H. and Kadlec, J.A. (1978) Wetlands and water quality. In Greeson, P.E., Clark, J.R. and Clark, J.E. (eds), *Wetland Functions and Values: the state of our understanding*, American Water Resources Association, Minneapolis, pp. 346–456.

Kaplan, W.A., Teal, J.M. and Valiela, I. (1977) Denitrification in salt marsh sediments; Evidence for seasonal temperature selection among populations of denitrifiers. *Microb. Ecol.*, **3**, 193–204.

Kaspar, H.F. (1983) Denitrification, nitrate reduction to ammonium, and inorganic nitrogen pools in intertidal sediments. *Mar. Biol.*, **74**, 133–9.

Koike, I. and Hattori, A. (1978) Denitrification and ammonia formation in anaerobic coastal sediments. *Appl. Environ. Microbiol.*, **35**, 278–82.

McLachlan, S.M. (1970) The influence of lake level fluctuation and the thermocline on water chemistry in two gradually shelving areas in Lake Kariba, Central Africa. *Arch. Hydrobiol.*, **66**, 499–510.

Madsen, J.D. (1991) Resource allocation at the individual plant level. *Aquatic Botany*, **41**, 67–86.

Mason, C.F. and Bryant, R.J. (1975) Production, nutrient content and decomposition of *Phragmites communis* Trin, and *Typha angustifolia* L. *J. Ecol.*, **63**, 71–95.

Melack, J.M. and Fisher, T.R. (1983) Diel oxygen variations and their ecological implications in Amazon floodplain lakes. *Arch. Hydrobiol.*, **98**, 422–42.

Melack, J.M. and Fisher, T.R. (1988) Denitrification and nitrogen fixation in an Amazon floodplain lake. *Verh. Internat. verein. Limnol.*, **23**, 2232–6.

Moss, B., Leah, R.T. and Clough, B. (1979) Problems of the Norfolk Broads and their impact on freshwater fisheries. *Proceedings of the First British Freshwater Fisheries Conference, University of Liverpool*, pp. 67–85.

Müller, M.M., Sundman, V. and Skugins, J. (1980) Denitrification in low pH spodosols and peats determined with acetylene inhibition method. *Appl. Environ. Microbiol.*, **40**, 235–9.

Nommik, H. (1956) Investigations of denitrification in soil. *Acta Agriculturæ Scandinavica*, **6**, 195–228.

Novitski, R.P. (1978) Hydrologic characteristics of Wisconsin's wetlands and their influence on floods, streamflow and sediment. In Greeson, P.E., Clark, J.R. and Clark, J.E. (eds), *Wetland Functions and Values: the state of our understanding*, American Water Resources Association, Minneapolis, pp. 377–88.

Oremland, R.W., Umberger, C., Culbertson, C.W. and Smith, R.L. (1984) Denitrification in San Francisco Bay intertidal sediments. *Appl. Environ. Microbiol.*, **47**, 1106–12.

Osborne, P.L. and Leach, G.J. (1983) Changes in the distribution of aquatic plants in a tropical swamp. *Environ. Conserv.*, **10**, 323–9.

Patrick, W.J. and DeLaune, R.D. (1976) Nitrogen and phosphorus utilisation by *Spartina alterniflora* in a salt marsh in Batararia Bay, Louisiana. *Estuar. Coast. Mar. Sci.*, **4**, 59–64.

Phillips, G.L., Eminson, D. and Moss, B. (1978) A mechanism to account for macrophyte decline in progressively eutrophicated freshwaters. *Aquat. Bot.*, **4**, 103–26.

Poth, M. (1986) Dinitrogen production from nitrite by a *Nitrosomonas* isolate. *Appl. Environ. Microbiol.*, **52**, 957–9.

Poth, M. and Focht, D.D. (1985) ^{15}N kinetic analysis of N_2O production by *Nitrosomonas europaea*: an examination of nitrifier denitrification. *Appl. Environ. Microbiol.*, **49**, 1134–41.

Priscu, J.C. and Downes, M.T. (1987) Microbial activity in surficial lake sediments of an oligotrophic and eutrophic lake, with particular reference to dissimilatory nitrate reduction. *Arch. Hydrobiol.*, **108**, 385–409.

Reddy, K.R., Patrick, W.H., Jr and Lindau, C.W. (1989) Nitrification–denitrification at the plant root sediment interface in wetlands. *Limnol. Oceanogr.*, **34**, 1004–13.

Riemer, D.M. and Toth, S.J. (1968) A survey of the chemical composition of aquatic plants in New Jersey. *New Jersey Agricultural Experimental Station, Rutgers University, Bull.* 820.

Richie, G.A.F. and Nicolas, D.J.D. (1972) Identification of the sources of nitrous oxide produced by oxidative and reductive processes in *Nitrosomonas europaea*. *Biochem. J.*, **126**, 1181–91.

Rose, D.J., McKersie, S.A., Fisher, P.J., Breen, P.F. and Bavor, H.J. (1987) Sewage treatment using aquatic plants and artifical wetlands. *Water*, September, 20–24.

Rother, J.A., Aziz, A., Hye Karim, N., and Whitton, B.A. (1988) Ecology of deep water rice fields in Bangladesh 4. Nitrogen fixation by blue-green algal communities. *Hydrobiologia*, **169**, 43–56.

Schimel, J.P., Firestone, M.H. and Kilham, K. (1984) Identification of heterotrophic nitrification in a Sierran forest soil. *Appl. Environ. Microbiol.*, **48**, 802–6.

Schipper, L.A., Cooper, A.B. and Dyck, W.J. (in press) Mitigating non-point source nitrate pollution by riparian zone denitrification. In Bognundi, I. and Kuzelka, R.D. (eds), *Nitrate Contaminaton: Exposure, consequences and control*, NATO Advanced Research Workshop, Nebraska, USA. Springer-Verlag, Berlin, Germany.

Schipper, L.A., Dyck, W.J., Barton, P.N. and Hodgkiss, P.D. (1989) Nitrogen renovation by denitrification in forest sewage land treatment systems. *Biological Wastes*, **29**, 181–7.

Seitzinger, S., Nixon, S., Pilson, M.E.Q. and Burk, S. (1980) Denitrification and N_2O production in nearshore marine sediments. *Geochem. Cosmochem. Acta*, **44**, 1853–60.

Sextone, A.J., Revsbech, N.P., Parkin, T.B. and Tiedje, J.M. (1985) Direct measurement of oxygen profiles and denitrification rates in soil aggregates. *Soil Sci. Soc. Amer. J.*, **49**, 645–51.

Sherr, B.F. and Payne, W.F. (1978) Effect of the *Spartina alterniflora* root rhizome system on salt marsh soil denitrifying bacteria. *Appl. Environ. Microbiol.*, **35**, 724–9.

Sherr, B.F. and Payne, W.F. (1981) The effect of sewage sludge on salt-marsh denitrifying bacteria. *Estuaries*, **4**, 146–9.

Slater, J.M. and Capone, D.G. (1987) Denitrification in aquifer soil and nearshore marine sediments influenced by groundwater nitrate. *Appl. Environ. Microbiol.*, **53**, 1292–7.

Smith, C.J. and DeLaune, R.D. (1983) Nitrogen loss from freshwater and saline estuarine sediments. *J. Environ. Qual.*, **12**, 514–18.

Sørensen, J. (1978a) Capacity for denitrification and reduction of nitrate to ammonia in a coastal sediment. *Appl. Environ. Microbiol.*, **35**, 301–5.

Sørensen, J. (1978b) Occurrence of nitric and nitrous oxides in a coastal marine sediment. *Appl. Environ. Microbiol.*, **36**, 807–13.

Sørensen, J., Jørgensen, T. and Brandt, S. (1988) Denitrification in stream epilithon: Season variation, in Gelbæk and Rabis Bæk, Denmark. *FEMS*, **53**, 345–54.

Sørensen, J., Tiedje, J.M. and Firestone, R.B. (1980) Inhibition by sulphide of nitric and nitrous oxide reduction by denitrifying *Pseudomonas fluorescens*. *Appl. Environ. Microbiol.*, **39**, 105–68.

Stewart, G.R. and Orebamjo, T.O. (1984) Studies of nitrate utilisation by the dominant species of regrowth vegetation of tropical West Africa: a Nigerian example. In Lees, J.A., Niel, S.M. and Rorison, I. (eds), *Nitrogen as an Ecological Factor*, Blackwell Scientific Publications, Oxford, pp. 167–88.

Tam, T.-Y. and Knowles, R. (1979) Effects of sulphide and acetylene on nitrous oxide reduction by soil and by *Pseudonomas aeruginosa*. *Can. J. Microbiol.*, **25**, 1133–8.

Tiedje, J.M., Sextone, A.J., Myrold, D.D. and Robinson, J. (1982) Denitrification: ecological niches, competition and survival. *Antonie van Leeuwenhoek*, **48**, 564–83.

Tiedje, J.M., Sexstone, A.J., Parkin, T.B., Revsbech, N.P. and Shelton, D.R. (1984) Anaerobic processes in soils. *Plant and Soil*, **76**, 197–212.

Tilton, D.L. and Kadlec, R.H. (1979) The utilisation of freshwater wetlands for nutrient removal from secondarily treated wastewater effluent. *J. Environ. Qual.*, **8**, 328–44.

Triska, F.J. and Oremland, R.S. (1981) Denitrification associated with periphyton communities. *Appl. Environ. Microbiol.*, **42**, 745–8.

Valiela, I., Teal, J.M., Volkmann, S., Shaufer, B. and Carpenter, E.J. (1978) Nutrient and particulate fluxes in a salt marsh ecosystem: Tidal exchanges and inputs by precipitation and groundwater. *Limnol. Oceanogr.*, **23**, 798–812.

Valiela, I. and Teal, J.M. (1979) The nitrogen budget of a salt marsh ecosystem. *Nature*, **280**, 652–6.

van der Linden, M.H.J.A. (1980) Nitrogen economy of reed vegetation in the Zuidelijk Flevoland Polder. I. Distribution of nitrogen among shoots and rhizomes during the growing season, and loss of nitrogen due to fire management. *Oceologia Plantarum*, **1**, 219–30.

van Raalte, C.D., Valiela, I., Carpenter, E.J. and Teal, J.M. (1984) Inhibition of nitrogen fixation in salt marshes measured by acetylene reduction. *Estuar. Coast. Mar. Sci.*, **2**, 301–5.

Vincent, W.F. and Downes, M.T. (1980) Variation in nutrient removal from a stream by watercress (*Nasturtium officinale* R.Br.). *Aquat. Bot.*, **9**, 221–35.

Vincent, W.F. and Downes, M.T. (1981) Nitrate accumulation in aerobic hypolimnia: relative importance of benthic and planktonic nitrifiers in an oligotrophic lake. *Appl. Environ. Microbiol.*, **41**, 565–73.

Viner, A.B. (1982) Nitrogen fixation and denitrification in sediments of two Kenyan lakes. *Biotropica*, **14**, 91–8.

Warwick, J. and Hill, A.R. (1988) Nitrate depletion in the riparian zone of a small woodland stream. *Hydrobiologia*, **157**, 231–30.

Whitton, B.A., Aziz, A., Francis, P., Rother, J.A., Simon, J.W. and Tahmida, Z.N. (1988) Ecology of deepwater rice fields in Bangladesh. 1. Physical and chemical environment. *Hydrobiologia*, **169**, 3–67.

Wijler, J. and Delwiche, C.C. (1954) Investigations on the denitrification process in soil. *Plant and Soil*, **5**, 155–69.

7 Nitrate Cycling in Marine Waters

N.J.P. OWENS
Plymouth Marine Laboratory, UK

7.1 INTRODUCTION

The oceans are the largest planetary reservoir of nitrate (excluding that bound in the lithosphere). Compared with other forms of nitrogen (most notably di-nitrogen), however, the quantities of nitrate dissolved in seawater are small (Table 7.1), nevertheless, the nitrate pool represents a central and fundamental part of the marine (and hence global) nitrogen cycle.

Nitrate enters the sea from the freshwater and terrestrial environment (Chapters 3—5), mainly through estuaries, although groundwater (Chapter 8) and atmospheric inputs (Chapter 2) are increasingly being recognised as important supplies. Although small in area compared with the overall marine system, estuaries are notable for their biological and chemical reactivity, and have the potential for modifying the supply of nitrate to the marine environment. This reactivity must be considered in any discussion of the marine nitrogen cycle, and estuaries must not be viewed simply as conduits between freshwater and marine waters, nor a simple admixture of the two. Estuaries are the most severely perturbed regions of the marine environment through anthropogenic activities (e.g. waste discharges, urbanisation, and transport).

Offshore from estuaries are the coastal seas. These are relatively simple to define topographically from depth criteria (e.g. those regions being confined to continental shelves). However, coastal seas may exhibit the biological and chemical characteristics of both estuarine and full oceanic regions. Depending upon the exact criteria used for their definition, coastal shelf seas represent approximately 10% of the total world ocean. These continental margins are also under extreme environmental pressure from human activities, and are an important food resource for a large part of the world population. Together with estuaries, there is increasing concern about the influence of nitrate in these environments.

Further still offshore are the deep oceans. Although these areas are generally under the least direct pressure from anthropogenic activities, the deep oceans are intimately linked to the coastal margins via complex trans-shelf transport mechanisms. There is an increasing realisation that the dynamic link between the deep ocean and more nearshore regions is of fundamental importance to the functioning of the coastal seas, and this is of particular relevance to nitrate cycling. Because of the overwhelming area and volume of the deep oceans, processes involving nitrate are of global significance.

In this chapter some features of importance relating to the cycling of nitrate in marine systems are considered. Much of the discussion will be concentrated on the estuarine and coastal waters, since these areas are those most directly influenced by human activity. However, as noted above, the deep oceans play an important role in the cycling of nitrate, and since the three domains are intimately linked, it is also appropriate to consider processes in the more offshore

Nitrate: Processes, Patterns and Management. Edited by T.P. Burt, A.L. Heathwaite and S.T. Trudgill
© 1993 John Wiley & Sons Ltd

Table 7.1 Marine nitrogen reservoirs (adapted by permission from Rosswall, 1983, SCOPE 21)

Units 10^{15} gN		%
Nitrate	570	2.5
Biomass	0.5	0.002
Dead organic matter	533–770	2.3–3.3
Dissolved N_2	22000	95
Other dissolved forms (e.g. ammonia)	8	0.03

regions. It should be noted that, in general, the same fundamental processes acting on nitrate occur in the marine environment as in terrestrial and freshwater environments; in the majority of cases, it is a question of scale where the important differences lie.

After a brief summary of the input of nitrate to the sea, the first part of the chapter is concerned with the characteristics of the various marine domains as they relate to the cycling of nitrate. Although exceptions to any 'typical' example of distribution and concentration of nitrate can always be found, an attempt has been made to provide examples of the likely levels of nitrate that occur in these environments. As in other environments, seasonal cycles are a common feature of nitrate in some regions of the marine system, and these are summarised in Section 7.4. Section 7.5 is concerned with sediments. The distinction between sediment and water column processes is more for convenience than a reality, but the wide physico-chemical differences between these 'phases' does, in part, at least, lead to a partitioning of nitrate cycling processes, although sediment-water column coupling is an important feature of marine systems. The final section, 7.6, is concerned with the historical trends in nitrate concentration. Clearly, this addresses the possible influence of human action but compared with other environments (e.g. freshwaters and groundwaters) the realisation of this possibility is relatively recent. Until only a few years ago, the philosophical view held by Euripedes, '... the sea washes all man's ills away', was also the practical approach, with the result that few long-term data are available on which objective judgements can be made.

It is assumed here that the fundamental processes of the aquatic nitrogen cycle are known. The processes are described in more detail in Chapter 5. Inorganic nitrogen is assimilated by algae for growth. The organic nitrogen so formed returns to the aquatic phase in the dissolved and particulate form, which is regenerated, principally by bacteria, via dissolved ammonium. Dissolved ammonium can then be regenerated to nitrate through the process of nitrification, thus completing the cycle. This is the typical nitrogen cycle that predominates in the well-oxygenated bulk of the ocean. Under certain conditions of low oxygen, dissolved nitrate can be utilised by a specialist group of bacteria which converts the nitrate to gaseous products (denitrification). Together, these processes combine to drive the cycle of nitrate in marine waters. There are important considerations of the physiology and biochemistry of the organisms involved, and, of course, higher organisms play an important role.

However, these details are not considered here. Good detailed discussions of the marine nitrogen cycle are provided by Carpenter and Capone (1983) and Blackburn and Sorensen (1988). It is also important to appreciate that although the cycle of nitrate is principally mediated through organisms, all processes in the marine system are dominated by the physics of the system. Reference to some of the more important physical oceanographic features has been made but these necessarily have received only a cursory treatment. A first-class introduction to physical oceanography is provided by Pickard and Emery (1990).

7.2 INPUTS OF NITRATE TO THE MARINE SYSTEM

There are two principal sources of nitrate to the marine system: rivers (via estuaries) and the atmosphere. To these must be added direct point-source discharges (e.g. sewage and industrial wastes through sea-outfalls and dumping) and groundwater inputs. Of these, the riverine sources are quantitatively the most significant on a global scale; however, any of the other sources may assume more importance on local, regional scales. The current input from rivers is of the order of $20-60 \times 10^{12}$ g NO_3-N per year (Table 7.2). A major difficulty arises in the estimation of the riverine input. In most cases riverine nitrate concentrations are measured at some point in the riverine system, invariably some distance from the sea. The input (or load) is calculated from the product of this concentration and the flow rate. While large effort may be involved to take account of the variability in the concentrations of nitrate and flow, the estimate remains a measure of the *gross* input since no account has been taken of the possible influence of the estuary in modifying this load (see below). Accepting this fundamental difficulty, it can be seen (Table 7.2) that the riverine inputs of nitrate calculated in 1963 differ by almost an order of magnitude from the most recent estimate. While part of this undoubtably reflects genuine increases in riverine nitrate concentrations, and thus possibly an increase in the *net* input to the sea, the estimates of Livingstone (1963), Van Bennekom and Salomons (1981) and Meybeck (1982) were intentionally made for unpolluted rivers. In contrast, the estimate of Walsh (1991) is an estimate for all rivers and includes the anthropogenic contribution. By characterising world rivers into three types — (1) pristine; (2) those perturbed by human activity through deforestation and Third World agricultural activities; (3) those draining heavily industrialised and intensively farmed catchments — and assuming nitrate concentrations of 10, 50 and 100 μmol NO_3-N l^{-1}, respectively, a world distribution of nitrate inputs to the marine system can be generated (Figure 7.1).

It should be noted that there is a marked global disparity between the distribution of the inputs of nitrate and freshwater to the marine system. For example, of the estimated global river discharge to the seas of 38.5×10^3 km^3 per year, 22% is accounted for by the world's three largest rivers — the Amazon, Zaire and Orinoco (Milliman and Meade, 1983) while these rivers account for only 3.9% of the estimated total nitrate input (16×10^{12} g N per

Table 7.2 Riverine inputs of nitrate to the marine environment

Flux ($\times 10^{12}$ g NO_3-N per year)	Comments	Source
7.13	1963: human impact minimised	Livingstone (1963)
3.7	Pristine rivers	Meybeck (1982)
3.5	Thinly populated areas	Van Bennekom and Salomons (1981)
3	Pristine	Wollast (1983)
24	Including anthropogenic influence; and contribution from ammonium	Wollast (1983)
6.0	Includes ammonium	Walsh (1991)
15.2	Includes anthropogenic sources; NO_3 only	This work[a] (Figure 7.1)

[a]Actual concentration data from rivers, where available, from Wollast (1983): if not, assumption of concentration as in text.
Flow data from Milliman (1991).

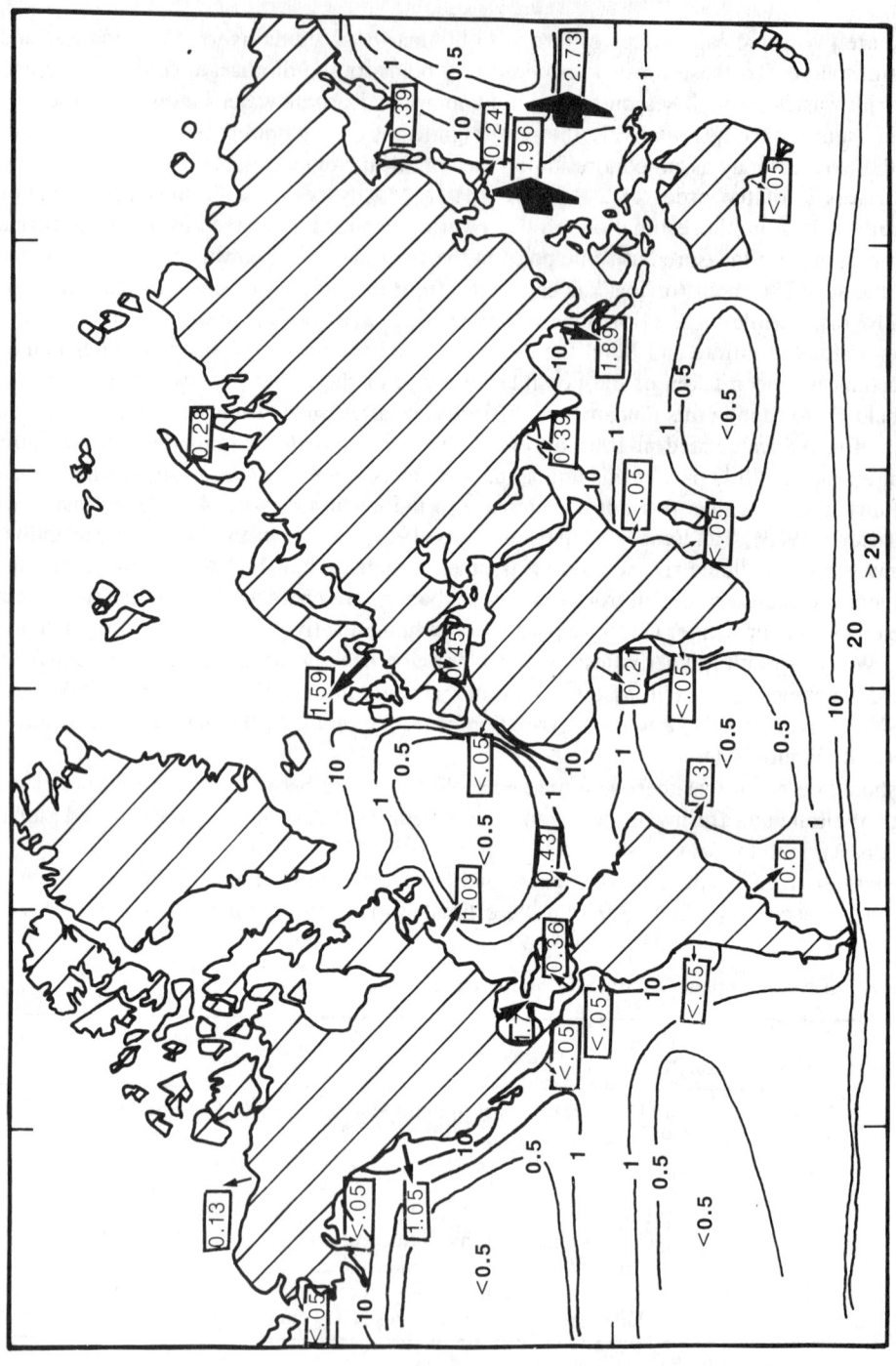

Figure 7.1 Global concentrations and inputs of nitrate to the sea. Contours (units: μmol NO_3-N l^{-1}) represent typical concentrations of nitrate in the surface waters. Figures within boxes indicate approximate inputs of nitrate from rivers (units: $\times$ 10^{12} g NO_3-N per year). Note that nitrate concentrations are greatest in sea areas bordering the continents with the exception of the Southern Ocean.

year). On the other hand, the rivers of south-east Asia and Oceania contribute approximately 25% of the river flow and 42% of the nitrate input; and industrialised Europe and North America contribute 8.3% of the river flow but 36% of the nitrate input. The consequences of these inputs on the marine recieving waters will be considered in detail below. However, compared with the total quantity of nitrate, this annual flux represents 2.8% of the standing stock of nitrate in the world ocean. Thus at the global scale, riverine inputs are an insignificant source of nitrate to the ocean. However, the riverine nitrate inputs are restricted to the coastal seas; thus the receiving volume is orders of magnitude smaller than the global oceanic volume. In these coastal seas, the riverine inputs of nitrate are important (see below). To a first approximation, riverine nitrate inputs represent approximately 25% of the total (i.e. including dissolved organic nitrogen) nitrogen inputs to coastal seas.

In contrast to the riverine inputs, which can be considered to be relatively discrete point-sources of nitrate, atmospheric inputs are more diffuse. A reasonable amount of information is available worldwide for these inputs and has been summarised in two recent reports (GESAMP, 1989, 1991; see also Prospero and Savoie, 1989 and Owens, Galloway and Duce, 1992). It is apparent (Table 7.3) that the atmospheric inputs of nitrate are quantitatively significant on the global scale, being of the same order as those derived from riverine inputs. Atmospheric inputs are regionally variable, with the largest nitrate loading being to the northern oceans. The most important source of this nitrate is continental, where anthropogenic activities contribute the largest source. The most significant aspect of these atmospheric inputs is that they represent a source of nitrate to fully oceanic systems, whereas, as noted above, the riverine inputs represent sources to restricted, coastal areas. Nevertheless, atmospheric inputs may also be important on a more localised scale. For example, the annual deposition of inorganic nitrogen ($\sim$50% of which is nitrate) to the North Sea is approximately 25% of the total riverine nitrogen inputs (Anon., 1987). Similar data are available for the Baltic Sea (Roenner, 1985).

7.3 PHYSICO-CHEMICAL CHARACTERISTICS OF THE MARINE ENVIRONMENT AND THE DISTRIBUTION AND CONCENTRATION OF NITRATE

7.3.1 ESTUARIES

Estuaries are difficult areas to define. Two definitions have stood the test of time, which describe the majority of estuaries. Ketchum (1951) defined an estuary as 'a region where

Table 7.3 Estimated atmospheric nitrate inputs to the oceans (units 10^{12} g N per year)

Deposition type	South Indian	North Atlantic Total	South Atlantic	North Pacific	South Pacific	North Indian
Wet		4.14	0.50	4.09	2.07	0.75
	1.24	12.80				
Dry		3.69	0.63	2.52	1.36	0.45
	0.72	9.37				
Subtotals		7.83	1.13	6.61	3.43	1.20
	1.96	22.17				

Source: GESAMP No. 38 (1989). Reproduced by permission.

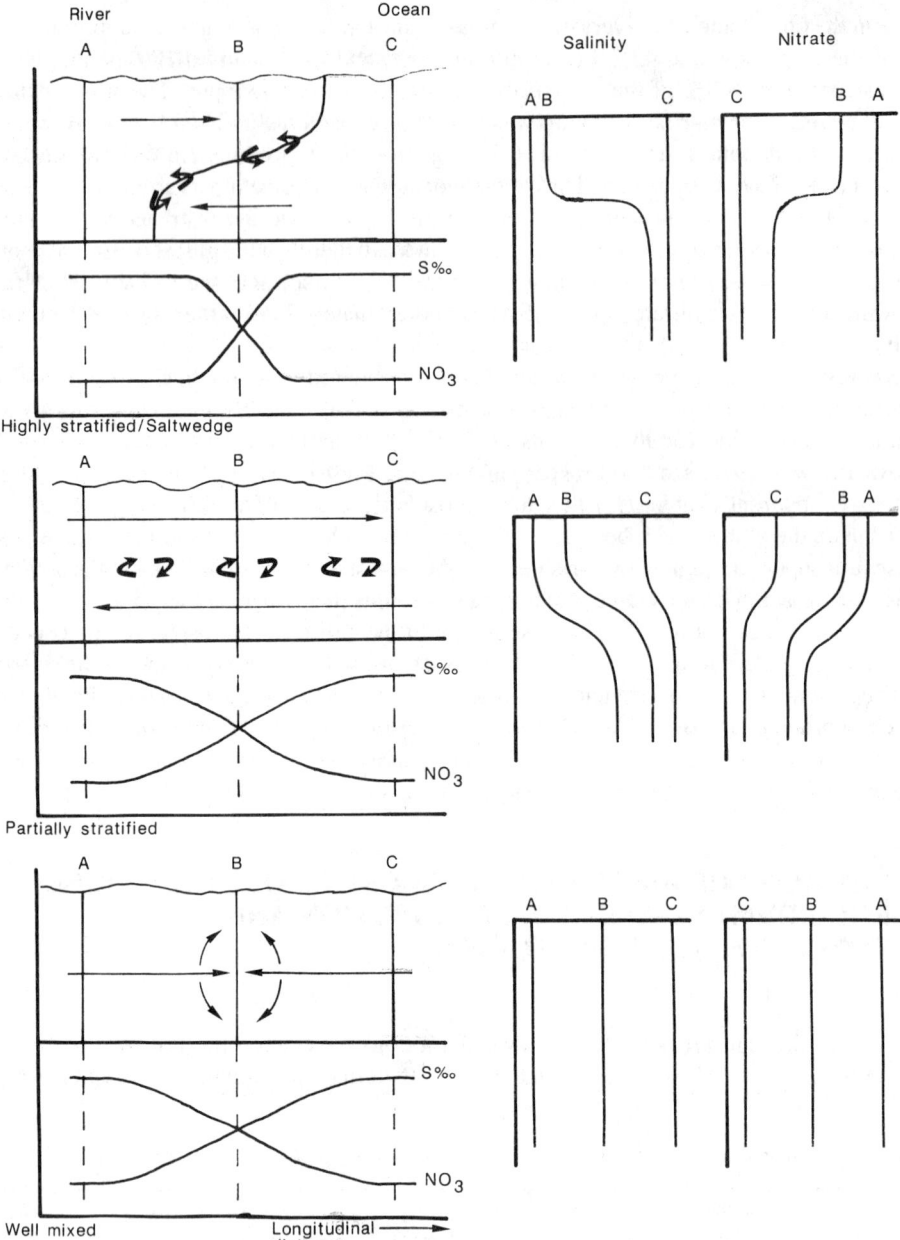

Figure 7.2 Schematic diagram representing idealised examples of the mixing that occurs between freshwater and seawater in estuaries, and the resulting effects on nitrate concentrations at various positions in the estuary. The left-hand panels for each estuary type (from top to bottom: highly stratified, partially stratified and well mixed) show longitudinal sections of salinity ($S\%_{oo}$) and nitrate (NO_3) concentrations in the surface waters from the river to the sea. The arrows represent the direction of water flow and the type of mixing that occurs. Thus, in the highly stratified estuary, salinity remains relatively low and nitrate relatively high in the upper estuary (in the region of position A), but salinity falls quickly, and nitrate concentrations decrease at some point (position B) within the estuary. The right-hand panels indicate the typical distributions of salinity and nitrate concentrations with depth at three locations (A, B and C) within the estuary

river water mixes with, and measurably dilutes, sea-water'. Although a largely accurate description, this definition includes open coastal waters which exhibit reduced salinity due to the influence of large riverine inputs (see below). These obviously non-estuarine regions were accounted for by the more refined definition of Pritchard (1952); that is: an estuary is 'a semi-enclosed sea or embayment where river water . . .'. However, this latter definition does not describe the negative or inverse estuary where the 'semi-enclosed' mixing region is an area of higher salinity than the open sea-water because evaporation (or sea-ice formation) results in freshwater removal that exceeds the input. The classical, positive estuary is typical of the majority of mid-latitude estuaries, whereas negative estuaries are usually confined to low-latitude, arid regions (e.g. the Red Sea) or polar latitudes (e.g. the Ross Sea). The physical features of estuaries are important in determining the concentrations and distribution of nitrate.

Estuaries can be classified into three general hydrodynamical categories (Figure 7.2). These features result from the balance of the freshwater and tidal flows, and in each case an estuary may change its category over periods of time ranging from days (in response to river spates), through weeks (in response to the neap-spring tide periodicity), to months (in response to seasonal variations in river flow). The hydrodynamical features of the estuary are fundamental in determining the horizontal and vertical distribution of nitrate within an estuary (Figure 7.2).

In a situation where the residence time of water in the estuary is short, relative to the timescale of biological activity, and there is only one source of nitrate (for example, a single riverine input), a simple nitrate−salinity dilution relationship results (Figure 7.3). This is known as 'conservative behaviour', and under such circumstances, there is no net removal or addition of nitrate. In contrast, 'non-conservative' behaviour can occur, and deviations from the idealised salinity−nitrate relationship results. Deviations above the salinity dilution will occur where there are point-source inputs within the estuary (for example, from industrial waste

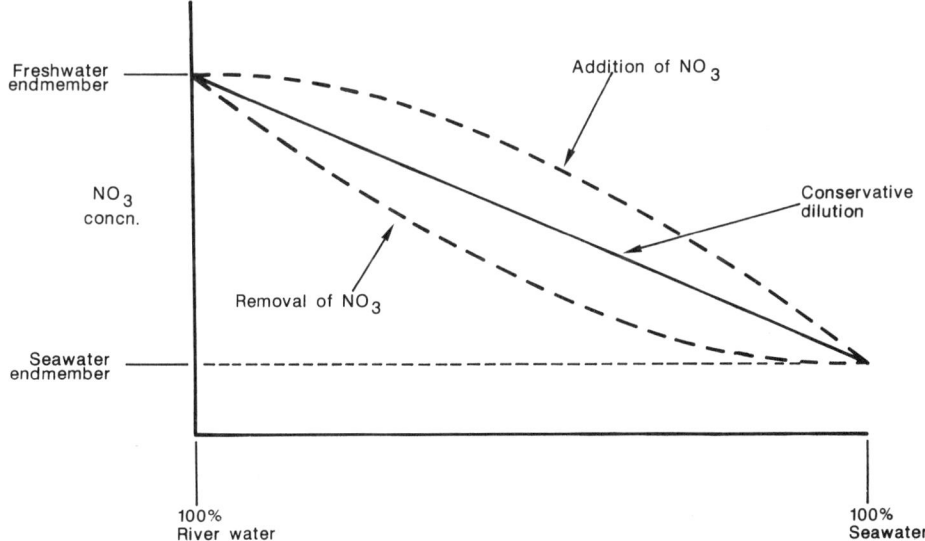

Figure 7.3 Idealised concentration of nitrate in the surface waters of an estuary at various salinities ranging from 100% river water to 100% seawater. The solid line represents the concentration that would be expected if there was no net removal or addition of nitrate within the estuary (conservative mixing). The dotted lines above and below the 'conservative dilution line' represent, respectively, the net addition or removal of nitrate within the estuary

inputs). However, deviations below the conservative dilution line are commonly observed and generally indicate biological removal processes occurring within the estuary. Furthermore, deviations from conservative behaviour are not always unequivocal if there are variations in the concentration of the nutrients in the end-members. It can be appreciated, for example, that if the concentration of nitrate in the freshwater increases (through a spate, for example) over a timescale shorter than the residence time of water in the estuary, a nitrate−salinity relationship will result which is above the dilution line measured after the spate has subsided; this could be mistakenly interpreted as an input of nitrate within the estuary. Thus, while nitrate−salinity relationships are a powerful indicator of the processes influencing nitrate concentrations in an estuary, they must be interpreted with care.

An example of non-conservative behaviour is shown in Figure 7.4. These data, for the Bristol Channel, a large estuary on the west coast of Britain, show two features. First, the filled circles indicate largely conservative behaviour. The data are from the main channel,

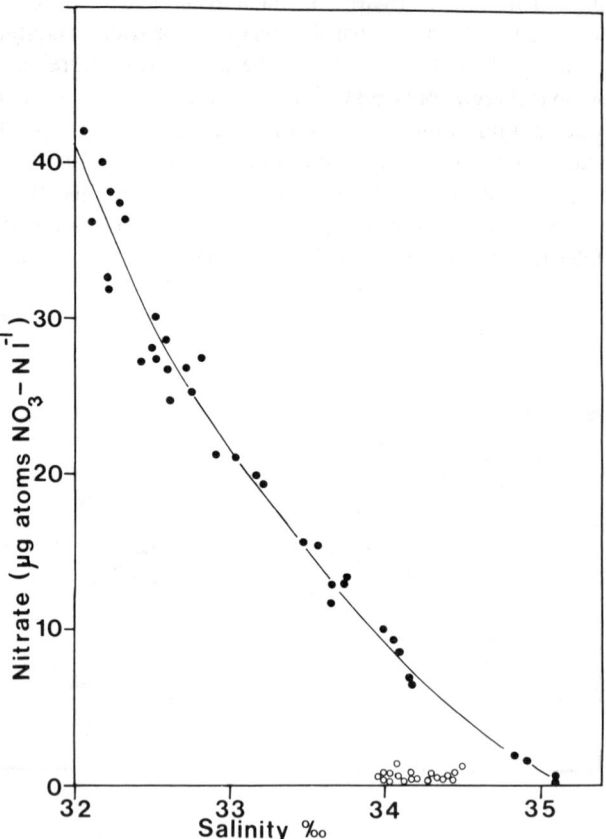

Figure 7.4 Actual concentration of nitrate at various salinities in the central Bristol Channel (filled circles) and Carmarthen Bay — a bay on the northern shore of the Bristol Channel (open circles) in August 1979. The central Bristol Channel data indicate essentially conservative behaviour of nitrate, whereas the Carmarthen Bay data indicate net removal (see text). Salinity of 35 is equivalent to 100% seawater (adapted from Mantoura *et al.*, 1988)

and the slight curvature indicates a small amount of removal of nitrate occurring there. Second, the open circles, obtained from an embayment removed from the main channel, indicate substantially depleted nitrate concentrations, that is, significant non-conservative behaviour (Mantoura et al., 1988). These low levels were brought about by removal of nitrate through utilisation by phytoplankton (Owens et al., 1986). The contrasting distributions were induced largely through differences in the two regions between the rate of plankton growth, relative to the residence time of water. Phytoplankton growth is restricted in the main channel because of low light levels in the water caused by high turbidity. Slow growth, combined with rapid flushing of the water, prevents a population of phytoplankton from establishing itself. Thus nitrate concentrations are predominantly under the control of physical factors, that is, mixing. Conversely, phytoplankton growth in the embayment is much more rapid because light levels there are significantly higher. This is brought about because particles in the water sink and are lost from the water column in the less tidally dynamic environment of the bay. Light can thus penetrate deeper into the water column, which, when combined with the reduced flushing rates, allows for a large population of phytoplankton to develop, which can strip nitrate from the water. If the favourable conditions persist for periods of several days, nitrate concentrations can become totally depleted, and possibly limit phytoplankton growth (Owens et al., 1986; Mantoura et al., 1988). On the other hand, a decrease in the flushing time of the bay (for example, caused by a heavy spate or spring tides) would reduce the favourable conditions for growth of phytoplankton, and the generally conservative distribution of nitrate would be re-established.

It can be appreciated from the above that it is an almost impossible task to allocate a 'typical' level of nitrate to an individual estuary, since the system is constantly changing in response to huge hydrodynamic influences coupled to biological modifications. Clearly, the upper and lower limits of the nitrate levels are governed by the concentrations of nitrate in the freshwater and marine end-members. Thus, it is generally true to say that estuaries are characterised by nitrate concentrations greater than those in marine waters, but less than those of freshwaters (see Chapter 5). It is not possible, however, to characterise estuaries into 'trophic' categories (e.g. oligotrophic, eutrophic) as are, commonly, freshwaters.

7.3.2 COASTAL AND SHELF WATERS

For the purposes of this discussion, these waters are defined as those between the land boundary and the edge of the continental shelf. The width of this zone waries considerably globally, from a few tens of kilometres (e.g. off Peru and the Iberian Peninsula) to hundreds (e.g. the Indonesian shelf, and off north-east North America); a global average is of the order of 75 km width and 135 m depth. As will be seen below, the topography of the shelf has a considerable influence on the levels and cycling of nitrate. In general, nitrate concentrations in shelf waters are the highest of the fully marine waters due to the high inputs of nitrate from the land and from regeneration processes occurring in sediments. The latter can be an important process since the relative shallowness of the water column provides for a large surface area of sediment to water volume ratio. Although shelf seas represent only about 10% of the surface area of the world ocean, they support about 90% of the world fishery. This is due in large part to the high levels of nutrients, predominantly nitrate, found there. Because of their proximity to land, some shelf seas are also subject to major impact from human activities.

In some areas, the gradation from the estuary to the shelf sea is indistinct, and the influence

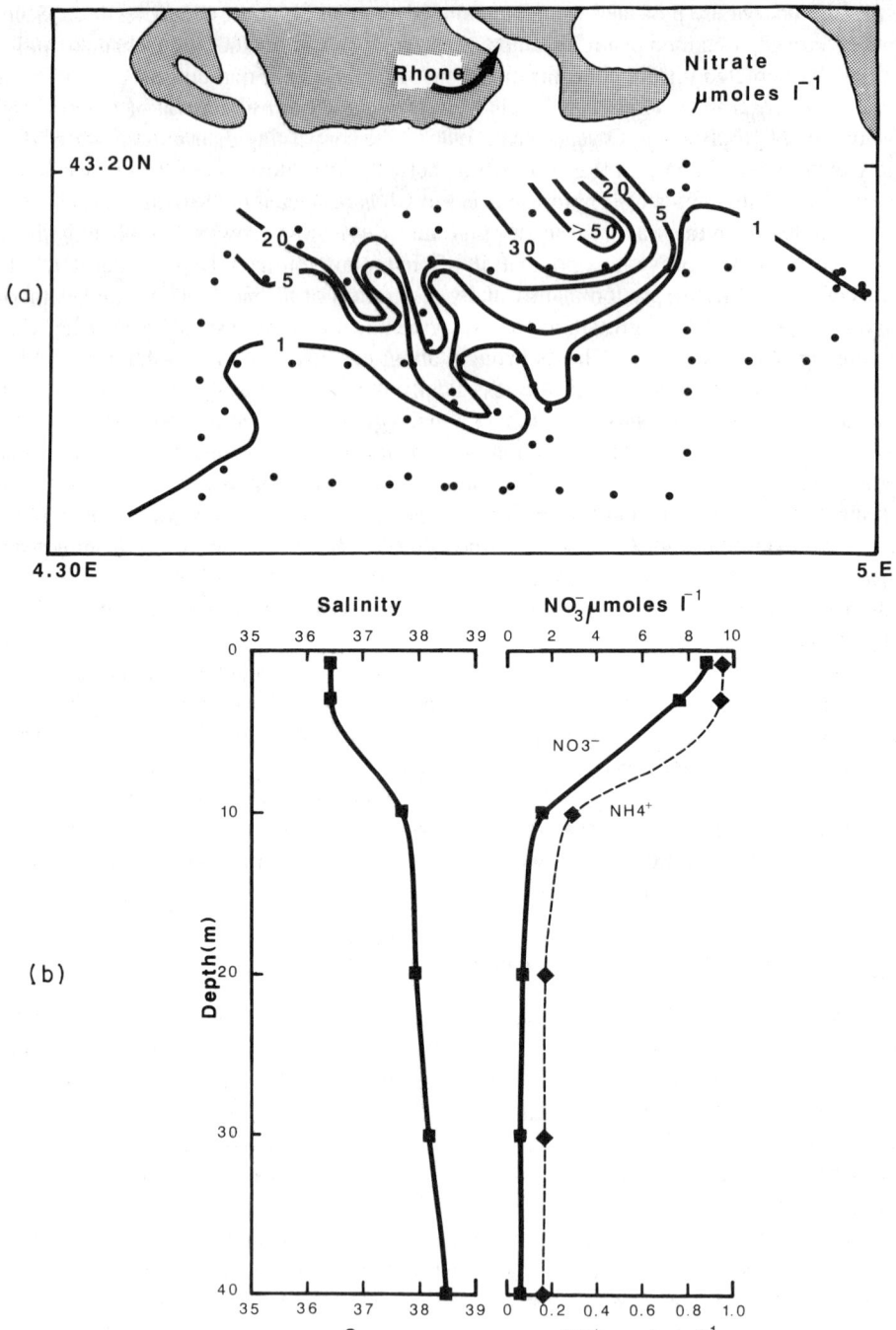

Figure 7.5 (a) Concentration of nitrate in the surface waters of the plume of the River Rhone, north-west Mediterranean Sea, January 1989. (b) Vertical distribution of salinity and nitrate (and ammonium) concentrations at a site within the plume, January 1989. (Figure 7.5(a) reproduced by permission from Woodward and Owens, 1989; Figure 7.5(b) reproduced from Owens *et al.*, 1989b)

of a riverine input can be detected many kilometres offshore as a marked area or 'plume' of low salinity. In the case of large rivers (for example, the Amazon) the plume can be detected hundreds of kilometres offshore. Under these circumstances, elevated nitrate concentrations can usually be observed, surrounded by water of low nitrate concentration (see Figure 7.5(a)). The boundary of the plume is usually abrupt, but subject to extreme, small-scale convolutions and eddies brought about by complex hydrodynamic mixing processes. The elevated nitrate concentrations are invariably restricted in depth to a few metres at the surface, since the low-salinity, riverine water is buoyant and spreads across the surface. This gives rise to a marked vertical distribution of nitrate (see Figure 7.5(b)). The physical features of plumes, and their mixing with seawater, are an important factor in the transport of dissolved and particulate material from the land, via rivers, to the sea.

An idealised distribution of nitrate along a transect across a continental shelf is shown in Figure 7.6 (considerable deviations from this ideal occur, and these will be considered later). From initially high concentrations close to the land, nitrate decreases rapidly. This is caused by a combination of physical dilution and biological utilisation. Apart from in plumes, nitrate concentrations are usually uniform from the surface to the bottom in these nearshore environments because strong physical mixing (tides) ensures a well-mixed water column. Small increases in nitrate concentrations at the bottom can sometimes be detected because

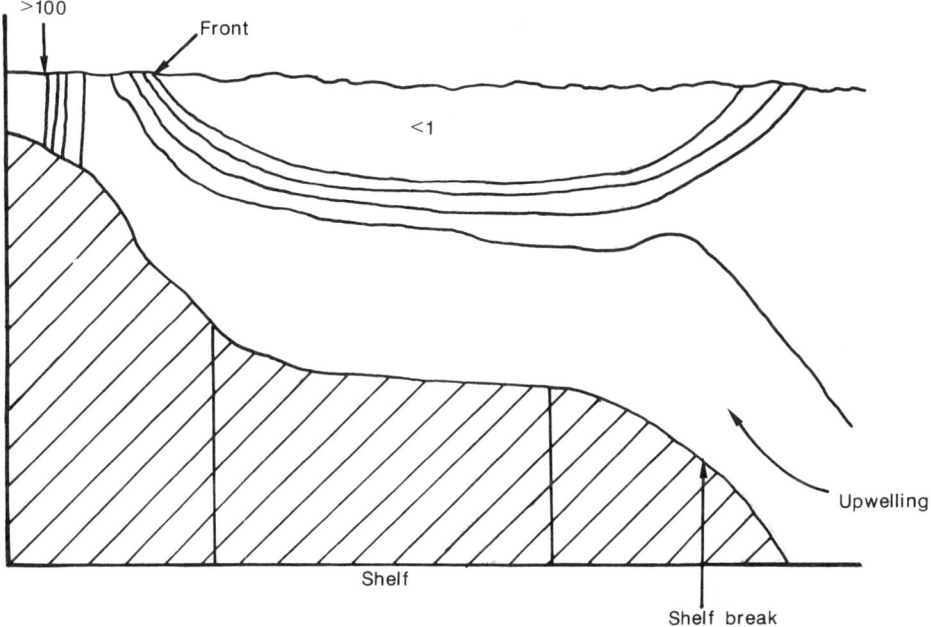

Figure 7.6 Schematic representation of the vertical distribution of nitrate concentrations from an estuary mouth across a continental shelf to the shelf break. Concentrations indicated (μmol NO$_3$ l^{-1}) are provided only as order of magnitude estimates. The important features are that nitrate concentrations are high and vertically uniform in the shallow very nearshore environments. Further offshore, nitrate concentrations decrease. At some distance offshore, seasonal vertical stratification may be established during the summer; a sharp discontinuity (front) in nitrate concentrations occurs at the boundary between these two regions. Nitrate concentrations in the surface waters are typically extremely low during periods of stratification, but overly deeper waters of higher nitrate concentrations. Upwelling of deeper, high nitrate concentration water may introduce nitrate to shallower waters at the shelf break.

of inputs from the sediments (e.g. Law and Owens, 1990). However, this input is usually quickly mixed throughout the water column. In the winter, or at high latitudes, this vertically uniform and generally decreasing concentration pattern continues across the shelf until full oceanic conditions persist.

During the summer months, particularly in temperate latitudes, where there are extremes of solar input, marked thermal stratification is set up caused by heating of the sea surface, which becomes relatively buoyant and tends to form two distinct layers. The boundary between the two layers (the thermocline) is invariably narrow and distinct, and temperatures may fall by several degrees over a vertical distance of 1 or 2 m. The formation of thermal stratification is of extreme importance in regulating the distribution of nitrate in shelf seas. The influence of thermal stratification can be seen in Figure 7.6 by the significant reduction in nitrate concentrations in the upper waters. This distribution is brought about because the upper, warm layer becomes essentially isolated from inputs of nitrate from deep, offshore waters by the strong density boundary imposed by the thermocline. This boundary is sufficiently 'robust' to prevent the normal mixing processes occurring. This, combined with biological removal of nitrate, forms an upper layer which progressively becomes more depleted in nitrate as the summer season progresses. Nitrate concentrations typically fall from many μmoles 1^{-1} to <10 nmoles 1^{-1} during the formation and development of stratified conditions (Brockmann, Billen and Gieskes, 1988; Woodward and Owens, 1990). These conditions give rise to a characteristic vertical profile of nitrate (Figure 7.6).

Whether an individual sea area becomes thermally stratified or not is brought about by a balance between the solar energy input (which promotes stratification) and vertical mixing (which tends to break down stratification), caused chiefly by tides, and, to a lesser extent, by wind. Also, for a given tidal or wind stress, the degree of vertical mixing imposed is strongly influenced by water depth. To a first approximation, all of these variables are relatively constant (and predictable) between years and for any sea area. Thus, it is possible to predict, in general, where seasonal thermal stratification will occur (Simpson and Hunter, 1974; Pingree and Griffiths, 1978). Figure 7.7 shows this distribution for the shelf waters around the UK. The correspondence between the areas where seasonal thermal stratification is expected and surface nitrate concentrations for a sub-set of this area (Figure 7.8) is good. The boundary between vertically well-mixed and stratified waters is extremely marked, and surface nitrate concentrations may change by an order of magnitude over a horizontal distance of less than 10 km. The 'front' between these two water masses while varying slightly in position from year to year and over a season, is relatively constrained in its position, for the factors outlined above.

A typical seasonal pattern, which is repeated each year, is shown in Figure 7.9. This illustrates the average vertical distribution of temperature, stability and nitrate for a site 20 miles south of Plymouth in the English Channel for the years 1901–75. As can be seen from Figure 7.7, thermal stratification should occur here, and indeed stability typically becomes established in early April. As the season develops and stability becomes more marked, nitrate concentrations decline, to become almost absent. Stratification breaks down in September, and there follows a return of high nitrate concentrations.

With reference to the transect offshore (Figure 7.6), the seasonally depleted nitrate concentrations in surface waters generally persist, with little change across the shelf to the shelf edge, and into the deep oceanic domain. However, nitrate concentrations in deeper waters tend to increase, particularly at the shelf edge. Here complex physical factors cause upwelling of deep, high nitrate concentration waters leading to an injection of nitrate (see Blanton, 1991,

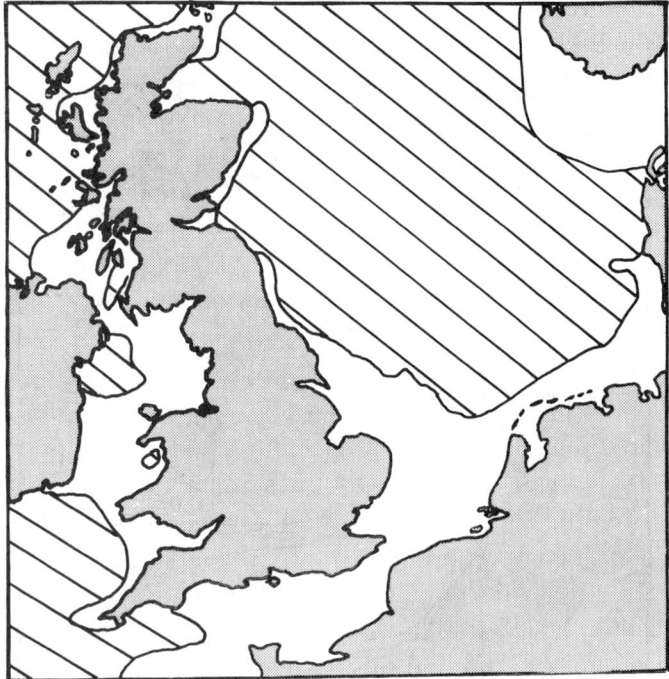

Figure 7.7 Sea areas around the UK where thermal stratification occurs during the summer months (shaded regions). The boundary between stratified and non-stratified water is the 'front' indicated in Figure 7.6 (adapted from Anonymous, 1981)

for a review of trans-shelf physical phenomena). These injections may introduce nitrate across the thermocline into surface waters.

While the simplified pattern above is a good description of many of the world's shelf seas, there are some notable exceptions. These are the coastal upwellings. An upwelling is the phenomenon where sub-surface (100–300 m depth) water is brought to the surface in response to the horizontal transport, offshore, of surface water. As can be seen to a limited extent from Figure 7.6 (but in greater detail below), subsurface oceanic waters are typified by high nitrate concentrations. Thus, upwelling introduces nitrate (and other nutrients) to the surface waters. This provides for a particularly fertile marine environment, and some of the world's largest and most economically important fisheries are based on upwelling systems (for a review of upwellings see Barber and Smith, 1981).

Although the detailed physical forcing of upwellings is complex, the underlying features are relatively straightforward. Their basis is the stable meteorological conditions that are established on the eastern boundary of the major oceans (that is, off the western coasts of North and South America and North and South Africa — Figure 7.10). Here, the generally persistent trade winds blow approximately parallel to the coastlines. This sets up a transport of surface water towards the equator, which because of the Coriolis force is deflected perpendicular (right in the northern hemisphere, to the left in the southern) to the coast. It is this offshore transport of surface water which is replaced by upwelling of deep water. This generalised description is known as 'Ekman pumping'. A particularly intense upwelling also occurs off the coast of Somalia and the Arabian peninsula as a result of the seasonally

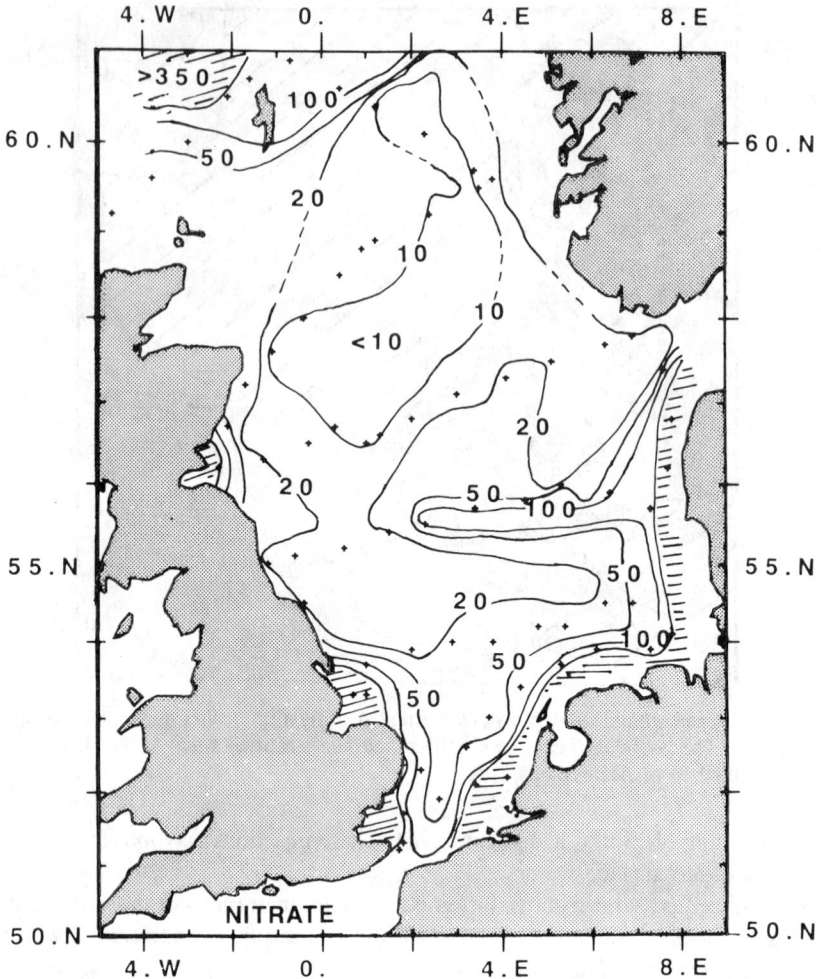

Figure 7.8 Distribution of nitrate (nmoles NO_3-N l^{-1}) in the surface waters of the North Sea, July 1987 (reproduced by permission from Woodward and Owens, 1990)

strong south-westerly monsoon. Although upwelling can occur along any coast should there be suitable alignment of coastline and winds (e.g. periodic upwelling occurs off the Iberian peninsula), it is the general persistence of the features and the length of coastline affected that make those areas shown in Figure 7.10 the most important; note the corespondence between Figure 7.10 and the levels of nitrate shown in Figure 7.1. Although the general meteorological conditions that induce upwellings usually persist for many months, local modifications do occur that alter the magnitude of upwelling, and commonly several individual 'upwelling events' may take place throughout the upwelling season. Because of the dependence of the fisheries of these areas on upwelling, their complete failure can lead to devasting crashes in the fishery. A well-known example of this is the occasional failure of the anchoveta fishery off Peru. A disturbance to the normal atmospheric circulation of the South Pacific occurs every 2–7 years, as a result of which the ocean surface temperature off the Peruvian coast

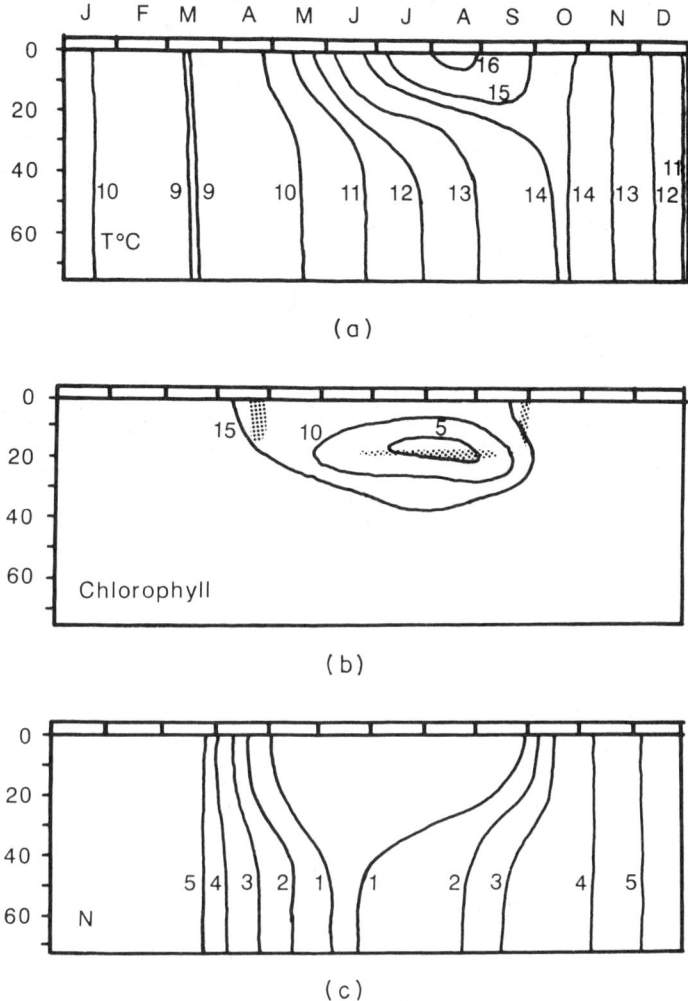

Figure 7.9 (a) Mean vertical temperature distribution over the year from 1903 to 1975 at a site in the western English Channel. (b) Vertical stability of the water column — lower numbers indicating stronger stability, stippled areas indicating concentrations of algae as measured by chlorophyll concentrations. (c) Mean vertical distribution of nitrate concentration (μmol NO_3-N l^{-1}) over the year from 1967 to 1977. Note the development of thermal stratification during the summer months, which coincides with the development of algae and concomitant reduction in nitrate concentrations (reproduced from Pingree, Maddock and Butler, 1977, by permission of Cambridge University Press)

suddenly rises by about 4°C, killing fish. The effect has become known as El Niño (The Christ Child) because of its near-coincidence to the Christmas period.

The typical effect of upwelling on nitrate concentrations is shown in Figures 7.11(a) and 7.11(b), which show transects for the coast of St Helena Bay, south-west Africa, and the Arabian Sea. The upwelling influence is particularly marked for the Arabian Sea, where the combination of the injection of nitrate and high levels of sunlight supports a particularly high rate of phytoplankton growth.

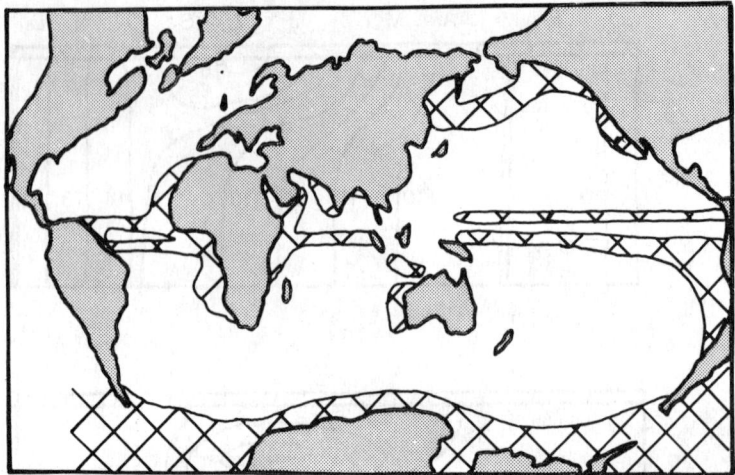

Figure 7.10   Global distribution of sea areas influenced by upwelling (hatched areas). The most extensive area subject to upwelling is the Southern Ocean. Strong upwelling occurs off the western boundary of the continents, and in a circumpolar band around the equator. These areas have an important influence on the concentrations of nitrate in the surface waters (reproduced from Barnes and Hughes, 1988, by permission of Blackwell Scientific Publications)

7.3.3   OCEANIC WATERS

Nitrate concentrations in the surface waters of fully oceanic systems are typically the lowest of all marine systems; indeed, over much of the world ocean nitrate is almost undetectable. In fact, it is only in recent years, with the introduction of new techniques (Garside, 1982), that exact concentrations of nitrate in many oceanic waters have been measurable; before this, nitrate concentrations were invariably reported as 'undetectable'. These low concentrations of nitrate are predominantly due to the great distances from land sources and the fact that nitrate removal processes in the upper layers are faster than the regeneration processes. The lowest nitrate concentrations are found in tropical and subtropical regions where there is a strong permanent thermocline. In these areas the supply of nitrate is restricted to weak tranport processes across the thermocline, and to atmospheric inputs. In tropical and subtropical systems nitrate concentrations are consistently about 10 nmole l^{-1}. Storm-induced inputs of rain and increased wind mixing may temporarily elevate nitrate concentrations up to about 50 nmole l^{-1} (Eppley and Renger, 1988; Eppley *et al.*, 1990). Temperate oceans exhibit the seasonal formation and destruction of the thermocline, thus in the summer period of phytoplankton growth, nitrate concentrations fall to low levels, while during the deep mixing of the water column during winter, nitrate concentrations are replenished.

The global distribution of nitrate in oceanic waters is shown in Figures 7.12(b) and 7.12(c). The low nitrate concentrations noted above are clearly seen in the surface waters of the central basins of the oceans. An important feature are the elevated nitrate concentrations around the equator (see also Figure 7.1), which are observed in all the oceans. This is caused by upwelling (see Figure 7.10) of subsurface water as a result of the divergence of the equatorial currents. The phenomenon is similar to the coastal upwelling described above, and is important in introducing nitrate to the surface waters of the equatorial regions.

It can be seen that nitrate concentrations are significantly elevated at depth, but there is not a continuous increase; rather, there are zones of differing nitrate concentration in the

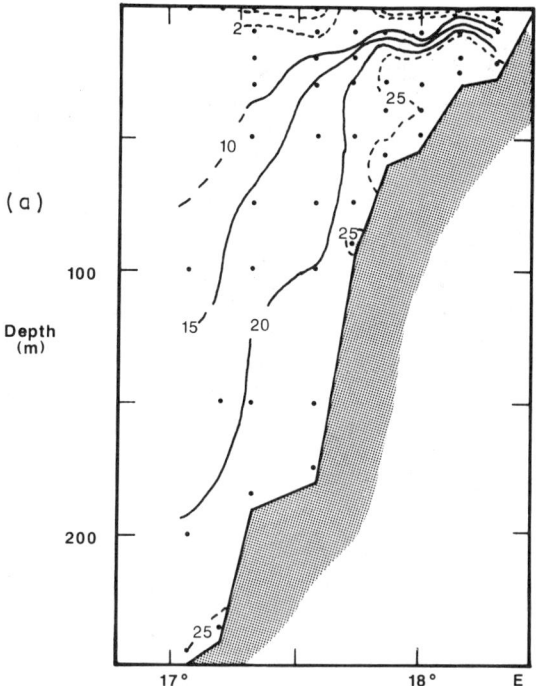

(a)

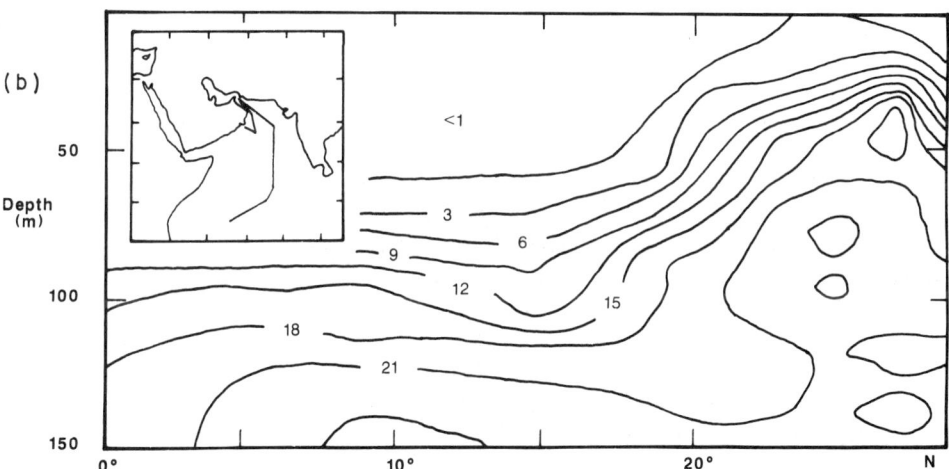

(b)

Figure 7.11   (a) Vertical distribution of nitrate (μmol NO_3-N l^{-1}) off St Helena Bay (west coast of South Africa) showing elevated nitrate concentrations introduced to shallow waters due to upwelling (from Bailey and Chapman, 1985). (b) Vertical distribution of nitrate in September 1986 along a longitudinal section in the Indian Ocean from the Equator to the Arabian Sea (see inset for location of transect). Strong upwelling evident at the northern end of the transect (reproduced from Mantoura *et al.*, 1992, by permission of Pergamon Press Ltd)

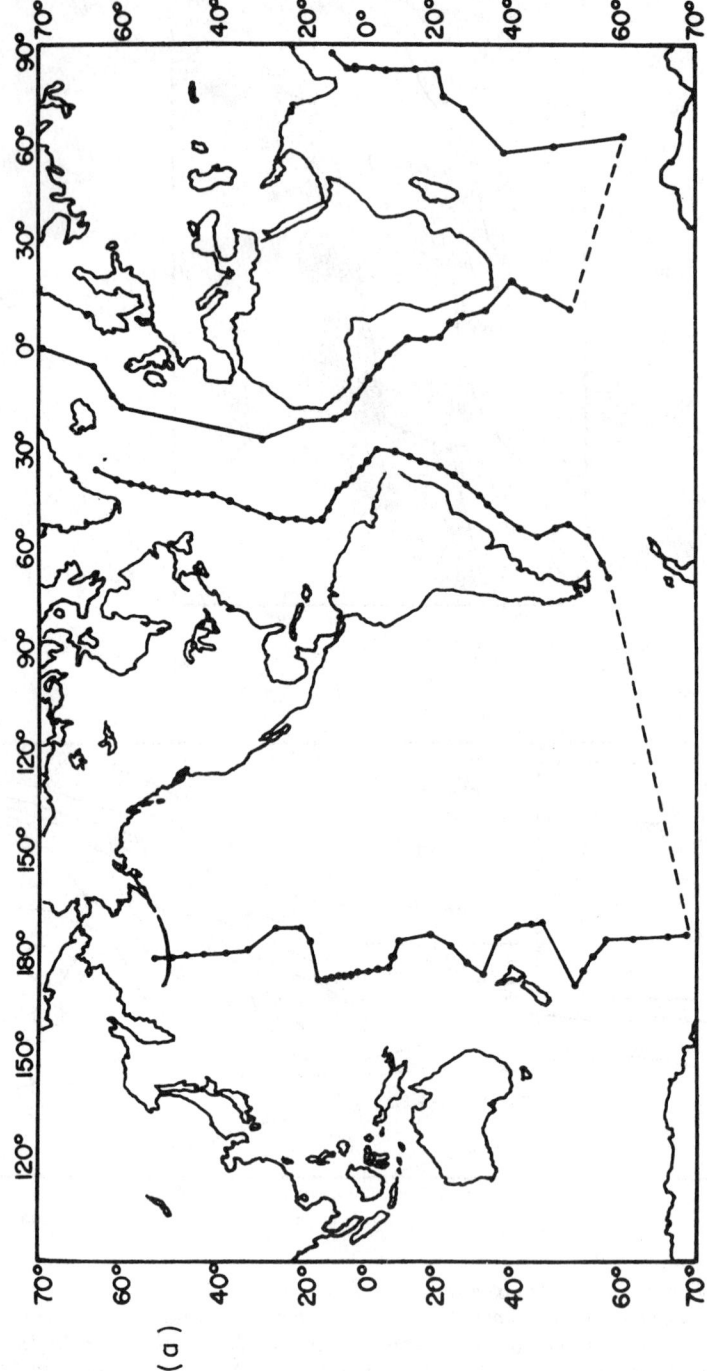

Figure 7.12 Global distribution of nitrate (μmol NO_3-N kg^{-1}); (1 l seawater is equivalent to approximately 1.025 kg). (a) Two vertical sections are shown with the track indicated.

(a)

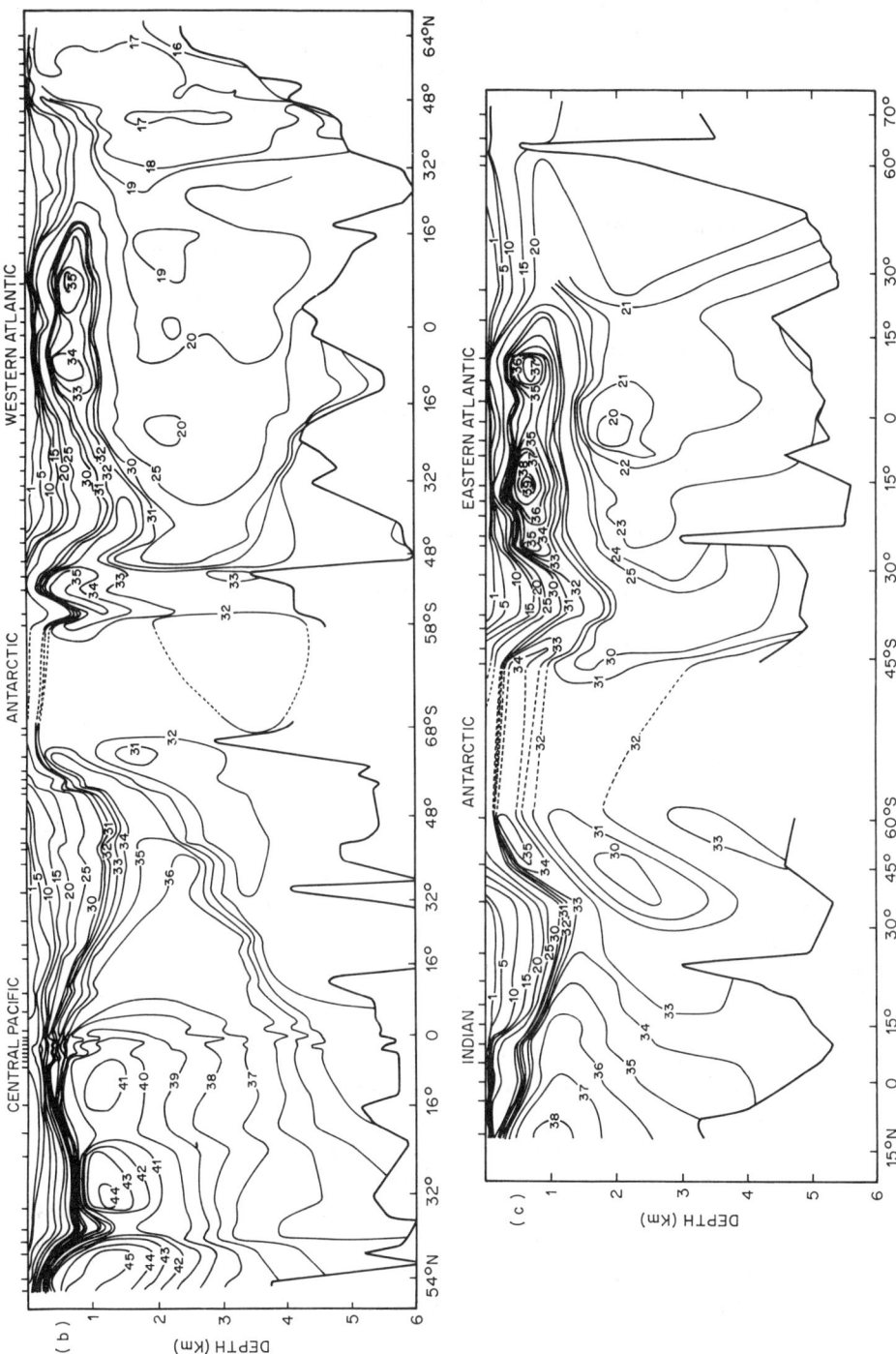

Figure 7.12 (*continued*) (b) Section running south through the Pacific Ocean to the Southern Ocean (where there is a break in the data coverage) and then northwards through the western Atlantic Ocean. (c) Section running south through the Indian Ocean and then northwards through the eastern Atlantic Ocean. Data from the GEOSECS (Geochemical Ocean Sections Study) programme conducted in the 1970s (reproduced by permission from Sharp, 1983)

oceans forming distinct 'lenses'. This zonal distribution is brought about by a combination of two factors: (1) production of nitrate by bacterial processes (see below) and (2) deep oceanic circulation (to understand the distributions shown in Figures 7.12(b) and 7.12(c) it is important to realise that all the oceans are intimately linked not only by surface currents but also by a deep water circulation — see Pickard and Emery, 1990). A feature of this deep oceanic circulation is that water masses of different origins are identifiable by their physical and chemical constituents. Thus, the water between about 1 and 4 km in the western Atlantic, with nitrate concentrations of between 17 and 25 μmol l^{-1}, is known as North Atlantic Deep Water and originates in high arctic regions. This water is dense (because of its cold temperature and high salinity) and flows southwards into the Southern Ocean. This is one of the two major sources of deep ocean water and is analogous to the more familiar surface ocean currents, although its driving force is caused by the physical properties of the water rather than the global wind fields, which predominantly drive the surface circulation. It can be seen that nitrate concentrations gradually increase in the direction of this flow. This is a result of the regeneration of nitrate from complex organic forms which had previously been incorporated into plant and animal matter, and which is transported in the mass water flow. Regeneration of nitrate is relatively slow and it can be appreciated that the nitrate concentrations will gradually increase with time. This introduces the concept of age of a water mass. Thus, to a first approximation, the 'older' the water mass, the higher are its constituent nitrate concentrations. The 'oldest' (about 4000 years since last at the surface) water masses are to be found in the central southern Pacific ocean and here the nitrate concentrations are the greatest. Broeker and Peng (1982) provide an excellent review of these aspects.

Other water masses are identifiable in Figures 7.12(b) and 7.12(c). One of the most important of these for oceanic circulation is the Antarctic water. This is the major source of deep water for the world ocean and is found at the bottom of all the oceans, and can be identified in the figures from its nitrate concentrations of between 30 and 34 μmol l^{-1}. Although nitrate concentrations increase with age, as noted above, and the 'oldest' waters contain the highest nitrate concentrations, the relationship is complicated by the fact that the initial (starting) concentration of nitrate in deep oceanic water is not everywhere the same. Thus the water mass clearly identifiable in the eastern Atlantic at about 1 km depth, Antarctic Intermediate Water, is relatively young, but contains high nitrate concentrations because its initial concentration is high.

A particularly interesting aspect of nitrate in oceanic waters influenced by upwelling is that the nitrate concentration is sometimes significantly lower than expected from a consideration of water masses and regeneration. The difference is due to denitrification occurring in the water column. Although denitrification is predominantly a process restricted to sediments (see Section 7.5), there are areas of the ocean where oxygen levels are sufficiently low to allow denitrification to take place (see Hattori, 1983). These are centred around upwellings, and result from the high (nitrate-supported) biological activity which stimulates organic matter production. The organic matter, once formed, sediments out of the surface waters, where bacterial degradation occurs. The combined effect of high degradation rates, which consume oxygen, and the hydrography leads to zones where oxygen becomes depleted or almost absent. Under such conditions denitrification can occur, and substantial losses of nitrate result. the phenomenon is present to a greater or lesser extent in all the major upwelling centres.

A large denitrification zone is present year-round in the Arabian Sea. This has as its basis extremely high phytoplankton production, stimulated by the south-westerly monsoon, which

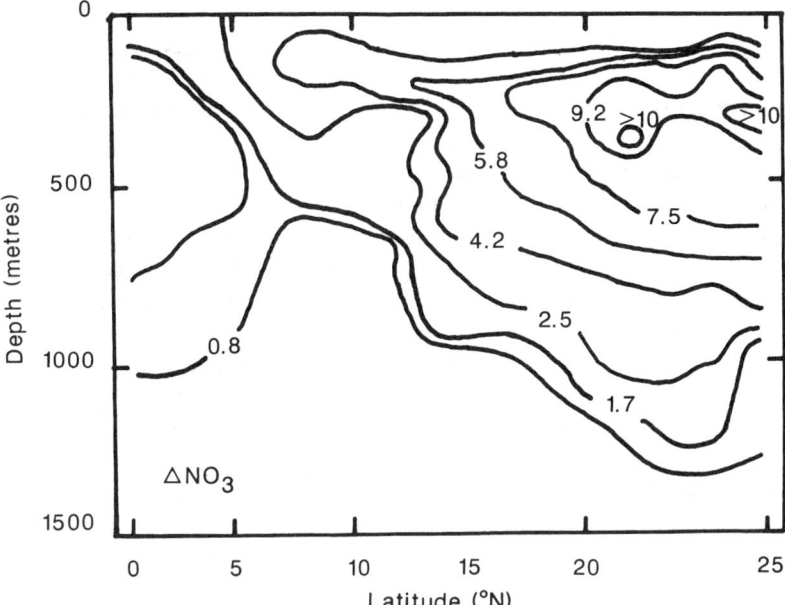

Figure 7.13 Vertical distribution of the 'nitrate anomaly' (-ve values, units: μmol NO_3-N l^{-1}) along the same transect in the Indian Ocean as shown in Figure 7.11(b). The 'nitrate anomaly' is the difference between the observed concentrations of nitrate and those expected from a knowledge of the source and age of the water mass. The large anomaly observed in the northern Indian Ocean is due to denitrification (reproduced from Mantoura *et al.*, 1993, by permission of Pergamon Press Ltd)

drives the strong upwelling noted above. The degree of the nitrate anomaly (i.e. the difference between the observed nitrate concentrations and those expected from the age of the water — Broeker and Peng, 1982) can be seen in Figure 7.13. Other important open-water denitrification areas are the eastern tropical north and south Pacific (Goering *et al.*, 1973; Codispoti and Packard, 1980), the Cariaco trench, Caribbean (Hattori, 1983), eastern tropical Atlantic (Goering, 1978), and the Black Sea (Goering *et al.*, 1973). These denitrification zones (and possibly others yet to be identified?) are important controls of the global nitrogen cycle.

A consequence of the high rates of denitrification is that high concentrations of nitrous oxide, a by-product of denitrification, are formed. Nitrous oxide is an important 'greenhouse gas' and interest in marine sources of nitrous oxide has increased recently with the realisation of 'global warming' (IPCC, 1990). Recent studies have shown, for example, that the Arabian Sea denitrification zone alone may contribute to a large proportion of the natural sources of nitrous oxide to the atmosphere (Law and Owens, 1990). It is possible that the marine sources have been severely underestimated.

It can be seen that the story is not a simple one, but a full coverage of deep oceanic circulation and its consequences for nitrate concentrations is outside the scope of this discussion. However, in principle, the cycle of nitrate in the deep ocean is rather like a 'roundabout' where nitrate is removed from surface waters by biological activity and converted to organic nitrogen. The organic nitrogen is removed from the surface waters either by sinking or entrainment in the deep circulation. Here, nitrate is gradually regenerated and, after sometimes several thousands of years, is transported back to the surface waters, where the cycle is repeated.

7.3.4 HIGH-LATITUDE SYSTEMS

It can be seen from Figures 7.12(b) and 7.12(c) that nitrate concentrations in the surface oceanic waters are greatest in high latitudes. This is primarily the result of the limited utilisation of nitrate by phytoplankton, although there are other contributory factors for the seas around Antarctica. Biological utilisation rates are reduced in these areas largely due to the limited availability of light. The major cause of the general lack of light is the severe restriction in solar radiation because of the high latitudes of these regions. But, just as these latitudes severely restrict light during the winter, the summer condition is one of greater light availability. However, two other factors restrict light availability further; ice-cover and turbulence. Ice cover is a major factor in both high-latitude regions and large areas of sea-surface are influenced (see, for example, Figure 7.14(a)). The effect of sea ice is to severely limit light penetration and reduce the amount of time suitable for phytoplankton growth. Coupled to this is the fact that these regions are frequently stormy (particularly the southern high latitudes) which produces a deep mixing of surface waters and carries phytoplankton cells into depths of water beyond which light cannot penetrate. Thus, although there may be almost continuous light during the summer there may still be a severe restriction on light availability. The combined effect is to prevent the full utilisation of nitrate with the result that nitrate concentrations remain generally high. Another suggestion which is receiving particular attention currently to explain the lack of utilisation of nitrate, is that algal growth is limited by the lack of iron, particularly in the Southern Ocean (Martin *et al.*, 1990a,b). As living organisms, phytoplankton require a variety of substances to suport their growth, including iron which is found in extremely low concentrations in the Southern Ocean. While there is some evidence to support this hypothesis, other observations are contradictory (de Baar *et al.*, 1990).

The average levels of nitrate in the surface waters of the Southern Ocean are around 20 μmol l^{-1}. It can be seen that there is an approximately circumpolar distribution bounded by the Antarctic convergence. This is a major oceanic 'front' bounding the warmer subtropical, less dense, and lower nitrate-containing waters to the north from the denser surface Antarctic water to the south. It is in this region that the northward-flowing Antarctic water dips below the subtropical water and there is a marked discontinuity in nitrate concentrations (this sinking Antarctic water is the source of the Antarctic Intermediate Water referred to above). Well to the south of the convergence there is a zone of upwelling which maintains the high levels of nitrate. Note that this upwelling is density driven rather than caused by the Ekman pumping outlined above. Reduced surface nitrate concentrations do occur in the Southern Ocean over small space scales (tens of kilometres). These result from short-lived accumulations of phytoplankton in areas where physical mixing is reduced, for example, around islands (e.g. Owens, Priddle and Whitehouse, 1991) or small eddies (Heywood and Priddle, 1987). Nitrate concentrations may be reduced to about 10 μmol l^{-1}, but this is only a temporary phenomenon.

Nitrate concentrations in the northern high latitude seas are lower than the Southern Ocean, being around 10−12 μmol l^{-1}. Like the Southern Ocean, nitrate concentrations remain high because of restricted biological activity (Figure 7.15).

An important feature unique to the seasonally ice-covered regions of both polar regions is the stimulation of high biological activity at the edge of the receding ice, during spring and summer. This area, the marginal-ice-zone (MIZ), exhibits marked variability in nitrate concentrations. The width of the MIZ is variable, but is of the order of 100−200 km 'seawards'

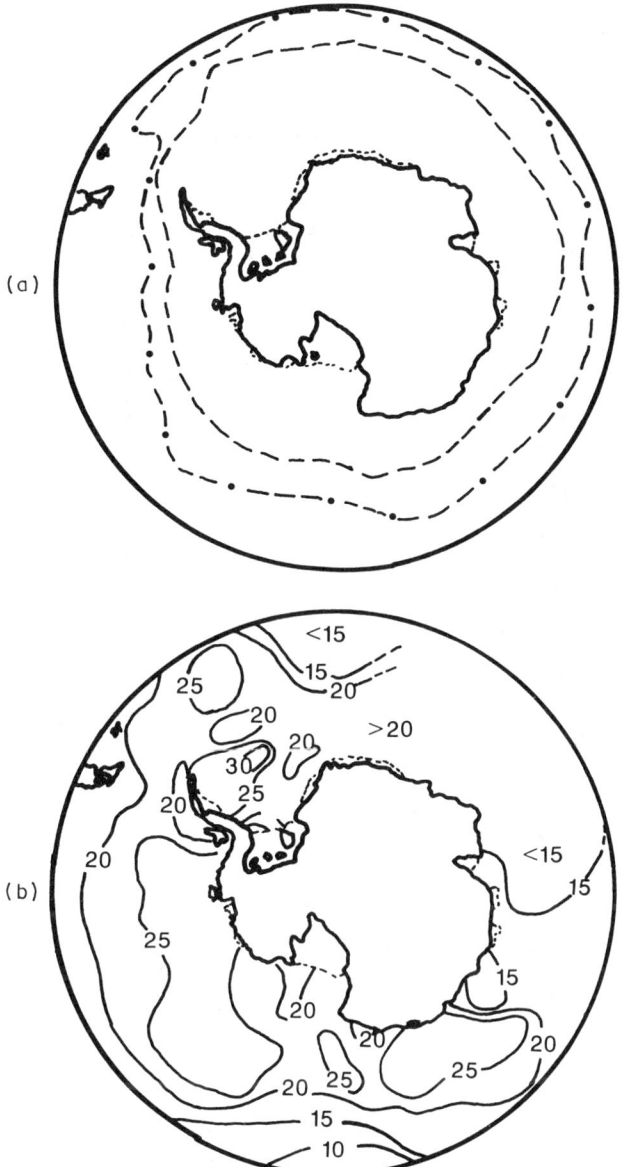

Figure 7.14 (a) The Southern Ocean showing the Antarctic Convergence (•– –•), the maximum extent of sea ice (– – –) and the minimum extent (- - - - -) close to the Antarctic continent. (b) Typical distribution of nitrate (μmol NO_3-n l^{-1}) (adapted from Foster, 1984)

from the ice-edge. The basis of the stimulated activity is the formation of a stable water column through the formation of low-salinity water from the ice-melt. This forms a surface layer of low density which severely restricts vertical mixing, thus increasing light availability to the phytoplankton. Under these conditions a dense ice-edge population of phytoplankton can be formed which can severely deplete the nitrate concentrations (for review of the MIZ system see Smith, 1987). Figure 7.16 shows a section through a MIZ in the Fram Strait (north-east

192

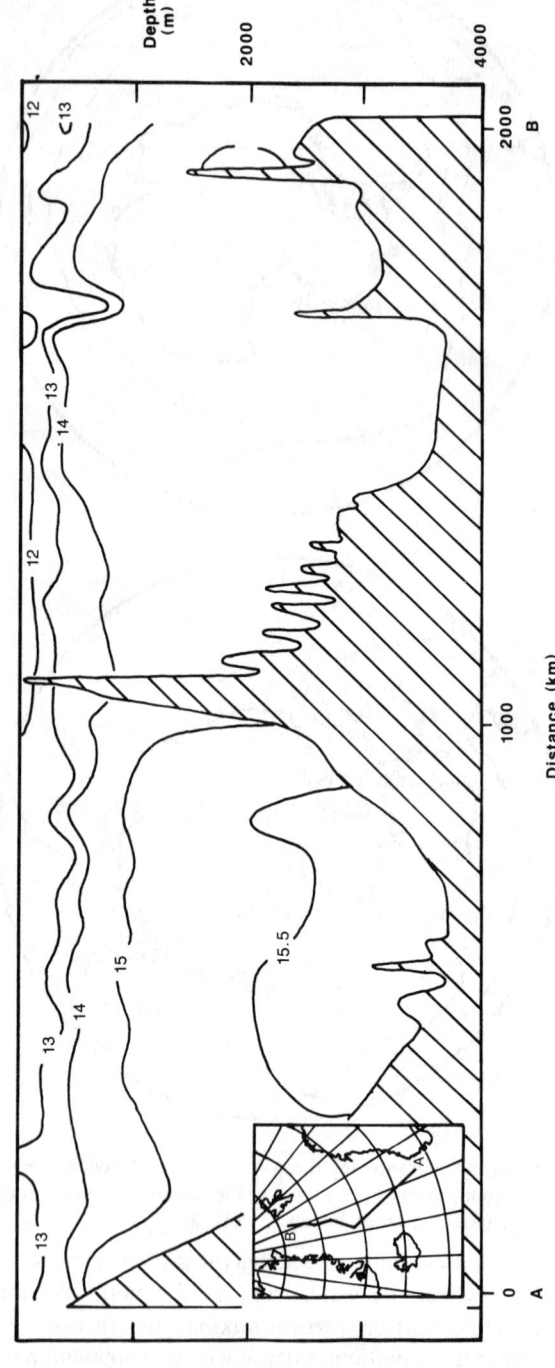

Figure 7.15 Vertical distribution of nitrate (μmol NO$_3$-N l^{-1}) in the Greenland Sea (see inset for location) (reproduced from Clarke *et al.*, 1990, by permission of Pergamon Press Ltd)

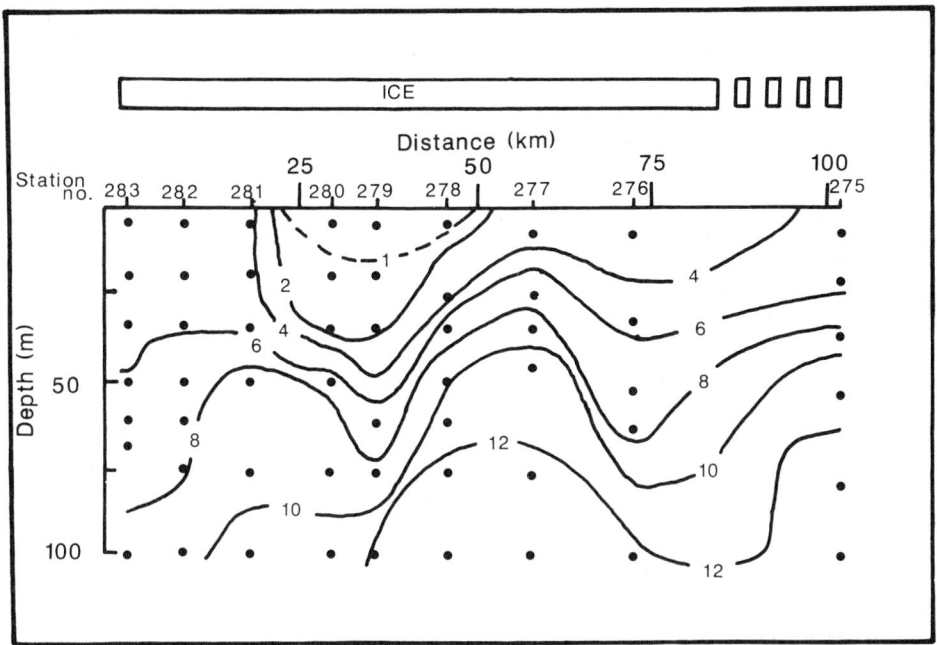

Figure 7.16 Distribution of nitrate (μmol NO_3-N l^{-1}) beneath seasonal sea ice and the marginal ice zone in the Fram Strait (Greenland Sea) (reproduced by permission from Smith and Kattner, 1989)

Greenland), and the reduced nitrate concentrations can be clearly seen. It is also interesting to note that nitrate concentrations are depleted under the ice, indicating that biological activity is not restricted to the ice-free waters. Also in this example, nitrate concentrations increase at the extreme edge of the ice. In other studies the reduced nitrate concentrations have been observed to extend many kilometres from the ice-edge. This phenomenon can lead to large areas of nitrate depletion. Upwelling of high nitrate-containing water occurs at the ice-edge and is important in maintaining a supply of nitrate to support the phytoplankton growth (Alexander and Niebauer, 1981; Buckley *et al.*, 1979). Phytoplankton growth at the ice-edge, supported by nitrate, is very important for the overall production of high-latitude systems.

7.4 SEASONAL CYCLES

Although seasonal variability has been referred to in the previous sections, it was thought worth summarising this in one section. Figure 7.17 is a schematic diagram bringing together all the marine areas and shows a generalised seasonal distribution of nitrate for surface waters.

The tropical and subtropical areas are the most stable marine environments with respect to nitrate. They are extremely oligotrophic, but can exhibit temporarily elevated concentrations through episodic atmospheric inputs and through periodic storm-induced mixing of deeper waters.

Temperate waters undergo very large variations in surface nitrate concentrations. The formation of the seasonal thermocline, which is the basis for the variability, was described above (Section 7.3.2). This cycle is of extreme importance for shelf seas since the

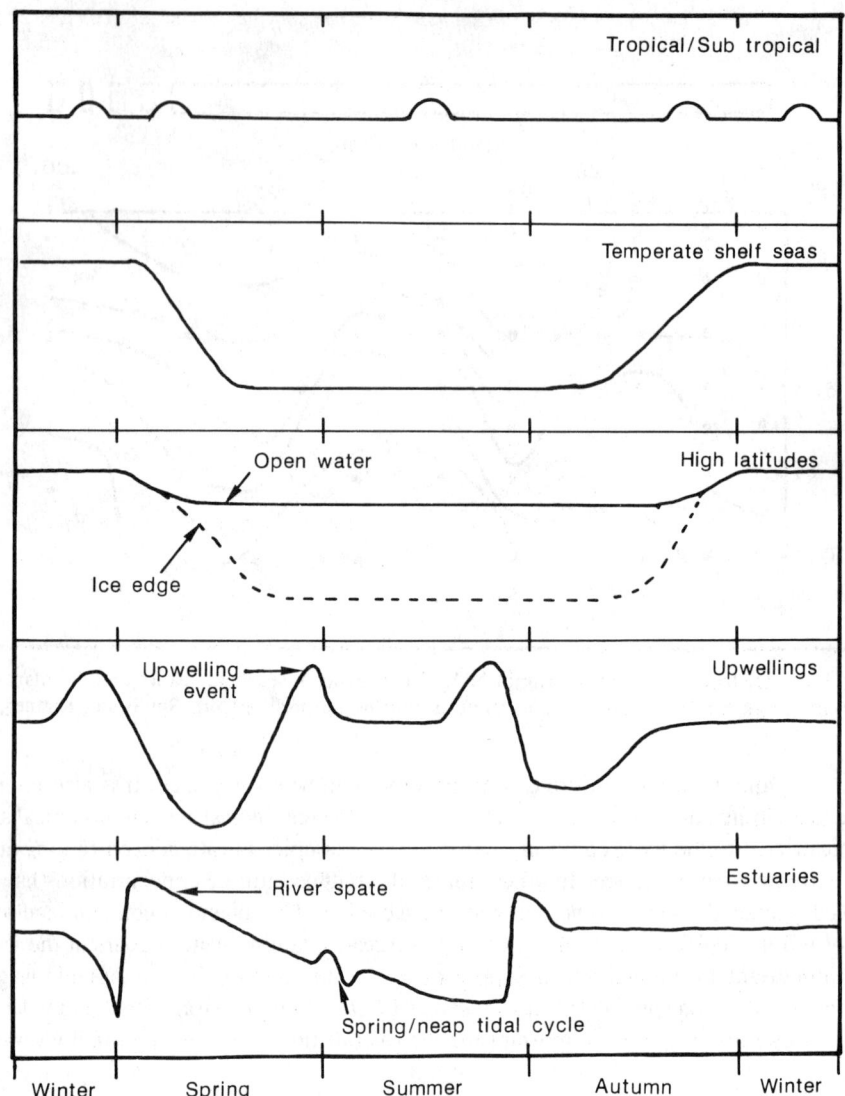

Figure 7.17 Schematic diagram showing the tyical seasonal variation in nitrate concentration in a range of marine environments

concentrations of nitrate fall to levels which may limit the growth of phytoplankton. Indeed, nitrate concentrations may fall to levels as low or lower than the highly oligotrophic central tropical gyres (e.g. Woodward and Owens, 1990). This has led to the concept of the limiting nutrient, and the possibility that nitrate introduced by human activity may lead to eutrophication. While eutrophication is a well-documented phenomenon in inland waters, the position is far less clear for marine waters (see Section 7.6).

Nitrate concentrations in the high-latitude areas are relatively stable over seasonal timescales. Slight depletion may be observed during summer months, particularly in the physically less dynamic areas. Major reductions in nitrate may occur, however, in the marginal ice-zones.

Extreme coastal areas in shallow water may exhibit temporary stability (e.g. Whitaker, 1982), in which case a cycle similar to the temperate shelf seas may be initiated. However, such events are ephemeral, in contast to the repeating cycle in temperate waters, and are brought about by local topography (e.g. Perissinotto and Duncombe-Rae, 1990) and short-term weather patterns.

Upwelling systems exhibit large fluctuations in surface nitrate concentrations. While the annual climatic features may be largely cyclical on an annual basis, that is, there are 'upwelling seasons' (e.g. Shannon and Pillar, 1986), whether upwelling occurs, or its intensity, is governed by more local effects. Thus upwelling 'events' take place which have the effect of introducing 'pulses' of nitrate into the surface water. Following these pulses, should upwelling decline, surface nitrate concentrations may fall, to be replaced some time later by a further pulse.

Estuaries also exhibit widely fluctuating seasonal cycles in nitrate concentrations. Because estuarine nitrate concentrations are dominated by riverine inputs, the seasonal variability in river nitrate concentrations, together with river flow, combine to cause rather unpredictable concentrations (Morris, Bale and Howland, 1981). Also, since estuaries are influenced greatly by the balance of seawater input, a cyclical variability may be observed caused by the regular spring-neap tidal cycle observed in many coastal regions. The residence time of water in some estuaries may be sufficiently long to allow phytoplankton populations to develop, and if this is seasonal, a depletion of nitrate may be observed (see Figure 7.4). Seasonal production of nitrate caused by nitrification has also been reported (Billen, 1975; Owens, 1986).

7.5 SEDIMENTS

Sediments are an important part of the cycling of nitrate in the marine environment. Their main role is predominantly one of involvement in the regeneration processes of the nitrogen cycle (see Chapter 5), but, as will be seen, they are also important sites leading to a loss of nitrate from the system. Because of their importance in the marine nitrogen cycle, sediments have received much attention. However, the majority of the work has been carried out on nearshore sediments due to the practical difficulties of working offshore. Thus much of the following has been derived from work describing the nearshore system.

Since the main reactions involving nitrate in sediments are mainly those of the bacterially mediated processes of nitrification and denitrification, it is useful to briefly consider these reactions. Nitrification is the oxidation of ammonium (NH_4^+) via nitrite (NO_2^-) to nitrate (NO_3^-). Although chemical nitrification can occur, the most important mediators responsible for the reaction in sediments are the autotrophic nitrifying bacteria. The most important feature of the nitrifiers, as far as marine sediments are concerned, is that they are aerobic organisms; thus nitrification can only occur in the presence of oxygen. Nitrification can lead to the formation of the gas nitrous oxide (N_2O), and this may be an important natural source of the gas on the global scale. However, although intimately linked with the processes considered in this section it is outside the scope of this discussion. Denitrification is the reduction of nitrate to di-nitrogen gas (N_2). This too is mediated by a diverse group of micro-organisms but their most important feature is that the reaction only occurs in the absence or near-absence of oxygen. Thus nitrification and denitrification are mutually exclusive. Denitrification is also a major route for the formation of nitrous oxide in the marine environment. Because of the gaseous end-products, the importance of denitrification lies in the fact that it provides the only route whereby nitrogen may be lost from the marine system. An alternative reduction of nitrate occurs in marine sediments which forms ammonium as the end-product; this is known as nitrate-reduction to distinguish it from denitrification. Nitrate reduction also requires anaerobic conditions. Both nitrification and denitrification occur to varying degrees in marine sediments, and it can be seen that the balance of the processes governs the levels of nitrate that may be present. Reviews on this complex system are provided by Kaplan (1983), Hattori (1983), Henriksen and Kemp (1988) and Koike and Sorensen (1988).

Because of the overriding influence of oxygen, the principal control of nitrification and denitrification is the oxygen status of the sediments. Assuming a fully aerated water column, oxygenated water will penetrate some distance into the sediment. A band of oxygenated sediment is thus created at the surface. Oxygen is consumed in the sediments by heterotrophic (mainly bacterial) activity, thus there is an oxygen demand on the sediment. The supply of oxygen to the surface sediments is usually sufficient to maintain their relatively high oxygen status. However, at greater depths oxygen penetration becomes increasingly restricted and the sediments become anoxic. Thus a band devoid of oxygen is found beneath the surface-oxygenated zone, and conditions are induced that allow both nitrification and denitrification to occur at the same site, but theoretically separated in the vertical axis.

Oxygen penetration is governed predominantly by two factors: the porosity of the sediments and their organic content. Coarse sediments (sands) allow a greater penetration of oxygen than muds, simply because of their physical structure, and this has an important influence on the distribution and cycling of nitrate in marine sediments. However, the more important feature is that of the organic content of the sediment. To a first approximation, the greater the organic content, the lower will be the oxygen status of the sediment. Generally, estuarine and nearshore sediments have a higher organic content than deep-sea sediments because of higher rates of sedimentation of organic matter. Thus, the nitrate content of nearshore sediments is usually lower than deep-sea sediments, primarily because of the balance between nitrification and denitrification (see Figure 7.18).

Note that the ratio of nitrification to denitrification appears to be related to water depth in these coastal sediments. This is brought about predominantly by the organic status of the sediments rather than water depth *per se*, and intuitively the relationship probably reaches an asymptote, perhaps somewhere at the shelf-break. However, there are insufficient data to establish this. These general considerations are, of course, an oversimplification, because differences in tidal scouring of sediments and sedimentation rates frequently prevent organic-rich sediments developing in nearshore environments. Nevertheless, the rates of nitrification and denitrification generally are highest in coastal environments, particularly estuaries and salt marshes, which are subject to high loads of organic sedimentation (Table 7.4).

The simplified model of a two-layered oxic and anoxic sediment described above is, however, frequently not observed, but rather, the highest rates of denitrification are commonly in the surface sediments (e.g. Kieskamp *et al.*, 1991; Law, Rees and Owens, 1991). This is brought about by the complex interaction of two factors. First, the simple oxygen profile described may be changed by the presence of anoxic micro-sites, close to the surface, caused by the presence of millimetre-sized pieces of organic matter. Second, the presence of fauna in the sediments drastically changes the surface area of the sediments and the overall oxic/anoxic status by the presence of burrows and bioturbation. The presence of animal burrows compresses the oxic/anoxic boundary and the activity of the animal introduces oxygen (through irrigation) and nutrient levels (through excretion). Thus the nutrient, and particularly oxygen profiles, are commonly significantly different from the ideal model (see Aller, 1988).

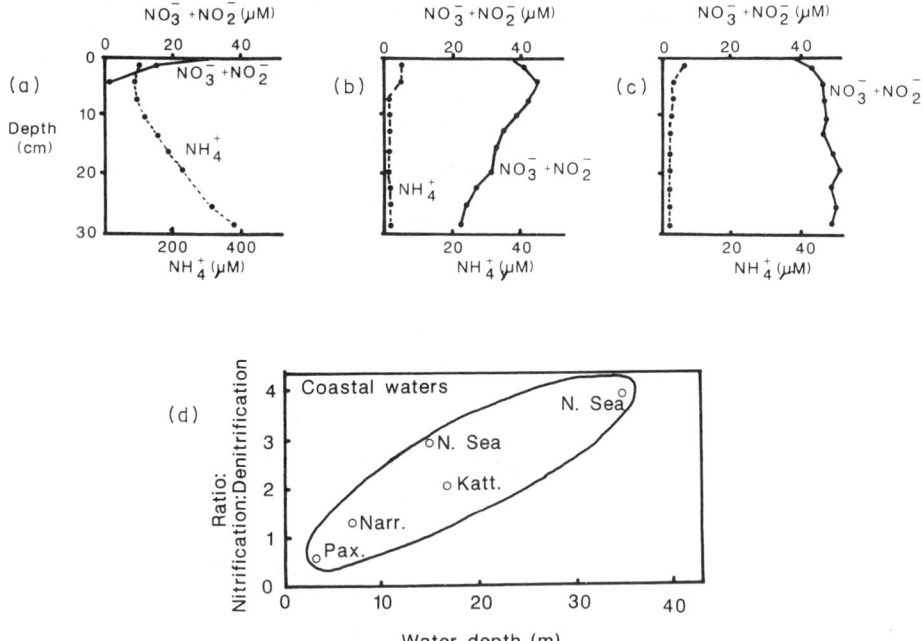

Figure 7.18 Top panels: distribution of nitrate (and ammonium) in sediments from various marine environments. (a) Typical organic-rich, shallow-water coastal sediment (Sagami Bay, Japan) showing steep decline in nitrate concentration due to active nitrate reduction. (b) Offshore site (Ogasawara Trench; 20°28′N, 143°20′E), with lower organic input exhibiting nitrification in the top 1−10 cm sediment, overlying a nitrate reduction zone. (c) Offshore site (western Pacific Ocean; 26°57′N, 142°55′E) with minimal organic input showing high nitrate throughout the sediment (nitrification) with no nitrate reduction (reproduced from Koike and Sorensen, 1988, by permission of Academic Press). Bottom panel (d): Ratio of nitrification to denitrification in sediments of various water depths (reproduced by permission from Henriksen and Kemp, 1988, SCOPE 33)

Also, the source of nitrate which is denitrified is not necessarily from the overlying water, but may be coupled to nitrification. For example, the rate of denitrification in the North Sea was closely related to the nitrate content in the surface 2 cm of sediment; however, this was not related to the overlying water concentration, but rather was controlled by the rate of nitrification in the sediment (Law and Owens, 1990). The percentage of the denitrification coupled to nitrification has been shown to vary between low levels of 7% (e.g. Nishio, Koike and Hattori, 1982) to 100% (e.g. Christensen and Rowe, 1984), but with a tendency towards the higher levels. Thus, maximal denitrification rates often are observed close to the zones of nitrification. All of these features make for a particularly complex ditribution of nitrification and denitrification in sediments.

It is not surprising that nitrification and denitrification exhibit marked seasonal variability. Examples are shown in Figure 7.19. Because of their biological controls, the processes are usually related to temperature (e.g. Hill, 1979), and thus highest rates are generally found during the summer. However, Figure 7.20 shows two peaks in denitrification rate. This may be a result of increased activity of benthic fauna, which have been suggested to be important in supplying nitrate to the sites of denitrification by increased irrigation (Aller, 1988), or the increase in the nitrate concentration of the overlying water.

Table 7.4 Rates of nitrate flux, nitrification and denitrification in marine sediments (μmol N m^{-2} h^{-1})

	Nitrate flux	Nitrification	Denitrification
Estuaries and salt marshes			
Bowden (1986)	−66 to 898		
Jorgensen and Sorensen (1988)	−420 to 1670		
Jorgensen and Sorensen (1985)	−670 to −2000		
Nishio *et al.* (1982)	−1145 to 1150		
Hill (1983)	−30 to −565		
Van Kessel (1977)	−2700		
Callender and Hammond (1982)	−15 to −60		
Seitzinger (1987)	−107 to 50		
Seitzinger *et al.* (1980)	15		
Seitzinger *et al.* (1984)	91		
Nedwell (1982)	−1		
Law *et al.* (1991)	−400 to −1600		
Hansen *et al.* (1981)		30 to 125	
Valiela and Teal (1979)		36 to 357	
Kaplan *et al.* (1979)		0 to 6	
Seitzinger *et al.* (1984)		10 to 130	
Macfarlane and Herbert (1984)		10 to 130	
Kaspar (1983)			6 to 12
Andersen *et al.* (1984)			0 to 150
Oren and Blackburn (1979)			7
Smith *et al.* (1983)			57
Valiela and Teal (1979)			71 to 214
Jorgensen and Sorensen (1988)			83 to 417
Jensen *et al.* (1988)			2 to 42
Seitzinger *et al.* (1980)			50
Jenkins and Kemp (1984)			77 to 89
Jorgensen and Sorensen (1985)			21 to 33
Nishio *et al.* (1982)			130 to 800

An important feature shown in Figure 7.20 is the relationship between the rate of denitrification and nitrate loading to the estuary from the riverine input. It can be seen that during the summer momths, when the inputs are at their lowest but denitrification rates are highest, all of the nitrate entering the estuary can be accounted for by denitrification. This has important implications for budgets of nitrogen entering the coastal waters and possible eutrophication effects, because riverine inputs are calculated prior to any possible modification by the estuary. Thus the current estimates of riverine inputs to coastal waters may possibly be in error. The weighted annual mean for the study shown, however, suggests that estuarine denitrification could account for only about 8% of the annual riverine input of nitrate. This value should be further reduced by the contribution made by the nitrification−denitrification couple, which was of the order of 50% (Law, Rees and Owens, 1991).

7.6 HISTORICAL ASPECTS AND EUTROPHICATION

Because of the undoubted increase in nitrate concentrations in rivers over recent decades, and the assumption that this must have led to a similar increase in nitrate concentrations in

Table 7.4 (*continued*)

	Nitrate flux	Nitrification	Denitrification
Law *et al.* (1991)			275 to 500
Kieskamp *et al.* (1991)			1 to 55
Coastal and shelf seas			
Henriksen *et al.* (1984)	−10 to −15		
Ennoksen and Samuelsson (1987)	−24 to 2		
Teague *et al.* (1988)	−112 to 79		
Sorensen (1984a)	−125 to −625		
Henriksen *et al.* (1981)	0 to 29		
Boynten and Kemp (1985)	−120 to 288		
Blackburn and Henriksen (1983)		33	
Seitzinger *et al.* (1984)		118	
Koike and Hattori (1978)		387 to 982	
Vanderborght and Billen (1975)		178 to 1430	
Sorensen i1978)			6 to 42
Henriksen *et al.* (1984)			40 to 45
Henriksen *et al.* (1981)			0 to 15
Koike and Hattori (1979)			21
Kaspar (1982)			13
Law and Owens (1990)			1 to 13
Offshore sediments			
Rittenberg *et al.* (1955)		2	
Bender *et al.* (1977)		3	
Sorensen (1984b)			0.3
Bender *et al.* (1977)			3.6
Wilson (1978)			10
Hattori (1983)			1 to 16
Devol (1991)			133

Note: negative fluxes indicate flux into the sediment from the water column.

coastal marine waters (but see above), there has been great concern expressed about the possibility of coastal eutrophication. Eutrophication of freshwaters has been a subject of debate for a considerable number of years. However, it is only recently that discussion has developed for marine systems. In part, the scientific debate has been stimulated by a very large increase in reports in the popular media. There is no doubt that there has been an increase in the number of 'exceptional algal blooms' *reported*. However, whether this increase is real or a product of observer effort (which obviously leads to an apparent increase) is not established. Certainly, algal blooms (periods of active growth leading to large biomass) are a feature of the natural marine environment, particularly in temperate waters. The real question is whether these have increased to the detriment of the marine environment, and if so, are they related to human activities? Unfortunately, because these questions have been addressed only recently, there are relatively few long-term data.

Unlike freshwaters, where phosphate is commonly assumed to be the nutrient that leads to problems of eutrophication (but see Chapter 5), attention is usually focused on nitrate in the marine system. This is because nitrate is usually the nutrient which becomes depleted first during periods of algal growth. As noted above, there are few long-term data to examine whether the concentrations of nitrate have increased in marine coastal waters. One of the

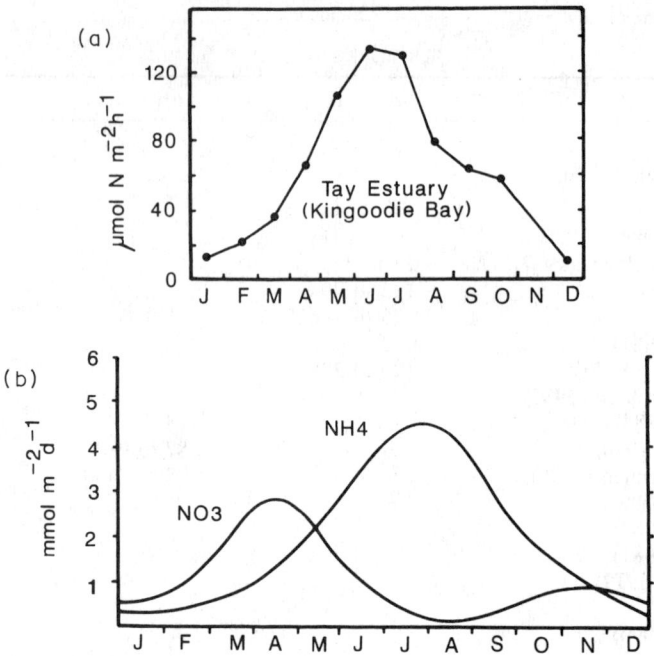

Figure 7.19 (a) Seasonal variation in the rate of sediment nitrification in an estuarine sediment (adapted from Macfarland and Herbert, 1984, and reproduced from Henriksen and Kemp, 1988). (b) Seasonal flux of nitrate (and ammonium) from shallow sediments from the Limfjorden, Denmark (reproduced from Blackburn, 1988, by permission of the American Society of Limnology and Oceanography)

most extensive data sets is from a study from the island of Heligoland, in the German Bight of the North Sea. The site is situated in an area exhibiting elevated nutrient concentrations (see Figure 7.8) from riverine inputs from much of the European coast. Figure 7.21 shows a marked increase in winter nitrate concentrations over the period 1962 to 1984. The summer period of reduced nutrients is clear, and these too appear to have increased over the period. These data together with marked changes in the composition of the phytoplankton (see Radach, Berg and Hagmeier, 1990) are strong evidence for eutrophication. However, these data are not entirely straightforward because the sampling site is situated in an area exhibiting complex hydrography (Otto et al., 1990) and changes in circulation and salinity leading to apparent nutrient increases cannot be excluded. Nevertheless, this example probably does indicate that eutrophication of nearshore waters is developing as a result of riverine nutrient inputs. Similar trends have been observed for other enclosed or semi-enclosed sea areas, (for example, the Baltic Sea (Rosenberg et al., 1990)).

Other data on the temporal distribution of numbers of algae in nearshore waters are available and are also strong evidence for eutrophication (e.g. Figure 7.22), although a direct causal link has yet to be established. A particular difficulty in establishing changes in the plankton is that changes over the timescale of interest (decades) are known to occur that can only have their origin in larger scale (e.g. meteorological, circulation) processes influencing whole ocean basins. Such trends are known from a long-term study known as the Continuous Plankton Recorder Survey (CPR) (Glover, 1967). This survey covers large areas of the north Atlantic Ocean and has shown marked fluctuations in the plankton, since the early 1940s, with common

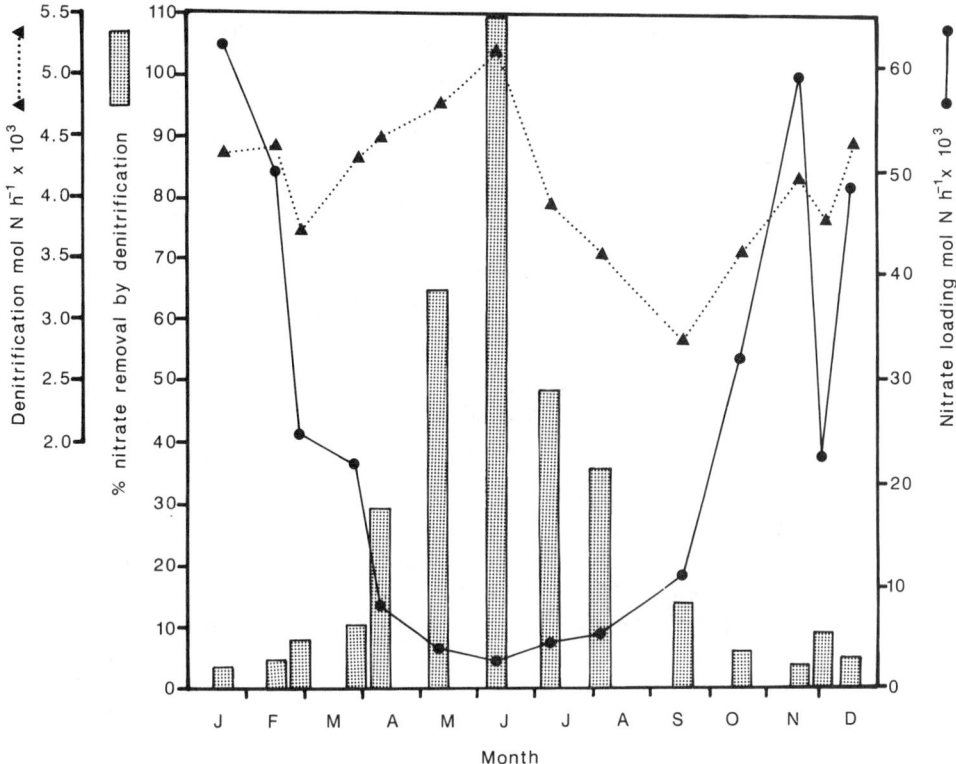

Figure 7.20 Seasonal variation in denitrification in estuarine sediments (Tamar estuary, south-west England). Also shown are the loading of nitrate to the estuary from the river, and the percentage of the loading calculated to be removed by denitrification (reproduced from Law, Rees and Owens, 1991, by permission of Academic Press Ltd)

trends frequently being observed over the whole region, including shelf sea areas such as the North Sea (Colebrook, 1982; Aebischer, Coulson and Colebrook, 1990). A particular example is provided by the alga *Phaeocystis*. This is a notable bloom-forming species which is particularly common in the southern North Sea. Under certain conditions of growth and weather patterns, large concentrations of foam are produced by this alga (Lancelot *et al.*, 1987) which can affect resort beaches. This species is considered to have increased in recent years as a result of eutrophication (Cadee, 1990; Lancelot *et al.*, 1987). Figure 7.22 shows that this alga has increased in recent years in nearshore waters, and it is reasonable to assume that this is a result of the known increases in nutrients at the same sites. However, *Phaeocystis* has shown a marked change in its occurrence over the whole north-east Atlantic from particularly high levels in the late 1940s through a decline in the mid-1970s followed by an increase in recent years. These changes cannot be a result of increases in nutrients since mid-ocean sites showed a very similar change (Figure 7.23). Also, the whole plankton community shows a similar trend (Owens *et al.*, 1989a). Thus it is possible that the observations shown in Figure 7.22 are a reflection of a more general, 'natural' phenomenon. However, the comparison between the data in Figures 7.22 and 7.23 is not entirely valid because the CPR survey poorly samples the nearshore environment. Nevertheless, this example serves to demonstrate the difficulties in establishing the existence or extent of eutrophication. Interesting

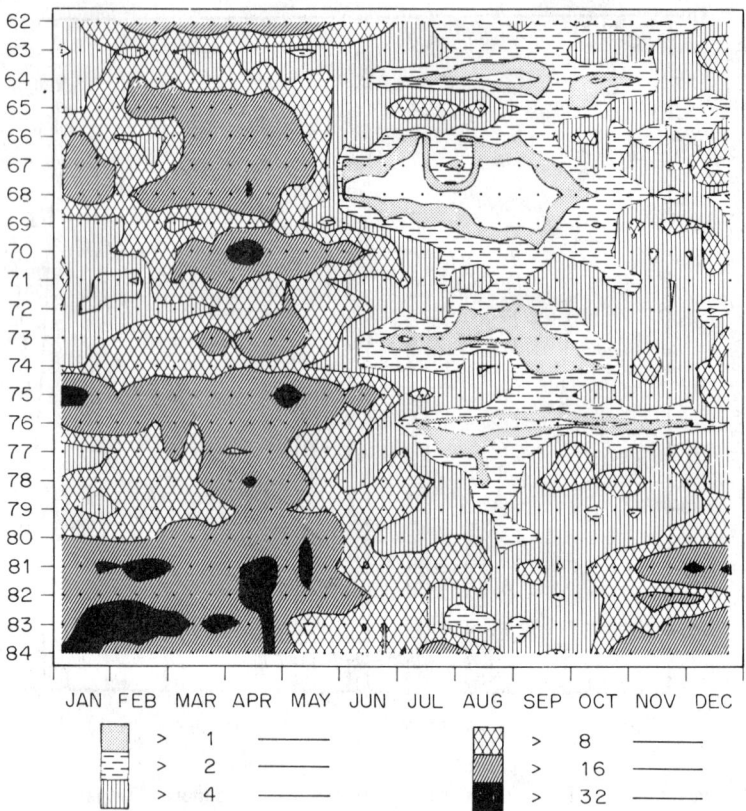

Figure 7.21 Seasonal variation in the concentration of nitrate (mean concentration over ten days; units: μmol NO_3-N l^{-1}) for the years 1962 to 1984 at a sampling location in the German Bight of the North Sea (reproduced from Radach, Berg and Hagmeier, 1990, by permission of Pergamon Press Ltd)

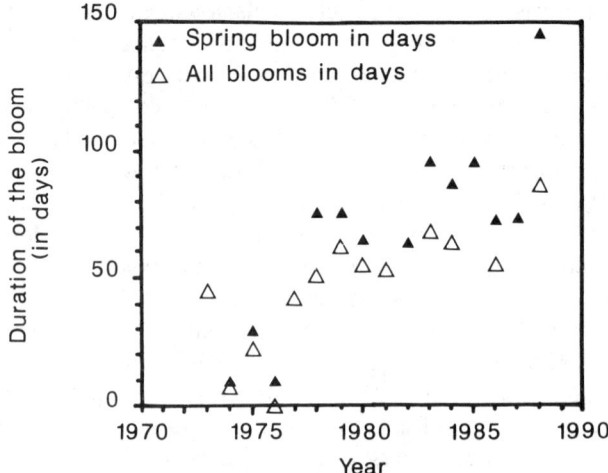

Figure 7.22 Duration of the spring bloom, and blooms occurring at other times of the year of the alga *Phaeocystis* sp. in the Marsdiep, The Netherlands (reproduced by permission from Cadee, 1990)

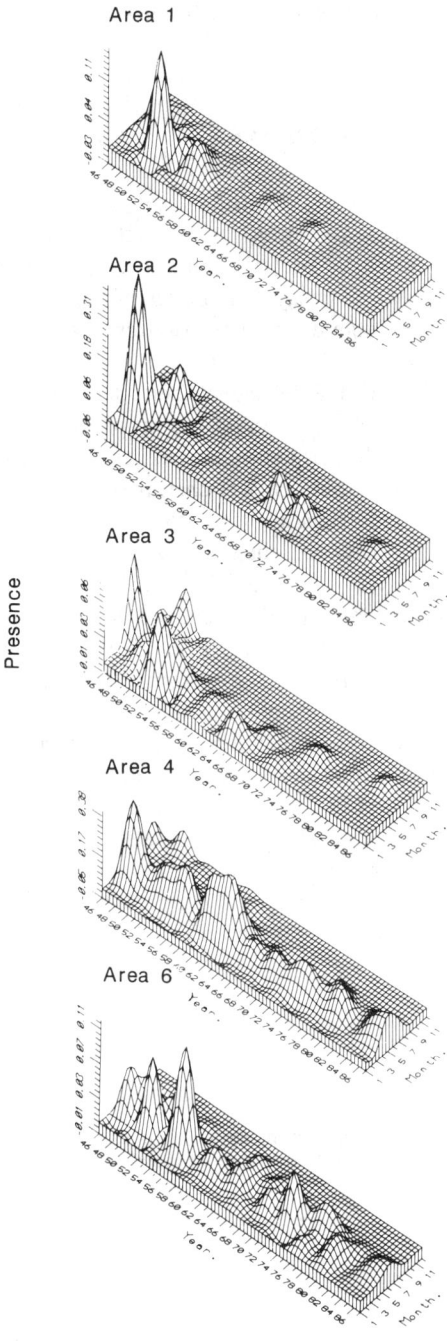

Figure 7.23 The occurrrence of *Phaeocystis* sp. in various regions of the north-east Atlantic and North Sea, for each month between the years 1946 and 1987 inclusive. Area 1: Atlantic $55°-63°$N, $11°-22°$W; Area 2: Atlantic $48°-55°$N, $11°-22°$W; Area 3: northern North Sea; Area 4: southern North Sea; Area 6: Irish Sea and western English Channel (reproduced from Owens *et al.*, 1989a, by permission of Cambridge University Press)

reviews addressing eutrophication questions have been provided by Reid *et al*. (1990) and
Smetacek *et al*. (1991).

7.7 SUMMARY AND CONCLUSIONS

It is clear from the above that the cycling of nitrate in marine systems is particularly diverse
and involves environmental driving forces as varied as global-scale meteorology to small-
scale variations in the distribution of bacteria. Nitrate cycling also encompasses human
activities, although it is probably true to say that human influence has probably had the least
effect on the marine natural cycle than in any other environment. Nevertheless, the possible
prospect of human-induced global climate change, which may have a profound effect on
oceanic circulation, cannot be excluded. Smaller-scale perturbations undoubtedly are occurring;
for example, eutrophication of enclosed sea-areas is probably a real phenomenon.

There is still much to learn. Although not discussed here, the measurement of nitrate in
the sea is based on relatively old-fashioned, wet-chemistry techniques. Refinements to the
techniques have led to a greater understanding of the nitrate cycle, but any major development
must wait for the development of new non-chemical (solid-state?) sensors (see Chapter 16).
On the global scale, the reasons why nitrate levels remain high in the Southern Ocean, for
example, is a current question. Although the subject is worth investigation in its own right,
because it will tell us much about the functioning of the first link in the chain of the marine
ecosystem, suggestions have been made that a fertilisation of the area with iron would stimulate
phytoplankton to utilise the nitrate, which could possibly reduce atmospheric carbon dioxide
levels and thereby ameliorate global warming. The validity of such a large 'eco-experiment'
requires a full understanding of the processes concerned. On a more local level, although
a worldwide problem, is the role of denitrification in reducing humans' inputs of nitrate to
the sea. It is possible that 'nitrate pollution' is moderated to some extent by the process,
but much more study is needed. And what if denitrification is a major sink for anthropogenic
nitrate inputs to the sea, could its stimulation lead to increases in nitrous oxide concentrations
in the atmosphere? Nitrous oxide is a potent 'greenhouse gas' and a human-induced increase
would be highly undesirable.

ACKNOWLEDGMENTS

The author thanks Andy Rees and Duncan Plummer for their assistance with the production of this
chapter. The Plymouth Marine Laboratory is a component of the UK Natural Environment Research
Council.

REFERENCES

Aebischer, N.J., Coulson, J.C. and Colebrook, J.M. (1990) Parallel long-term trends across four marine
    trophic levels and weather. *Nature*, **347**, 753−5.
Alexander, V. and Niebauer, H.J. (1981) Oceanography of the Bering Sea ice-edge zone in spring.
    Limnol. Oceanogr., **26**, 1111−25.
Aller, R.C. (1988) Benthic fauna and biogeochemical processes in marine sediments: the role of burrow
    structures. In Blackburn, T.H. and Sorensen, J. (eds), *Nitrogen Cycling in Coastal Marine
    Environments*, Wiley, Chichester, pp. 301−38.

Andersen, T.K., Jensen, M.H. and Sorensen, J. (1984) Diuranl variation of nitrogen cycling in coastal, marine sediments. *Mar. Biol.*, **83**, 171–6.

Anonymous (1981) *Atlas of the Seas around the British Isles*, Ministry of Agriculture Fisheries and Food, HMSO, London.

Anonymous (1987) *Quality status of the North Sea — Summary*, Second International Conference on the Protection of the North Sea, HMSO, London.

Bailey, B.W. and Chapman, P. (1985) The nutrient status of the St. Helena Bay region in February 1979. In Shannon L.V. (ed.), *South African Ocean Colour and Upwelling Experiment*, Sea Fisheries Research Institute, Cape Town, pp. 125–45.

Barber, R.T. and Smith, R.L. (1981) Coastal upwelling ecosystems. In Longhurst, A.R. (ed.), *Analysis of Marine Systems*, Academic Press, London, pp. 31–68.

Barnes, R.S.K. and Hughes, R.N. (1988) *An Introduction to Marine Ecology*, Blackwell, Oxford.

Bender, M.L., Fanning, K.A., Froelich, P.N., Heath, G.R. and Maynard, V. (1977) Interstitial nitrate profiles and oxidation of sedimentary organic matter in the Eastern Equatorial Atlantic. *Science*, **198**, 605–9.

Billen, G. (1975) Nitrification on the Scheldt Estuary (Belgium and the Netherlands). *Est. Coast. Mar. Sci.*, **3**, 79–89.

Blackburn, T.H. (1988) Benthic mineralisation and bacterial production. In Blackburn, T.H. and Sorensen, J. (eds), *Nitrogen Cycling in Coastal Marine Environments*, Wiley, Chichester, pp. 175–90.

Blackburn, T.H. and Henriksen, K. (1983) Nitrogen cycling in different types of sediments from Danish waters. *Limnol. Oceanogr.*, **28**, 477–93.

Blackburn, T.H. and Sorensen, J. (1988) *Nitrogen Cycling in Coastal Marine Environments*, Wiley, Chichester.

Blanton, J.O. (1991) Circulation processes along oceanic margins in relation to material fluxes. In Mantoura, R.F.C., Martin, J.-M. and Wollast, R. (eds), *Ocean Margin Processes in Global Change*, Wiley, Chichester, pp. 145–63.

Bowden, W.B. (1986) Gaseous nitrogen emissions from undisturbed terrestrial ecosystems: an assessment of their impacts on local and global nitrogen budgets. *Biogeochem.*, **2**, 249–79.

Boynten, W.R. and Kemp, W.M. (1985) Nutrient regeneration and oxygen consumption by sediments along an estuarine salinity gradient. *Mar. Ecol. Prog. Ser.*, **23**, 45–55.

Brockmann, U., Billen, G. and Gieskes, W.W.C. (1988) North Sa nutrients and eutrophication. In Salomons, W., Bayne, B.L., Duursma, E.K. and Forstner, U. (eds), *Pollution of the North Sea: an Assessment*, Springer-Verlag, Berlin, pp. 348–89.

Broeker, W.S. and Peng, T.-H. (1982) *Tracers in the Sea*, Columbia University, New York.

Buckley, J.R., Gammelsrod, T., Johanessen, J.A., Johanessen, O.M. and Roed, L.P. (1979) Upwelling: oceanic structure at the edge of the Arctic pack ice in winter. *Science*, **203**, 165–7.

Cadee, G.C. (1990) Increase of *Phaeocystis* blooms in the westernmost inlet of the Wadden Sea, the Marsdiep, since 1973. In Lancelot, C., Billen, G. and Barth, H. (eds), *Eutrophication and Algal Blooms in North Sea Coastal Zones, The Baltic and Adjacent Sea Areas: Predictions and assessment of preventative actions*, CEC Water Pollution Research Reports, 12, Guyot, Brussel, pp. 105–12.

Callender, E. and Hammond, D.E. (1982) Nutrient exchange across the sediment–water interface in the Potomac River Estuary. *Est. Coast. Shelf Sci.*, **15**, 395–413.

Carpenter, E.J. and Capone, D.G. (1983) *Nitrogen in the Martine Environment*, Academic Press, New York.

Christensen, J.P. and Rowe, G.T. (1984) Nitrification and oxygen consumption in northwest Atlantic deep-sea sediments. *Deep-Sea Res.*, **42**, 1099–1116.

Clarke, R.A., Swift, J.G., Reid, J.L. and Koltermann, K.P. (1990) The formation of Greenland Sea deep water: double diffusion or deep convection? *Deep-Sea Res.*, **37**, 1385–1424.

Codispoti, L.A. and Packard, T.T. (1980) Denitrification rates in the eastern tropical South Pacific. *J. Mar. Res.*, **38**, 453–77.

Colebrook, J.M. (1982) Continuous plankton records: phytoplankton, zooplankton and environment, North-east Atlantic and North Sea, 1958–1980. *Oceanol. Acta*, **5**, 473–80.

de Barr, H.J.W., Buma, J.G.J., Nolting, R.F., Cadee, G.C., Jacques, G. and Treguer, P.J. (1990) On iron limitation of the Southern Ocean: experimental observations in the Weddell and Scotia Seas. *Mar. Ecol. Prog. Ser.*, **65**, 105–22.

Devol, A.H. (1991) Direct measurement of nitrogen gas fluxes from continental shelf sediments. *Nature*, **349**, 319–21.

Ennokson, B. and Samuelsson, M.O. (1987) Nitrification and dissimilatory ammonium production and their effects on nitrogen flux over the sediment–water interface in bioturbated coastal sediments. *Mar. Ecol. Prog. Ser.*, **36**, 181–9.

Eppley, R.W. and Renger, E.H. (1988) Nanomalar increase in surface layer nitrate concentration following a small wind event. *Deep-Sea Res.*, **35**, 1119–25.

Eppley, R.W., Garside, C., Renger, E.H. and Orellana, E. (1990) Variability of nitrate concentration in nitrogen-depleted subtropical surface waters. *Mar. Biol.*, **107**, 53–60.

Foster, T.D. (1984) The marine environment. In Laws, R.M. (ed.), *Antarctic Ecology*, Volume 2, Academic Press, London, pp. 345–371.

Garside, C. (1982) A chemiluminescent technique for the determination of nanomolar concentrations of nitrate and nitrite in seawater. *Mar. Chem.*, **11**, 159–67.

GESAMP (Joint Group of Experts on the Scientific Aspects of Marine Pollution) (1989) *The Atmospheric Input of Trace Species to the World Ocean*, Rep. Stud. 38, WMO, Geneva.

GESAMP (Joint Group of Experts on the Scientific Aspects of Marine Pollution) (1991) *Global Changes and the Air–Sea Exchange of Chemicals*, Rep. Stud. 48, WMO, Geneva.

Glover, R.S. (1967) The continuous plankton recorder survey of the North Atlantic. *Symp. Zool. Soc. Lond.*, **19**, 189–210.

Goering, J.J., Richards, F.A., Codispoti, L.A. and Dugdale, R.C. (1973) Nitrogen fixation and denitrification in the ocean: biogeochemical budgets. *Proc. Symp. Hydrogeochem. Biogeochem.*, **2**, 12–27.

Goering, J.J. (1978) Denitrification in marine systems. In Schlessinger, D. (ed.), *Microbiology — 1978*, American Society of Microbiology, Washington, DC, pp. 357–61.

Hansen, J.I., Henriksen, K. and Blackburn, T.H. (1981) Seasonal distribution of nitrifying bacteria and rates of nitrification in coastal marine sediments. *Microb. Ecol.*, **7**, 297–304.

Hattori, A. (1983) Denitrification and dissimilatory nitrate reduction. In Carpenter, E.J. and Capone, D.G. (eds), *Nitrogen in the Marine Environment*, Academic Press, New York, pp. 191–232.

Henriksen, K. and Kemp, W.M. (1988) Nitrification in estuarine and coastal marine sediments. In Blackburn, T.H. and Sorensen, J. (eds), *Nitrogen Cycling in Coastal Marine Environments*, Wiley, Chichester, pp. 207–49.

Henriksen, K., Hansen, J.I. and Blackburn, T.H. (1981) Rates of nitrification, distribution of nitrifying bacteria and nitrate fluxes in different types of sediment from Danish waters. *Mar. Biol.*, **61**, 299–304.

Henriksen, K., Jensen, A. and Rasmussen, M.B. (1984) Aspects of nitrogen and phosphorus mineralization and recycling in the northern part of the Danish Wadden Sea. *Neth. J. Sea Res.*, **10**, 51–69.

Heywood, R.B. and Priddle, J. (1987) Retention of phytoplankton by an eddy. *Cont. Shelf Res.*, **7**, 937–55.

Hill, A.R. (1979) Denitrification in the nitrgen budget of a river ecosystem. *Nature*, **281**, 291–2.

Hill, A.R. (1983) Denitrification: its importance in a river draining an intensively cropped watershed. *Agric. Ecosys. Environm.*, **10**, 47–62.

IPCC (Intergovernmental Panel on Climate Change (1990) *Climate Change The IPCC Scientific Assessment*, Houghton, J.T., Jenkins, G.J. and Ephraums, J.J. (eds), Cambridge University Press, Cambridge.

Jenkins, M.C. and Kemp, W.M. (1984) The coupling of nitrification and denitrification in two estuarian sediments. *Limnol. Oceanogr.*, **29**, 609–19.

Jensen, M.H., Andersen, T.K. and Sorensen, J. (1988) Denitrification in coastal bay sediment: regional and seasonal variation in Aarhus Bight, Denmark. *Mar. Ecol. Prog. Ser.*, **48**, 155–62.

Jorgensen, B.B. and Sorensen, J. (1985) Seasonal cycles of O_2, NO_3^- and SO_4^{2-} reduction in estuarine sediments: the significance of a NO_3^- reduction maximum in the spring. *Mar. Ecol. Prog. Ser.*, **24**, 65–74.

Jorgensen, B.B. and Sorensen, J. (1985) Two annual maxima of nitrate reduction and denitrification in estuarine sediment (Norsminde Fjord, Denmark). *Mar. Ecol. Prog. Ser.*, **48**, 147–54.

Kaplan, W.A. (1983) Nitrification. In Carpenter, E.J. and Capone, D.G. (eds), *Nitrogen in the Marine Environment*, Academic Press, New York, pp. 139–90.

Kaplan, W.A., Valiela, I. and Teal, J.M. (1979) Denitrification in a salt marsh ecosystem. *Limnol. Oceanogr.*, **24**, 726–34.

Kaspar, H.F. (1982) Denitrification in marine sediment: measurement of capacity and estimate of *in situ* rate. *Appl. Env. Microbiol.*, **43**, 522—7.

Kaspar, H.F. (1983) Denitrification, nitrate reduction to ammonium, and inorganic nitrogen pools in intertidal sediments. *Mar. Biol.*, **74**, 133—9.

Ketchum, B.H. (1951) The exchanges of fresh and salt water in tidal estuaries. *J. Mar. Res.*, **10**, 19—38.

Kieskamp, W.M., Lohse, L., Epping, E. and Helder, W. (1991) Seasonal variation in denitrification rates and nitrous oxide fluxes in intertidal sediment of the western Wadden Sea. *Mar. Ecol. Prog. Ser.*, **72**, 145—51.

Koike, I. and Hattori, A. (1978) Denitrification and ammónia formation in anaerobic coastal sediments. *Appl. Env. Microbiol.*, **35**, 278—82.

Koike, I. and Hattori, A. (1979) Estimates of denitrification in sediments of the Bering Sea Shelf. *Deep-Sea Res.*, **26**, 409—15.

Koike, I. and Sorensen, J. (1988) Nitrate reduction and denitrification in marine sediments. In Carpenter, E.J. and Capone, D.G. (eds), *Nitrogen in the Marine Environment*, Academic Press, New York, pp. 251—73.

Lancelot, C., Billen, G., Sournia, A., Weisse, T., Colijn, Fl., Veldhuis, M.W.J., Davies, A. and Wassman, P. (1987) *Phaeocystis* blooms and nutrient enrichment in the continental coastal zones of the North Sea. *Ambio*, **16**, 38—46.

Law, C.S. and Owens, N.J.P. (1990) Denitrification and nitrous oxide in the North Sea. *Neth. J. Sea Res.*, **25**, 65—74.

Law, C.S., Rees, A.P. and Owens, N.J.P. (1991). Temporal variability of denitrification in estuarine sediments. *Est. Coast. Shelf Sci.*, **33**, 37—56.

Livingstone, D.A. (1963) Chemical composition of rivers and lakes. In Fleischer, M. (ed.), *Data of Geochemistry*, US Geological Survey, Washington, DC, pp. 1—61.

Macfarlane, G.T. and Herbert, R.A. (1984) Dissimilatory nitrate reduction and nitrification in estuarine sediments. *J. Gen. Microbiol.*, **130**, 2301—8.

Mantoura, R.F.C., Owens, N.J.P. and Burkill, P.H. (1988) Nitrogen biogeochemistry and modelling of Carmarthen Bay. In Blackburn, T.H. and Sorensen, J. (eds), *Nitrogen Cycling in Coastal Marine Environments*, Wiley, Chichester, pp. 415—41.

Mantoura, R.F.C., Law, C.S., Owens, N.J.P., Burkill, P.H., Woodward, E.M.S., Howland, R.J.M. and Llewellyn, C.A. (1993) Nitrogen biogeochemical cycling in the northwestern Indian Ocean. *Deep-Sea Res.*, in press.

Martin, J.H., Fitzwater, S.E. and Gordon, R.M. (1990a) Iron deficiency limits phytoplankton growth in Antarctic waters. *Global Biogeochem. Cyc.*, **4**, 5—12.

Martin, J.H., Gorden, R.M. and Fitzwater, S.E. (1990b) Iron in Antarctic waters. *Nature*, **345**, 156—8.

Meybeck, M. (1982) Carbon, nitrogen and phosphorus transport by world rivers. *Am. J. Sci.*, **282**, 401—50.

Milliman, J.D. (1991) Flux and fate of fluvial sediment and water in coastal seas. In Mantoura, R.F.C., Martin, J.-M. and Wollast, R. (eds), *Ocean Margin Processes in Global Change*, Wiley, Chichester, pp. 69—89.

Milliman, J.D. and Meade, R.H. (1983) World-wide delivery of river sediment to the oceans. *J. Geol.*, **98**, 966—76.

Morris, A.W., Bales, A.J. and Howland, R.J.M. (1981) Nutrient distributions in an estuary: evidence of chemical precipitation of dissolved silicate and phosphate. *Est. Coast. Shelf Sci.*, **12**, 205—16.

Nedwell, D.B. (1982) Exchange of nitrate, and the production of bacterial nitrate reduction, between seawater and sediment from a UK saltmarsh. *Est. Coast. Shelf Sci.*, **14**, 557—66.

Nishio, T., Koike, I. and Hattori, A. (1982) Denitrification, nitrate reduction, and oxygen consumption in coastal and estuarine sediments. *Appl. Env. Microbiol.*, **43**, 444—50.

Oren, A. and Blackburn, T.H. (1979) Estimation of sediment denitrification rates at *in situ* nitrate concentrations. *Appl. Environ. Microbiol.*, **37**, 174—6.

Otto, L., Zimmerman, J.T.F., Furnes, G.K., Mork, M., Saetre, R. and Becker, G. (1990) Review of the physical oceanography of the North Sea. *Neth. J. Sea Res.*, **26**, 161—238.

Owens, N.J.P. (1986) Estuarine nitrification: a naturally occurring fluidized bed reaction? *Est. Coast. Shelf Sci.*, **22**, 31—44.

Owens, N.J.P., Cook, D., Colebrook, M., Hunt, H. and Reid, P.C. (1989a) Long term trends in the occurrence of *Phaeocystis* sp. in the north-east Atlantic. *J. Mar. Biol. Ass.* UK, **69**, 813—21.

Owens, N.J.P., Galloway, J.N. and Duce, R.A. (1992) Episodic atmospheric nitrogen deposition to oligotrophic oceans. *Nature*, **357** (6377), 397—9.

Owens, N.J.P., Mantoura, R.F.C., Burkill, P.H., Howland, R.J.M., Pomroy, A.J. and Woodward, E.M.S. (1986) Nutrient cycling studies in Carmarthen Bay: phytoplankton production, nitrogen assimilation and regeneration. *Mar. Biol.*, **93**, 329—42.

Owens, N.J.P., Priddle, J. and Whitehouse, M.J. (1991) Variations in phytoplanktonic nitrogen assimilation around South Georgia and in the Bransfield Strait (Southern Ocean). *Mar. Chem.*, **35**, 287—304.

Owens, N.J.P., Rees, A.P., Woodward, E.M.S. and Mantoura, R.F.C. (1989b) Size fractionated primary production and nitrogen assimilation in the north-western Mediterranean Sea during January 1989. In Martin, J.-M. and Barth, H. (eds), *EROS 2000: European River Ocean System, First Workshop on the North Western Mediterranean Sea*. CEC Water Pollution Report, 13, Guyot, Brussels, pp. 126—35.

Perissinotto, R. and Duncombe-Rae, C.M. (1990) Occurrence of anitcyclonic eddies on the Prince Edward Plateau (Southern Ocean); effects on phytoplankton biomass and production. *Deep-Sea Res.*, **37**, 777—93.

Pickard, G.L. and Emery, W.J. (1990) *Descriptive Physical Oceanography*, Pergamon, Oxford.

Pingree, R.D. and Griffiths, D.K. (1978) Tidal fronts on the shelf seas around the British Isles. *J. Geophys. Res.*, (Oceans and Atmospheres), **83**, 4615—22.

Pingree, R.D., Maddock, L. and Butler, E.I. (1977) The influence of biological activity and physical stability in determining the chemical distributions of inorganic phosphate, silicate and nitrate. *J. Mar. Biol. Ass. UK*, **57**, 1065—73.

Pritchard, D.W. (1952) Estuarine hydrography. *Advanc. Geop.*, **1**, 243—80.

Prospero, J.M. and Savoie, D.L. (1989) Effect of continental sources on nitrate concentrations over the Pacific Ocean. *Nature*, **339**, 687—9.

Radach, G., Berg. J. and Hagmeier, E. (1990) Long-term changes of the annual cycles of meteorological, hydrographic, nutrient and phytoplankton time series at Heligoland and at LV Elbe 1 in the German Bight. *Cont. Shelf Res.*, **10**, 305—28.

Reid, P.C., Lancelot, C., Gieskes, W.W.C., Hagmeier, E. and Weichart, G. (1990) Phytoplankton of the North Sea and its dynamics: A review. *Neth. J. Sea Res.*, **26**, 295—331.

Rittenberg, S.C., Energy, K.O. and Orr, W.L. (1955) Regeneration of nutrients in sediments of marine basins. *Deep-Sea Res.*, **3**, 23—45.

Roenner, U. (1985) Nitrogen transformations in the Baltic proper: denitrification counteracts eutrophication. *Ambio*, **14**, 134—8.

Rosenberg, R., Elmgren, R., Fleischer, S., Jonsson, P., Persson, G. and Dahlin, H. (1990) Marine eutrophication case studies in Sweden. *Ambio*, **19**, 102—8.

Rosswall, T. (1983) The nitrogen cycle. In Bolin, B. and Cook, R.B. (eds), *The Major Biogeochemical Cycles and their Interactions*, Wiley, Chichester, pp. 46—50.

Seitzinger, S.P. (1987) Nitrogen biogeochemistry in an unpolluted estuary: the importance of benthic denitrification. *Mar. Ecol. Prog. Ser.*, **41**, 177—86.

Seitzinger, S.P., Nixon, S.W. and Pilson, M.E.Q. (1984) Denitrification and nitrous oxide production in a coastal marine ecosystem. *Limnol. Oceanogr.*, **29**, 73—83.

Seitzinger, S.P., Nixon, S.W., Pilson, M.E.Q. and Burke, S. (1980) Denitrification and N_2O production in near-shore marine sediments. *Geochim. Cosmochim. Acta*, **44**, 1853—60.

Shannon, L.V. and Pillar, S.C. (1986) The Benguela ecosystem. 3. Plankton. *Oceanogr. Mar. Biol. Ann. Rev.*, **24**, 65—170.

Sharp, J.H. (1983) The distribution of inorganic nitrogen and dissolved and particulate organic nitrogen in the sea. In Carpenter, E.J. and Capone, D.G. (eds), *Nitrogen in the Marine Environment*, Academic Press, New York.

Simpson, J.H. and Hunter, J.R. (1974) Fronts in the Irish Sea. *Nature*, **250**, 404—6.

Smetacek, B., Bathmann, U., Nothig, E.-M. and Scharek, R. (1991) Coastal eutrophication: causes and consequences. In Mantoura, R.F.C., Martin, J.-M. and Wollast, R. (eds), *Ocean Margin Processes in Global Change*, Wiley, Chichester, pp. 251—79.

Smith, C.J., Delaune, R.D. and Patrick, W.H. Jr (1983) Nitrous oxide emission from Gulf Coast wetlands. *Geochim. Cosmochim. Acta*, **47**, 1805—14.

Smith, W.M. Jr (1987) Nutrient regeneration and denitrification in low oxygen fjords. *Deep-Sea Res.*, **34**, 983—1006.

Smith, W.O. Jr and Kattner, G. (1989) Inorganic nitrogen uptake by phytoplankton in the marginal ice zone of the Fram Strait. *Rapp. P.-v. Reun. cons. int. Explor Mer.*, **188**, 90−7.

Sorensen, J. (1978) Capacity for denitrification and reduction of nitrate to ammonia in a coastal marine sediment. *Appl. Env. Microbiol.*, **35**, 301−15.

Sorensen, J. (1984a) Seasonal variation and control of oxygen, nitrate, and sulfate respiration in coastal marine sediments. In Klug, M.S. and Reddy, C.A. (eds), *Current Perspectives in Microbial Ecology*, Am. Soc. Microbiol, Washington, DC, pp. 447−53.

Sorensen, J. (1984b) Denitrification in a deep-sea sediment core from the eastern equatorial Atlantic. *Limnol. Oceanogr.*, **29**, 653−7.

Teague, K.G., Madden, C.J. and Day, J.W. Jr (1988) Sediment−water oxygen and nutrient fluxes in a river-domianted estuary. *Estuaries*, **11**, 1−9.

Valiela, I. and Teal, J.M. (1979) The nitrogen budget of a salt marsh ecosystem. *Nature*, **280**, 652−6.

Van Bennekom, J.J. and Salomons. W. (1981) Pathways of organic nutrients and organic matter from land to ocean through rivers. In Burton, J.D., Eisma, D. and Martin, J.-M. (eds), *River Input to the Ocean System*, SCOR, Rome, pp. 33−51.

Van Kessel, J.F. (1977) Factors affecting the denitrification rate in two water−sediment systems. *Wat. Res.*, **11**, 259−67.

Vanderborght, J.P. and Billen, G. (1975) Vertical distribution of nitrate concentration in interstitial water of marine sediments with nitrification and denitrification. *Limnol. Oceanogr.*, **20**, 953−61.

Walsh, J.J. (1991) Importance of continental margins in the marine biogeochemical cycling of carbon and nitrogen. *Nature*, **350**, 53−5.

Whitaker, T.M. (1982) Primary production of phytoplankton off Signy Island, South Orkneys, the Antarctic. *Proc. R. Soc., Lond. Ser. B*, **214**, 169−89.

Wilson, T.R.S. (1978) Evidence for denitrification in aerobic pelagic sediments. *Nature*, **274**, 354−6.

Wollast, R. (1983) Interactions in estuaries and coastal waters. In Bolin, B. and Cook, R.B. (eds), *The Major Biogeochemical Cycles and their Interactions*, Wiley, Chichester, pp. 385−407.

Woodward, E.M.S. and Owens, N.J.P. (1989) The influence of the River Rhone upon the nutrient fluxes of the Golfe du Lion. In Martin, J.-M. and Barth, H. (eds), *EROS 2000: European River Ocean System, First Workshop on the North Western Mediterranean Sea*, CEC Water Pollution Report, 13, Guyot, Brussels, pp. 79−86.

Woodward, E.M.S. and Owens, N.J.P. (1990) Nutrient depletion studies in offshore North Sea areas. *Neth. J. Sea Res.*, **25**, 57−63.

Part II

SPATIAL AND TEMPORAL PATTERNS OF NITRATE TRANSFER

8 Nitrate in Groundwater

T.P. BURT
School of Geography, University of Oxford
and
S.T. TRUDGILL
Department of Geography, University of Sheffield

8.1 NITRATE AS A GROUNDWATER CONTAMINANT

8.1.1 INTRODUCTION

Nitrate is the pollutant most commonly identified in groundwater (Freeze and Cherry, 1979). Nitrate contamination of aquifers is becoming widespread for a whole variety of reasons, notably because of increasing use of inorganic fertilisers, ploughing of old grassland and disposal of organic material (farmyard manure, slurry, sewage sludge) on or beneath the land surface. Though nitrate (NO_3^-) is the main form in which nitrogen occurs in groundwater, dissolved nitrogen may also be present as ammonium (NH_4^+), nitrite (NO_2^-), nitrogen (N_2), nitrous oxide (N_2O) and as organic nitrogen.

Figure 8.1 shows the sources and pathways of nitrate in groundwater. Nitrate in groundwater may originate from organic matter or ammonium. These are converted into nitrate by the processes of mineralisation (decomposition of organic matter into inorganic ammonium) and nitrification (oxidation of ammonium to form nitrate). These are aerobic processes which commonly take place above the water table. Nitrate may be lost from groundwater by denitrification which, because it takes place in an anaerobic environment, is most likely under saturated conditions (i.e. below the water table). Denitrification is principally a biologically mediated process, though it may occur chemically or biologically. NO_3 is first reduced to NO_2 and then to N_2O. Complete reduction results in N_2, though often a trace of N_2O is left; these remain in solution until degassing can take place. Chemodenitrification can occur if the redox potential of the groundwater is low enough (Freeze and Cherry, 1979, p. 414). Biological denitrification may be limited in groundwater because of the lack of a carbon substrate. Its operation in groundwater systems is little known though much debated since it is potentially an important nitrate loss mechanism (see Section 8.2.3 below). Sprent (1987, p. 25) notes that some bacteria can reduce nitrate to ammonium. Being a cation, this ammonium may then be adsorbed by clay minerals present in the rock.

The division between 'soil water' and 'groundwater' is necessarily somewhat arbitrary. Nitrogen cycling in soils is described in Chapters 3 and 4 in some detail. In Chapter 9, the drainage of nitrate through soil to streams is considered; this is essentially an aspect of hillslope hydrology, the dominant flow direction being horizontal (downslope) rather than vertical. Here, we must necessarily concentrate both on vertical percolation through the unsaturated zone (initiated by vertical drainage through the soil profile) and on lateral flow within the saturated zone. However, the distinction between subsurface flow in deep soils and in shallow

Nitrate: Processes, Patterns and Management. Edited by T.P. Burt, A.L. Heathwaite and S.T. Trudgill
© 1993 John Wiley & Sons Ltd

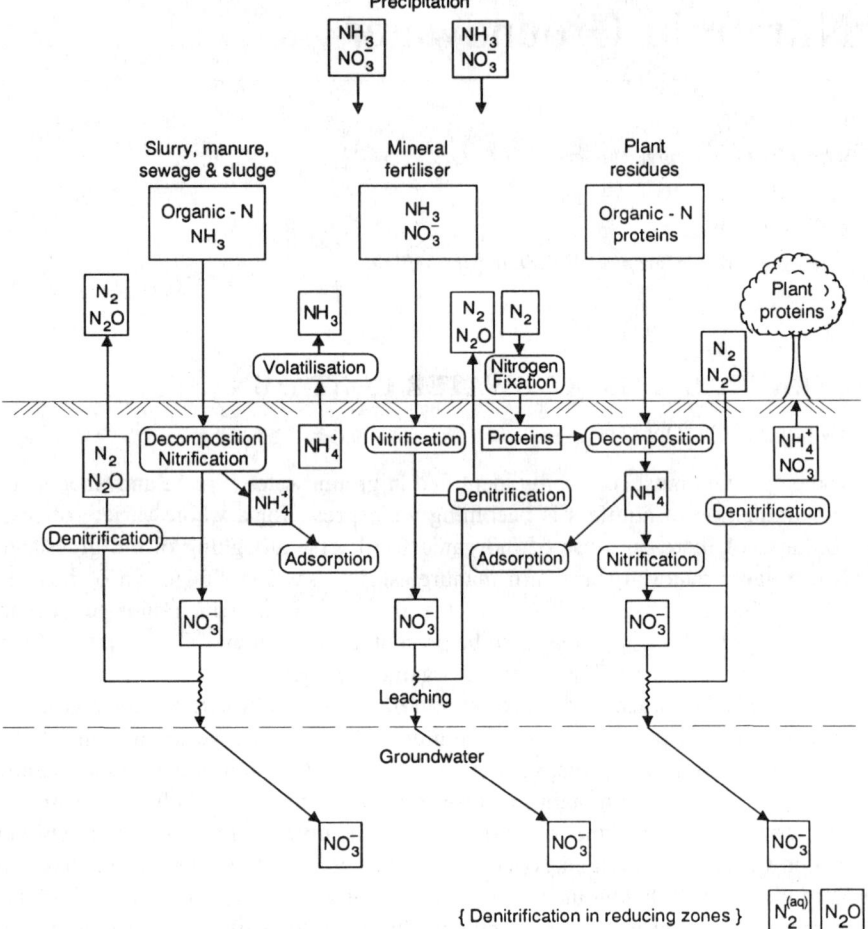

Figure 8.1 Sources and pathways of nitrate in groundwater (adapted from Freeze and Cherry, 1979)

aquifers can sometimes be hard to make, so that there is inevitably some overlap between this and the next chapter. Finally, it should be pointed out that soil scientists have, naturally enough, concentrated their efforts on vertical drainage in soils, especially within the root zone. In both Chapter 9 and here we refer to soil water flow where it is relevant to do so.

8.1.2 AQUIFER PROPERTIES

Two types of flow occur through rock: laminar flow through the rock matrix and turbulent flow through large fractures. Because flow through a porous medium is complex, attempts to estimate flow using simple measures of pore geometry have met with little success. Darcy's Law (1856) shows that the flow of water through a porous medium is proportional to the driving force acting on the liquid (i.e. the hydraulic gradient) and to the ability of the medium to transmit the liquid (i.e. the hydraulic conductivity). Darcy was not concerned with flow through individual pores, referred to as the microscopic level, but rather with flow at the

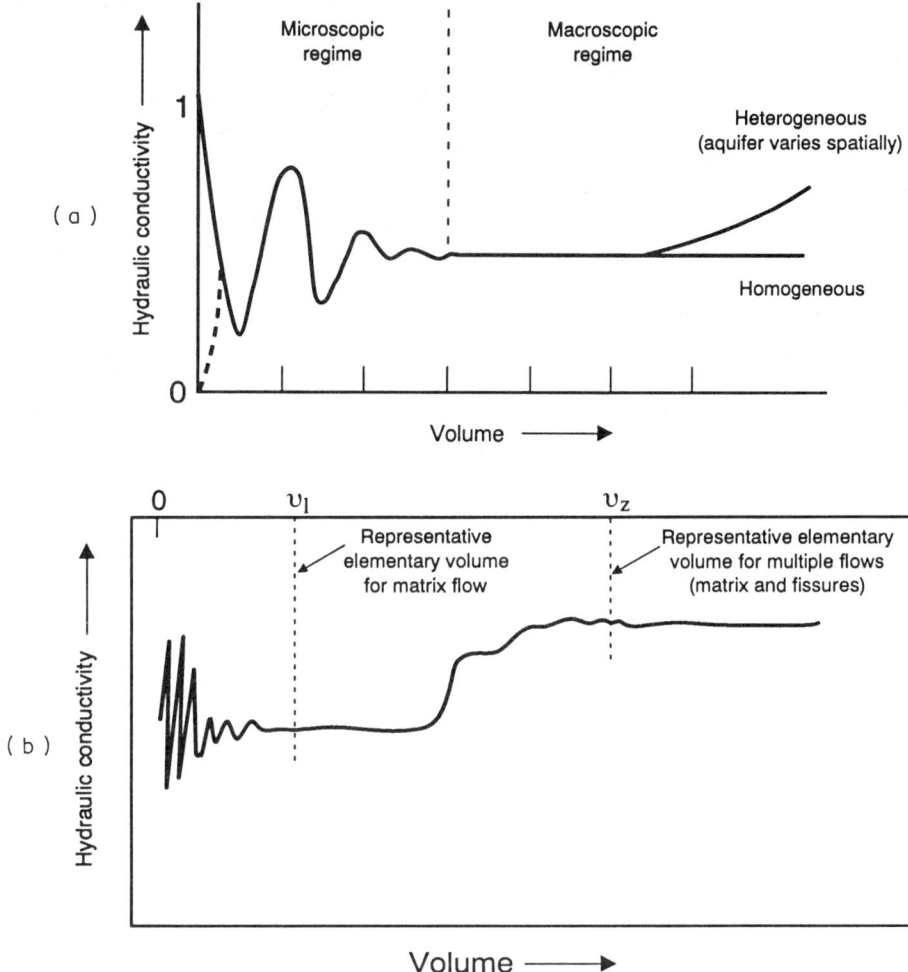

Figure 8.2 Diagrams illustrating representative elementary volumes for (a) homogenous, porous rock and (b) porous rock with fractures (adapted from Domenico and Schwartz, 1990)

macroscopic scale. Darcy experimented with a volume of sand that was large with respect to a single pore but small with respect to the space within which significant variations of macroscopic properties may be anticipated (Domenico and Schwartz, 1990). At the microscopic level, the size and tortuousity of pores varies widely; as volume increases, variability decreases markedly until the point where there are no longer any variations with the size of sample. Hence Darcy sought the smallest possible sample that exhibited an acceptable level of homogeneity, as shown in Figure 8.2(a). At larger scales, some measure of heterogeneity is reintroduced since bedrock properties vary spatially. Though variance will increase with scale, it may still be possible to define a mean value of hydraulic conductivity with acceptable accuracy.

Hydraulic conductivity is maximal when all pores are filled with water; this is referred to as the saturated hydraulic conductivity. For a given porosity, pore size is the principal factor controlling hydraulic conductivity. Thus a massive rock containing only a few large fractures may have a much higher hydraulic conductivity than a porous but fine-grained rock whose pores are all of very small radius. For fractured rocks, it is likely that as the volume increases beyond that which is representative of the macroscopic properties of the rock matrix, the flow parameters will start to vary again before becoming constant once more for large volumes of rock (Domenico and Schwartz, 1990). Thus the representative volume may exist on more than one scale, this being shown schematically in Figure 8.2(b). Not surprisingly, the saturated hydraulic conductivity of many rocks is determined by its secondary (or bulk) porosity rather than by the primary porosity of the matrix. This is illustrated in Figure 8.3. Each rock type is represented by two fields, of which the one to the left of the diagram shows the primary porosity and pore size. The second field shows the total rock porosity and the range of size of the secondary voids. There is very little change in porosity between primary and secondary fields, but the secondary pore size is two to three orders of magnitude larger; it is this latter variable that mainly determines the saturated hydraulic conductivity of a rock (Smith, Atkinson and Drew, 1976). In the zone of saturation, flow is likely to remain Darcian, even in a well-fissured rock, since the hydraulic gradient is usually low, though clearly the highest *individual* pore velocities are associated with the largest fractures (see Section 8.1.3).

As a rock becomes progressively unsaturated, so its hydraulic conductivity is reduced, often by several orders of magnitude. As the degree of saturation is reduced, flow is confined to

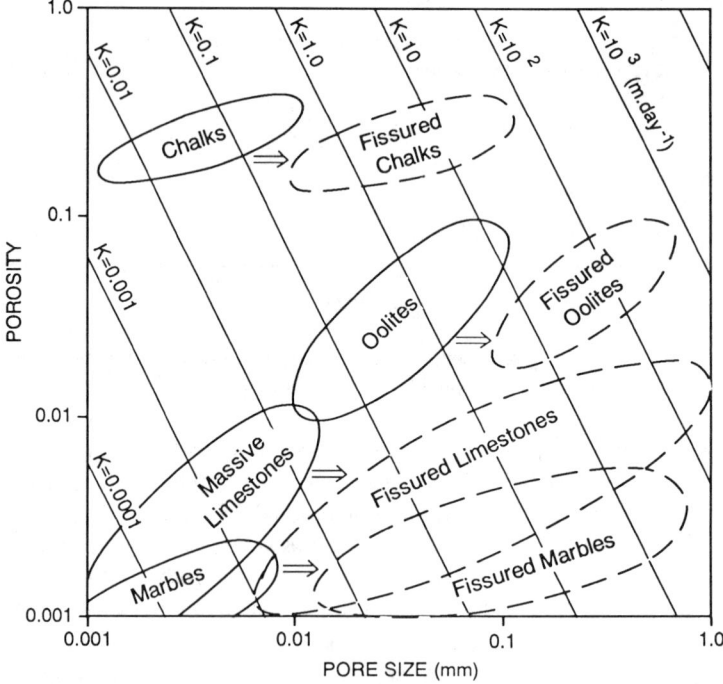

Figure 8.3 Primary and secondary porosity, pore size and hydraulic conductivity (K) of carbonate rocks (adapted from Smith, Atkinson and Drew, 1976). Contour values of K are based on the assumption that the rock behaves as a bundle of straight, parallel capillary tubes. Solid lines indicate primary conductivities of the rock matrix; dashed lines are secondary or total conductivity

smaller pores and to less of the pore space, and conductivity decreases. The largest pores drain first. This may leave the smaller pores which remain full of water poorly interconnected. Thus, drainage of large fractures is very likely to cause a severe reduction in flow and the hydraulic conductivity can fall by several orders of magnitude (Figure 8.3). If the vertical flux in an unsaturated fractured rock is less than the saturated hydraulic conductivity of the matrix, the flow will tend to remain in the smaller sized pores within the matrix and flow will be slow. As the flux increases, the water content of the rock matrix will become higher and some flow will then take place within larger pores. In other words, though the rock as a whole remains unsaturated, the capillary suction in the rock matrix has fallen sufficiently that larger pores (but only those which are of capillary size) can now hold water; these provide pathways for more rapid flow. If the flux is greater than the saturated hydraulic conductivity of the matrix, the matrix will saturate. Some water must then enter the fractures (those cavities too large to function as capillaries) where flow is likely to be much more rapid than through the small pores.

Recent research on cracked clay soils has shown that this is not the only mechanism whereby large fractures can conduct flow rapidly down through the unsaturated zone (Beven and Germann, 1982). In soils where the rainfall rate exceeds the infiltration rate of the soil matrix, excess water will enter the structural voids (or 'macropores') and water may move quickly down through the soil profile, even though the matrix remains unsaturated. The depth of movement is determined by the rate of input and by losses of water absorbed into the soil matrix through the walls of the macropore. The magnitude of absorption losses into the soil depends on the matric suction of the soil water, which itself is a function of soil moisture content. Where soil macropores connect with fractures in the rock or where drainage from the soil exceeds the infiltration rate of the rock matrix, percolating water may enter large rock fissures and may therefore by-pass the rock matrix.

The important point is that, depending on rock type, the downward movement of water through the unsaturated zone may or may not be uniform. This has particular implications for the speed with which nitrate is carried down to the zone of saturation.

8.1.3 SOLUTE TRANSPORT PROCESSES

Considerations of solute movement in porous media often begin with Fick's first law which refers to solute flux in relation to solute gradient (Nye and Tinker, 1977). Even in static water, solutes diffuse along the concentration gradient from high to low concentration. If the water itself is moving, convection also occurs: that is, the mass transport of water and any solute dissolved in it from one location to another. However, the solute will move down any concentration gradient simultaneously with the convection process, so the two components of solute movement (diffusion and convection) must be combined. A further complication in a porous medium is that some portions of water (and its solutes) travel faster than others, in large pores and/or by more direct routes (Burt and Trudgill, 1985). The combined movement of convection and diffusion is referred to as dispersion (or, more formally, hydrodynamic dispersion). Dispersion causes a zone of mixing to develop between a fluid of one composition that is adjacent to or being displaced by a fluid of another composition (Domenico and Schwartz, 1990).

Dispersion can be demonstrated by the displacement of water by solute in a simple column apparatus (Figure 8.4(a)). For a saturated rock with initially no solute present, simple convection (or 'piston flow') for a solute pulse would yield a breakthrough curve as shown

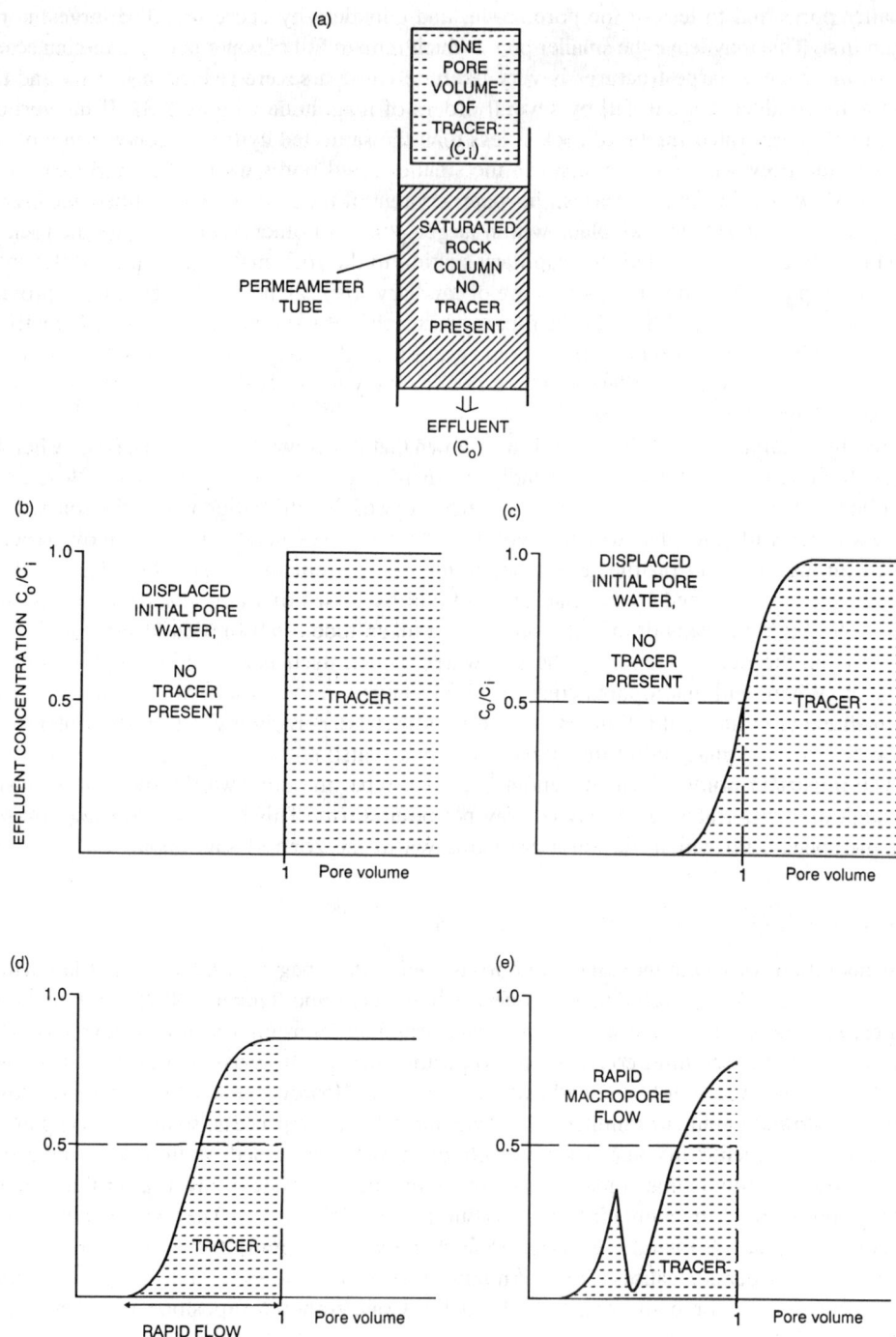

Figure 8.4 Movement of solute through a saturated rock sample (adapted from Burt and Trudgill, 1985)

in Figure 8.4(b) when one pore volume has been displaced. The effect of dispersion is to spread the solute mass beyond the region it would normally occupy due to convection alone, with more rapid faster appearance in the faster flow paths and some tailing in the slower, more tortuous paths (Figure 8.4(c)). This pattern can be greatly exaggerated by the presence of large fractures which cause the solute to arrive in the outflow considerably in advance of one pore volume or even giving initial peaks of solute (Figures 8.4(d) and (e)).

Burt and Trudgill (1985) have identified three flow domains in soils. These are equally applicable to flow in rock and provide a basis for understanding the movement of nitrate:

(1) Immobile water which is tightly held in the smallest pores; these pores remain saturated at all times. This water does not appear in drainage water, but it is important because nitrate may diffuse into and out of the domain from or to the more mobile water.

(2) Mobile or displaced water, equivalent to capillary or matric water, and is held at intermediate tensions (which vary with rock type). Nitrate concentrations may build up in this domain depending on the pattern of displacement from above and on the exchange of nitrate with immobile water via diffusion.

(3) Transient flow transmitted rapidly through large fissures. This flow will be maintained provided that the supply rate exceeds infiltration losses into the rock matrix.

8.2 NITRATE MOVEMENT IN THE UNSATURATED ZONE

8.2.1 EXTRACTION AND ANALYSIS OF BOREHOLE WATER

A substantial proportion of the total research effort on nitrate in UK groundwater continues to be dedicated to the unsaturated zone (Parker, Young and Chilton, 1991). The research has been based primarily on the extraction of interstitial pore water from rock cores by centrifugal techniques or pressure extraction. A full description of the extraction and analysis of pore water is given, *inter alia*, in Young, Hall and Oakes (1976). Typically, extracted pore water may be analysed for nitrate, nitrite, ammonium, chloride, calcium and tritium. In addition, rock samples may be tested for evolved carbon dioxide, carbohydrate content and bacterial contamination.

Chloride is widely regarded as a stable, conservative, non-reactive ion, not subject to biological transformation. Chloride proiles can therefore represent a useful tool in the interpretation of solute transport in the unsaturated zone. The transport of nitrate is likely to be similar to that of chloride except that denitrification or nitrification within an aquifer cannot be discounted (Whitelaw and Rees, 1980). The determination of tritium (^{3}H) profiles aids the evaluation of solute transport mechanisms in the unsaturated zone because of its unique temporal distribution in rainfall and infiltration. It can thus be used as an analogue for non-reactive solutes of similar physico-chemical behaviour to nitrate whose temporal patterns of input are less easy to estimate. Many tritium profiles show a distinct peak originating from fallout in rainfall in 1963−5 following atmospheric testing of thermonuclear weapons (Geake and Foster, 1989). Despite attenuation because of radioactive decay and the effects of hydrodynamic dispersion, this peak is still evident in water moving downwards in the unsaturated zone (see next section and Figure 8.3). However, given the assumptions and possible sources of error involved in evaluating the tritium mass balance (particularly uncertainties over the tritium content of rainfall), profiles must be interpreted with some caution (Foster and Smith-Carington, 1980).

8.2.2 RATES OF DOWNWARD MOVEMENT OF NITRATE

Interpretation of pore water profiles requires a detailed understanding of flow mechanisms and dispersion effects in the unsaturated zone. In the UK, three main aquifers have been studied: Cretaceous Chalk, the Jurassic limestones, and the Triassic Sherwood (or Bunter) sandstone (Rodda, Downing and Law, 1976).

8.2.2.1 Cretaceous Chalk

The Chalk of south-east England is the major aquifer in the UK, providing over 40% of the total groundwater abstracted. In recent years, there has been mounting concern about the increasing concentrations of nitrate in water suplied from boreholes and spring in the Chalk; hence the Chalk is the most studied UK aquifer in this respect. The Chalk is a fine-grained limestone frequently traversed by joints and fissures. The matrix (intergranular) porosity is high (0.25–0.5) but the pores are so small (mostly 0.25–0.5 μ – radius) that the hydraulic conductivity is very low (below 10^{-2} d^{-1}; Foster and Smith-Carington, 1980 – see also Figure 8.3). These pores remain full of water almost all the time but the average annual rate of downward movement of the water is very low, approximately 0.9 m a^{-1} (Wellings and Bell, 1980). In addition to these micropores, the Chalk contains a large number of fissures of varying sizes (Reeves, 1979). These can contain water either only when the matric tension is low or, more rarely, if the percolation rate from the soil above exceeds the hydraulic conductivity of the rock matrix. Except in valleys, depth to the water table is often in the range 20–50 m. Given very high nitrate concentrations in the upper part of the unsaturated zone and the slow rate of downward percolation, it is expected that nitrate concentrations in chalk groundwater supplies will increase steadily over the next few decades (see below).

There is no simple universal model for solute transport through the unsaturated zone of the Chalk (Parker, Young and Chilton, 1991). Two mechanisms have been put forward to explain observed nitrate and tritium profiles. Marked tritium peaks have been taken to indicate that most of the infiltration is moving slowly downwards by intergranular flow with very little dispersion. Microfissures and the largest pores of the chalk (0.01–0.2 mm diameter) account for the slow mode of downward tritium movement. Such pores are unlikely to represent more than 1% of the Chalk by volume and probably produce a hydraulic conductivity of less than 0.01 m per day (Foster and Smith-Carington, 1980). Foster (1975) suggested that the main mechanism of recharge was flow through fissures. After intense rainfall, the infiltration capacity of the rock matrix is likely to be exceeded, suctions fall concomitantly and positive potentials develop locally, allowing horizontal groundwater flow to the macrofissures (larger than 0.2 mm in aperture). Diffusion between the moving water in the fissures and the relatively static pore water retards the downward movement of isotopes and solutes so that a virtually non-dispersive solute movement can still occur. This would, however, only hold for low-velocity flow through a dense network of microfissures and for solutes with a high diffusion coefficient. When flow takes place in larger fissures, rates may be too rapid to allow significant difussion exchange with the pore water and solutes may by-pass the normal slow mode of solute transport.

Geake and Foster (1989) conclude that the relative importance of each flow mechanism may vary spatially and with depth; moreover, at any one site, the mode of water flow can vary temporally with excess rainfall intensity and antecedent moisture conditions. Where the matrix hydraulic conductivity is relatively high, intergranular flow will predominate. There is straightforward 'piston flow' displacement of water and least dispersion of the tritium peak, implying that nitrate will itself move downwards without significant dispersion. With much lower matrix hydraulic conductivity, more flow must move through fissures and solutes become more dispersed down the profile. Dispersion is implied where there has been more downward movement of solute than would be expected if piston flow was occurring and where there has been an apparent flattening and broadening of the original solute peak. As a result, there are higher concentrations than expected in the deeper part of the profile (Foster and Smith-Carington, 1980; Parker, Young and Chilton, 1991).

Isotope and solute profiling in the Chalk has been reviewed recently by Geake and Foster (1989) and by Parker, Young and Chilton (1991). Figure 8.5(a) shows a tritium profile from the unsaturated zone of the Chalk at a site in Hampshire where the tritium peak, moving slowly downwards with time, has been preserved despite attenuation because of radioactive decay and the effects of hydrodynamic dispersion. Figures 8.5(b) and (c) show the nitrate profile from an adjacent borehole and the land use since 1949 for the field in which both the tritium and nitrate boreholes were drilled. Young, Hall and Oakes (1976) postulate that the large releases of nitrate accompanying the ploughing of ley (temporary) grassland should be recognisable as major peaks in nitrate concentration in interstitial water. A comparison of the land-use record with the vertical nitrate profile suggests that the peaks at 30−32 m, 20 m, 14 m, 9 m and 2 m may correspond to periods of ploughing after leys in 1948/9, 1955/6, 1960/1, 1965/6 and 1972/3, respectively. This chronology would seem to correspond well with the 1963/4 tritium peak at 8−9 m, which represents a mean downward movement of about 0.8 m a^{-1}.

The preservation of tritium peaks at many sites demonstrates that the bulk of the nitrate transport through the unsaturated zone of the Chalk, by whatever means, is slow. Evidence shows that downward flow takes place with varying degrees of dispersion. At those sites where solutes and isotopes are moving downwards without significant dispersion, all the pore water to the base of the tritium peak can be assumed to be post-1963 in age. Given this assumption, nitrate leaching losses from Chalk soils can be seen to have increased steadily since the mid-1960s. The very high nitrate concentrations in the upper part of the unsaturated zone imply that nitrate concentrations in Chalk groundwater supplies will continue to rise slowly for many years (Geake and Foster, 1989).

8.2.2.2 Jurassic (Cotswold or Lincolnshire) limestones

These are found to the west of the Chalk and form an important aquifer in south-central and eastern England. They are thin formations (20−35 m) with moderate intergranular porosity (0.1−0.25) but with very small pores, giving very low values for matric hydraulic conductivity (0.5−5 × 10^{-4} m d^{-1}). The limestone, however, is cut by a large number of bedding plane fractures and joints along which the bulk of the groundwater flow occurs. Groundwater levels increase rapidly following the first effective infiltration in late autumn and early winter. This immediate respose to recharge suggests a low storage aquifer with rapid groundwater flow (Smith-Carington et al., 1983). It suggests too that slow percolation through the rock matrix, a feature of the Chalk aquifer, is not very important in this limestone. Since the movement of nitrate down to the water table is so rapid, the history of leaching losses from the soil is not preserved in unsaturated zone profiles. Despite the low rate of recharge (100−250 mm annually), it is likely that the effects of any change in land use, and in the amount of nitrate leached, will be rapidly transmitted to the saturated zone.

8.2.2.3 Triassic (Sherwood) sandstone

This aquifer is located in the Midlands of England and comprises a series of medium- to fine-grained red sandstones underlain by the so-called Pebble Beds, a coarse pebbly sandstone and conglomerate which forms the main part of the aquifer. The unsaturated zone varies from more than 60 m deep beneath higher ground to less than 10 m deep in valleys. Groundwater levels show no substantial seasonal variation. Measurement of the moisture content and physical properties of sandstone cores from investigation boreholes suggests that the rate of downward percolation of water through the unsaturated zone is about 2 m per year (the mean annual rate of recharge is 360 mm a^{-1}). Hence there is a time lag of some 5–30 years before infiltration reaches the water table (STW, 1988). Investigations in the Hatton catchment showed that nitrate concentrations in the unsaturated pore waters vary from 0.5 to 50 mg NO_3-N l^{-1}, reflecting the different land uses (Figure 8.6). Very high leaching losses occur from arable land (60 kg N ha^{-1}) and from grazed grassland which is heavily fertilised (up to 150 kg N ha^{-1}), losses being equivalent to 40% of inputs in both cases.

8.2.3 CLIMATIC AND LAND-USE CONTROLS

Parker, Young and Chilton (1991) have reviewed the relationship between land use and nitrate leaching. Where it can be assumed that the nitrate and tritium in pore water profiles is moving down without dispersion, then it is possible to interpret the history of recent nitrate leaching losses in relation to cultivation history (e.g. Figure 8.5) and mass balance calculations may be made to quantify the amount of nitrate leached. In the discussion below, the word 'equivalent' indicates that most of the nitrate leached does not come directly from the nitrogen applied in the previous season. Most of the applied nitrogen which is not taken up by the crop, or that which is returned to the soil in plant residues or as in dung and urine, will become incorporated into the soil organic matter and microbial biomass. It is this nitrogen which is subsequently mineralised to leachable nitrate (STW, 1988; Dowdell and Webster, 1980).

Many investigations under arable land have revealed marked peaks in nitrate concentration in the upper part of the unsaturated zone; Figures 8.5 and 8.6 show typical results. Maximum concentrations range from 20 to 70 mg NO_3-N l^{-1} in the upper 10 m (well in excess of EC limits) compared with typical values of the order of 5–20 mg NO_3-N l^{-1} at depth. Foster, Cripps and Smith-Carington (1982, table 1) estimated annual leaching losses since 1963 in the range 40–70 kg N ha^{-1}, equivalent to between 26% and 78% (mean 49%) of the N fertiliser applied. Note that one site listed by Foster, Cripps and Smith-Carington (1982) is excluded here because of the complication that grassland was ploughed during the study period. The ploughing out of ley grassland is a particularly important source of nitrate, resulting in an estimated leaching loss of 280 kg ha^{-1} over the three years following ploughing and 150 kg ha^{-1} in the first year alone. Parker, Young and Chilton (1991) note that current estimates of the average leaching of nitrate under arable land are equivalent to about 25% of applied nitrogen. This lower figure may relate in part to the indirect benefits of the recent moves to autumn-sown cereals and to the use of minimal cultivation techniques, both of which might reasonably be hoped to reduce leaching losses in winter.

Apart from the ploughing up of grassland, many studies in the 1970s concluded that nitrate leaching losses were low under grassland. However, in many cases such studies were confined to unimproved or unfertilised grassland, and recent studies have shown that nitrate concentrations found under grazed grassland are often greater than those found under

223

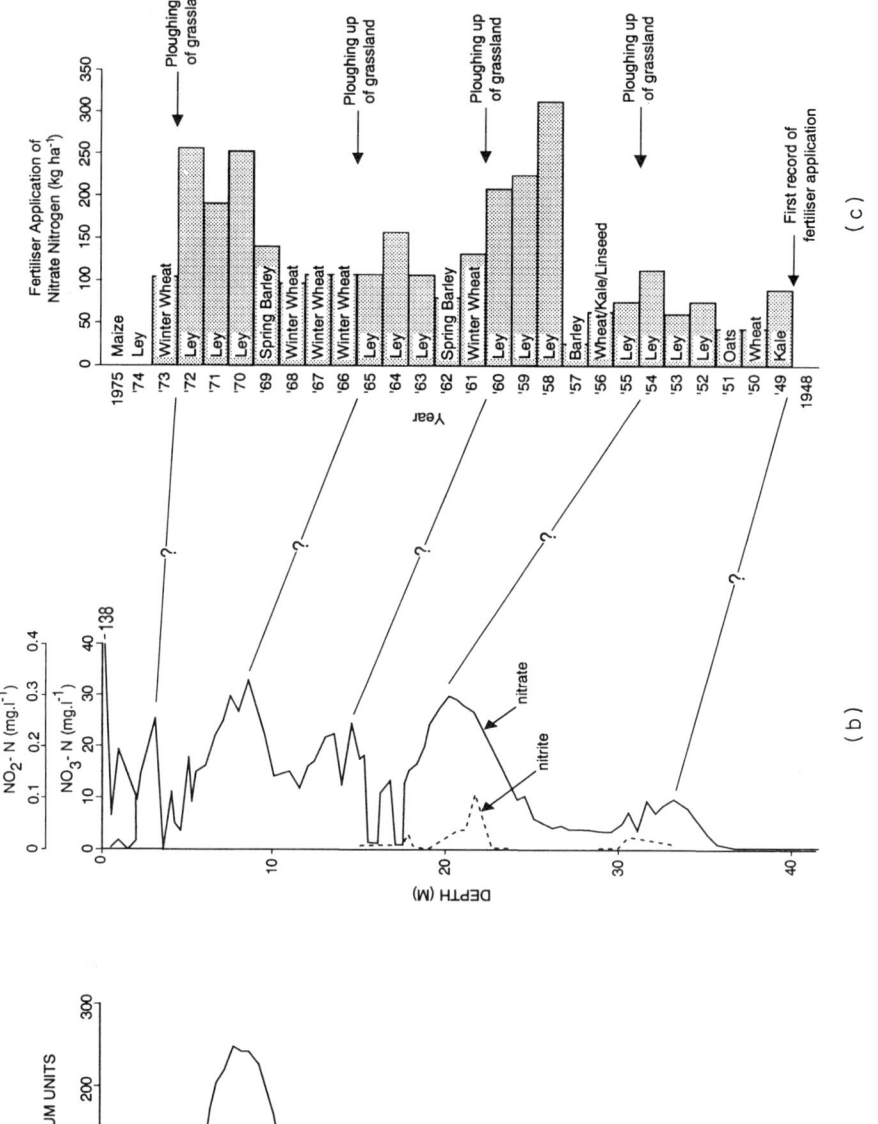

Figure 8.5 Tritium (a) and nitrate (b) profiles, together with land use since 1949 for land in the vicinity of the boreholes (c) for a site on the Chalk near Winchester (based on Young, Hall and Oakes, 1976, by permission of WRc)

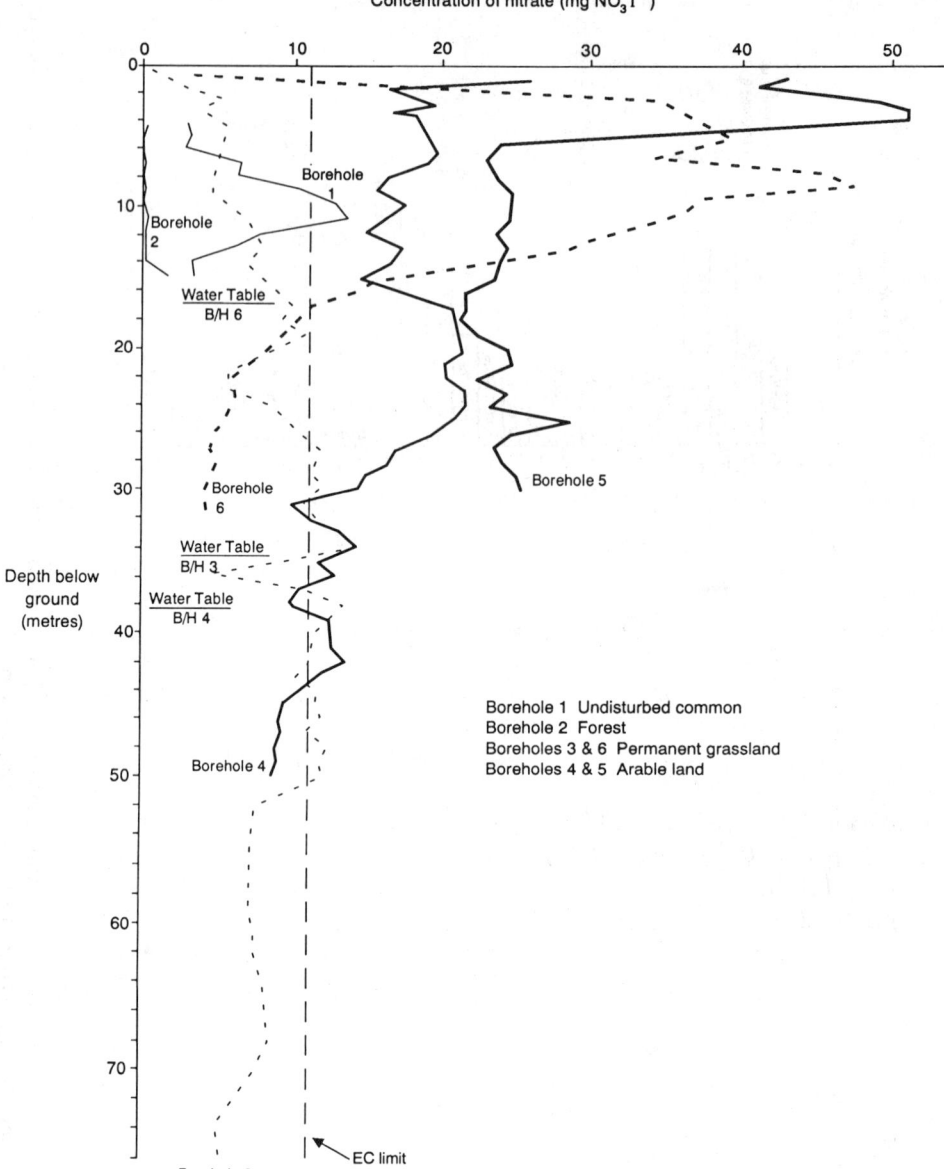

Figure 8.6 Nitrate profiles of cored boreholes in the Hatton catchment study area (reproduced by kind permission of the National Rivers Authority, Severn-Trent Region)

intensively managed arable land (Parker, Young and Chilton, 1991). Grassland production responds more to higher rates of fertiliser application than do most other crops. Optimum rates of application are in the range 350–450 kg N ha^{-1} a^{-1} and have increased considerably over the last 30 years. Ryden, Ball and Garwood (1984) emphasised the possibility of significant leaching from intensively managed, grazed grassland; they estimated an equivalent leaching loss of 160 kg N ha^{-1} a^{-1} from an application of 420 kg N ha^{-1} a^{-1}, significantly greater than that beneath a comparable cut grass sward from which most of the nitrogen is removed in hay or silage. Parker, Young and Chilton (1991) conclude that mean leaching losses are between 75 and 150 kg N $ha^{-1} a^{-1}$ from intensively managed grazed grassland, equivalent to at least 20% of the applied nitrogen. Such losses are likely to give rise to nitrate concentrations above the EC maximum. As noted above, leaching losses from cut grass systems are low, typically around 10 kg N $ha^{-1} a^{-1}$, except where large amounts of fertiliser are added (applications over 250 kg N $ha^{-1} a^{-1}$), because most of the nitrogen is removed when the grass is cut.

8.2.4 EVIDENCE FOR N-CYCLING AND NITRATE LOSS

The transport of nitrate downwards through the unsaturated zone is likely to be similar to that of chloride and tritium, unless the processes of nitrification or denitrification take place, in which case the concentration of nitrate will change independently of any dispersive effects. Such processes cannot be discounted since the presence of both the required bacteria and suitable carbon substrates has been shown. Whitelaw and Rees (1980) found significant bacterial populations throughout the whole of the unsaturated zone of the Chalk. Nitrate-reducing and ammonium-oxidising bacteria were detected in substantial numbers at various horizons. Denitrification was identified beneath permanent grassland whereas the unsaturated zone beneath fertiliser arable land was essentially aerobic and little (if any) attentuation of nitrate by denitrification was occurring. They concluded that nitrate seems certain to be involved in biological reactions in the unsaturated zone of the Chalk, although quantification of the rate and extent of such reactions was not possible. Young, Hall and Oakes, (1976) took the occurrence of small concentrations of nitrite at various points within the Chalk to be an indication of biological activity at depth (Figure 8.5(b)). Moreover, the content of organic carbon in the form of carbohydrate was more than sufficient to support biological activity within the aquifer. Once again, however, they were unable to establish the extent to which denitrification has occurred or could take place. The general implications of such studies are that denitrification rates are low, since the unsaturated zone is predominantly aerobic, so that biological reduction of nitrate concentrations is likely to be minimal. However, in aquifers like the Chalk, with slow percolation rates through the unsaturated zone, significant rates of denitrification could influence the pattern of decreasing nitrate concentration with depth.

8.3 NITRATE MOVEMENT IN THE SATURATED ZONE

The distinction is often made between unconfined and confined groundwater because of hydraulic differences between the flow of water under pressure in a confined aquifer and the flow of free, unconfined groundwater. In the case of unconfined groundwater, the upper boundary of the zone of saturation is the water table; its shape tends to follow that of the overlying topography, albeit in more subdued form. Recharge to an unconfined aquifer takes place directly by percolation from the land surface above. In the case of a confined aquifer, the upper boundary of the water body is formed by an overlying, less permeable bed which prevents direct recharge from above. Recharge to a confined aquifer can only take place at those locations where the aquifer is unconfined; hence the distinction drawn above is, to some extent, an arbitrary one (Ward and Robinson, 1990).

In order to forecast nitrate concentrations from groundwater sources, research has been carried out to determine the input, migration and attentuation of nitrate within the saturated zone. The input of nitrate to the saturated zone via percolation was considered in the previous section. Research on the saturated zone has included detailed catchment investigations in unconfined aquifers to establish the origin and distribution of nitrate within the saturated zone and to understand the factors controlling the concentration of nitrate in groundwater supplies (whether pumped supplies or spring-fed). Investigations in confined aquifers have sought to understand the origin and security of low-nitrate groundwater; these studies have necessarily included consideration of the transition zone between the unconfined and truly confined parts of the aquifer (Parker, Young and Chilton, 1991).

8.3.1 UNCONFINED AQUIFERS

The precise pattern of nitrate leached from soils into the aquifer below is rarely known, but lysimeter studies suggest that the highest nitrate concentrations occur in autumn and early winter, this nitrate being derived from mineralisation of the soil biomass. Later in the winter, concentrations tend to fall as the nitrate available for leaching becomes less (see Chapters 3 and 9). Nitrate concentrations in groundwater increase immediately and rapidly with the first infiltration of soil drainage water into the aquifer in autumn. Both groundwater levels and nitrate concentration rise to a peak in late winter; thereafter both decline through the spring and summer until further recharge takes place. Such marked seasonal fluctuations tend to happen even in aquifers with a generally deep unsaturated zone, indicating that rapid recharge to the saturated zone must take place locally, most usually in valleys where depth to the water table is least. The dispersion of nitrate from the autumn peak in infiltration to the winter peak in the saturated zone may happen for several reasons (Smith-Carington et al., 1983): dilution by pre-existing groundwater, diffusion exchange with immobile pore water and slow movement of a component of the infiltrating water through the upper part of the permanently saturated zone.

The distribution of nitrate within the saturated zone depends on several factors: the depth of the unsaturated zone, the saturated hydraulic conductivity of the aquifer, the degree to which any pumping induces vertical groundwater flow, and lithological variations within the aquifer. Parker, Young and Chilton (1991) report the case of the Sherwood Sandstone in Yorkshire (England): the unsaturated zone is thin (about 10 m), encouraging rapid recharge, and deep penetration of nitrate into the sandstone is thought to occur as a result of groundwater abstraction from boreholes drilled down to the bottom of the aquifer. The occurrence of major fissuring in a rock and the presence of less permeable beds may block or attentuate the downward movement of polluted water. The Chalk limestone is generally more fissured in its upper levels, so that significantly higher nitrate concentrations occur than in the less-fissured rock below. In the Lincolnshire Limestone, the Kirton Cementstones appear to act as an important aquiclude between the upper and lower limestone beds, creating chemical stratification between the two aquifers. Where they occur, faults allow transfer of groundwater between the two limestone aquifers, and the boreholes themselves may act in a similar way (Smith-Carington et al., 1983).

The occurrence of stratified groundwater can cause uncertainty in terms of the effectiveness of water sampling from boreholes. The nitrate content of pumped samples can vary according to borehole depth, length of solid casing, intersection of fissure flow, the proportion of inflow from various depths in the aquifer and the degree to which changes in the pumping regime

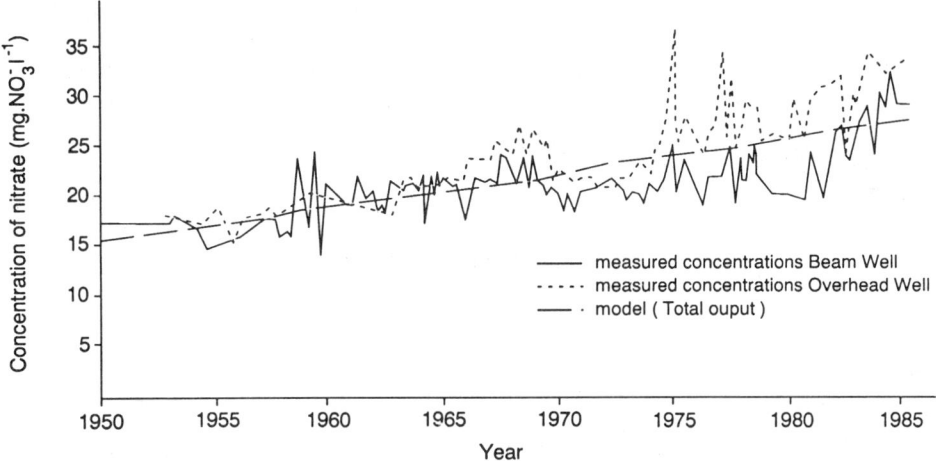

Figure 8.7 Nitrate concentrations for two wells in the Hatton catchment, 1950−1986 (reproduced by kind permission of the National Rivers Authority, Severn-Trent Region)

cause water to be withdrawn from different levels in the aquifer. For the early detection of diffuse pollution from agricultural sources, sampling from shallow depths in the aquifer will be most effective (Parker, Young and Chilton, 1991). Sampling from boreholes which fully penetrate an aquifer is suspect since samples are likely to represent an integrated value of concentrations of contaminants for the thickness of saturated aquifer that has been penetrated. By contrast, interstitial water extracted from cores recovered from beneath the water table may reflect more accurately the composition of water at various levels (Young, Hall and Oakes, 1976). Water taken from neighbouring boreholes may well yield different nitrate concentrations, although exhibiting similar trends, if the water is drawn from different depths in the aquifer. An example of this is given in Figure 8.7.

The concentration of nitrate in a groundwater sample from beneath a particular investigation site is a function not only of the concentrations entering the saturated zone by vertical drainage but also of the additions to the groundwater in an up-gradient direction. This laterally flowing water may be of higher or lower nitrate content so that at a given site the infiltrating groundwater may become more concentrated or more dilute. If vertical mixing within the saturated zone is minimal, inputs of differing concentrations to the water table along a flow path may remain distinct for some distance, leading to vertical variations in concentration below the water table (Young, Hall and Oakes, 1976). Figure 8.8 shows a section through the Chalk parallel to the groundwater flow lines; a single borehole has been drilled into the saturated zone. The section shows that the unsaturated zone is of variable thickness. Young, Hall and Oakes (1976) postulate that the higher concentrations of nitrate at 3 m depth below the water table demonstrates the earlier arrival of higher concentrations of nitrate in interstitial water from an area of reduced unsaturated zone thickness upgradient of the borehole. Young, Hall and Oakes (1976) used their model to predict the rate at which nitrate already in the unsaturated zone and future surface inputs would move down to the water table in response to infiltration. Estimates of the expected nitrate concentrations in groundwater were obtained from a catchment model in which infiltrating water and nitrate were routed through the

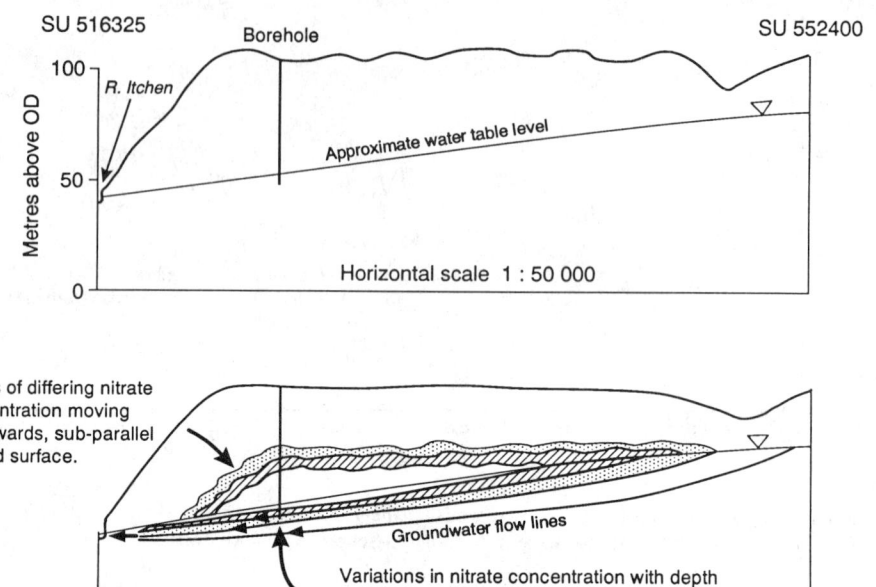

Figure 8.8 Section parallel to groundwater flow through the Chalk near the Itchen valley, Hampshire, England (above) and sketch showing postulated mechanism responsible for variation in nitrate concentration with depth below water table at the borehole (from Young, Hall and Oakes, 1976, reproduced by permission of WRc)

saturated zone of the aquifer. A vertical flow model was used to estimate the nitrate flux across the water table from 1950 to 2030 with the assumption that past agricultural practice and infiltration rates would be repeated cyclically in the future. A downward migration rate through the unsaturated zone of 1 m per annum was assumed. Because of the large difference in the hydraulic conductivities between the saturated and unsaturated zones, times of transit through the saturated aquifer were estimated to be small in comparison, having a maximum of about 5 years. Model predictions of the nitrate concentration of water reaching the water table in the vicinity of the borehole and for the mean nitrate concentration in groundwater beneath the borehole are shown in Figure 8.9. The steady rise in groundwater concentration from 1970 onwards is the result of contributions from up-gradient zones where the unsaturated thickness is less than at the borehole. The large depth of the unsaturated zone at the borehole site means that the increase in nitrate concentration of water entering the saturated zone at that point lags considerably behind the overall response of the groundwater.

8.3.2 CONFINED AQUIFERS

To date, the impact of agricultural pollution on confined aquifers in the UK has been small and groundwater in these aquifers is generally characterised by low or zero nitrate concentrations (Parker, Young and Chilton, 1991). The importance of such aquifers is not the volume of water which such sources provide but the fact that they are an important source of low-nitrate water which can be used to blend with polluted supplies from unconfined aquifers. This is especially important in the east of England where supplies of low-nitrate

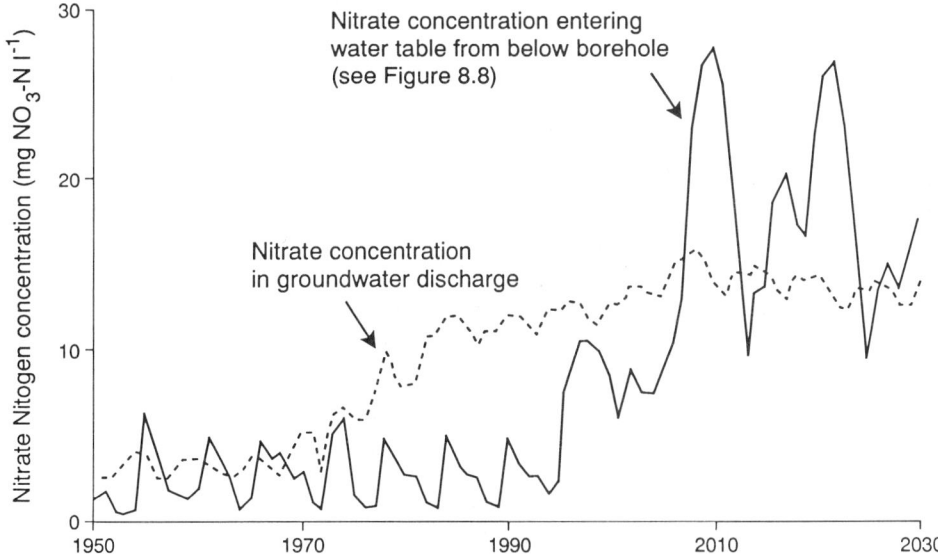

Figure 8.9 Model predictions of nitrate concentration in water reaching the water table for the borehole site shown in Figure 8.8 (from Young, Hall and Oakes, 1976, reproduced by permission of WRc)

water for blending are increasingly rare. Not surprisingly, the principal focus for research has been the security of such supplies. A major concern is that the significant pollution noted in recent years in the unconfined sections of certain aquifers will migrate into the confined strata. In some cases, the flow direction is naturally down-dip from the unconfined outcrop at higher elevation to the confined strata at lower elevation. In other cases, pumping may establish a hydraulic gradient towards the confined layer.

A front of groundwater contaminated by agrochemicals has migrated into the confined section of the Lincolnshire Limestone (Figure 8.10); tritium in the pore water proves this water to be recent in origin. Nitrate and dissolved oxygen diminish to negligible concentrations some 10 km from the edge of the aquifer outcrop. Dissolved oxygen concentrations gradually fall due to bacterial demand and also as a result of oxidation of finely disseminated pyrite and other ferrous minerals (Edmunds and Walton, 1983). This reduction in dissolved oxygen apparently provides a stimulus for the process of denitrification. Pore water samples from a borehole some 4 km from the outcrop contain significant nitrate which demonstrates that diffusion from the fissure water to the limestone matrix occurs (Parker, Young and Chilton, 1991). Much lower nitrate concentrations were found in pore water at greater distances into the confined aquifer. However, significant tritium concentrations were found in this water, which indicates that the water is of modern (post-1963) origin and should theoretically have contained nitrate also. The absence of this associated nitrate implies that denitrification is occurring. Smith-Carington *et al.* (1983) argue that the decreasing nitrate concentrations across this zone could not be the result of bacterial denitrification, since the relevant bacteria, even if present in sufficient numbers, would be unlikely to denitrify in the presence of dissolved oxygen. They argue that the major attenuation of nitrate in the marginally aerobic zone, and the relatively small increases in the decade 1969−79 (Edmunds and Walton, 1983), may be due primarily to transient dilution resulting from the diffusion of nitrate into the rock matrix. Bacterial denitrification would be more likely further down-gradient, beyond the point where

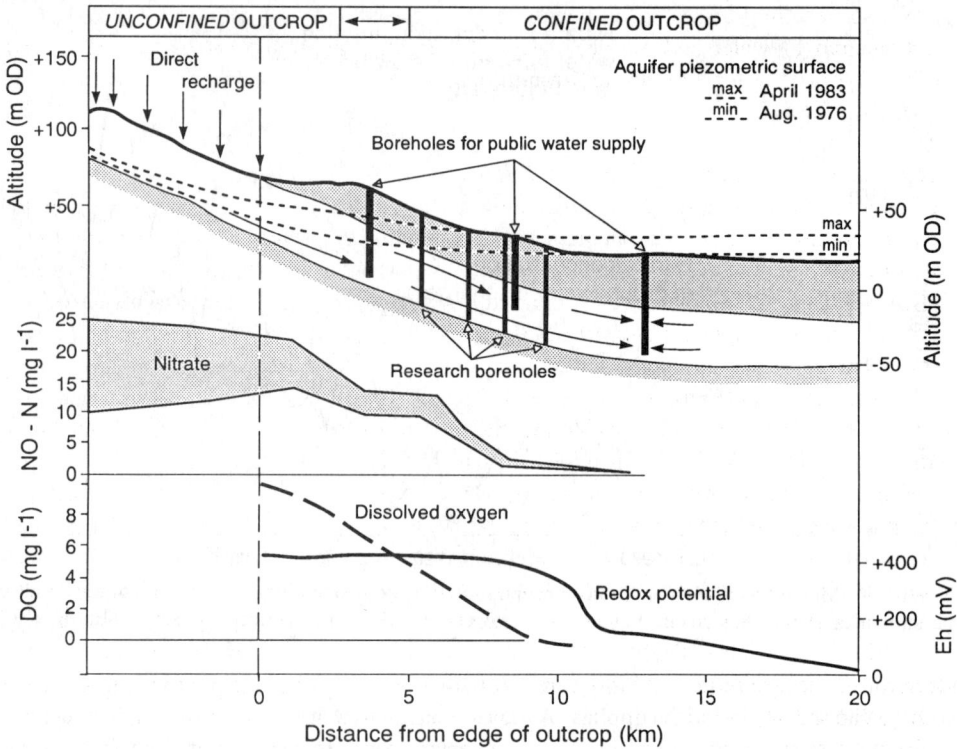

Figure 8.10 Down-dip hydrogeological and hydrochemical section through the Lincolnshire Limestone of south Lincolnshire (after Smith-Carington *et al.*, 1983; Parker, Young and Chilton, 1991)

there is a sharp reduction in Eh below 400 mV (Sprent, 1987, p. 61); the decrease in sulphate concentrations beyond 18 km suggests reduction via anaerobic microbial respiration. However, the parallel decrease of dissolved osygen and nitrate concentrations does imply that denitrification processes are also active in the marginally aerobic zone upstream of the Eh boundary. This might occur because of deoxygenation of micropore water, despite the availability of oxygen close by in the larger fissures; this situation is also thought to occur in soils (Umarov, 1990). A similar pattern to that shown in Figure 8.10 is reported by Parker, Young and Chilton (1991) for the Sherwood Sandstone aquifer. Despite 10 years of abstraction, there has been virtually no movement of nitrate polluted water from the recharge area of the outcrop into the confined aquifer.

There is obviously still much uncertainty about the potential role of denitrification in confined aquifers. Nevertheless, a continued supply of low-nitrate water from these aquifers depends on the effectiveness of the processes of oxygen removal and nitrate reduction.

8.3.3 MODELLING STUDIES

In the UK, the mathematical model developed by the Water Research Centre has been extensively used to model nitrate movement in groundwater (Oakes, 1982, 1987). The groundwater catchment is divided by a grid into square cells, with an edge length of 0.25 km. The simulation involves in three stages with the calculation of:

(1) Nitrate leaching losses from the soil zone to the underlying aquifer;
(2) Transport of water and nitrate through the unsaturated zone to the water table; and
(3) Transport of water and nitrate through the saturated zone to pumping wells or springs (DOE, 1986).

The model is run in two parts: in the first, the movement of water through the aquifer is calculated; in the second part, the quantity of nitrate carried by the water is estimated. Leaching losses of nitrate are calculated from a simple set of rules which identify different types of crops with the average annual loss expressed as a percentage equivalent of the nitrate applied (see Table 12.2). These rules, derived from evidence such as that presented in Section 8.2, have been the subject of some criticism (Addiscott and Powlson, 1989) and it is likely that future versions of the model will incorporate physically based leaching models such as that of Addiscott and Whitmore (1991) and Whelan (1992). The fissure flow and molecular diffusion processes described in Section 8.2 are incorporated into the unsaturated zone component of the model. Nitrate which has reached the water table is assumed to move horizontally towards springs and pumping wells at velocities and in directions determined by groundwater gradients. A necessary prerequisite to running the nitrate model is to run the corresponding groundwater flow model to provide the groundwater gradients (Oakes and Pontin, 1976). To calculate the diluting effect of recharge water mixing with water from the saturated zone it is necessary to assume an effective depth of flow in the aquifer over which this mixing occurs (DOE, 1986). The model has been used extensively to predict future nitrate trends in groundwater and to assess the impact of land-use changes on those trends (DOE, 1986; STW, 1988). In the case of the Hatton catchment study (STW, 1988), the model results have been combined with a financial analysis to indicate the benefits and costs of particular solutions (see also Chapter 12).

8.4 PATTERNS OF NITRATE DISCHARGE FROM AQUIFERS

8.4.1 ANNUAL REGIME

Discharge from aquifers may be via springs, diffuse seepage into streams and lakes, or from pumped wells. Depending on the nature of the aquifer, peak groundwater discharge may lag significantly behind precipitation inputs. There is a continuum of rainfall−runoff responses: cavernous (karst) limestones and thin aquifers provide the most rapid, peaked groundwater hydrographs while deep aquifers like the Chalk yield the most subdued and lagged hydrographs. Deep aquifers such as the Chalk have low flows only slightly below the mean flow because the response time for discharging water in storage is comparable to the time between wet and dry seasons. In such aquifers, there will be a long lag before a discharge increase occurs in response to the reasonal recharge, with a slow rise to the peak groundwater discharge and a low rate of recession (Burt, 1992). However, even in a deep aquifer like the Chalk, groundwater discharge is noticeably seasonal. One result of this is that many headwater streams are seasonally ephemeral. Prolonged summer discharge lowers the water table to such an extent that the source of the river migrates several kilometres downstream. Autumn recharge eventually raises the water table and the intermittent ('bourne') stream reappears, though perhaps not until several months after the main period of rainfall.

The seasonal variation in groundwater discharge is in many cases paralleled by a similar fluctuation in nitrate concentration; usually the two are positively correlated (see also Sections

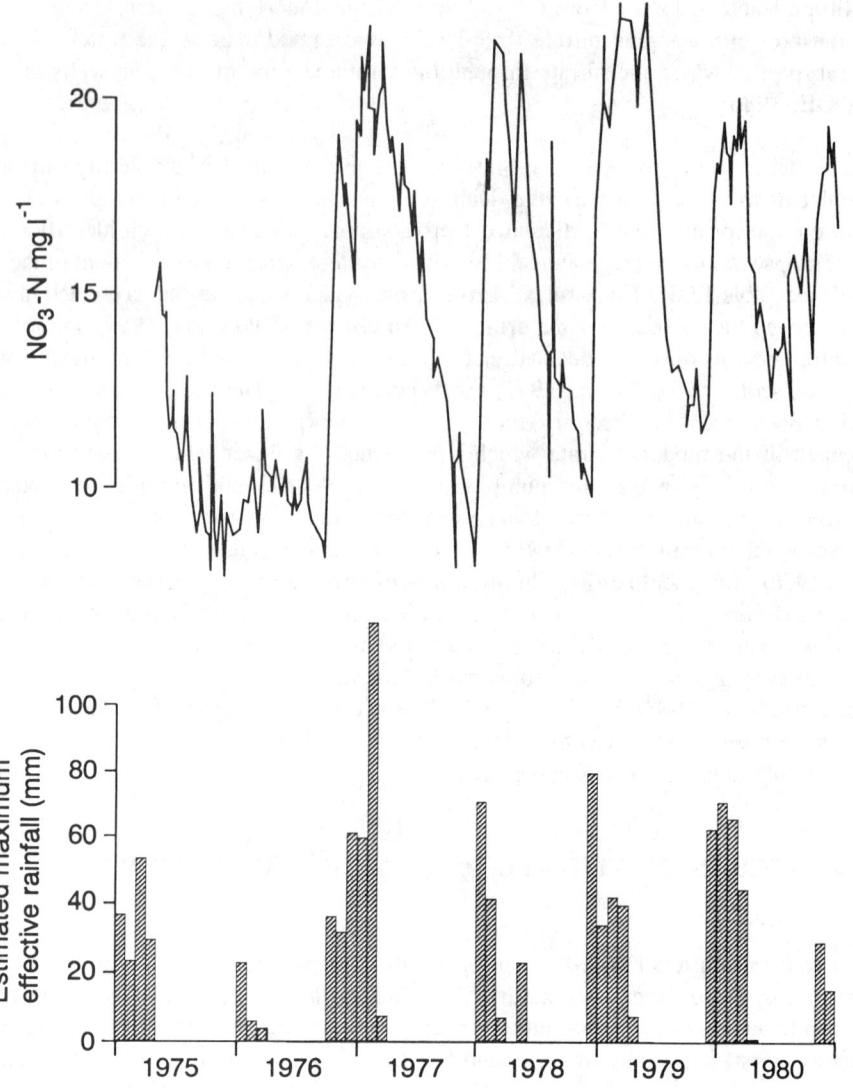

Figure 8.11 Nitrate concentrations from a public supply well in the Lincolnshire Limestone (after Smith-Carington *et al.*, 1983)

10.3 and 12.4). Groundwater discharge and nitrate concentration often increase sharply with the first recharge to the water table in late autumn or early winter; both then rise to peak in middle to late winter and thereafter both decline through the spring and summer until further recharge takes place. Figure 8.11 provides an example of marked seasonal fluctuations in the nitrate concentration of borehole water in the Lincolnshire Limestone. The sharp increases in nitrate concentration in the autumn, particularly after the drought of 1976, are especially notable. By contrast, in the previous winter, which was remarkably dry, minimal leaching of soils took place and there was little increase in the nitrate concentration of groundwater during that winter as a result.

As noted above (Section 8.3.1), the attenuation of nitrate from a sharp autumn peak in infiltration to a winter peak in discharge may happen for several reasons, notably because of dilution by pre-existing groundwater and dispersion as the infiltrating water percolates through the rock. Peak discharge often precedes peak nitrate concentration because the flood peak propagates more rapidly through the aquifer than the movement of the water itself and its solute load; dispersion may further delay the nitrate peak. The relationship between percolation and groundwater discharge is shown schematically in Figure 8.12 (which is based in part on Figure 12.6).

Results from Haigh and White (1986) and from Roberts (1987) show a decrease in the nitrate concentration of soil drainage water through the winter. The seasonal pattern of groundwater discharge is based on results from Burt *et al.* (1983), Burt and Arkell (1987) and Haycock (1991). A strongly seasonal response is seen even for a deep aquifer like the Chalk: over most of the aquifer, recharge is considerably delayed by the great depth of the unsaturated zone, but in the valleys, where depth to the water table is much less, recharge to the water table is rapid and the linkage between infiltration and discharge is clear.

Many examples exist in the published literature which show the annual nitrate regime for groundwater-fed rivers. For example, Casey and Clarke (1979) examined weekly nitrate samples for the river Frome which drains a Chalk catchment in southern England. Using multiple regression analysis they found that weekly nitrate concentrations over an 11-year period could be explained most successfully using a linear trend to describe the long-term increase and a cosine function to describe the seasonal variation. Discharge was a less useful independent variable. Although discharge was closely related to the seasonal variation in nitrate concentration when both were based on weekly data averaged over several years ($r = 0.858$), the individual discharges and nitrate concentrations were much less closely related, because both can vary considerably and independently from week to week. There is further discussion of the annual nitrate regime in rivers in Sections 10.3.1 and 12.4.

8.4.2 STORM HYDROGRAPHS

Except for cavernous limestones, it is unusual for aquifers to provide flood hydrographs. If there are no significant sources of overland flow, floods may be generated by groundwater. Otherwise high discharges are generated by surface and near-surface runoff processes and groundwater provides the prolonged recession flow associated with the falling limb of the hydrograph.

A number of studies of hillslope hydrology have identified the occurrence of double-peaked storm hydrographs (Dunne, 1978). In addition to an immediate runoff response at the time of the rainfall, a second hydrograph peaking several days after the rainfall input can occur in some basins, especially in the winter when soil moisture deficits are minimal. The second peak is entirely subsurface flow. Such hydrographs are usually the major volumatric response and may also provide the peak discharge. Burt and Butcher (1985) have analysed the occurrence of such hydrographs in catchments where permeable soils overlie impermeable bedrock. The delayed hydorgraph has also been identified as a time of maximum nitrate concentration (Burt *et al.*, 1983; Burt and Arkell, 1987). Given the coincidence of high discharge and high nitrate concentration, this means that such hydrographs can represent the major period of nitrate loss from a catchment (see also Section 9.4). Haycock (1991) has identified a similar runoff response for a small basin draining the Cotswold Limestone (a southern version of the Lincolnshire Limestone discussed above). The results in Figure 12.6 (p. 360 below) show

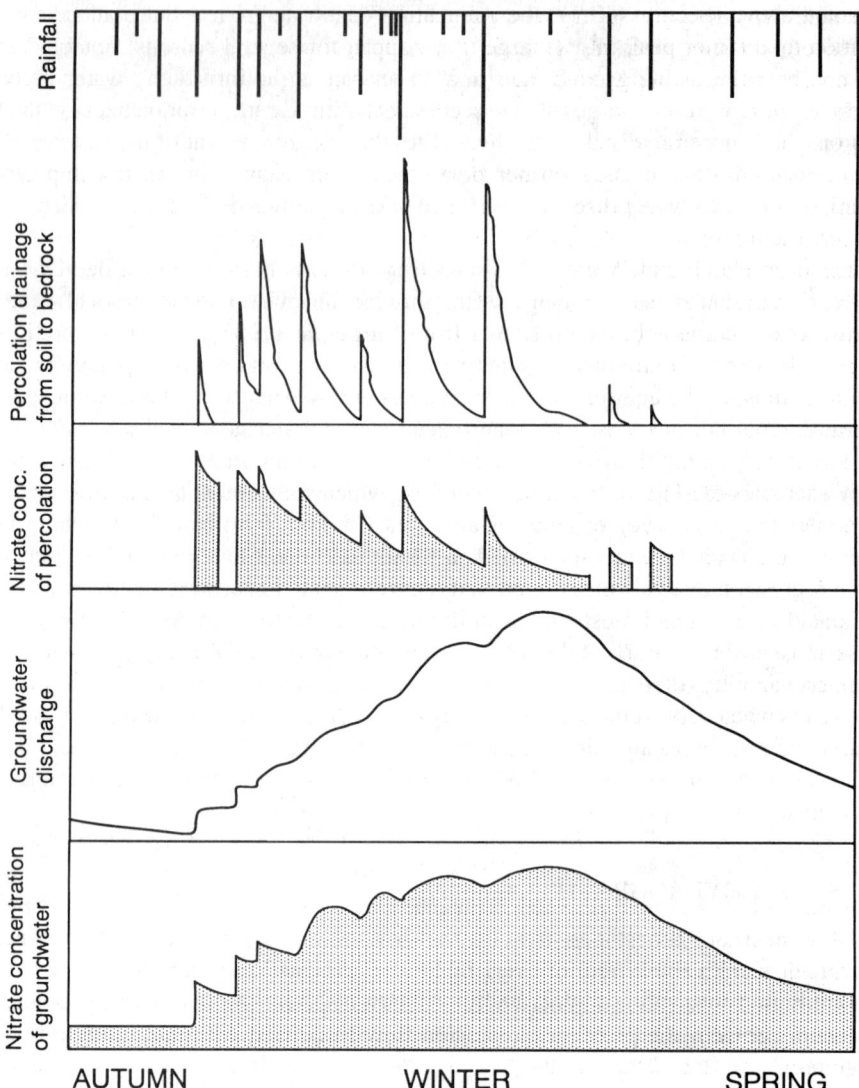

Figure 8.12 Schematic diagram showing percolation drainage into and discharge out of an aquifer. Also shown are the nitrate concentrations of percolation and discharge waters

that a thin aquifer can produce individually recognisable storm hydrographs and that each delayed hydrograph is associated with enhanced nitrate leaching. It is clear that deep throughflow in permeable soil and shallow groundwater flow may produce identical hydrological responses, and the distinction between the two is therefore necessarily an arbitrary one.

8.4.3 LONG-TERM TRENDS

Many groundwater sources in the UK have shown a steady rise in nitrate concentration since the 1960s, which must reflect post-war changes in farming practice. Smith-Carington *et al.*

(1983) note that there are several possible causes of a rising trend but that their relative importance is difficult to assess. If agricultural changes are slow, it is to be expected that nitrate concentrations will also respond slowly. However, in aquifers with a deep unsaturated zone, the effects of even relatively sharp increases in nitrate leaching associated with modern farming methods may be greatly attenuated by slow percolation through the unsaturated zone. Even where nitrate reaches the saturated zone quickly, diffusion of nitrate into the immobile storage could lead to an attenuation of surface-derived nitrate in the mobile fissure water.

In many cases, detailed sampling of groundwater and rivers in the UK began in 1975; the limited amount of data and the unusual meteorological sequence at the beginning of the period make trend analysis difficult. Johnes (1990) examined the nitrate concentrations for the river Windrush which drains an area of Cotswold Limestone in the upper Thames basin. For thin aquifers, where there is intimate linkage between recharge and discharge, it is possible to distinguish between the effects of climate and land-use change in relation to the changing nitrate concentration of groundwater. Using partial correlation analysis, Johnes (1990) showed that the annual mean nitrate concentration of the Windrush was most successfully explained by a long-term trend. The year number (*Anno Domini*) was used in the regression to denote the time trend; this variable alone accounted for 73% of the explained variance in nitrate concentration. Discharge, though significantly correlated with nitrate concentration, alone accounted for only 20% of the explained variance in annual mean nitrate concentration. A small joint effect (7%) indicated that higher nitrate concentrations had occurred in later, wetter years. These results mirror those found by Burt *et al.* (1988) for the Slapton Wood catchment and suggest that changes in land use provide a more significant explanation of changing nitrate concentrations over recent times than do climatic conditions. Further discussion of long-term trends is to be found in Section 10.3.3.

8.5 CONCLUSIONS: PREVENTION OR CURE?

Considerable quantities of freshwater are stored naturally in permeable strata. In England and Wales, the most important aquifers are the Chalk and the Sherwood (Triassic) sandstone which together yield about 60% of the groundwater abstracted; groundwater supplies 35% of the water used for public supply. Where groundwater is available in sufficient quantities, it is generally preferred as a source of water because of its low cost, good quality, relatively constant temperature and the fact that it is not so susceptible to pollution as surface storage (Rodda, Downing and Law, 1976). However, as the evidence reviewed above demonstrates, the security of groundwater is now under threat. Over the last two decades, nitrate concentrations in some groundwater sources have risen to levels which may intermittently or continuously exceed the EC limit of 11.3 mg NO_3-N l^{-1} (e.g. Figure 8.5). It is such nitrate concentrations, and the upward trend, which are of concern to the UK water industry, which has initiated a programme of field and laboratory investigations designed to determine the extent of nitrate contamination in the unsaturated and saturated zones of the principal aquifers; to evaluate the mechanisms and rates of movement of nitrate derived from the land surface; and to estimate future trends in groundwater nitrate concentrations (Royal Society, 1983). Investigations have revealed large concentrations of nitrate in the unsaturated zones of the principal aquifers. The evidence indicates that this nitrate is moving slowly downwards towards the main groundwater body. The slow rate of percolation and the great depth of the unsaturated zone mean that maximum nitrate concentrations in many supplies may not occur until well the twenty-first century (Figure 8.7). By then, the nitrate concentration of groundwater will be well above the current EC maximum acceptable concentration in some cases.

Two possible strategies are available to control nitrate concentrations in potable waters obtained from aquifers. Curative engineering solutions include blending of waters from different sources, installation of treatment processes for nitrate removal from water supplies (Chapter 13) and the relocation of wells within the confined sections of an aquifer. Recently much more attention has been given to preventative measures which tackle the problem at source (Chapter 12). In the Hatton Catchment Study (STW, 1988) a computer simulation model (Oakes, 1987) was used to assess options for the control of nitrate in water supply. The simulations showed that, unless there is a change in land-use practice, groundwater in the Hatton area will exceed the EC nitrate limit early next century. Blending supplies could defer the impact, but it will eventually be necessary to install a treatment plant to remove nitrate. Chemical treatment could be avoided if land use in the catchment is modified to reduce the amount of nitrate being leached. In the Hatton area it would be relatively cheap to provide blending and treatment; even so, the costs are not significantly less than solutions involving land-use controls. In other areas where favourable engineering solutions are not available, options involving a modification of land use might well offer clear financial advantages. In the Hatton catchment, a small 'local protection zone' was favoured, largely on practical grounds, as the means to control land use. This would impose stringent land-use controls on a small area in the vicinity of the borehole. Catchment-wide restrictions would be more expensive to implement and police; though they have a lower financial impact on individual farmers, the total cost can be greater.

The Hatton Catchment Study did not consider the social implications of the possible land-use changes, though these could be significant. Nor did it take account of wider issues of agricultural and environmental policy, the costs and benefits which may be associated with these or the question of who should pay compensation to farmers for loss of income. DOE (1988) also favoured a local protection zone policy and concluded that it would be in the national interest to compensate farmers for income lost in complying with the restrictions imposed in a local protection zone. Though water treatment may be the marginally cheaper option, other considerations, such as more strategic use of funds already being used to take farmland out of production, may favour the land-use change option. On this basis, the UK has introduced a pilot Nitrate Sensitive Area scheme (Section 12.3.3) aimed at controlling land use around selected boreholes. This scheme marks the first concerted attempt by government to use agricultural restrictions to prevent the pollution of groundwater supplies. This theme is explored in greater detail in Chapter 12.

REFERENCES

Addiscott, T. and Powlson, D.S. (1989) Laying the ground rules for nitrate. *New Scientist*, 29 April, 28–9.

Addiscott, T. and Whitmore, A.P. (1991) Simulation of solute leaching in soils of differing permeabilities. *Soil Use and Management*, **7**, 94–102.

Beven, K.J. and Germann, P. (1982) Macropores and water flow in soils. *Water Resources Research*, **18**, 1311–25.

Burt, T.P. (1992) The hydrology of headwater catchments. In Calow, P. and Petts, G.E. (eds), *Rivers Handbook*, Volume 1, Chapter 1, Blackwell, Oxford, pp. 3–28.

Burt, T.P. and Arkell, B.P. (1987) Temporal and spatial patterns of nitrate losses from an agricultural catchment. *Soil Use and Management*, **3**, 138–43.

Burt, T.P. and Trudgill, S.T. (1985) Soil properties, slope hydrology and spatial patterns of chemical denudation. In Richards, K.S., Arnett, R.R. and Ellis, S. (eds), *Geomorphology and Soils*, Allen and Unwin, London, 13–36.

Burt, T.P., Butcher, D.P., Coles, N. and Thomas, A.D. (1983) The natural history of the Slapton Ley Nature Reserve. XV: Hydrological processes in the Slapton Wood catchment. *Field Studies*, **5**, 731–52.

Burt, T.P., Arkell, B.P., Trudgill, S.T. and Walling, D.E. (1988) Stream nitrate levels in a small catchment in south west England over a period of 15 years (1970–85). *Hydrological Processes*, **2**, 267–84.

Casey, H. and Clarke, R.T. (1979) Statistical analysis of nitrate concentrations from the River Frome (Dorset) for the period 1965–76. *Freshwater Biology*, **9**, 91–7.

Darcy, H. (1856) *Les Fontaines Publique de la Ville de Dijon*, Victor Dalmont, Paris.

DOE (1986) *Nitrate in water*. Department of the Environment Pollution Paper 25, HMSO, London.

DOE (1988) *The Nitrate Issue*, Department of the Environment, HMSO, London.

Domenico, P.A. and Schwartz, F.S. (1990) *Physical and Chemical Hydrogeology*, Wiley, Chichester.

Dowdell, R.J. and Webster, C.P. (1980) A lysimeter study using nitrogen-15 on the uptake of fertiliser nitrogen by perennial ryegrass swards and losses by leaching. *Journal of Soil Science*, **31**, 65–75.

Dunne, T. (1978) Field studies of hillslope flow processes. In Kirkby, M.J. (ed.), *Hillslope Hydrology*, Wiley, Chichester, pp. 227–93.

Edmunds, W.M. and Walton, N.R.G. (1983) The Lincolnshire Limestone — hydrogeochemical evolution over a ten-year period. *Journal of Hydrology*, **61**, 201–11.

Foster, S.S.D. (1975) The Chalk tritium groundwater anomaly — a possible explanation. *Journal of Hydrology*, **25**, 159–65.

Foster, S.S.D. and Smith-Carington, A.C. (1980) The interpretation of tritium in the Chalk unsaturated zone. *Journal of hydrology*, **46**, 343–64.

Freeze, R.A. and Cherry, J.A. (1979) *Groundwater*, Prentice-Hall, Englewood Cliffs, NJ.

Geake, A.K. and Foster, S.S.D. (1989) Sequential isotope and solute profiling in the unsaturated zone of British Chalk. *Hydrological Sciences Journal*, **34**, 79–95.

Haigh, R.A. and White, R.E. (1986) Nitrate leaching from a small, underdrained, grassland, clay catchment. *Soil Use and Management*, **2**, 65–70.

Haycock, N.E. (1991) *Riparian Land as Buffer Zones in Agricultural catchments*, Unpublished DPhil thesis, Oxford University.

Johnes, P.J. (1990) *An Investigation of the Effects of Land Use upon Water Quality in the Windrush Catchment*, Unpublished DPhil thesis, Oxford University.

Nye, P. and Tinker, B. (1977) *Solute Movement in the Soil–Root System*, Blackwell, Oxford.

Oakes, D.B. (1982) Nitrate pollution of groundwater resources — mechanisms and modelling. In Zwirnmann, K.-H. (ed.), *Non-point Nitrate Pollution of Municipal Water Supply Sources: Issues of Analysis and Control*, IIASA Collaborative Proceedings series CP-82-S4, IIASA, Laxenburg, Austria, pp. 207–30.

Oakes, D.B. (1987) *Prediction of Groundwater Nitrate Concentrations*, Water Research Centre Publication PRU 1566-M, Medmenham.

Oakes, D.B. and Pontin, J.M.A. (1976) *Mathmatical Modelling of a Chalk Aquifer*, Water Research Centre Technical Report 24, Medmenham.

Parker, J.M., Young, C.P. and Chilton, P.J. (1991) Rural and agricultural pollution of groundwater. In Downing R.A. and Wilkinson, W.B. (eds), *Applied Groundwater Hydrology*, Clarendon Press, Oxford, pp. 149–63.

Reeves, M.J. (1979) Recharge and pollution of the English Chalk: some possible mechanisms. *Engineering Geology*, **14**, 231–40.

Roberts, G. (1987) Nitrogen inputs and outputs in a small agricultural catchment in the eastern part of the United Kingdom. *Soil Use and Management*, **3**, 148–54.

Rodda, J.C., Downing, R.A. and Law, F.M. (1976) *Systematic Hydrology*, Newnes-Butterworth, London.

Royal Society (1983) *The Nitrogen Cycle of the United Kingdom*, the Royal Society, London.

Ryden, J.C., Ball, P.R. and Garwood, E.A. (1984) Nitrate leaching from grassland. *Nature*, **311**, 50–54.

Smith, D.I., Atkinson, T.C. and Drew, D.P. (1976) The hydrology of limestone terrains. In Ford, T.D. and Cullingford, C.H.D. (eds), *The Science of Speleology*, Academic Press, New York, pp. 179–212.

Smith-Carington, A.K., Bridge, L.R., Robertson, A.S. and Foster, S.S.D. (1983) *The nitrate pollution problem in groundwater supplies from Jurassic limestones in central Lincolnshire*, Institute of Geological Sciences Report 83/3, HMSO, London.

Sprent, J.I. (1987) *The Ecology of the Nitrogen Cycle*, Cambridge University Press, Cambridge.

STW (1988) *The Hatton Catchment Study*, Severn-Trent Water, Birmingham.

Umarov, M.M. (1990) Biotic sources of nitrous oxide in the context of the global budget of nitrous oxide. In Bouwman, A.F. (ed.), *Soils and the Greenhouse Effect*, Wiley, Chichester, pp. 263–8.

Ward, R.C. and Robinson, M. (1990) *Principles of Hydrology*, 3rd edition, McGraw-Hill, New York.

Wellings, S.R. and Bell, J.P. (1980) Movement of water and nitrate in the unsaturated zone of Upper Chalk near Winchester, Hants., England. *Journal of Hydrology*, **48**, 119–36.

Whelan, M.J. (1993) *A Physically Based Model of Soil Nitrate Leaching*, Unpublished PhD thesis, Leeds University, in preparation.

Whitelaw, K. and Rees, J.F. (1980) Nitrate-reducing and ammonium-oxidising bacteria in the vadose zone of the Chalk aquifer of England. *Geomicrobiology Journal*, **2**, 179–87.

Young, C.P., Hall, E.S. and Oakes, D.B. (1976) *Nitrate in groundwater — studies on the Chalk near Winchester, Hampshire*. Water Research Centre Technical Report 31, Medmenham.

9 Nitrate Losses from Agricultural Land

A.C. ARMSTRONG

ADAS Soil and Water Research Centre, Cambridge

and

T.P. BURT

School of Geography, University of Oxford

9.1 INTRODUCTION — SCALE AND DIMENSIONS

At all scales, the movement of nitrate is intimately associated with the movement of water. Because it is highly soluble, nitrate is almost everywhere carried in solution, and the measurement of nitrate fluxes requires a consideration of the water fluxes that carry it. However, the same is not true for all forms of nitrogen, so that any consideration of the total nitrogen budget must also include both the gaseous phase and the binding and release of nitrogen as relatively immobile soil organic matter. Nevertheless, the first stage in the consideration of measuring nitrate fluxes is frequently the definition of the water fluxes. The measurement and theory of water movement in soils is a well-researched area, and many reviews are available (for example, in Childs, 1969; Marshall and Holmes, 1988). The movement of water in agricultural soils is particularly well documented as it relates to the artificial control of soil water by irrigation or drainage, topics which are reviewed, for example, by ILRI (1973) and Smedema and Rycroft (1983).

The study of water and nitrate movement can be undertaken at a variety of scales and dimensions. Commonly, studies are carried out at one of three scales which also effectively introduce three separate dimensions to the problem: profile, slope or experimental plot, and catchment. These are, in effect, one-, two- and three-dimensional studies, and the choice of the relevant framework for any study will depend on that which is the most appropriate to the problem of interest, and will bring with it a degree of complexity. Nevertheless, certain typical situations occur frequently, corresponding to each of these scales.

Profile studies are common, both for the study of the processes, for field observations, and as generalisations of large areas. Typical among these studies are those of the movement of nitrate to groundwater, in which the rate of movement of nitrate through the soil profile to the underlying aquifer is the main focus of interest. Frequently, then, the one-dimensional profile is considered a representation of the processes occurring over a large area, which may, for example, be a grid cell in a wider model. Studies of the agricultural use of fertiliser also frequently follow this approach, being concerned with the utilisation of the applied fertiliser by the plant and the rate of loss to the wider environment.

Two-dimensional studies of nitrate delivery to rivers often require a consideration of whole slopes or of experimental plots. Typically, both surface and subsurface flow patterns are considered, so that the movement of water and nitrate in both the horizontal and lateral

Nitrate: Processes, Patterns and Management. Edited by T.P. Burt, A.L. Heathwaite and S.T. Trudgill
© 1993 John Wiley & Sons Ltd

directions are included in the analysis. Lastly, the study of nitrate movement in whole catchments often requires the consideration of three-dimensional situations, involving the profile processes, the lateral movement through the soil and over the surface, and its transformation along stream lines.

Increasing the dimensionality of the study increases the scale. There is thus a general increase in the level of aggregation as the process goes from simple profile-based studies to complete catchment studies. However, it is also necessary to recognise that studies can also be carried out over a whole range of temporal scales. As with the choice of spatial scale, so the choice of temporal scale also imposes a set of perspectives. Thus, for example, studies of plant physiological function frequently consider time scales of the order of hours and days. Studies of the mechanisms of nitrate movement on hillslopes frequently consider whole storm events, and as such are concerned with a temporal period of the order of a few to tens of days. On the other hand, studies of nitrogen balances concerned with water quality and the management of land in order to meet environmental objectives, particularly within the agricultural context, require data over whole years.

Soil studies, including those of soil nitrate, need to be aware of these scale considerations, because each spatial and temporal scale brings with it a perspective. There is always a danger in following the reductionist argument, that systems must be studied at the smallest level, and that large scales must be derived from an aggregation of small-scale studies. While it remains true that all phenomena are governed by the basic rules of chemistry and physics that operate at the smallest level, it is unwise to attempt to resolve every problem from small-scale rules. At each level, appropriate generalisations need to be made, and the complexity of the landscape system must preclude the possibility of ever predicting the macroscopic behaviour of systems by repeated application of microscopic rules.

9.2 PROFILE STUDIES

9.2.1 NITRATE TRANSPORT PROCESSES

The classical problem in soil physical and chemical studies is to determine the fluxes of water and solutes one-dimensionally in the vertical direction. This situation is the easiest to describe theoretically, the easiest to reproduce in the laboratory, and the easiest to instrument in the field. Furthermore, it has proved to be a very reliable tool, since the results derived from single-profile analyses have appeared to represent the behaviour of quite large areas.

The reason for this satisfactory behaviour of the one-dimensional approximation is that the two main vectors of transport of both water and solutes, which are the downward movement under gravity and the upwards movement through evapotranspiration, both act essentially normally to the soil surface. For this reason, it is possible to simplify the system and suggest that the three major routes for nitrate movement out of a profile are:

(1) Upwards, crop uptake and gaseous loss.
(2) Sidewards, via surface/subsurface flow.
(3) Downwards, as profile leaching.

This chapter is concerned with the last two of these, but we must remain aware of the potential for gaseous interactions, both inputs and outputs, which are fully discussed in Chapter 3.

The balance between vertical and lateral flow on a site depends on its hydrology, and in particular on the nature of the soil. In free-draining soils, movement is predominantly in the vertical direction, and nitrate studies are typically concerned with the downward movement of solute through the base of the profile. However, on impermeable soils there is effectively no downward movement below the base of profile, and losses are predominantly horizontal. Consequently, studies of such soils inevitably require a consideration of lateral losses, which introduces a two-dimensional element. Many soils occupy a position intermediate between the two extremes of being free-draining or impermeable, and experience a mixture of both vertical and lateral movement, the balance between the two varying with the time of the year. Many soils experience periods of lateral flow during winter waterlogging, but can be considered to have simple vertical flows during the summer period.

Water movement within a soil profile can be subdivided into three simple portions: saturated flow, unsaturated flow, and by-pass flow.

Saturated flow is the simplest and best understood. Under saturated conditions beneath the water table, water flows in response to the hydraulic gradient, following Darcy's law. Prediction of fluxes normally requires a two-dimensional analysis to identify the hydraulic gradient, and is well treated in many texts (e.g. Childs, 1969; ILRI 1973). If, however, attention is focused on the mid-drain (wettest case) situation, then the movement of water is essentially vertical. In this situation, both the fluxes through the water table and the position of the water table can generally be predicted using simple analyses. It should also be noted that, in general, soils are anaerobic below the water table. Consequently, the chemical reactions governing the production of nitrate are different from those on unsaturated, aerated soils above the water table. The position of the water table is thus a major variable controlling the nitrogen economy of a site.

Unsaturated flow follows a similar relationship, in that water and hence solutes move in response to a potential gradient. For vertical flow, this movement is normally described by the Richards equation:

$$\frac{\delta\theta}{\delta t} = \frac{\delta}{\delta z}\left(K(\theta)\,\frac{\delta\psi}{\delta z}\right) + \frac{\delta K}{\delta z}$$

(Marshall and Holmes, 1988, Chapter 4) where

   θ is the volumetric water content,
   K the hydraulic conductivity,
   z the vertical direction upwards from an arbitrary base,
   ψ the soil water potential,
   t is time.

In general, this is much slower than saturated flow because for most soils the hydraulic conductivity drops very rapidly with decreasing soil moisture content. Because of this, unsaturated water flow has long residence times and the soil water solutions are often near equilibrium.

Unsaturated flow can be either downwards in response to the gravitational component of the potential or upwards in response to evapotranspirative and capillary forces. Consequently, it is possible for both water and solutes to move in both directions. In a global context, the total water balance, given by rainfall minus evapotranspiration, can be downward leaching (as in UK) or upward accumulation of salts on the soil surface, the semi-arid problem of salinisation.

By-pass flow occurs wherever there is preferential movement of water along specific pathways. Frequently this involves structural fissures and cracks, while the soil peds are unsaturated, and is related to the phenomenon of macroporosity. Within a saturated soil, preferential flow may also occur along channels which can be termed pipe-flow. By-pass flow is rapid, and so has the possibility for only limited interaction with the soil mass. It thus tends to be low in solutes, particularly in those originating from the soil, such as nitrate. The phenomenon of by-pass flow has only recently been the subject of much study, and it is not described by classical soil physics. Nevertheless, it is an important component of water movement in many natural soils. Attempts to describe this flow component mathematically have been made by, *inter alia*, Hoogmoed and Bouma (1980), Beven and Germann (1981), and Trudgill *et al.* (1983), and the movement of solutes by this process by Smettem, Trudgill and Pickles (1983), and Jarvis and Leeds Harrison (1987a,b) and Jarivs *et al.* (1991).

Although the detailed consideration of the flow paths is necessary for the study of short-term movement of water, for many studies it is possible to simplify the approach. Thus, for example, for the calculation of the annual leaching budget to an underlying aquifer it is the total flux that is important, and this is given by the annual effective rainfall. In these circumstances, the concept of piston flow, with the water from each year displacing that from the preceding down the profile, is generally adequate.

9.2.2 METHODS OF STUDYING NITRATE MOVEMENT IN THE PROFILE

Several methods are available for measuring the movement of water and solutes in the profile. The measurement of lateral water movement is generally considered a two-dimensional problem and is discussed in the next section. The problem is to measure and intercept the vertical, largely unsaturated, water movement. The techniques have recently been reviewed by Ballif and Muller (1990) and Addiscott (1990). In general there is no single preferred technique, as each method involves different degrees of effort, and each is, to some extent, unsatisfactory. The choice of any one method therefore remains to a degree a compromise between what is ideal and what is practical. However, the choice is also, to some extent, dictated by the soil type.

Direct sampling of soil and soil water is perhaps the simplest technique. Samples are taken (for example, by coring) and the amount and chemical composition of the water within them determined by laboratory methods. This method is, however, both labour intensive and destructive. For this reason it cannot be used routinely to track movements down a profile, although it does give a useful reference set of data at a single time. The major problem identified by Addiscott (1990) is that of interpretation: it will describe the location of nitrate in the profile but not the fluxes. Nevertheless, the method has given useful results in unsaturated rock above aquifers, notably in the chalk of south-east England. In this rock, matrix flows occurs through very small fissures (1.0−1.5 μm diameter), in which water and its solute load moves uniformly downwards (Wellings 1984a). The pattern of nitrate concentrations down the soil profile and into the underlying unsaturated aquifer can thus be closely related to the sequence of land uses at the surface (Wellings 1984a; Ryden, Ball and Garwood, 1984).

Porous ceramic cup samplers (Figure 9.1) extract water from the soil by applying a suction, so that solute concentrations in the water can be analysed using normal laboratory methods. There are, however, problems with their use, both in their installation, which inevitably involves some degree of disruption, and in the interpretation of the results they provide. Porous cup samplers have been used extensively since their introduction by Briggs and McCall (1904), and recent reviews are given by, *inter alia*, Hansen and Harris (1975), Barbee and Brown (1986), Wood (1973) and Grossman and Udluft (1991). The largest unresolved problem relates

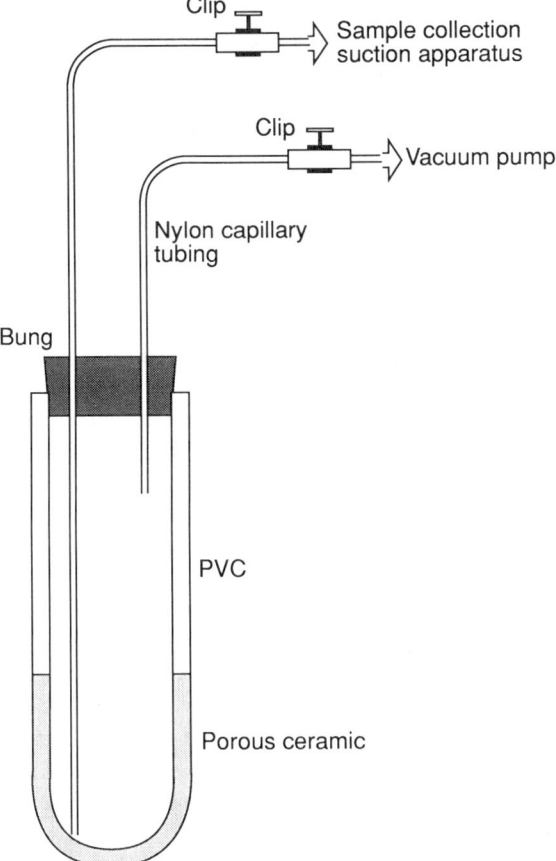

Figure 9.1 Porous cup samplers

to the nature of the soil pores that are sampled, and the degree to which the water sampled is representative of all the water in the soil. The size of the pores from which the water is drawn are defined by the tension applied. For applied suction, s, the largest pore to remain full of water has diameter, r:

$$r = 2\gamma/pgs$$

where p is the density of water,
     g is the acceleration due to gravity, and
     γ is water surface tension.

Consequently, applying a given tension to a soil releases water from a given range of pore sizes. (This technique is the basis of the laboratory determination of pore size distribution by examining the water released at successively higher tensions.) It is impossible to apply more than 1 atmosphere of tension, and, in practice, rarely more than 0.8 bar. This imposes a limit to the minimum size of pore that can be sampled. When water is extracted using a porous cup, then water is released from those pores whose diameters lie between those emptied between the initial and the applied suction. Water from smaller pores will not be drawn to the sampler, and no water will be drawn from pores that were initially empty. Consequently, this method invariably fails to sample water moving in macropores. If properties of the water are affected by the size of pore in which it resides (which is quite feasible, since, in general, water is less mobile in smaller pores, and will thus be nearer to equilibrium with dissolved salts), then the sample removed by a porous cup might systematically bias the results. Although this problem has been recognised for some time, no practical field solution has been suggested and adopted. The limitations of sampling are particularly problematic where macropore flow occurs. Porous cups only sample the soil matrix and cannot give figures for preferential paths. However, the problem is likely to be least on sandy soils where most water is in the large pores and matrix flow dominant.

The use of ceramic samplers also requires an independent estimation of the water flux which needs to be multiplied by the concentrations to give the nitrate fluxes. Occasionally these water fluxes are calculated from the potential gradient measured with tensiometers, using, for example, the Zero Flux Plane Method (e.g. Wellings, 1984a). This gives a calculated flux, not a direct measurement. For some studies, notably those based on annual budgets, it is possible to estimate the total flux from the water balance:

$$\text{Drainage} = \text{rainfall} - \text{evapotranspiration}$$

Porous cups are, however, the only method that can be used routinely within an agricultural cultivation system, and the majority of results that discuss the effects of various cultivation systems, cropping patterns, etc. are derived using this technique (e.g. Shepherd and Lord, 1990; Ballif and Muller, 1990).

Lysimeters, by contrast, give direct measurement of fluxes (Figure 9.2). In essence, they consist of an isolated block of soil, in which movement is constrained to the downward direction. This block of soil can either be repacked soil material or else an undisturbed monolith excavated from the soil. In either case, a free water surface is created at the base by suspending it over gravel or a void. This in itself causes problems, as the very act of creating that free surface introduces an artificial lower boundary to the soil sample, whereas in the field there would be a suction to a water table below. In order to answer this limitation, some workers have suggested applying suction to the base of lysimeters.

Distrubed lysimeters, effectively large laboratory columns, are very useful for process studies (for sample, the study of breakthrough curves and for studies of plant nutrition) but have an uncertain relationship with field conditions. A considerable body of French work using disturbed lysimeters is summarised in Ballif and Muller (1990).

Undisturbed lysimeters are much more difficult to construct (Belford, 1979), so are expensive in time and effort. All involve some degree of disturbance, although the aim is to minimise it. Flow round the sides is particularly problematical. Because the sample is taken directly from the field situation, the question is raised as to the degree to which it is representative of the field from which it is taken. This raises the common soil problem of defining representative soil units, often termed the Representative Elementary Volume (REV). It is by no means clear that this volume is enclosed within an average lysimeter, even those as large as the 0.8 m diameter and 1.2 m deep as used by Cannell *et al.* (1980a), in the case of structured soils. Nevertheless, there have been many significant studies using this technique.

One particularly relevant study is the Rothamsted drain gauges, established in 1870 by Lawes, Gilbert and Warrington (1881), and continued to the present day. Reports of the work have been published by Voelcker (1884), Miller (1906) and Russell and Richards (1920),

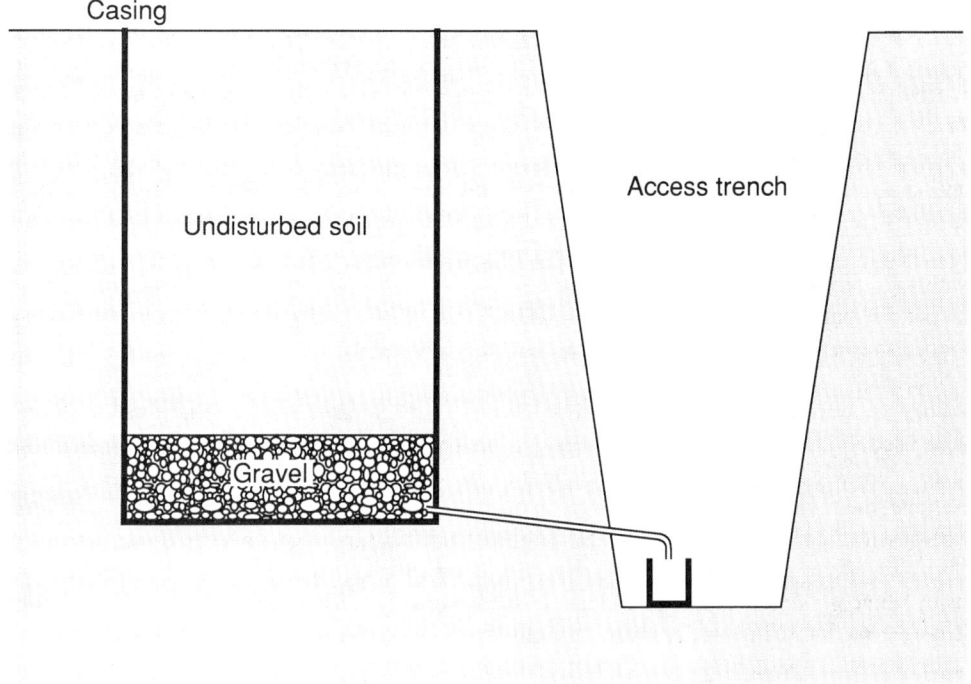

Figure 9.2 Lysimeter installation

and the long-term data set recently reviewed by Addiscott (1988). The lysimeters consisted of a series of soil monoliths, each 1/1000 of an acre in area, lined with brick, undermined and supported by perforated iron plates. Three replicates, 20, 40, and 60 inches deep, were constructed, although the results of the 40-inch lysimeter are considered unreliable due to leakage. Although the soils of these lysimeters have remained fallow since their initial construction, they have shown a continued leakage of nitrate due to the breakdown of soil organic matter. Addiscott (1988) demonstrated that the leaching rate could be predicted by a combination of an exponential decline and a consideration of the annual rainfall amounts. Initial rates of nitrate losses from the soil were about 45 kg ha^{-1}, with the exponential decline showing a half-life of 41 years. These results indicate that uncropped and uncultivated soils will continue to lose nitrates for a very long time, as the soil organic nitrogen is only slowly exhausted by microbial breakdown.

Many subsequent workers have used lysimeters as the basis for studies of crop performance and nitrate movement. These, for example, include the studies of Cannell *et al.* (1980a), which demonstrate both the potential for detailed crop and soil performance studies afforded by this technique and the complexity of the installation required. Thus in order to undertake experiments in which the soil water regime is controlled within the lysimeters it was necessary both to restrict the inputs of rain through a movable artificial shelter and to allow for irrigation (Cannell *et al.*, 1980b). In order to provide adequate replication, 64 lysimeters (32 replicates each of two soil types) were installed. To provide records of (and hence to control) the soil water regime, each lysimeter was provided with a neutron probe access tube, and tensiometers installed through the side walls of the monoliths. Edge effects around the lysimeters were minimised by surrounding them with soil in which the same crop was grown. This facility was then used to study the effects of short periods of waterlogging on both winter cereals (Cannel *et al.*, 1980b; Belford, Cannell and Thomson, 1985) and also a variety of other crops (Cannell *et al.*, 1985; Belford *et al.*, 1980; Cannell and Belford, 1980). Several of these experiments have yielded information about the nitrogen economy of the installation. Thus, for example, Cannell *et al.* (1980b) have indicated that waterlogged soils exhibit a greater degree of denitrification, and hence a lower loss of nitrate by leaching, when compared to similar soils without waterlogging. Later studies by Belford, Cannell and Thomson (1985), however, showed no significant effect of waterlogging on nitrate leaching loss. A large proportion of the nitrate leached (85%), however, was lost before the onset of winter waterlogging, when nitrate concentrations in the soil solution were high. These studies show the pattern that is repeated throughout most British studies of nitrate leaching, with high nitrate concentrations in the autumn, declining during the winter as the readily available nitrate is exhausted.

These examples illustrate the use and flexibility of lysimeter studies. Many other authors have used similar techniques, with varying degrees of sophistication. A great deal of our current knowledge of the nitrogen dynamics of soils derived from these studies (see, for example, the review of earlier studies given by Wild and Cameron, 1980). However, lysimeters are not without their problems, and the degree to which these problems can be solved places a limit on the quality of our knowledge. The most critical problem is that of matching the lower boundary to the natural conditions. In sandy soils, lysimeters break the vertical continuity of pores, so that plants suffer from drought stress, whereas in clay soils lateral flow which is important in the natural state is not reproduced, and so clay lysimeters can suffer from continual waterlogging, leading to structural collapse. Nevertheless, there is a general consensus that lysimeters represent the most accurate method for measuring nitrate leaching from field soils, and other techniques are often calibrated and cross-checked against them.

^{15}N studies

In addition, it is possible to use, in conjunction with any of these methods, isotopically labelled nitrogen. The naturally occurring stable isotope of nitrogen (^{15}N) can be applied to the soil (either enriched or depleted) in a fertiliser, and the proportion of the isotope in any output (e.g. in the leachate or in the crop) used to measure the proportions of the original application in that output. Normally this involves the use of mass spectrometry to determine the isotopic proportions (Nason and Myrold, 1991). In this way, quantitative estimates of the pathways involved can be derived. This technique has been widely used for small-scale experiments (see, for example, Barraclough, 1988, 1991; Geens *et al.*, 1991), but the expense involved is usually prohibitive for field-scale studies.

9.3   TWO-DIMENSIONAL FIELD SCALE STUDIES — BOUNDED EXPERIMENTAL PLOTS

9.3.1   DEFINITION OF STUDY SITES

For saturated soils, movement in both the lateral as well as the vertical sense is a major component of the fluxes of both water and nitrate. For clay soils, which become effectively impermeable at depth, a well-defined hydrological system allows the interception and measurement of fluxes with some degree of success. Studies of nitrate transport in field soils thus generally require the measurement of the water movement carrying that transport, to determine the total nitrate flux:

$$\text{Total solute flux} = \Sigma \, (\text{flow . concentration})$$

Hence, the field problem is one of sampling, catching and measuring the water and the concentrations of nitrate within it. Although any system for intercepting water inevitably introduces a boundary into the soil which itself causes a distortion of the flow paths, this effect can normally be kept small. For this reason, most studies consider fairly large blocks of land: medium-scale experimental plots or whole hillslopes typically of the order of $0.1-1.0$ ha in area. The instrumentation is the same as for most hydrological studies, particularly for the saturated phase.

Some studies of two-dimensional unsaturated flow have, nevertheless, been undertaken. Generally, these have used a collection of tensiometers coupled with porous cup samplers. The two-dimensional array involved is often extremely complex to install (as, for example, in field methods described by Hoover, 1987). The problems of applying this instrumentation are similar to those given above the one-dimensional case, so are not elaborated further.

The study of saturated water and solute flow from whole areas of land tends to adopt a common design, in which a portion of land is isolated (for example, by interceptor ditches or drains or impermeable barriers), and the outflows from this land collected by a further set of interceptor drains or ditches. In agricultural soils these may include both artificial drains and gutters and troughs to intercept flows near the surface. Such isolated blocks of soil have also become known as 'large plot lysimeters'. Figure 9.3 (after Armstrong and Garwood, 1991) illustrates the kind of installation necessary for these studies.

Such studies have several advantages. First, they are related to standard hydrological investigations, and are based on well-established techniques for the establishment of the fluxes. Second, they are normally sufficiently large for 'normal' agricultural practices to be carried out on them, and they thus contrast markedly with the very special cultivations that most lysimeters and small-plot experiments receive. Third, by offering a block of land of a moderate size they are a potential solution to the problem of small-scale variability: as long as there are no significant trends or soil boundaries within the plot areas, they are normally sufficiently large to encompass the majority of the soil variability that is so problematic when dealing with small samples. The converse is that such isolated plot studies are very costly to install and run, and for this reason their number is limited. The isolated-plot experiment is further limited by the need to define the lower boundary for water movement, and therefore the majority of these installations have been in clay soils which become effectively impermeable at depth.

Once a flow has been intercepted, there is then the standard problem of measuring that flow. Common solutions include collection of the total flux, tipping buckets, weirs and flumes plus head-measuring device, and electronic measurement of within-pipe flows. These are all standard hydrological tools which are well documented elsewhere (e.g. Gregory and Walling, 1973; Goudie, 1990).

9.3.2 SAMPLING STRATEGIES

Once the water flux-measuring techniques have been defined, then the second problem is that of sampling the water. Unfortunately, no satisfactory continuously operating nitrate concentration probe is available, so the only practicable way to record concentrations is to

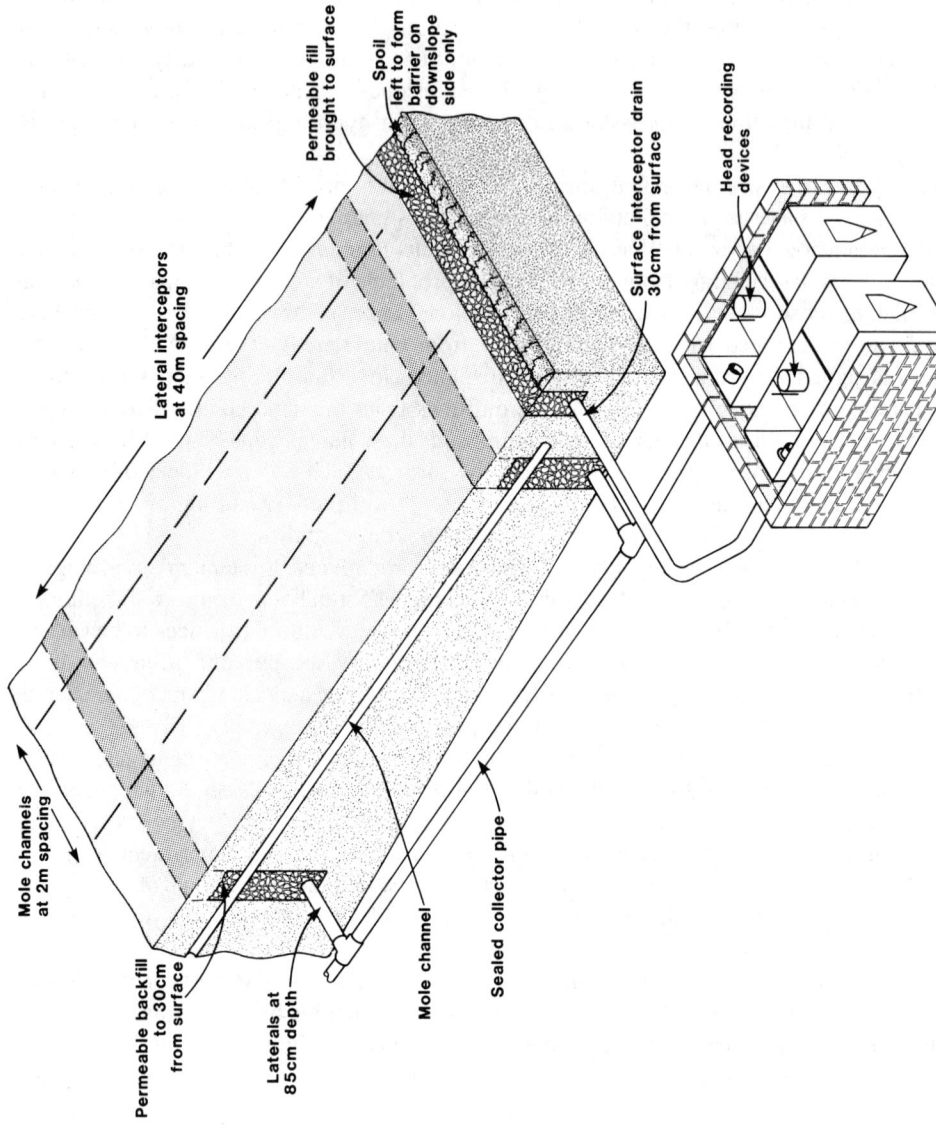

Figure 9.3 Discharge and sampling times from a flow-proportional sampler

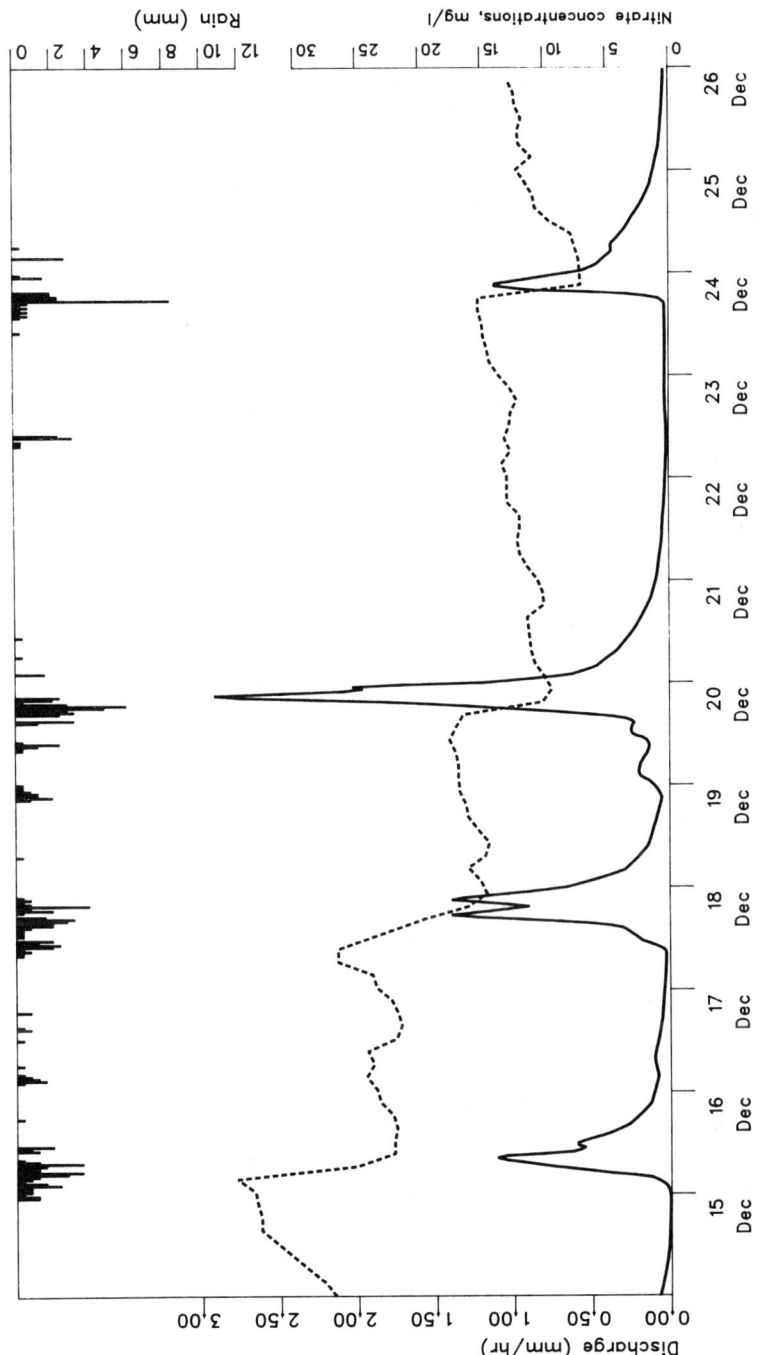

Figure 9.4 Discharges (solid line) and nitrate concentrations (dotted line) from the Brimstone Farm drainage experiment, December 1989 (after Rose *et al.*, 1991, Fig. 3)

take samples and analyse them in the laboratory. The results that are obtained thus depend on the adoption of a suitable water-sampling strategy. Choice of this strategy depends on the observation that at the field scale flows vary rapidly within short periods, and so, as a rule, do nitrate concentrations. This is illustrated by Figure 9.4, which shows the pattern of discharges through an agricultural drain and the nitrate concentrations for four runoff events. Hydrograph peaks commonly last for no more than 2 days, and these are accompanied by simultaneous rapid shifts in nitrate concentrations. This variation in nitrate concentration is most marked where fissures and macropores lead to preferential flow.

Consequently, choice of a suitable time step for repeated samples can be difficult. Where the sampling intervals are large compared to the scale of changes in the concentrations, then it is not clear that the concentrations recorded are relevant for the larger part of the flow in that interval. On the other hand, if the sampling interval is set too low, then the number of samples soon becomes very large, and the analytical load can become prohibitive. The problem is further compounded where automated water samplers are used. These commonly have only a limited number of sample bottles that can be filled, which imposes a further limit on the sampling frequency. The configuration of a typical problem is thus to devise a sampling strategy which will result in no more than 24 samples in a 7-day period.

The commonly suggested (though much less commonly adopted) solution is to establish a flow-proportional sampling strategy. The philosophy behind this strategy is that, if the rate of change of nitrate concentration is some function of the rate of change of flow rate, and observing that most flow events in small-scale studies are of short duration, then the choice of sampling interval inversely proportional to the rate of flow will generate sufficient information to establish the variation in nitrate intensity in an unbiased fashion. In addition, the individual samples in a flow proportional sample all relate to the same volume of flow, irrespective of the flow rate. Flow-proportional samples are thus frequent at high flows and less frequent at low ones.

Implementation of anything other than a simple regular sampling system requires the intervention of a field-based intelligent system, either a data logger or microcomputer, which can interrogate a flow sensor and then use that information to trigger a sampler. Because of the problems of implementing such equipment in a remote field site away from mains electricity (and often in environments uncongenial to electronic equipment) this sort of strategy has not been commonly adopted for field studies.

One such system has been implemented at an ADAS FDEU site by Charlesworth and Chisholm (in preparation). Flows through a weir are converted to a head measurement via a float. The head displacement is converted to an electrical signal by use of a counterweighted string which turns a rotary potentiometer. The signal from these potentiometers is returned to a data logger and, at the same time, output to a portable, battery-operated PC. This computer examines the flows and then, when sampling conditions are indicated, sends out a trigger signal to a water sampler and also to the data logger, which records the occurrence of a sampling event. The data recovered from the data logger thus include not only all the flow data but also the timing of all the water samples taken. The results of using such an approach are shown in Figure 9.5; data taken from the logger record indicated both flow and the time that samples were taken.

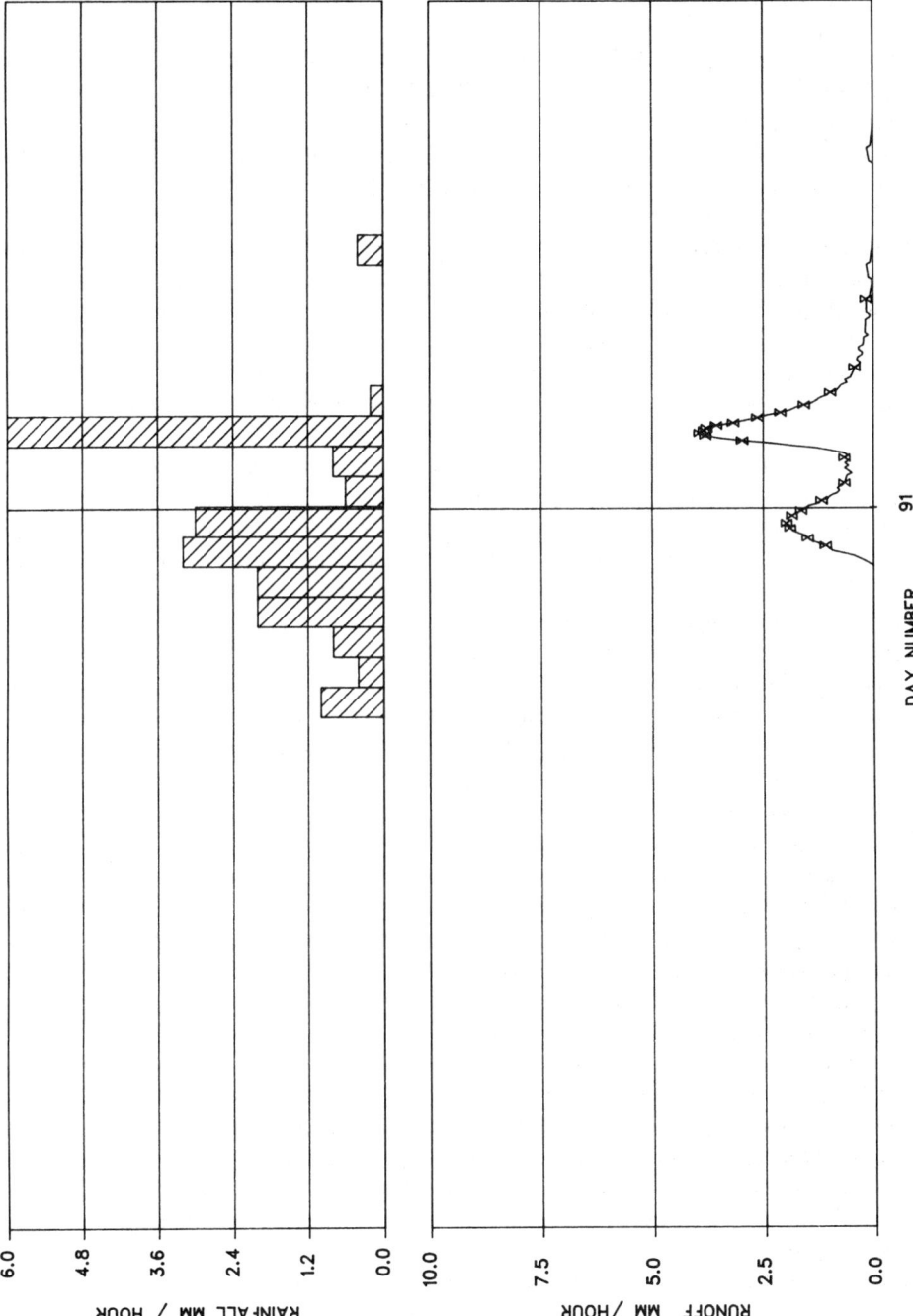

Figure 9.5 Isolated plot experiment (reproduced by permission from Armstrong and Garwood, 1991, Fig. 3)

9.3.3 FIELD STUDIES

Examples of the study of nitrate leaching from whole-field systems are infrequent. However, in recent years a number of such experiments have been undertaken in the UK with funding from the Ministry of Agriculture: at Brimstone Farm, at Cockle Park and at North Wyke. In each of these, a series of hydrologically isolated plots have been used to identify both the flows of water the nitrate fluxes under a number of imposed agronomic conditions. Such experiments are, however, not unique, and a number of similar experiments (for example, Vinten, Howard and Redman, 1991; Arlot and Dutetre, 1990; Hood, 1986; Barraclough, Hyden and Davies, 1983) have been concerned to establish the nitrate balance on a variety of soil conditions. Nevertheless, these three MAFF experiments form an important sequence, and are described here to illustrate both the techniques and the results that can be obtained. In addition, these experimental sites have the important advantage that the nitrate movement studies have been undertaken in conjunction with intensive hydrological and agronomic studies.

The Brimstone Farm experiment, in Oxfordshire, is probably the most comprehensively studied site in UK. Initially set up to study the interaction between the drainage needs and performance of winter cereals grown under conditions of minimal tillage, it includes detailed studies of soil water regime, water loss, soil nutrient loss and crop performance (Cannell *et al.*, 1984). The soil of the site is a pelo-stagnogley of the Denchworth series.

This experiment comprises 20 plots, of which 10 were drained using mole drains at 0.6 m depth and 2 m spacing intercepted by permeable fill over pipe drains at 0.9 m depth and 46 m spacing. The remaining 10 were left undrained. In addition, the plots were subjected to two agronomic treatments, conventional ploughing and reduced tillage. Soil water status was, in addition, measured by tensiometry from permanent instrument pits (Howse and Goss, 1982) as well as by neutron probe, open auger hole and piezometric techniques. Water leaving the plots was intercepted at the surface by a gutter, within the plough layer by a trench infilled with gravel and drained by a permanent pipe; and, where appropriate, the flow through the drainage system was collected. Plots were isolated across the slope by deep (1.1 m) drains and up and down the slope by polythene barriers. All flows leaving the plots were measured by a combination of V-notch weir and stage-recording system. Water samples were taken from the points at which the flows entered the measuring chambers and subjected to laboratory analyses (Harris *et al.*, 1984). In addition, detailed studies were made of the growth and development of the crop (Ellis *et al.*, 1984).

Nitrate leaching results from the experiment show considerable variation from year to year. Table 9.1, extracted from Goss *et al.* (1988), shows the mean annual leaching losses of nitrate through the drainage systems, subdivided in to the autumn and spring periods. The spring period was defined as that following the first top-dressing application of fertiliser nitrogen. These results show that the predominant nitrogen leaching loss is through the winter, following the decay of crop residues. A general relationship is observed between the volume of drainage water and the amount of nitrate exported. Spring losses were almost directly related to the amount of rainfall falling after the application (Goss *et al.*, 1990).

Nitrate concentrations in the drainage waters from Brimstone Farm have also shown the same overall decline through the winter as has been reported from lysimeter work (Section 9.2). In general, concentrations are high at the start of drainflow in the autumn, and follow an approximately exponential pattern of decay throughout the winter, which is particularly apparent if plotted against cumulative drainage (e.g. Goss *et al.*, 1988). This pattern, of high concentrations in the autumn following the first heavy rainfall, has been repeatedly observed

Table 9.1 Autumn and spring nitrate nitrogen losses through mole and pipe drain systems from the Brimstone farm experiment. (Data from Goss *et al.* 1988, reproduced by permission of Elsevier Applied Science Publishers Ltd)

Year	Crop	Autumn fertiliser applied (kg ha^{-1})	Winter NO$_3$ loss (kg ha^{-1})	Winter drain flow (mm)	Spring fertiliser applied (kg ha^{-1})	Spring NO$_3$ loss (kg ha^{-1})
1979	W. wheat	17	40.1	163	117	0.0
1980	W. wheat	24	54.3	219	130	1.0
1981	W. wheat	0	11.1	198	149	5.6
1982	W. wheat	24	33.6	198	149	0.8
1983	W oats	30	42.2	183	110	25.0
1984	W. wheat	17	4.2	73	223	0.0
1985	Oilseed rape	46	34.0	144	239	10.3
1986	W. wheat	0	45.7	227	130	0.7

at a number of sites (Rose *et al.*, 1991). Data from the Brimstone Farm site also show short-term variation in concentration, a period of dilution with each major hydrograph event being preceded by a small rise in concentration (see Figure 9.4).

Similar experimental work at *Cockle Park* (Armstrong, Shaw and Wilcockson, 1983; Armstrong, 1984) followed a comparable methodology on three unreplicated plots, and reported the contrast between arable residues and a subsequent grass crop. Early results have been shown to be anomalous. Initially, they suggested that there was little difference in nitrate content between the surface runoff from undrained plots and the drainflow from the drained plots. This was accompanied by the observation that peak flow rates were lower on the undrained plots than on the drained. As neither of these observations have been replicated at other sites, these anomalous results are now considered to reflect the short-term effect of a subsoiling operation which was intended to remove a plough pan. Following the subsoiling, the undrained plots experienced a condition in which water entered the loosened soil and moved down the topographic slope at the base of the subsoiling, and by moving through the soil this water leached nitrates, and so acquired chemical characteristics very similar to that moving through the soil to the drains. By contrast, it is most commonly observed that the dominant route of water loss from most clay soils in the UK in their undrained state is across the surface. Surface flow does not have the opportunity to interact with the soil nitrate, and therefore has low concentrations of nitrate. After a period of grass, recent observations at the same site have suggested a more normal response (R. Hodgkinson, FDEU, pers. comm). These results thus indicate the overriding importance of cultivation and land-management effects on the hydrology and solute balance of agricultural land.

An experiment of similar complexity and on a soil like that at Brimstone Farm, but conducted in the context of a grazed sward, at *North Wyke*, Devon, has been reported by Armstrong and Garwood (1991). The installation consists of 12 1 ha plots. Six of the plots are drained by mole drains at 55 cm depth and 2 m spacing over permanent pipes at 40 m spacings. All drain and surface flows are measured as they leave the plots. Water tables are measured by continuous meters within each of the plots. The plots are subject to two levels of nitrogen fertiliser application (200 and 400 kg N ha^{-1}) and some plots were also reseeded. All plots are continuously grazed, with stocking rate adjusted to maintain a sward height of 5−6 cm. Detailed measurements of grass production and animal production are also maintained.

Water-table data demonstrate that the drainage has achieved its fundamental aim, of reducing waterlogging. A rapid drainflow response reflects the importance of macropore flow through the soil, particularly the fissuring associated with mole channel formation. Measured times of water travel from the surface to the mole channels are of the order of 30 minutes, and travel times within the mole channel are between 30 and 60 minutes (Hallard and Armstrong, 1992). These values explain the time difference between peak rainfall and peak drain flows on the drained plots. Travel times for surface runoff are generally less than 30 minutes, so the observed difference in time to concentration between drained and undrained plots is of the order of $1-1\frac{1}{2}$ hours. Consequently, for this soil, it is suggested that drainage has the effect of slightly reducing and delaying hydrograph peaks, and reducing potential flood risk downstream. Drainage has thus changed the dominant route of water movement from the drained plots from surface flow to drainflow, and thus imposes a change in the hydrological processes on the two sets of plots. On *undrained plots* during the field capacity period the undrained land remains saturated, with water tables close to the surface. Rainfall infiltrates and moves through the soil on these plots only very slowly, so surface runoff is the dominant process. On *drained plots*, however, the water table is lower, so that rainfall infiltrates and moves through the soil to the drains relatively rapidly. Only for short periods of time does the water table on the drained plot rise sufficiently to restrict vertical water movement, and so generate short periods of surface runoff. Consequently, although the overall volume of water leaving the site is changed only slightly by drainage, the route of movement is changed significantly.

These differences in soil water regime lead, in turn, to differences in the quality of water leaving the plots. Mean nitrate concentrations in the water leaving the site are given in Table 9.2, from Armstrong, Scholefield and Hallard (1990). Drained plots are aerobic, and nitrogen-mineralisation rates are high. Consequently, the concentration of nitrate in the drainage water is high, falling from a peak in excess to 50 mg NO_3-N l^{-1} in October to 20 mg NO_3-N l^{-1} in April. Anaerobic, waterlogged conditions on undrained plots restrict the supply of nitrogen and, consequently, the concentration of nitrate in the surface flows from the undrained plots are much lower. A combination of these techniques and detailed studies of the agronomy, together with some direct measurement of the gaseous losses, allowed Garwood (1988) to compile a preliminary nitrogen balance sheet for this site (Table 9.3).

Table 9.2 Mean monthly nitrate concentrations in drain and surface water flows, North Wyke, mg l^{-1} N

| | 200 kg N ha^{-1} old sward | | | 400 kg N ha^{-1} old sward | | | 400 kg N ha^{-1} reseed | | |
| | Undrained | Drained | | Undrained | Drained | | Undrained | Drained | |
	Surface	Surface	Drain	Surface	Surface	Drain	Surface	Surface	Drain
August	7.7	10.3	14.3	17.1	0.0	44.7	4.3	1.5	16.4
September	13.7	8.6	14.8	29.3	0.0	48.5	9.6	4.9	17.5
October	11.6	10.3	15.6	24.0	12.1	51.4	6.3	6.9	17.6
November	5.1	7.7	12.8	19.3	9.0	44.0	3.4	3.9	16.3
December	3.8	3.1	12.1	17.8	8.0	39.9	4.6	2.2	16.5
January	2.1	2.7	8.9	9.2	6.6	31.8	3.3	1.9	12.7
February	1.3	3.0	6.0	4.5	5.1	22.9	3.5	3.5	8.9
March	3.7	6.3	7.6	14.2	7.2	22.1	8.7	6.1	11.8
April	2.1	4.5	6.6	8.7	4.7	21.0	4.0	3.4	9.8
May	1.4	0.0	7.8	6.4	0.0	26.1	4.9	0.0	8.8

Table 9.3 Provision annual nitrogen balance sheet for drained (D) and undrained (UD) grazed grass swards at North Wyke, Devon (after Garwood, 1988)

| | Permanent grass | | | | Reseed | |
	UD	D	UD	D	UD	D
N input	200	200	400	400	400	400
Animal output (LWG)	24	27	28	30	27	34
Leached	20	56	48	187	24	74
Denitrified	90	56	111	84	115	81
Ammonia loss	30[a]	30	60[a]	69	101	102
Stored in soil OM	72	64	48	45	137	128
Total N accounted for	236	233	295	415	404	430

[a]Estimated.

The results from these studies demonstrate that leaching of nitrate from clay soils is the result of two processes: the release of nitrate and the movement of water through the soil to remove it. Where either of these processes is not operating then the loss of nitrate is small. Thus on undrained land which remains saturated for much of the winter, organic matter is not converted to soluble nitrate but lost gaseously, and, in addition, water loss is predominantly across the surface. By contrast, on undrained land, aerobic processes convert organically bound nitrogen to nitrate, which is then removed by water moving through the soil. In addition, it can be observed that where the supply of nitrogen is low, as, for example, in the low-fertilizer plots at North Wyke, then the leaching losses of nitrate are also lower. The estimation of leaching loss from agricultural areas thus requires the combined consideration of both agricultural practice and soil hydrology.

9.4 THREE-DIMENSIONAL FIELD-SCALE STUDIES — UNBOUNDED HILLSLOPE UNITS IN NATURAL CATCHMENTS

9.4.1 STATEMENT OF THE PROBLEM

With bounded plots the area of the study is known accurately and problems of leakage are minimised. Where nitrate loss from a 'natural' hillslope is to be studied, the fact that the plot is not an isolated block, separated from the rest of the catchment by an impermeable barrier, may create several problems. In the absence of a barrier, the boundary of the plot must be defined by a topographical survey of the ground surface. This approach may be acceptable in steep catchments, but the direction of flow may not correspond with surface topography in areas of gentle relief. Problems of 'unknown' leakage will be compounded where there is a substantial component of groundwater discharge; the area contributing to a spring or to a length of river bank may be very difficult to define in a groundwater-fed catchment unless detailed surveys of water-table elevation or water-tracing experiments are undertaken. On low-angle clay slopes, the haphazard connectivities of macropores might also render topographic boundaries somewhat uncertain. In most cases, unbounded subcatchment units will be irregular in plan; only rarely (e.g. the area draining to a tile drain) can regular unbounded plots be defined.

Given the lack of an impermeable barrier, it is very difficult to measure outflow discharge from an unbounded plot accurately. Even where there is a spring at the foot of the slope, this does not ensure that all the outflow emerges at that point. Moreover, weirs installed to intercept spring flow may suffer from leakage below and around the structure. Therefore measurements of spring discharge may provide a good indication of temporal patterns of hillslope discharge but are likely to yield *under*estimates of the total discharge leaving the slope unit. Measurement of stream discharge above and below the section of channel where the slope drains into the stream may be helpful, but any additions of discharge between the two points must include contributions from the opposite side of the valley, which introduces a further complication into the discharge calculations (Anderson and Burt, 1978; Burt and Arkell, 1986).

Where percolation losses to groundwater are negligible, it is possible to estimate slope discharge by applying Darcy's law. This requires good knowledge of soil water conditions, particularly the elevation and slope of the water table, depth to the impermeable horizon, and the hydraulic conductivity of the soil. If these are known, discharge may be calculated at the downslope boundary of a given area of hillslope (Anderson and Burt, 1978; Burt and Butcher, 1985). From this, the nitrate flux may be calculated if the nitrate concentration of soil water is known. In practice, very few studies of nitrate leaching have successfully achieved this distributed approach (the study of Haycock, 1991, being a notable exception) and the hydrological processes responsible for nitrate leaching on unbounded hillslope plots have had to be inferred from observations made at the foot of the slope. McDonnell (1990) provides an excellent example of how to design a field experiment to study the distributed nature of slope runoff and solute leaching.

While properly structured experiments are easily achieved in bounded-plot studies (see Section 9.3 above), this is rarely possible when seeking to compare unbounded subcatchment units unless they are very small. In most cases, any subcatchment unit (running from stream channel to drainage divide) will contain a range of features (topography, soil type, etc.) within it, so that it becomes difficult to formalise comparisons between slopes. Under such circumstances, controlled experiments are difficult to perform and replication becomes near-impossible. Studies of whole hillslope segments are often crucial, however, because bounded plots may occupy too small an area to be representative of the catchment as a whole. Their isolated nature may also cause problems in this respect.

9.4.2 TEMPORAL PATTERNS

Two timescales are important in the study of runoff generation on hillslopes: seasonal variations in soil moisture content, which lead directly to seasonal fluctuations in subsurface discharge, and episodic storm events, when both surface and subsurface storm flow may be observed. Several studies of hillslope hydrology and nitrate leaching have been conducted within the Slapton Wood catchment (Burt *et al.*, 1983; Burt and Butcher, 1985; Burt and Arkell, 1987); these studies are introduced here to illustrate the temporal patterns of hillslope flow which may occur.

Only 1% of the annual runoff in the Slapton Wood stream is quickflow (Troake and Walling, 1973). Not all of this is surface runoff: some is rapid subsurface flow from saturated areas close to the stream or through macropores. The remaining 99% is entirely generated from within the soil since the bedrock is impermeable. Much of this subsurface flow occurs in winter in the form of delayed hydrographs which peak up to several days after the rainfall, with high flows lasting for as long as two weeks (Burt and Butcher, 1985; see also Section 10.3.2). These large delayed hydrographs occur only in winter when the soil water deficit has been recharged (Burt, 1988). The generation of the delayed hydrograph is associated with the development of a deep zone of saturation which builds up above the impermeable bedrock. The zone of saturation extends upslope from the stream and, during periods of high subsurface runoff, it is continuous right onto the plateau areas above the steeper valley side slopes. Drainage from the plateau appears to be especially important in generating the largest delayed hydrographs. Much of the drainage is through large hillslope hollows into which subsurface drainage off the plateau converges (Burt and Butcher, 1985).

Delayed hydrographs are times of major nitrate loss since both discharge and nitrate concentrations are high (Burt and Arkell, 1987; see also Figure 9.6). For example, in the dry winter of 1983−4, four such hydrographs accounted for 70% of the total nitrate loss from December to February inclusive (about 55% of the annual nitrate loss). Saturation at the base of the soil horizon is produced by moving soil water down from the root zone, leaching nitrate as it does so. In addition, growth of the zone of saturation towards the soil surface may cause more effective removal of nitrate from the soil than can take place with unsaturated flow (Burt *et al.*, 1983). McDonnell (1990) draws a similar conclusion: he argues that as 'new' soil water backs up into the soil profile, it mixes with the much larger volume of water stored in the soil matrix. Accordingly, water leaving the base of the slope appears 'old', despite the fact that the subsurface flow is driven by infiltration down cracks in the soil as that 'new' water initially by-passes the soil matrix. Although the leaching process at Slapton is now understood in general terms, confirmation of the exact mechanism is lacking. It has proved difficult to install sufficient numbers of deep piezometers to allow the downslope displacement of old, nitrate-rich soil water to be observed.

As expected, the delayed peaks of discharge and nitrate concentration observed at the basin outlet are mirrored at the springs. Figure 9.6 shows that the discharge response of a spring is much smoother than that of the stream, because the latter involves surface as well as subsurface flow. Indeed, the discharge record for the spring shows that an extended period of high flow at the basin outlet (24−29.1.84) can be separated into two distinct peaks at the spring. Nitrate concentration rises at much the same time as discharge at the spring. In the stream, although coincident increases do also sometimes take place, it is more usual for the rise in nitrate concentration to lag behind the delayed discharge response. Thus, not all slopes respond in an exactly similar manner, even during the same event. Where nitrate lags behind discharge, this may show that the subsurface flood wave is propagated more quickly downslope than the water itself, although this has not been confirmed by direct observation. However, other explanations are also possible, and it may simply be that the later drainage has a higher concentration because of its longer residence time.

Even in soils dominated by macropore flow, solute leaching can be an efficient process (McDonnell, 1990). Haigh and White (1986) described losses of nitrate from a clay soil under grazed grassland at Wytham, near Oxford. The soil is drained by tile drains 0.8 m deep at a spacing of 30 m with mole drains at 25−40 cm depth, with a spacing of 2.2 m. Kneale (1986) showed that runoff from the slope occurs as intermittent subsurface flow between late autumn and spring. No surface runoff was observed but rainfall prompted a large and rapid response from the drain at the base of the slope whenever the soil was wet. Since the hydraulic conductivity of the soil matrix is only $1-2$ mm h^{-1}, macropore flow is easily initiated (Kneale and White, 1984). Kneale (1986) assumed that flow took place vertically through the soil until the macropore water table was reached, then by lateral flow through a saturated channel to the drain. Haigh and White (1985) argued that nitrate losses were large because rainfall thoroughly mixes with the soil solution of the A horizon due to its small

258

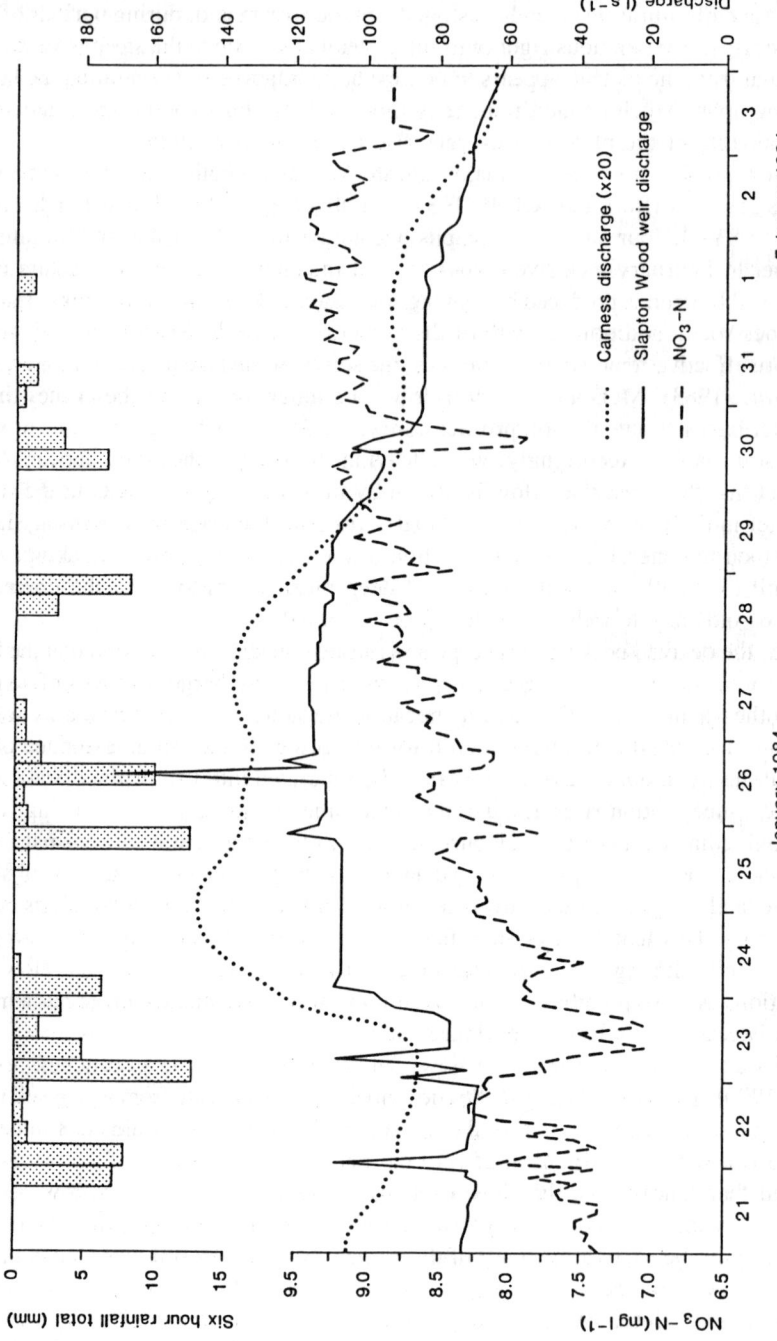

Figure 9.6 Hydrographs and nitrate chemographs for the Slapton Wood catchment, January 1984 (from Burt and Arkell, 1987, *Soil Use and Management*, reproduced by permission of Blackwell Scientific Publications)

crumb structure. This water then moves rapidly to the drains through macropores in the B horizon. Though Kneale (1986) found no evidence of a perched water table in the B horizon, Haigh (1985) shows that one can develop — not necessarily long enough for lateral flow to occur but sufficient for infiltrating water to mix with water stored in the soil matrix. The Wytham stream is fed both by drained and undrained land (Kneale, 1986). Subsequent work on the same stream (Burt, unpublished data) has shown that runoff events in autumn and early winter were all associated with increases in stream nitrate concentration at (or soon after) the discharge peak; a similar pattern is seen for tile drainage. In midwinter, a decrease in stream nitrate concentration is characteristics at the flood peak, but even then tile drains can sometimes show sharp increases in nitrate concentration.

The nitrate concentration in the drainage falls through the winter, suggesting a gradual decline in the nitrate available for leaching (Haigh and White, 1986). Even so, losses in storms can be large, and concentrations may remain very high in drier winters. This shows that losses of nitrate from grazed clay soils can be relatively large compared to other sites; the predominance of macropore flow in clay soils does not seem to limit this loss significantly (see also Haycock and Burt, in press).

Anderson and Burt (1978) and McDonnell (1990) calculate subsurface flows on hillslopes using the method described briefly above (Section 9.4.1). Haycock (1991) has extended this approach to calculate nitrate fluxes draining from a shallow limestone aquifer through permeable floodplain soils to a stream channel (Section 12.4). Given the relatively high hydraulic conductivity of the soil ($K_{sat} = 0.012$ cm s^{-1}), Haycock's calculations show that the amounts of subsurface flow through the floodplain soil, when the water table is high, are quite sufficient to account for increases in stream discharge along the stream section. Given the large quantities of flow involved, the decrease in nitrate concentration observed as water moves across the floodplain must represent a real loss, probably by denitrification, and not a dilution effect caused by mixing with floodplain water. Chloride concentrations in the subsurface flow exhibit only a small decrease in concentration as water moves through the floodplain soil.

9.4.3   SPATIAL PATTERNS

Two factors are likely to be most influential in controlling nitrate losses from hillslopes: land use and topography. These factors may combine to cause variable inputs of runoff and nitrate along the stream channel. Burt and Arkell (1986) determined the amount of nitrate being contributed from subcatchment areas in the Slapton Wood catchment. Stream discharge and nitrate concentrations were measured at a number of points along the stream so that inputs to each reach could then be estimated. Reaches are classified depending on whether they are fed by a hillslope hollow or whether the valley-side slopes are rectilinear (in plan and profile). Land use on the plateau tends to be more intensive at the top of the valley; valley-side slopes are grassed except for the bottom two reaches, where they are wooded.

Table 9.4 shows runoff contributions to the Slapton Wood stream at a time of modest subsurface flow. The discharge generated per unit area of flow (dQ/A) shows that the valley-side slopes are more productive. This may be because they are steeper, and because they drain only small areas of plateau. The large tributary hollows (Carness and Eastergrounds), despite the importance of convergent flow processes, contain large areas distant from the stream which seem not to be important in slope drainage. Figures for discharge per unit length (dQ/L) show that hollows are important sources of discharge.

Table 9.5 gives the nitrate load for each stream reach: once again the valley-side slopes

Table 9.4 Flow contributions in the Slapton Wood catchment (26.2.84)

Site	Q	Reach type	dQ	dQ/A	dQ/L	dQ/(AL)
Above Carness	2.01	H	4.06	1.97	12.88	6.26
Below Carness	6.07	H	2.61	12.71	12.42	60.51
Below Valleyside hollow	8.68	V	6.70	4.64	1.59	1.10
Above Eastergrounds	15.38	H	5.83	2.51	15.3	7.94
Below Eastergrounds	20.21	V	6.50	7.29	2.06	2.31
Basin Outlet	26.71					

Abbreviations:
H = hillslope hollow, V = valley-side slope.
dQ = change in stream discharge over stream reach ($l\ s^{-1}$).
A = hillslope area draining to stream reach (m^2).
L = length of stream reach (m).
Units:
dQ/A = $l\ s^{-1}\ m^{-2} \times 10^{-5}$
dQ/L = $l\ s^{-1}\ m^{-1} \times 10^{-2}$
$dQ/(AL)$ = $l\ s^{-1}\ m^{-1}\ m^{-2} \times 10^{-7}$

Table 9.5 Nitrate inputs to the Slapton Wood stream (26.2.84)

Site	NO$_3$-N	Load	Reach type	dL	dLD/A	d/LD/A	dL/(A.L)
Above Carness	12.3	24.72	H	71.72	3.16	1.37	6.03
Below Carness and Valleyside hollow	11.1	96.35	V	54.37	3.76	0.13	0.89
Above Eastergrounds	9.8	150.72	H	43.30	2.24	1.38	7.12
Below Eastergrounds	9.6	194.02	V	39.69	4.45	0.13	1.41
Basin outlet	8.8	233.71					

Abbreviations:
H = hillslope hollow, V = valley-side slope.
Load = nitrate load at gauging point ($mg\ s^{-1}$).
dL = change in nitrate load over steam reach ($mg\ s^{-1}$).
A = hillslope area draining to stream reach (m^2).
L = length of stream reach (m).
Units:
dL/A = $mg\ s^{-1}\ m^{-2} \times 10^{-4}$
dL/L = $mg\ s^{-1}\ m^{-1} \times 1$
$dL/(A.L)$ = $mg\ s^{-1}\ m^{-1}\ m^{-2} \times 10^{-7}$

have higher unit area contributions, but the hollows form major point sources. In regard to controlling farm practices within the basin to limit nitrate losses to the stream, all fields which abut the valleyside slopes must be involved and not just those immediately upstream of hollows. Burt and Arkell (1986, 1987) concluded that, in relation to topography, runoff and nitrate are contributed in a relatively uniform manner. Thus, variations in land use seem to be of more importance in determining nitrate levels along the stream (Figure 9.7).

9.5 MODELLING APPROACHES

Because of its importance, there have been many attempts to model the loss of nitrogen from

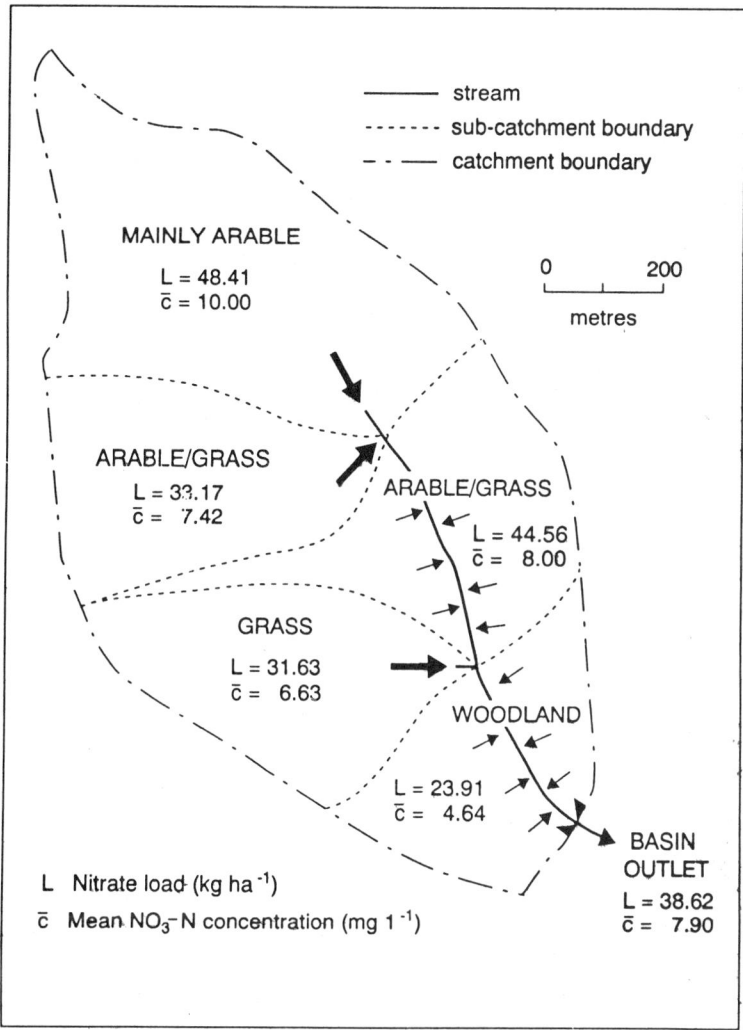

Figure 9.7 The spatial pattern of nitrate leaching in the Slapton Wood catchment, south Devon, as indicated by dilution gauging experiments. Large arrows indicate point sources of nitrate (springs and tributary streams); small arrows imply non-point inputs (subsurface seepage) for that section of the stream (after Burt and Arkell, 1987)

soils. The review that follows is an attempt to identify some of the more important threads in a fast-growing literature, and to point the way for the interested reader.

Modelling is essentially a problem-solving activity, and so one way to classify the various models is to consider the nature of the problem they address. A fundamental division can perhaps be made between those models that attempt to define the components of the annual nitrogen budget (frequently motivated by agricultural needs) and those concerned to represent the details of nitrogen movement within the profile. It is, of course, possible to use detailed models to estimate annual fluxes, and so there is a continuity between the two types of model.

A second subdivision in modelling activity is between those concerned with a mechanistic description of the soil physical state, and those that adopt stochastic viewpoint (Addiscott and Wagenet, 1985). Because they adopt a somewhat different perspective, stochastic models are treated separately.

9.5.1 NITROGEN BUDGET MODELS

A major thread in nitrate research has been the need to predict the nitrogen fertiliser requirements of agricultural crops, with the twin aims of improving crop productivity and, by limiting over-application of fertiliser, reducing the nitrate leaching losses. For many agricultural purposes the problem is formulated in terms of the total leaching loss during a winter, so that the residual nitrate available to the crop, and hence the fertiliser requirement, can be estimated.

Burns (1975) has suggested that for British conditions the fraction of nitrate leached (f) from the rooting zone of the soil can be given by a simple function of the drainage volume:

$$f = \left(\frac{P}{P+V_m} \right)^x$$

where P is the drainage volume (cm) and V_m is the field capacity moisture content by volume. The exponent x depends on the assumed distribution of nitrogen within the soil profile. However, the Burns model has been found to behave only moderately successfully for many soils (e.g. Barraclough, Hyden and Davies, 1983). Consequently, Addiscott has developed a model which calculates a daily nitrogen balance for the soil throughout the winter period. this model makes the important distinction between mobile and immobile water (Addiscott, 1977; Addiscott and Whitmore, 1987, 1991).

In order to establish an effective model the soil is divided into a series of layers, and the vertical movement of water in both directions is determined from a consideration of drainage and evapotranspiration. Incoming rain displaces water and nitrate through the mobile phase only, and this displacement is followed by nitrate movement between the mobile and immobile phases. The equilibriation between the phases does not occur completely, representing the 'holdback' by the intra-aggregate diffusion. The model also considers the generation of nitrate by mineralisation and the balance also takes into account the uptake by the crop.

Modelling nitrate movement is further complicated in grazed grass soils by animals, which consume herbage that contains nitrogen taken up into the plant, but then return it, in the form of urine and faeces, in a highly concentrated but spatially non-uniform way. Models of the nitrogen mass balance in grass include that of Scholefield et al. (1991). This model attempts to define the total components but not the day-to-day simulation of balances. Simulation of daily totals in a growing grass crop have, however, been attempted by Hansen (1986, 1987) but the models here becomes so complex as to be virtually unusable for all except experimental situations.

9.5.2 NITRATE MOVEMENT MODELLING

Full modelling of solute movement within a soil requires numerical solutions to the Richards equation for water flow in unsaturated soils, together with the convection−dispersion equation for chemical transport (Wagenet, 1990). In addition, the modelling of nitrogen within a soil profile also requires the modelling of the transformation of nitrogen species (nitrification, denitrification, mineralisation, immobilisation, plant uptake and absorption) within the flow domain (Hutson and Wagenet, 1991). As a consequence, models are complicated, require a large number of parameters and are often difficult to verify. Nevertheless, a number of models have been constructed, such as those of Barraclough (1989), Hutson and Wagenet (1991), and Bergstrom and Jarvis (1991). For example, Kaluarachchi and Parker (1988) present a computational scheme based on finite-element techniques for the solution of the relevant system of equations. However, the computational effort involved, and the parameterisation required, preclude the use of this sort of analysis in all except the most demanding circumstances.

9.5.3 STOCHASTIC MODELLING

More recently, several workers, considering the problems of soil spatial variability, have suggested that the predictions of water and solute movement in soils are best approached from a stochastic viewpoint. Such studies require a consideration of both the stochastic properties of the soil and its effect on transport processes. Thus, for example, Jury (1982) has made the extremely useful suggestion that the 'breakthrough' curve observed in many laboratory studies represents the distribution of travel times through the soil column, leading to a probabilistic interpretation of the leaching phenomena. Use of the transfer function model has been further developed by White (1989), White and Magesan (1991) and Vauclin (1990).

Lastly, models developed in the petroleum geology field have suggested that flows through spatially random structures are dominated by routes of high movement, and that large portions of the structures may contribute only slightly to the total flux. Although the application of such models to soils and nitrate has not yet been undertaken, these results are suggestive, and again indicate the probable importance of macropores and other discrete routes of movement.

ACKNOWLEDGEMENTS

The contribution of ACA to this chapter has been funded by the Ministry of Agriculture, Fisheries and Food. Thanks are due to Eunice Lord and Mark Shepherd of ADAS Soil Science for their comments on the manuscript.

REFERENCES

Addiscott, T.M. (1977) A simple computer model for leaching in structured soils. *Journal of Soil Science*, **28**, 554−63.
Addiscott, T.M. (1988) Long-term leakage of nitrate from bare unmanured soil. *Soil Use and Management*, **4**, 91−5.
Addiscott, T.M. (1990) Measurement of nitrate leaching: a review of methods. In Calvet, R. (ed.), *Nitrates, Agriculture, Eau*, INRA, Paris, pp. 157−68.
Addiscott, T. and Wagenet, R.J. (1985) Concepts of solute leaching in soils: a review of modelling approaches. *Journal of Soil Science*, **36**, 411−24.
Addiscott, T.M. and Whitmore, A.P. (1987) Computer simulation of changes in soil mineral nitrogen and crop nitrogen during autumn, winter and spring. *Journal of Agricultural Science, Cambridge*, **109**, 141−57.
Addiscott, T.M. and Whitmore, A.P. (1991) Simulation of solute leaching in soils of different permeabilities. *Soil Use and Management*, **7**, 94−102.

Anderson, M.G. and Burt, T.P. (1978) The role of topography in controlling throughflow generation. *Earth Surface Processes and Landforms*, **3**, 331–44.

Arlot, M.-P. and Dutetre, A. (1990) Les mesures sur eaux de drainage: application à l'étude du lessivage des nitrates sur le site experimental de La Jaillière. *Perspectives Agricoles*: Nitrates: pour concilier agriculture performant et qualité des eaux, Tire-à-part des No. 144 et 145, Fevrier, Mars, pp. 45–50.

Armstrong, A.C. (1984) The hydrology and water quality of a drained clay catchment, Cockle Park, Northumberland. In Burt, T.P. and Walling, D.E. (eds), *Catchment Experiments in Fluvial Geomorphology*, Geo Books, Norwich, pp. 153–68.

Armstrong, A.C. and Garwood, E.A. (1991) Hydrological consequences of artificial drainage of grassland. *Hydrological Processes*, **5**, 157–74.

Armstrong, A.C., Scholefield, D. and Hallard, M. (1990) Soil water status, drainage, and the production of grassland. *Papers of the British Grassland Society Second Research Conference*, British Grassland Society, Hurley.

Armstrong, A.C., Shaw, K. and Wilcockson, S.J. (1983) Field drainage and nitrogen leaching: some epxerimental results. *Journal of Agricultural Science, Cambridge*, **101**, 253–5.

Ballif, J.L. and Muller, J.C. (1990) Les bougies poreuses et les lysimetres. *Perspectives Agricoles*: Nitrates: pour concilier agriculture performant et qualité des eaux, Tire-à-part des No. 144 et 145, Fevrier, Mars, pp. 24–33.

Barbee, G.C. and Brown, K.W. (1986) Comparison between suction and free-drainage soil solution samplers. *Soil Science*, **141**, 149–54.

Barraclough, D. (1988) Studying mineralization/immobilization turnover in field experiments: The use of Nitrogen-15 and simple mathematical analysis. In Jenkinson, D.S. and Smith, K.A. (eds), *Nitrogen Efficiency in Agricultural Soils*, Elsevier, New York, pp. 409–17.

Barraclough, D. (1989) A usable mechanistic model of nitrate leaching. I. The model. *Journal of Soil Science*, **40**, 543–54.

Barraclough, D. (1991) The use of mean pool abundances to interpret ^{15}N tracer experiments. I. Theory. *Plant and Soil*, **131**, 89–96.

Barraclough, D., Hyden, M.J. and Davies, G.P. (1983) Fate of fertilizer nitrogen applied to grassland. 1. Field leaching results. *Journal of Soil Science*, **34**, 483–97.

Belford, R.K. (1979) Collection and evaluation of large soil monoliths for soil and crop studies. *Journal of Soil Science*, **30**, 363–73.

Belford, R.K., Cannell, R.Q., Thomson, R.J. and Dennis, C.W. (1980) Effects of waterlogging at different stages of development on the growth and yield of peas (*Pisum sativum* L.). *Journal of the Science of Food and Agriculture*, **31**, 857–69.

Belford, R.K., Cannell, R.Q. and Thomson, R.J. (1985) Effects of single and multiple waterloggings on the growth and yield of winter wheat on a clay soil. *Journal of the Science of Food and Agriculture*, **36**, 142–56.

Berstrom, L. and Jarvis, N.J. (1991) Prediction of nitrate leaching losses from arable land under different fertilization intensities using the SOIL-SOILN models. *Soil Use and Management*, **7**, 79–85.

Beven, K. and Germann, P. (1981) Water flow in soil macropores. II. A combined flow model. *Journal of Soil Science*, **32**, 15–29.

Briggs, L.J. and McCall, A.G. (1904) An artificial root for inducing capillary movement of soil moisture. *Science*, **20**, 566–69.

Burns, I.G. (1975) An equation to predict the leaching of surface-applied nitrate. *Journal of Agricultural Science, Cambridge*, **85**, 443–54.

Burt, T.P. (1988) Seasonality of subsurface flow and nitrate leaching. *Catena*, Supplement, **12**, 59–65.

Burt, T.P. and Arkell, B.P. (1986) Variable source areas of stream discharge and their relationship to point and non-point sources of nitrate pollution. *IAHS Publication* **157**, 155–64.

Burt, T.P. and Arkell, B.P. (1987) Temporal and spatial patterns of nitrate losses from an agricultural catchment. *Soil Use and Management*, **3**, 138–43.

Burt, T.P. and Butcher, D.P. (1985) Topographic controls of soil moisture distribution. *Journal of Soil Science*, **36**, 469–86.

Burt, T.P., Butcher, D.P., Coles, N. and Thomas, A.D. (1983) The natural history of the Slapton Ley Nature Reserve. XV: Hydrological processes in the Slapton Wood catchment. *Field Studies*, **5**, 731–52.

Cannell, R.Q. and Belford, R.K. (1980) Effects of waterlogging at different stages of development on the growth and yield of winter oilseed rape (*Brassica napus* L.). *Journal of the Science of Food and Agriculture*, **31**, 963–5.

Cannell, R.Q., Belford, R.K., Gales, K. and Dennis, C.W. (1980a) A lysimeter system used to study the effect of transient waterlogging on crop growth and yield. *Journal of the Science of Food and Agriculture*, **31**, 105–16.

Cannell, R.Q., Belford, R.K., Gales, K., Dennis, C.W. and Drew, R.P. (1980b) Effects of waterlogging at different stages of development on the growth and yield of winter wheat. *Journal of the Science of Food and Agriculture*, **31**, 117–32.

Cannell, R.Q., Belford, R.K., Blackwell, P.S., Govi, G. and Thomson, R.J. (1985) Effects of waterlogging on soil aeration and on root shoot growth and yield of winter oats (*Avena sativa* L.). *Plant and Soil*, **85**, 361–73.

Cannell, R.Q., Goss, M.J., Harris, G.L., Jarvis, M.G., Douglas, J.T., Howse, K.R. and Le Grice, S. (1984) A study of mole drainage with simplified cultivation for autumn-sown crops on a clay soil. 1. Background, experiment and site details, drainage systems, measurement of drainflow and summary of results, 1978–80. *Journal of Agricultural Science, Cambridge*, **102**, 539–59.

Charlesworth, S. and Chisholm, J. (in preparation) A versatile and intelligent system for flow related water sampling. ADAS Field Drainage Experimental Unit, Trumpington, Cambridge.

Childs, E.C. (1969) *An Introduction to the Physical Basis of Soil Water Phenomena*, Wiley, London.

Ellis, F.R., Christian, D.G., Bragg, P.L., Henderson, F.K.G., Prew, R.D. and Cannell, R.Q. (1984) A study of drainage with simplified cultivation for autumn-sown crops on a clay soil. 3. Agronomy, root and shoot growth of winter wheat 1978–80. *Journal of Agricultural Science, Cambridge*, **102**, 539–59.

Garwood, E.A. (1988) Water deficiency and excess in grassland. In Wilkins, R.J. (ed.), *Nitrogen and Water Use by Grassland*, AFRC Institute for Grassland and Animal production, Hurley, pp. 24–41.

Geens, E.L., Davies, F.P., Maggs, J.M. and Barraclough, D. (1991) The use of mean pool abundances to interpret ^{15}N tracer experiments. II. application. *Plant and Soil*, **131**, 97–105.

Goss, M.J., Colbourne, P., Harris, G.L. and Howse, K.R. (1988) Leaching of nitrogen under autumn-sown crops and the effects of tillage. In Jenkinson, D.S. and Smith, K.A. (eds), *Nitrogen Efficiency in Agricultural Soils*, Elsevier, New York, pp. 269–82.

Goss, M.J., Howse, K.R., Harris, G.L. and Colbourn, P. (1990) The leaching of nitrates after spring fertilizer application and the influence of tillage. In Merckx, R. Vereecken, H. and Vlassak, K. (eds), *Fertilization and the Environment*, Leuven University Press, pp. 20–25.

Goudie, A.S. (1990) *Geomorphological Techniques*, 2nd edition, Unwin Hyman, London.

Gregory, K.J. and Walling, D.E. (1973) *Drainage Basin Form and Process*, Edward Arnold, London.

Grossman, J. and Udluft, P. (1991) The extraction of soil water by the suction-cup method: review. *Journal of Soil Science*, **42**, 83–93.

Haigh, R.A. (1985) *Water balance and water quality studies in an underdrained clay soil catchment*, Unpublished DPhil thesis, Oxford University.

Haigh, R.A. and White, R.E. (1986) Nitrate leaching from a small, underdrained, grassland, clay catchment. *Soil Use and Management*, **2**, 65–70.

Hallard, M. and Armstrong, A.C. (1992) Observations of water movement to and within mole drainage channels. *Journal of Agricultural Engineering Research*, **52**, 309–15.

Hansen, E.A., and Harris, A.R. (1975) Validity of soil-water samples collected with porous ceramic cups. *Soil Science Society of America, Proceedings*, **39**, 528–36.

Hansen, G.K. (1986) *HEJMDAL: analser og simulering af afgroders vækst vand-og kulstofbalance* (HEJMDAL — analyses and simulation of crop growth, water and carbon balance), Institut for landbrigets Plantekultur Hojbakkegard, Kobenhavn.

Hansen, G.K. (1987) A summary of a simulation model, Hejmdal for water use and growth of crops. In Feyen, J. (ed.), *Agriculture: Simulation models for cropping systems in relation to water management*, EUR 10869, pp. 73–90.

Harris, G.L., Goss, M.J., Dowdell, R.J., Howse, K.R. and Morgan, P. (1984) A study of mole drainage with simplified cultivation for autumn-sown crops on a clay soil. 2. Soil water regimes, water balances and nutrient losses in drain water, 1978–80. *Journal of Agricultural Science, Cambridge*, **102**, 539–59.

Haycock, N.E. (1991) *Riparian land as buffer zones in agricultural catchments*, Unpublished DPhil thesis, University of Oxford.

Haycock, N.E. and Burt, T.P. (in press) The role of floodplain soils in reducing the nitrate concentration of subsurface runoff: a case study in the Cotswolds, England. *Hydrological Processes*.

Hood, A.E.M. (1976) Nitrogen, grassland and water quality in the United Kingdom. *Outlook on Agriculture*, **8**, 320–27.

Hoogmoed, W.B. and Bouma, J. (1980) A simulation model for predicting infiltration into cracked clay soil. *Soil Science Society of America Journal*, **44**, 458–61.

Hoover, J.R. (1987) Instrumentation system for determining watershed hydrologic characteristics. *Transactions, American Society of Agricultural Engineers*, **30**, 1051–6.

Howse, K.R. and Goss, M.J. (1982) Installation and evaluation of permanent access pits which permit continuity of measurement in cultivated soils. *Experiemtal Agriculture*, **18**, 267–76.

Hutson, J.L. and Wagenet, R.J. (1991) Simulating nitrogen dynamics using a deterministic models. *Soil Use and Management*, **7**, 74–8.

ILRI (International Institue for Land Reclamation and Improvement) (1973) *Drainage Principles and Applications*, Publication 16, ILRI, Wageningen, The Netherlands.

Jarvis, N.J. and Leeds-Harrison, P.B. (1987a) Modelling water movement in drained clay soil. I. Description of the model, sample output and sensitivity analysis. *Journal of Soil Science*, **38**, 487–98.

Jarvis, N.J. and Leeds-Harrison, P.B. (1987b) Modelling water movement in drained clay soil. II. Application of the model in Evesham series clay soil. *Journal of Soil Science*, **38**, 499–509.

Jarvis, N.J., Jansson, P.-E., Dick, P.E. and Messing, I. (1991) Modelling water and solute transport in macroporous soil. I. Model description and sensitivity analysis. *Journal of Soil Science*, **42**, 59–70.

Jury, W.A. (1982) Simulation of solute transport using a transfer function model. *Water Resources Research*, **18**, 363–8.

Kaluarachchi, J.J. and Parker, J.C. (1988) Finite element modelling of nitrogen species transformation and transport in the unsaturated zone. *Journal of Hydrology*, **103**, 249–74.

Kneale, W.R. (1986) The hydrology of a sloping, structured clay soil at Wytham, near Oxford, England. *Journal of Hydrology*, **85**, 1–14.

Kneale, W.R. and White, R.E. (1984) The movement of water through cores of a dry (cracked) clay loam grassland topsoil. *Journal of Hydrology*, **67**, 361–5.

Lawes, J.B., Gilbert, J.H. and Warrington, R. (1881) On the amount and composition of rain and drainage water collected at Rothamsted. Parts I, II, and III. *Journal of the Royal Agricultural Society of England*, **17**, 241–79; **17**, 311–35; **18**, 1–70.

Marshall, T.J. and Holmes, J.W. (1988) *Soil Rhysics*, Cambridge University Press, Cambridge.

McDonnell, J.J. (1990) A rationale for old water discharge through macropores in a steep, humid catchment. *Water Resources Research*, **26**, 2821–32.

Miller, N.H.J. (1906) The amount and composition of the drainage through unmanured and uncropped land. *Journal of Agricultural Science, Cambridge*, **1**, 280–303.

Nason, G.E. and Myrold, D.D. (1991) ^{15}N in soil research: appropriate application of rate estimation procedures. *Agriculture, Ecosystems and Environment*, **34**, 427–41.

Rose, S.C., Harris, G.L., Armstrong, A.C., Williams, J.R., Howse, K.R. and Tranter, N. (1991) The leaching of agrochemicals under different agricultural land uses and its effect on water quality. In Peters, E. and Walling, D.E. (eds), *Sediment and stream water quality in a changing environment: Trends and explanations*, IAHS Publication **203**, 249–57.

Russell, E.J. and Richards, E.H. (1920) The washing out of nitrates by drainage from uncropped and unmanured land. *Journal of Agricultural Science, Cambridge*, **9**, 22–43.

Ryden, J.C., Ball, R.P. and Garwood, E.A. (1984) Nitrate leaching from grassland. *Nature*, **311**, 50–53.

Scholefield, D., Lockyer, D.R., Whitehead, D.C. and Tyson, K.C. (1991) A model to predict transformation and losses of nitrogen of Nitrogen in the UK pastures grazed by beef cattle. *Plant and Soil*, **132**, 165–77.

Shepherd, M.A. and Lord, E.I. (1990) Effect of crop rotation and husbandry on nitrate loss by leaching from sandland. In Calvet, R. (ed.), *Nitrates, Agriculture, Eau*, INRA, Paris, pp. 381–7.

Smedema, L.K. and Rycroft, D.W. (1983) *Land Drainage: planning and design of agricultural drainage systems*, Batsford, London.

Smettem, K.R.J., Trudgill, S.T. and Pickles, A.M. (1983) Nitrate loss in soil drainage waters in relation to by-passing flow and discharge on an arable site. *Journal of Soil Science*, **34**, 499–509.

Troake, R.P. and Walling, D.E. (1973) The hydrology of Slapton Wood stream. A preliminary report. *Field Studies*, **3**, 719–40.

Trudgill, S.T., Pickles, A., Smettem, KR.J. and Crabtree, R.W. (1983) Soil water residence time and solute uptake. 1. Dye tracing and rainfall events. *Journal of Hydrology*, **60**, 257−79.

Vauclin, M. (1990) Modelisation des transfers dans les sols non-satures: approach deterministique ou stochastique? In Calvet, R. (ed.), *Nitrates, Agriculture, Eau*, INRA, Paris, pp. 169−79.

Vinten, A.J.A., Howard, R.S. and Redman, M.H. (1991) Measurement of nitrate leaching losses from arable plots under different nitrogen input regimes. *Soil Use and Management*, **7**, 3−14.

Voelcker, A. (1984) On the composition of waters of land-drainage. *Journal of the Royal Agricultural Society of England*, **2**, 132−65.

Wagenet, R.J. (1990) Quantitative prediction of the leaching of organic and inoranic solutes in soil. *Philosophical Transaction of the Royal Society of London, Series B*, **329**, 321−30.

Way, J.T. (1856) On the composition of the waters of land-draining and of rain. *Journal of the Royal Agricultural Society of England*, **17**, 123−62.

Wellings, S.R. (1984a) Recharge of the Upper Chalk aquifer at a site in Hampshire, England. 1. Water balance and unsaturated flow. *Journal of Hydrology*, **69**, 259−73.

Wellings, S.R. (1984b) Recharge of the Upper Chalk aquifer at a site in Hampshire, England. 2. Solute Movement. *Journal of Hydrology*, **69**, 275−85.

White, R.E. (1989) Prediction of nitrate leaching from a structured clay soil using transfer functions derived from externally applied or indigenous solute fluxes. *Journal of Hydrology*, **107**, 31−42.

White, R.E. and Magesan, G.N. (1991) A stochastic-empirical approach to modelling nitrate leaching. *Soil Use and Management*, **7**, 85−94.

Whitemore, A.P. and Addiscott, T.M. (1986) Computer simulation of winter leaching losses of nitrate from soils cropped with winter wheat. *Soil Use and Management*, **2**, 26−30.

Wild, A. and Cameron, K.C. (1980) Soil nitrogen and nitrate leaching. In Tinker, P.B. (ed.), *Soils and Agriculture*, Blackwell, Oxford, pp. 35−70.

Wood, W.W. (1973) A technique using porous cups for water sampling at any depth in the unsaturated zone. *Water Resources Research*, **9**, 486−8.

10 Nitrate in Surface Waters

P.J. JOHNES
Department of Environmental and Evolutionary Biology, University of Liverpool
and
T.P. BURT
School of Geography, University of Oxford

10.1 BACKGROUND INFORMATION

10.1.1 GENERAL CONCERNS

Recent reports by the Organisation for Economic Co-operation and Development (OECD, 1985, 1987) have commented upon significant increases in nitrate pollution over the last 20 years. Concern has focused on rivers in the developed world where high concentrations of instream nitrate result through the delivery of nitrogen from intensive agricultural practices and from the discharge of high-concentration point sources of nitrogen such as sewage effluent. High levels of nitrate in drinking water are becoming a problem in some regions, such as Western Europe (Roberts and Marsh, 1987; ECETOC, 1988). Such concerns have, for example, led the European Community (EC) to adopt the World Health Organisation (WHO) recommendations on acceptable concentrations of nitrate in water intended for human consumption (WHO, 1970). Recently, this control has been extended to regulate the concentration of nitrate in all waters which are, or may be, abstracted for the supply of drinking water, affording more general protection to the aquatic environment. The legislative control of nitrate concentrations in water is more fully reviewed in Chapter 14.

Meybeck, Chapman and Helman (1989) present a worldwide analysis of nitrate concentrations in a range of rivers, demonstrating the variability of nitrate concentrations in surface waters. The data presented suggest that the nitrate problem is most serious in Western Europe, with the global median nitrate concentration in surface water, excluding Europe, at 0.25 mg NO_3-N l^{-1}, while the European median concentration is 4.5 mg NO_3-N l^{-1}. The OECD (1990) concurs, mentioning nitrate increases in The Netherlands, Belgium, France, Italy, and Sweden; in the United Kingdom, for nine out of 25 rivers sampled, mean nitrate concentrations in the period 1981−5 were found to be higher than in any previous period. Smith and Alexander (1985) present a review of water quality in 298 North American rivers, identifying a widespread increase in surface farmed agricultural regions. The OECD (1990) note that non-point source pollution in the USA constitutes 80% of the total nitrogen load reaching its surface waters.

Clearly, then, the greatest increases in nitrate concentrations in surface waters seem to be occurring in Western Europe and North America, with the implication from these various surveys that nitrate pollution is less serious in Asia, Africa and South America. In this chapter, since the majority of published data are presented for surface waters in the first two regions, the discussion will be centred on nitrate in surface waters in industrialised rather than

Nitrate: Processes, Patterns and Management. Edited by T.P. Burt, A.L. Heathwaite and S.T. Trudgill
© 1993 John Wiley & Sons Ltd

developing countries. It may be, however, that nitrate concentrations in surface waters elsewhere in the world are also increasing, and in some cases may also be approaching levels which would be a cause for concern. One problem with selective surveys of large rivers is that the magnitude of the nitrate problem may be underestimated: major rivers in continental areas are sustained by large volumes of low-nitrate water from the headwaters so that inputs of nitrate from tributaries in the middle and lower parts of the catchment are easily diluted in the main river channel. Thus, more local problems of nitrate pollution may easily be overlooked.

10.1.2 NITROGEN CYCLING IN RIVERS AND LAKES

Heathwaite (Chapter 5, this volume) has described nitrogen cycling in surface waters in some detail. Our purpose here is to emphasise the salient points from her review which are relevant to a survey of nitrate concentrations in river basins. Inputs may be from point or non-point sources. Effluent from sewage-treatment works is an important point source of nitrogen delivery to aquatic environments, and in urbanised basins may constitute a significant proportion of the total nitrogen load in a river. Equally important as point sources of nitrogen in surface waters are the discharge of silage liquor, liquid runoff from stored manures and dairy wastes, particularly in predominantly agricultural basins. However, the greatest loss of nitrogen from agricultural land to adjacent surface waters is as non-point or diffuse inputs (though a good deal may be channelled into rivers via tile drains or open ditches). Within the surface water body, instream cycling of nitrogen will determine the proportion of the total nitrogen load present as nitrate at any given time. Nitrogen is lost to the sea, and to the atmosphere through denitrification. Abstraction of water for public supply, while less important in volumetric terms, is of interest because of legislation controlling maximum acceptable nitrate concentrations in abstracted waters.

Significant decreases in nitrate concentration in a surface water body may result from assimilation of nitrate by algae, and by instream and riparian macrophytes, especially in the upper and middle reaches of a river, and in the littoral fringe and the euphotic zone of standing waters. Eventual decomposition of plant and animal matter results in increases in the nitrate concentration as organically bound nitrogen is mineralised to ammonium and then, if the dissolved oxygen concentration in the water is adequate, oxidised to form nitrate. Algae will have a strong but localized influence on nitrogen concentrations in surface waters. In standing waters, blooms of algal production in response to increases in the availability of nutrients in the water column can have a significant and spatially variable impact on water quality, both horizontally and vertically. Macrophytes will also effect changes in standing water quality, particularly where the littoral fringe is extensive. The influence that macrophytes may exert on the nitrate concentration of flowing water is also spatially variable (Pinay et al., 1990). The river continuum (Vannote et al., 1980) and nutrient spiralling (Newbold et al., 1981; Elwood et al., 1983) concepts incorporate the idea of longitudinal linkages: unidirectional water flow processes promote the mixing of materials from upstream to downstream zones. In headwater tributaries, channels are small and macrophytes may easily interact with stream water; lower down, the hydraulic radius (the ratio of channel cross-section area to the length of the wetted perimeter) is larger and a smaller proportion of the water body interacts with the riparian and instream vegetation. The net result is longer nutrient spiralling lengths and more stable nutrient concentrations.

Odum (1989) notes that the character of a river changes from source to mouth: not only does discharge increase, but community metabolism, species composition and diversity changes as well. In unpolluted headwater tributaries, respiration often exceeds primary productivity and stream biota are then dependent on allochthonous nutrient sources, usually as leaf matter or dissolved organic matter washed in from terrestrial ecosystems (or sometimes from adjacent lakes). Lower down, streams are wider and often less well shaded, allowing higher rates of primary production, with greater species density. In the lowest reaches of a river, high velocity and increased turbidity in the water body from suspended solids reduce the potential for macrophyte colonisation, with a decrease in nitrogen input to the river through instream primary production. This continuum concept thus suggests that, in an unpolluted river basin, the dominant source of instream nutrients will vary from terrestrial in the headwaters to aquatic production plus upstream delivery in the middle reaches, and back to terrestrial sources plus upstream delivery in the lowest reaches. However, the continuum concept is usually applied to rivers with headwater zones in forested landscapes, notably in the USA. Haycock (1991) has questioned whether the continuum concept should be applied to rivers with intensively farmed headwaters, as in southern England, where headwater tributaries may be just as highly polluted as the main river (see also Chapter 12). Low-nitrate waters, derived from headwater catchments, may not be available to dilute sewage and other effluents derived from urban areas further downstream, especially in those regions where agriculture is most intensive.

10.1.3 NITROGEN SPECIATION

Nitrate concentrations in surface waters are strongly dependent not only on the surrounding land in the catchment, but also on instream cycling of nitrogen species, except at small-basin scales. These species consist of nitrite ($NO_2 - N$), ammonium (ionised, $NH_4 - N$; un-ionised, $NH_3 - N$) and organic nitrogen, in addition to nitrate ($NO_3 - N$). The predominant form present is determined by environmental conditions in the water body, particularly pH, temperature and oxygen availability (Brady, 1984; O'Neill, 1985), coupled with plant uptake, and mineralisation rates of labile organic nitrogen. Indirectly, season can be said to be the key contol of the speciation balance, regardless of the total nitrogen concentration in the water body.

In the predominantly agricultural catchment of the River Windrush, a headwater tributary of the River Thames, Johnes (1990) demonstrated that nitrate dominates the annual nitrogen loading. The highest nitrate concentrations coincided with periods of highest discharge during the winter. However, organic nitrogen was also important, constituting almost 40% of the annual total nitrogen load, the proportion increasing in a drier year, with organic nitrogen contributing both to on-site and downstream inorganic nitrogen pollution (see also Johnes and Burt, 1991). In a study of nitrogen transport in surface runoff from a variety of land uses in the Slapton Ley catchment, south Devon, Heathwaite, Burt and Trudgill (1990) demonstrated the value of nitrogen speciation analyses. They found that more than 90% of the nitrogen lost from heavily grazed grassland was in the form of ammonium, but that this proportion varied according to the land-use type and the fertiliser and grazing practice in each field. Heathwaite and Johnes (in press) discuss the collection of nitrogen flux data for surface waters with respect to sample collection, preservation and analysis; their sampling strategy is summarised in Figure 10.1.

Despite the potential importance of ammonium and organic nitrogen, the majority of aquatic nutrient studies have focused on nitrate concentrations in surface waters, rather than on all nitrogen species; accordingly, the remainder of this chapter will assess the spatial and temporal patterns of nitrate alone.

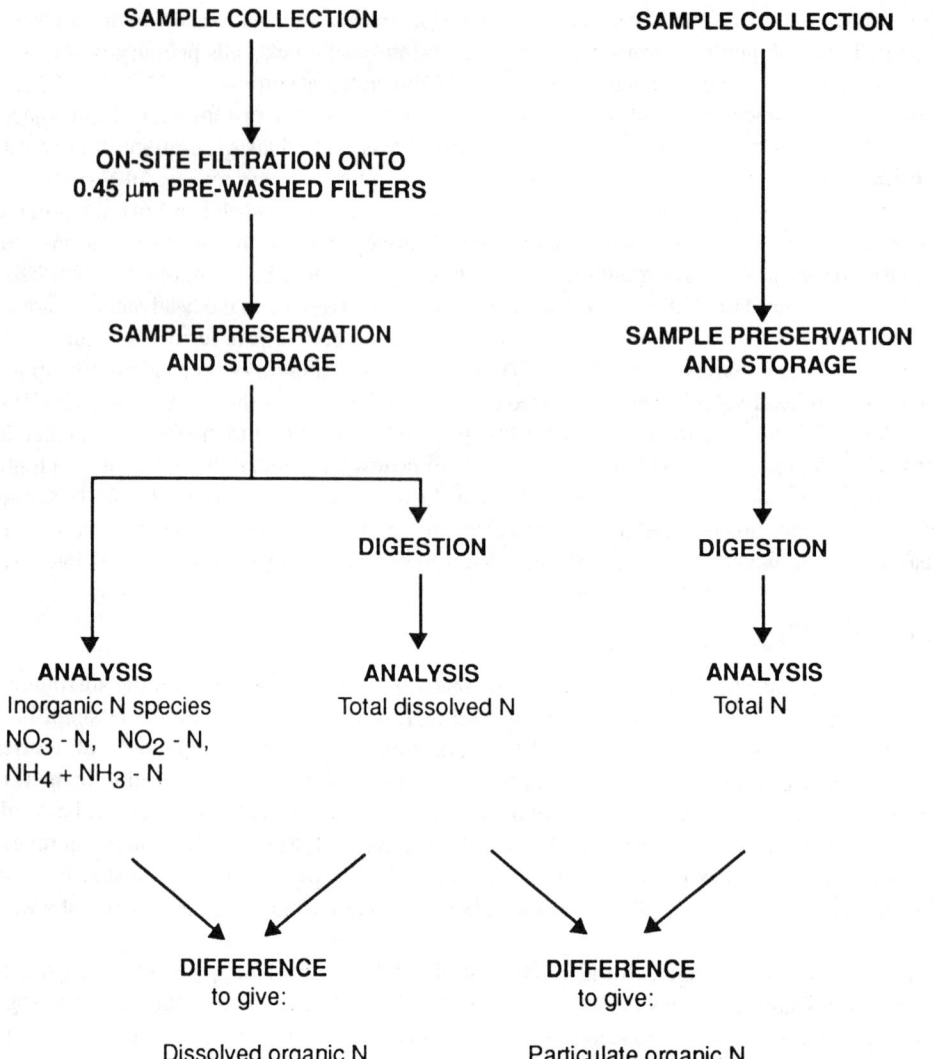

Figure 10.1 General sampling strategy for determination of nitrogen concentrations

10.1.4 DATA COLLECTION AND ANALYSIS

Traditionally, studies of nitrate in surface waters have concentrated on temporal variation in surface water nitrate concentrations (Section 10.3). These studies have been based on water quality time series collected from a limited number of sampling points in the research basin. The data present an integration of the hydrological processes within the flow domain at that point. However, variation in both time and space must be studied (Falkenmark and Chapman, 1989). Studies of spatial variation in land use can give an indication of the spatial patterns of suspended sediment and nutrient concentrations in water bodies (Section 10.2). Mosaic interpretation of these data rather than interpolation between points allows a pictorial representation of patterns to be presented which can then be used to identify controlling factors

(Walling and Webb, 1978). However, to date, less attention has been paid to spatial variations in nitrogen in the aquatic environment, than to temporal patterns, though recent spatial studies include Burt and Arkell (1987), Kesner and Meentemeyer (1989) and Johnes (1990).

Since the concentration and load of nitrate in any surface water is dependent on both temporal and spatial variation in source, delivery mechanism and instream cycling processes, any study aiming to assess fluxes in nitrate concentrations in surface waters will clearly need to monitor changes over a variety of temporal and spatial scales. A sampling strategy employed to provide representative nitrate concentration and load data for suface waters must meet both the statistical requirement for a regular long-term monitoring and the hydrological requirement for more intense, flexible sampling over periods of storm discharge (Walling and Webb, 1982).

Collection of nitrate data can be through on-site, instantaneous determination of nitrate concentrations and discharge or on-site discharge measurement and sample collection for subsequent laboratory determination of nitrogen concentrations. Solute concentrations may be monitored using continuous-flow probes linked to automatic data-logging equipment. However, while this has been suitable for the monitoring of physical and some chemical variables, ion-selective electrodes for on-site monitoring of nitrate concentrations have been found to be less reliable. More recently, the use of ion-exchange resins has gained support, but these are still under development (Edwards, A., MLURI, pers. comm.). Alternatively, nitrogen concentrations may be more reliably determined through sample collection and subsequent laboratory analysis. However, changes in the concentrations of determinands between collection and analysis result from a variety of physical, chemical and biological processes. The nature and rate of these reactions may be such that the samples on analysis are totally unrepresentative of conditions at the time of sample collection.

A variety of changing conditions in the sample container may influence nitrogen speciation. An increase of $20°C$, for example, will result in a pH change of 0.1 (O'Neill, 1985). The speciation of nitrogen is heavily dependent on the pH of the sample. At high temperatures or high pH, or both, significant losses of ammonia can occur through volatilisation. At low pH, nitrogen may be lost in gaseous form as N_2. Between pH 3 and pH 9 the inorganic nitrogen species will be in equilibrium with no net loss, but any change in pH may result in transformation from one species to another. Therefore, rapid freezing or thawing of samples is undesirable if the samples are to be used for subsequent nitrogen speciation analysis or, indeed, if the absolute concentration of any particular species is required (Heathwaite and Johnes, in press).

Organisms in the water sample may decompose, nitrify or denitrify nitrogen compounds. Hence it is vital that biological activity in the sample is retarded. One standard method of achieving this is to freeze the sample, but, as noted above, this is undesirable (Batley and Gardiner, 1977; Giesy and Briese, 1978; Berg, 1982). The most widely recommended technique for the preservation of water samples for nitrogen speciation analysis is the addition of a biocidal reagent such as mercuric chloride ($HgCl_2$) with storage at $4°C$ (e.g. Jenkins, 1968; German Chemists Association, 1981; Suess, 1982). Heathwaite and Johnes (in press) advocate the addition of 1 ml of 40 mg l^{-1} $HgCl_2$ to each 100 ml sample in the sample bottle prior to collection. This allows sufficient mercuric chloride to stop biological action, but produces a low enough concentration in the sample to prevent interference in standard colorimetric analytical procedures for the determination of nitrogen species concentrations.

Polyethylene or glass bottles are universally recommended for the collection of water samples for nitrogen speciation and suspended sediment analysis (Heathwaite and Johnes (in press)). Polyethylene bottles are particularly suitable for field work, since these do not break. New bottles should be washed with 10% hydrochloric acid prior to use to prevent sorption of suspended matter and nitrogen compounds onto the surface of the bottle. This should be repeated each time the bottles are used for sample collection. It also prevents release of amines from the polyethylene which could cause nitrogen contamination.

Samples collected by hand should be filled completely to eliminate air and hence inhibit aerobic biological action. Automatic water samplers should be protected from rain, snow and dust with an insulated plastic cover. The sample lines from each sampler to the river should be regularly checked to prevent blockage either by macrophytes in the river or by algal growth inside the sample lines. The lines should be made as short as possible to minimise the contact time between sample and line and to remove bends in which water could stagnate, promoting algal growth.

In rivers, samples should be collected from a well-mixed site to ensure a representative sample. Samples should be taken away from the boundaries of the river and, to avoid collection of surface films or the entrainment of bottom deposits, at mid-depth. A number of authors recommend collection at least 30 cm from the bottom or surface, where possible (Standing Committee of Analysts, 1980; Suess, 1982; Hunt and Wilson, 1986). In standing waters, samples should be collected at a range of depths, or an integrated depth sample collected at a range of sites within the water body in order to obtain a representative sample. Alternatively, where there are clear inflow and outflow streams, a standing water body may be considered as a mixed-reaction vessel, and samples collected from the main inflow and outflow waters.

10.1.5 LOAD CALCULATION FROM CONCENTRATION DATA

The protocol presented in Figure 10.1 does not deal with sampling frequency, since this is largely determined by basin size, accessibilty and the aims of the sampling programme. Nevertheless, the accuracy of any load calculation is largely dependent upon the extent to which sample concentrations are representative of environmental conditions. Thus sampling frequency is of primary importance. This is especially so in the calculation of sediment-associated nutrient loads, because sediment load increases greatly with discharge: the rapid suspended sediment response during storm hydrographs requires a greater sampling frequency than is necessary for nitrate- and nitrite-nitrogen (which are more conservative) in order to be fully representative of temporal variation during extreme events (Walling and Webb, 1981a). The minimum acceptable sampling interval will be largely dictated by the size, reactivity and flow duration characteristics of the drainage basin. For the River Windrush (363 km²) daily sampling with hourly sampling during flood events was found to be sufficient (Johnes, 1990). In contrast, in smaller drainage basins such as the Slapton catchment in south Devon (46 km²) minimum sampling frequency for determination of nitrate concentrations was found to be daily sampling with 15-minute sampling during periods of storm discharge (Heathwaite, Burt and Trudgill, 1990).

Inadequate sampling frequency is a key problem in calculating nitrogen loading from historical records of water quality and discharge. Nitrate data are often only available on a sporadic basis and interpretation of load predictions based on such data should take this into account. Loads can be calculated in two ways: interpolation and extrapolation (e.g. Walling and Webb, 1982; Ongley, 1987). These vary in both precision and accuracy and suitability for use with different data sets.

Interpolation is the process of inserting in a series an intermediate number or quantity ascertained by calculation from those already known. Interpolation of instantaneous loadings to produce an annual load or yield for a basin is an acceptable procedure provided that sampling is sufficiently frequent to minimise the error attached to unmonitored periods. Marsh (1980), Walling and Webb (1982) and Hagebro, Bang and Somer (1983) have assessed the relative accuracy and precision of such interpolations. Various loading equations are presented in Walling and Webb (1982). *Extrapolation* involves calculation, based on a known series of data, of the trend preceding or following the known series. It is particularly attractive in drainage-basin studies where sampling instruments undergo mechanical failure, for periodic sampling strategies, or where sampling frequency is insufficient for an acceptable level of accuracy. There is considerable debate over the accuracy and precision of the extrapolation procedures available, particularly for the calculation of suspended sediment load (e.g. Walling and Webb, 1981a; Ferguson, 1986; Labadz, 1988). The technique most widely favoured involves extrapolation from a log-linear rating relationship between suspended sediment load and discharge, and this approach may also be applied to nutrients. In instances where nitrogen concentration data are scarce, data may be estimated from the rating curve using a continuous record of discharge. Ferguson (1986) shows the need for a correction to be made in order to prevent an underestimation of loads when using a log-linear rating curve.

10.1.6 CONSIDERATIONS OF TIME AND SPACE

As in any geographical analysis, the simultaneous portrayal of system behaviour in time and space presents problems. In the rest of this chapter we present spatial analyses for both small and large drainage basins, and then separate out periodic and episodic variations over time for different locations. The final section deals with modelling initiatives which, in the case of distributed models at least, allows some integration of time and space in a single analysis.

One further aspect relating to the spatial analyses deserves mention: an interesting interpretation of such results may be gained by invoking the ergodic hypothesis. This suggests that, under certain circumstances, sampling in space can be equivalent to sampling in time and that space−time transformations are possible. In our case, comparisons between nitrate losses from woodland, grassland and arable fields at one point in time may resemble the change which has occurred in many basins as farming has become more intensive over the last half century.

10.2 SPATIAL VARIATIONS OF NITRATE WITHIN RIVER BASINS

Mapping spatial variations of discharge and solute transport within a river basin presents certain difficulties. Falkenmark and Chapman (1989) argue that hydrological processes fall into two types, depending on whether the direction of water movement is mainly vertical (e.g. rainfall) or horizontal (e.g. river discharge). Horizontal flows are generally quantified by calculating the discharge (i.e. the volume of water per unit time crossing a channel cross-section perpendicular to the flow direction; its dimension is L^3T^{-1}). For a solute such as nitrate, the product of the discharge and the solute concentration gives a mass transport or load (MT^{-1}). For horizontal transport, the discharge at a given point does not relate to conditions at the measurement point but represents the integrated effect of the entire catchment upstream of the section. It follows that changes in horizontal transport should not be investigated areally but along the path of movement (Kovacs, 1989). Thus, detailed sampling along a river channel is needed in order to map variations in discharge and solute transport, especially with regard to the sharp changes in concentration and load which may occur downstream of point sources or tributary junctions, and isoline maps derived from widely spaced samples must be viewed with some caution.

Most studies of the spatial pattern of nitrogen concentration and load within river networks have been made in relatively large drainage basins (> 100 km^2); detailed mapping within small (> 10 km^2) basins is rare despite the large number of plot studies which have been conducted (see Chapter 9).

10.2.1 VARIATIONS WITHIN SMALL BASINS

Only a few studies have concentrated on drainage basins with areas of less than 10 km^2 (e.g. Lowrance, Todd and Asmussen, 1984; Jacobs and Gilliam, 1985). At this scale it may be possible to aggregate the results of plot studies, to relate them to patterns of nitrate transport at the basin outlet, and to ascribe the pattern of nitrate loss within the basin to the distribution of land use and topography. Haycock (1991) has used spatial sampling of stream and spring water to demonstrate the protection afforded by nutrient retention zones on floodplains and to show the magnitude of instream nitrate losses during summer (see Chapter 12, Section 12.4).

Burt and Arkell (1987) studied the spatial pattern of runoff and nitrate loss within the Slapton Wood catchment, South Devon. Subcatchment divisions and land use within this 0.94 km^2 catchment are shown in Figure 10.2, together with the spatial pattern of nitrate loss. Two controls of nitrate loss were likely to be most relevant: land use (woodland, arable and pasture) and topography (especially the contrast between the tributary valleys, 'Carness' and 'Eastergrounds', and other valley-side slopes). To determine the amount of nitrate being contributed from differrent parts of the catchment, stream discharge was measured by dilution gauging at a number of points along the stream. This exercise was repeated a number of times so that an average contribution for each reach of the stream could be estimated. Frequent sampling along the channel allowed the spatial pattern of nitrate concentration in the stream to be established.

Table 10.1 shows the annual budget of nitrate inputs from the source areas defined in Figure

Table 10.1 Annual budget of nitrate inputs from source areas in the Slapton Wood catchment (adapted from Burt and Arkell, 1987)

Source area	Mean annual discharge (l s^{-1})		Area (ha)		Mean NO$_3$−N conc. (mg l^{-1})	Nitrate load (kg)	(kg ha^{-1})
1. Headwaters	4.4	(30)	28.5	(30)	10.0	1380	48.4
2. Valley-sides	2.9	(20)	16.6	(18)	8.0[a]	740	44.6
3. Carness	2.9	(20)	20.6	(22)	7.4	680	33.2
Flume	10.2	(70)	65.7	(70)	8.7	2800	39.6
4. Eastergrounds	2.9	(20)	19.3	(21)	6.6	610	31.6
5. Slapton Wood	1.5	(10)	8.9	(9)	4.6[a]	210	23.9
Basin outlet	14.6	(100)	93.9	(100)	7.9	3626	38.6

[a] Estimated from nitrate load and discharge.
Figures in parentheses are percentages for those of the basin outlet.
Source area numbers accord with subcatchments shown in Figure 10.2.

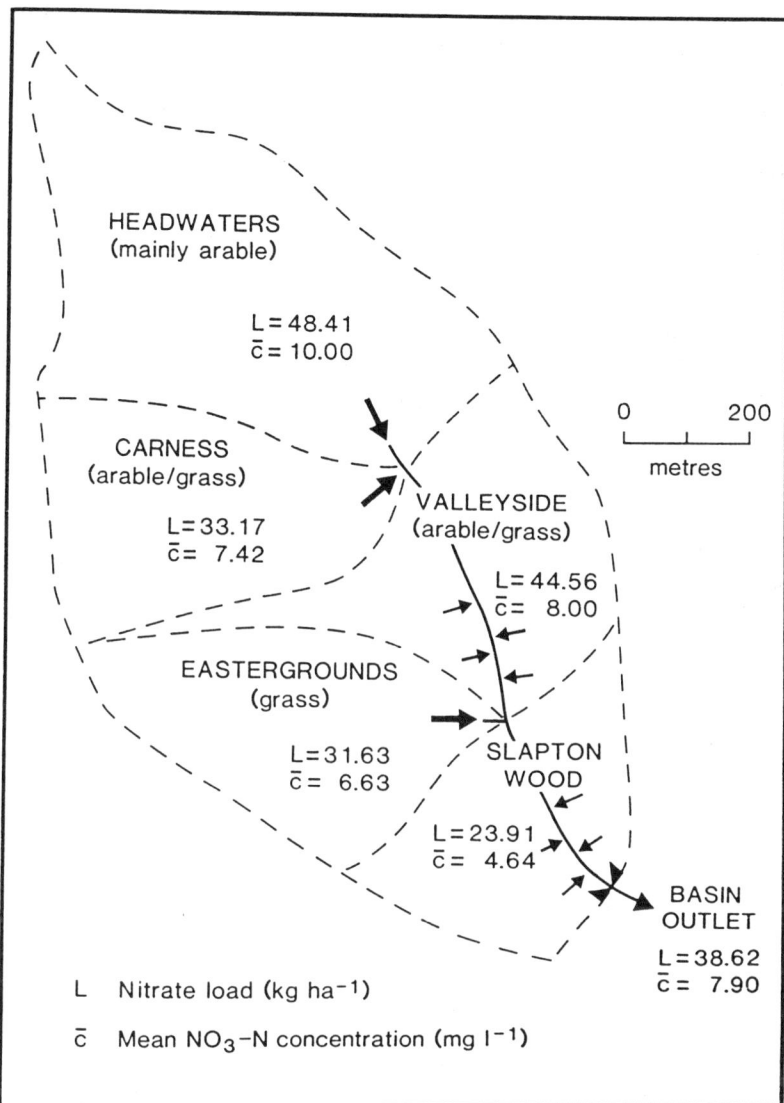

Figure 10.2 The spatial pattern of nitrate leaching in the Slapton Wood catchment, south Devon, as indicated by dilution gauging experiments. Large arrows indicate point sources of nitrate (springs and tributary streams); small arrows imply non-point inputs (subsurface seepage) for that section of the stream (after Burt and Arkell, 1987)

10.2. As expected, the largest nitrate losses per unit area (and the highest concentrations) are derived from the mainly arable headwaters. The two major subcatchments, Carness and Eastergrounds, are important point sources of discharge and nitrate, but since they include large areas distant from the stream which contribute little to runoff, their nitrate loss per unit area is less than might be expected. The wood provides the smallest input of nitrate; this might have been less were it not for an arable field at the top of the wood on its south side. The results suggest that all slopes which abut the steeper valley-side slopes, but especially those within the tributary valleys, are sensitive locations with respect to nitrate loss (Burt and Arkell, 1986). Nevertheless, land use seems to be the main control of nitrate losses from the basin. The spatial variation in nitrate losses from wood, grass and arable land provides an analogy for changes in the nitrate concentration of streams in the Slapton catchments, and elsewhere, during the twentieth century (see Section 10.4.3).

10.2.2   VARIATIONS IN LARGE BASINS

Few studies have detailed spatial variation in nitrogen concentrations at a large-basin scale (>100 km^2), linking these to the spatial distribution of nitrogen source areas within the catchment. Walling and Webb (1978) have highlighted the importance of spatial data as a means of identifying key land use and geological controls of water quality. Spatial variations in nitrate concentrations in the Exe basin have been described by Walling and Webb (1985) and for one of the tributaries by Webb and Walling (1985). Recently, more attention has been paid to the value of spatial analysis as an important tool in basin management and, with the advent of Geographical Information Systems (GIS) capable of handling greater quantities of data, spatial patterns in water quality in larger basins are now being analysed (e.g. Kesner and Meentemeyer, 1989; Johnes, 1990). Examples of spatial variation in nitrate concentrations in large basins are presented below.

In a three-year study based in the catchment of the *River Windrush* (363 km^2), a tributary of the River Thames, the land-use distribution in the catchment was mapped and compared with variations of nitrogen concentrations along the river channel (Johnes, 1990). This use of spatial analysis in the Windrush has permitted identification of key nutrient export zones in the catchment contributing to instream nitrogen loading, and identified sensitive areas of the Windrush catchment where management strategies to reduce nitrogen loading on the river might be most profitable (see Section 10.4.3).

The results of the 41 water quality surveys conducted along the Windrush are summarised in the form of box-whisker diagrams for nitrate and total nitrogen in Figures 10.3(a) and 10.3(b), respectively. The upstream shaded box represents the River Dikler, the downstream shaded box represents the Sherborne Brook, both key tributaries of the River Windrush. The Sherborne Brook is predominantly spring-fed, with surface water nitrogen concentrations being largely determined by land use over the highly reactive limestone aquifer in this catchment. Thus, while the river corridor in the Sherborne is lightly grazed grassland, a low-nutrient export crop, the high nitrogen leaching rates under intensively farmed cereals on the limestone result in high nitrate and total nitrogen concentrations in the aquifer. As a result, the nitrate concentrations in the Sherborne Brook are higher than in the River Windrush. The River Dikler receives a lower proportion of its flow from groundwater than either the Windrush in its headwaters, or the Sherborne Brook, and so the nitrate concentration of the Dikler is lower than that of the Windrush, and causes a decrease in Windrush nitrate at the confluence. These results indicate the relative contribution of nitrate from groundwater, and the importance of intensive cereal cropping over the aquifer in determining nitrate concentrations in the Windrush.

The total nitrogen data summarised in Figure 10.3(b) confirm the general patterns observed both for nitrate (Figure 10.3(a)) and total organic nitrogen concentrations in the Windrush. The lower concentrations of total nitrogen at site 2 may be attributed to the greater contact time between nutrients in the water body and aquatic biota in the artifical lake at that site.

Overall, there is a downstream decrease in concentration from the headwaters to the top of the Lias Clay floodplain at site 8. Mean total nitrogen concentrations at all sites exceed the EC nitrate limit of 11.3 mg $NO_3 - N$ l^{-1}. Of course, total nitrogen includes all nitrogen species, Nevertheless, the total reserve of nitrogen in the water column provides the potential for nitrate concentrations to exceed the limit through instream cycling processes, although a number of organic nitrogen compounds present in the total nitrogen pool will be refractory, and therefore less available for such cycling.

Nitrate concentrations throughout the *upper Thames basin* show marked spatial patterns which can be related to land use and major rock types in the basin. The geology of the upper Thames basin is shown in Figure 10.4, and the spatial patterns in nitrate concentrations are illustrated in Figure 10.5. The nitrate data are mean concentrations based on data from at least 15 samples in the 30-month period, 1984—6.

Rock types in the upper Thames basin play a key role in determining the proportion of river flow deriving from surface and groundwater sources. In the north and south, in the areas of predominantly Jurassic (Oolite) Limestone and Chalk, most of the river flow comes from groundwater; surface water quality in these regions therefore reflects nitrate concentrations in the aquifers in each zone (see Chapter 8). The main channel of the Thames flows through the Oxford Clay zone, where a greater proportion of the surface flow is delivered through surface and near-surface runoff. As a result, non-point and point sources of nitrogen in this part of the catchment have a more immediate impact upon surface water quality than in the Jurassic Limestone zone, where there is a mean time lag of five years between transfer from the land surface to the groundwater, or the Chalk zone, where the time lag is at least 40 years (see Section 8.2). This impact is broadly shown in Figure 10.5. The lowest nitrate concentrations are found in surface waters in the Chalk where nitrate pollution has yet to reach the saturated zone. The highest concentrations occur in those areas where the rivers have their sources within the clay vales or the Jurassic Limestone zone; in both areas the impact of leaching from intensively farmed land is soon seen in local surface waters. The clay vales are often overlooked as important source areas of nitrate, but land drainage allows intensive farming on such land and nitrate losses may be high (see Chapter 9). Particularly high concentrations of nitrate occur along the River Cherwell, a major tributary of the Thames, with its confluence at Oxford; these result in part from point source discharge of sewage effluent to this river due to extensive urban expansion in this subcatchment. Similar point source pollution is noted below the large town of Swindon; inputs of nitrogen from its sewage-treatment works influence nitrate concentrations throughout the upper Thames basin.

10.2.3 NATIONAL PATTERNS OF NITRATE CONCENTRATION: GREAT BRITAIN

A number of studies have mapped nitrate patterns in the rivers of Great Britain. Roberts and Marsh (1987) used data from the Department of the Environment's Harmonised Monitoring Scheme. However, the 250 or so sites selected in the scheme are located primarily on the major rivers. Data collected from large rivers provide only an aggregate measure of water quality in the area upstream and cannot give a detailed representation of the countrywide variation in solute levels. Additional information must be obtained from small- and medium-sized basins where solute levels will reflect that of runoff from a local source area. Such a sampling strategy will minimise (though never totally prevent) the influence of point source inpute from sewage-treatment works and so highlight relations between land use and water

280

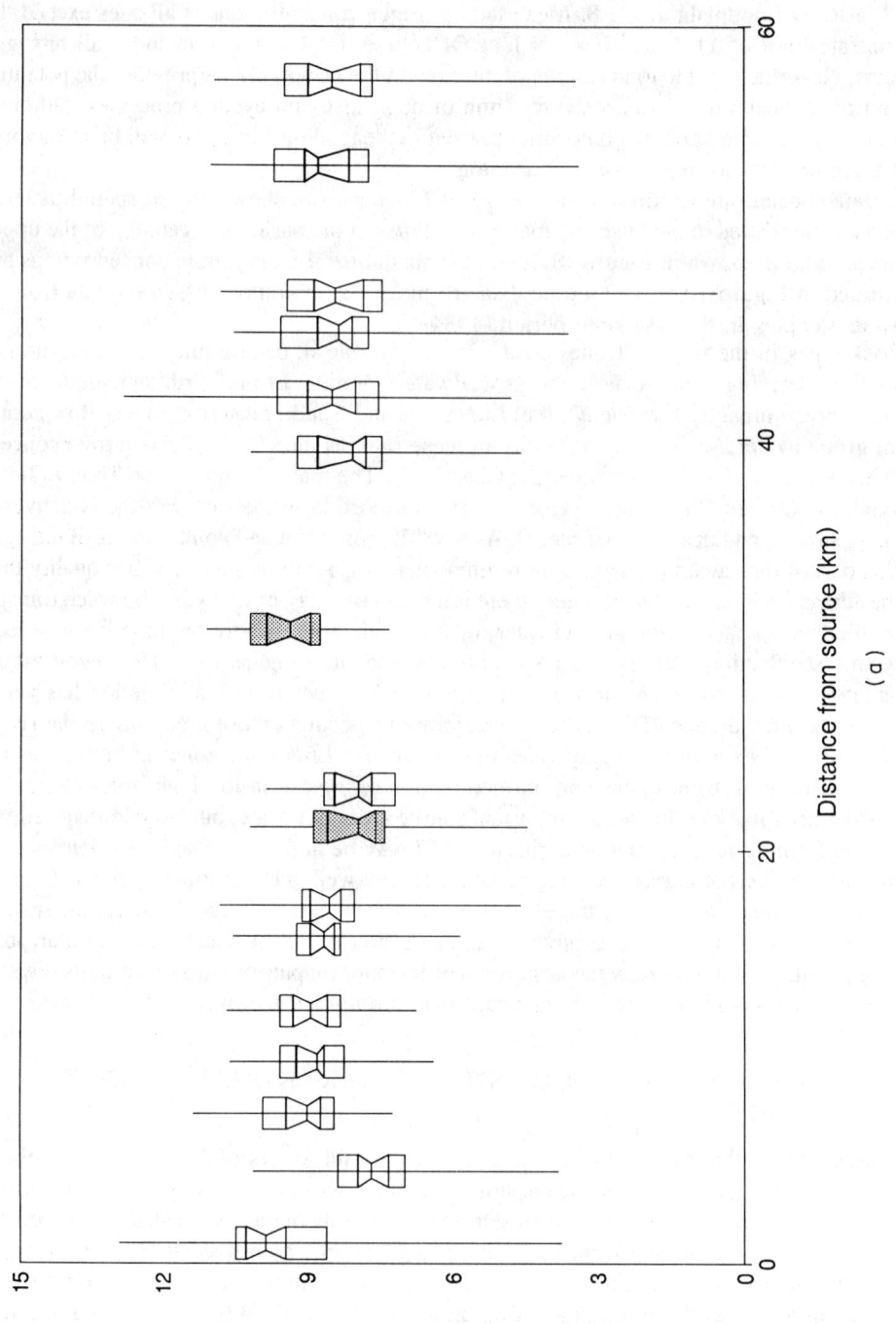

(a)

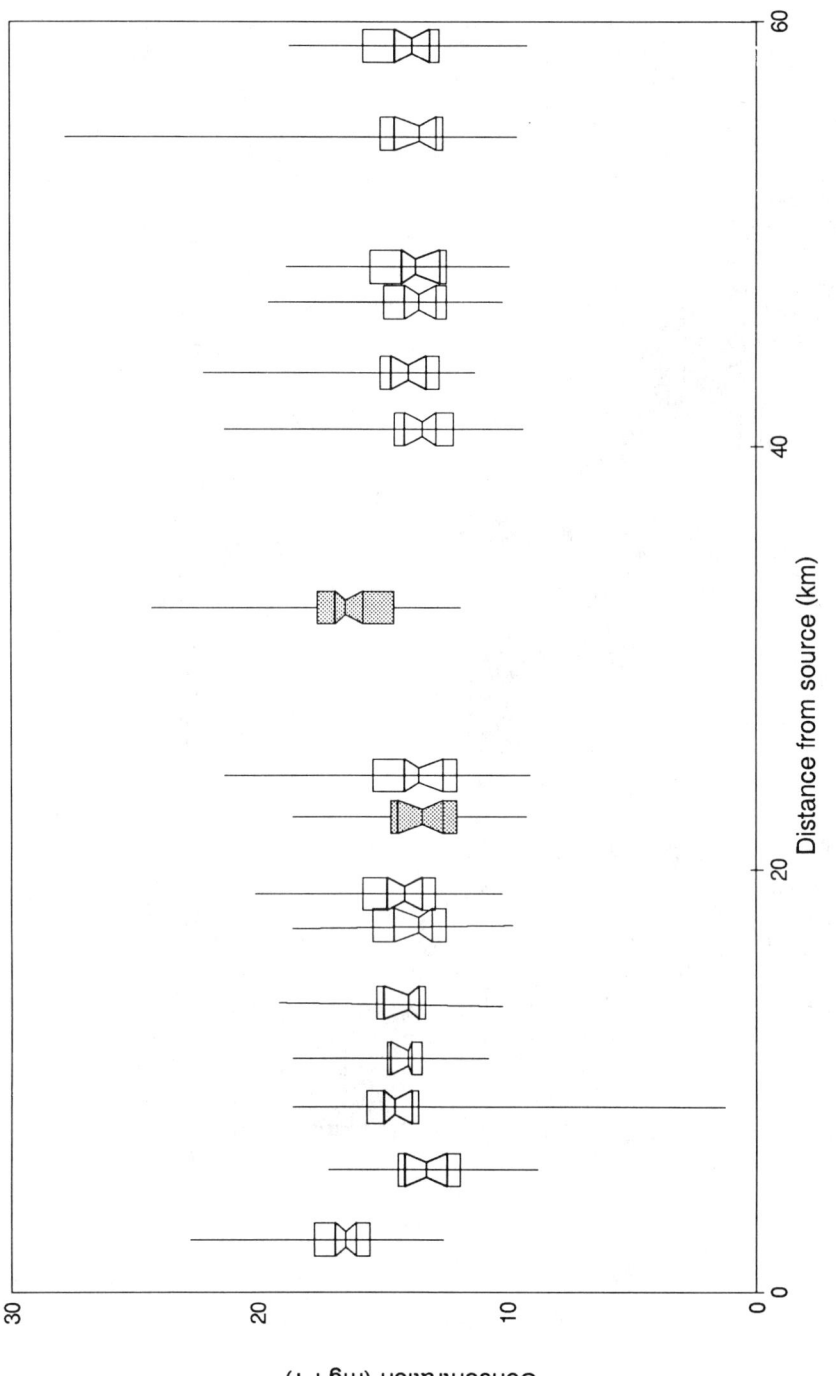

Figure 10.3 Box-whisker plots of (a) nitrate concentrations and (b) total nitrogen concentrations along the River Windrush. Boxplots show the median, confidence interval, interquartile range and range for each site

282

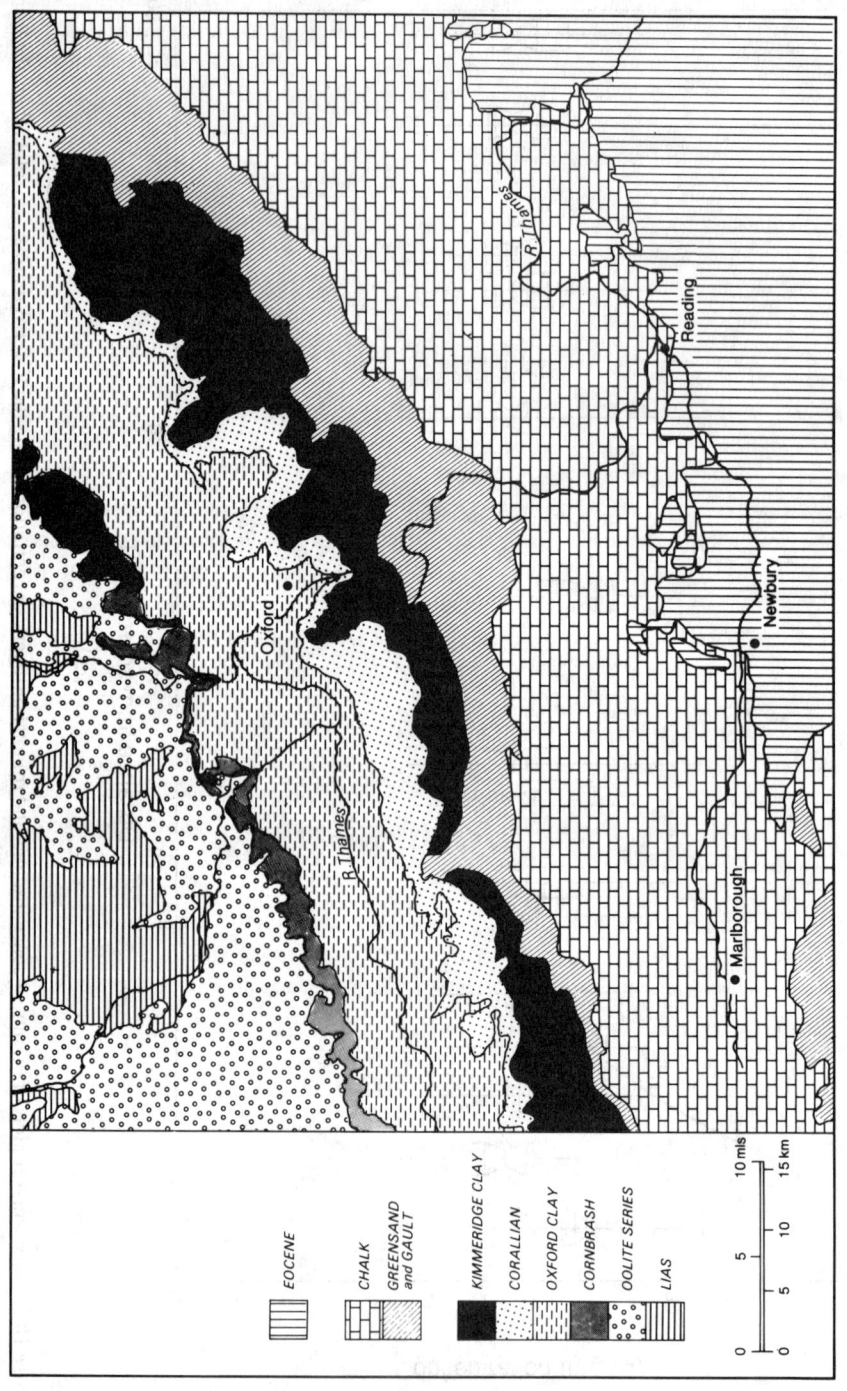

Figure 10.4 Geology of the upper Thames basin

EOCENE

CHALK

GREENSAND
and GAULT

KIMMERIDGE CLAY

CORALLIAN

OXFORD CLAY

CORNBRASH

OOLITE SERIES

LIAS

10 mils

15 km

0 5 5 10 10

Reading

Newbury

Marlborough

Oxford

R Thames

R Thames

quality. Accordingly, Walling and Webb (1981b) used water quality data collected by the regional water authorities from small- and medium-sized basins to augment information from the Harmonised Monitoring network. Betton (1990) followed a similar course; her survey of nitrate concentration in some 900 British streams and rivers is the most comprehensive and up to date available (see also Betton, Webb and Walling, 1991).

The maps shown in Figures 10.6(a) and 10.6(b) show the spatial patterns in mean nitrate concentration and nitrate load determined by Betton (1990). A clear distinction is apparent between upland Britain, where mean concentrations are below 2.5 mg $NO_3 - N\ l^{-1}$, and lowland regions, where concentrations are considerably higher. In upland areas, nitrate levels are very low because of the combination of high rainfall and runoff, undisturbed land use, and low rates of mineralisation (nitrogen is conserved in organic form in upland soils). In lowland areas, nitrate concentrations increase partly because of decreased runoff, but also because of the increasing intensity of agricultural activity. The traditional view that the highest nitrate levels are found in East Anglia is too simple; many areas in the Midlands and south of England also have intensive agriculture. Moreover, since grassland is now being managed much more intensively, with fertiliser applications equalling or exceeding those for arable land, the old distinction between intensive arable in the east and extensive grassland in the west is now untenable (if it were ever true). Thus, most of the rivers in lowland England have mean nitrate concentrations above 5.5 mg $NO_3 - N\ l^{-1}$ and, as discussed above, in some streams mean levels are very much higher than this. Even so, the highest nitrate loads do tend to be in the east, though by no means exclusively so. Because more runoff produces higher nitrate concentrations, intensive agriculture in the wetter lowlands to the west may be just as likely to generate high nitrate levels as in the drier areas to the east. Indeed, it may be that there is now less difference between nitrate levels in the rivers of East Anglia compared with those further west than there used to be.

Of particular interest are the lower nitrate concentrations in the extreme south-east, in a belt from Dorset to Kent, with small areas of much reduced agricultural activity (New Forest and the Weald). This area is dominated by streams draining the Chalk. At present, these streams have low nitrate concentrations, but predictions suggest that nitrate concentrations here will rise very steeply over the next few decades as nitrate pollution percolates slowly down through the unsaturated zone of the chalk aquifer to reach the water table (see Chapter 8). The area of highest concentration includes not just outcrops of shallow, fissured aquifers such as the Sherwood Sandstone and the Oolitic Limestones but also large areas of clay vale. As noted above (Section 10.2.2), these areas are often overlooked but may constitute major sources of nitrate loss in a large catchment.

The map of mean nitrate load on receiving waters shown in Figure 10.6(b) indicates that loadings of 15 to 30 kg $ha^{-1}\ a^{-1}$ are typical of large parts of central England. Loads are very much lower in Wales and Scotland because of very low nitrate concentrations, and in the south-east because of low annual runoff. The highest losses of nitrate take place where both concentrations and runoff volumes are relatively high: by implication, in areas of intensive agriculture in the wetter parts of the lowlands.

284

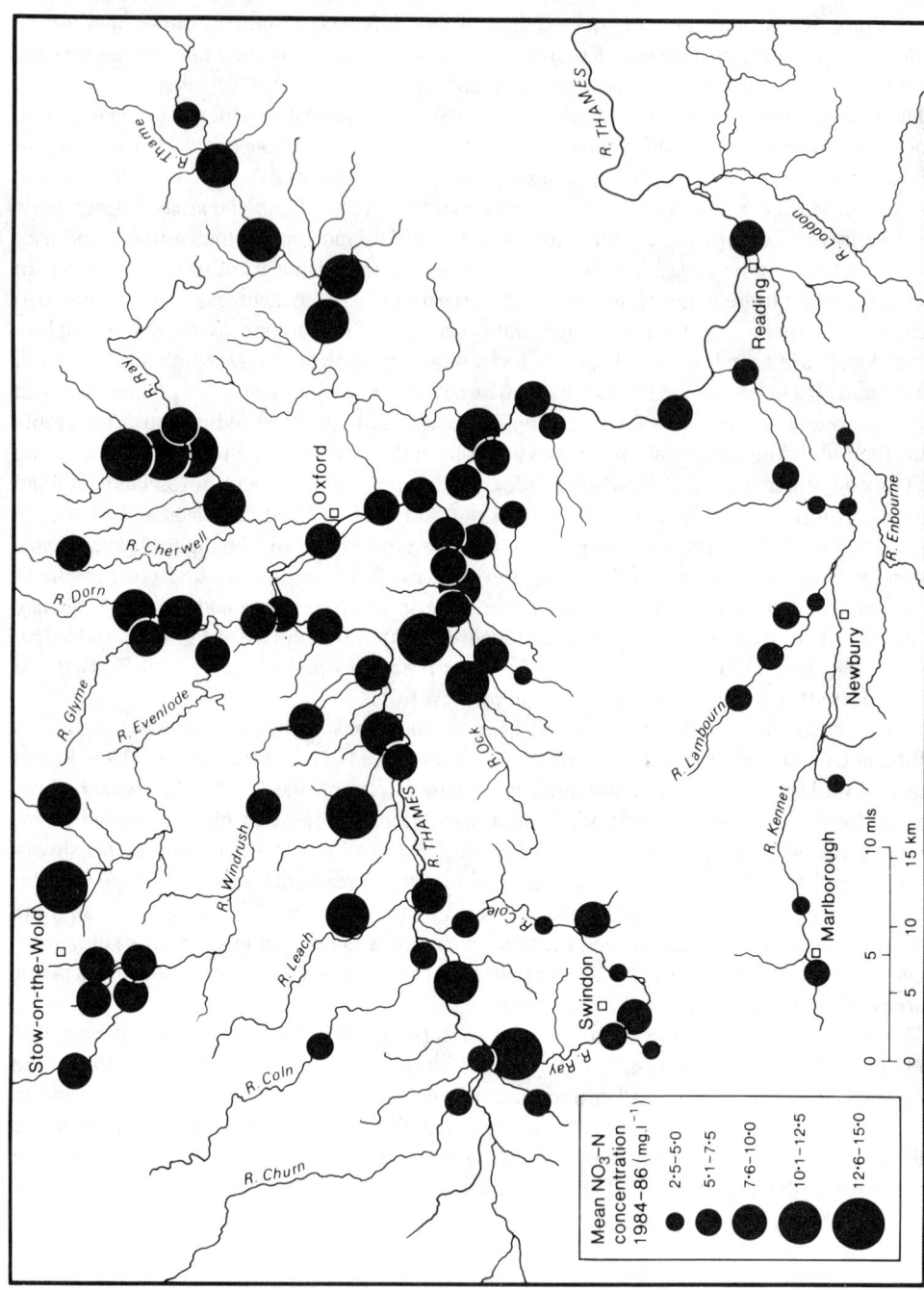

Figure 10.5 Spatial patterns in nitrate concentrations in the upper Thames basin (data provided by the former Thames Water Authority)

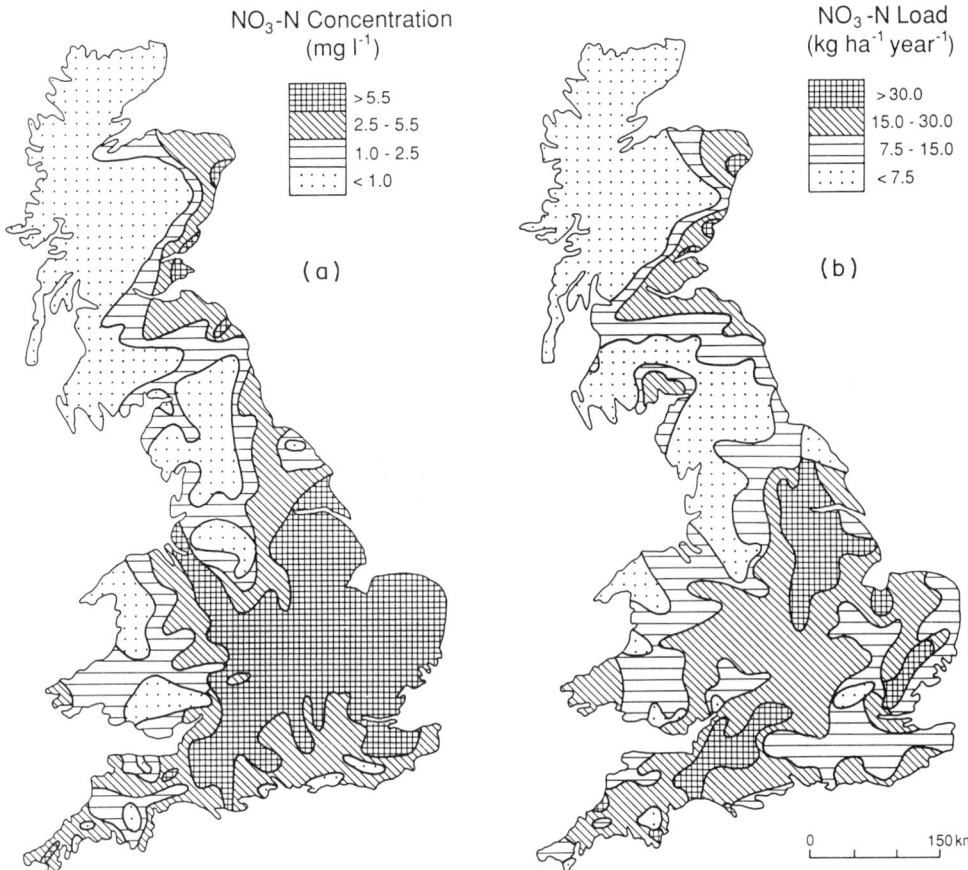

Figure 10.6 (a) Mean nitrate concentrations and (b) mean nitrate loads in British rivers (from Betton, Webb and Walling, 1991, reproduced by kind permission of the authors)

10.3 TEMPORAL VARIATIONS

10.3.1 ANNUAL REGIME

A within-year cycle of nitrate concentration and load is evident for many rivers although, as Walling and Webb (1984) demonstrate for the Exe (see below), this need not be the case at all sites, even within the same basin. Figure 10.7 shows discharge and nitrate concentrations (daily values) at Worsham on the River Windrush, a tributary of the Thames, over two years (Johnes, 1990). The relationship between discharge and nitrate concentration is immediately evident from the general correspondence between the graphs ($r=0.59$), although over short periods discharge and nitrate patterns may differ widely. Peak nitrate concentrations occur in the winter months with higher concentrations in wetter winters when levels may exceed the EC limit for short periods. The smooth seasonal variation in nitrate concentration noted in Figure 10.7 is repeated in Figure 10.8, which shows nitrate concentrations (monthly samples) for the period April 1973 to August 1990 for the same site.

In studying such annual nitrate regimes, two approaches have been taken: to quantify the

286

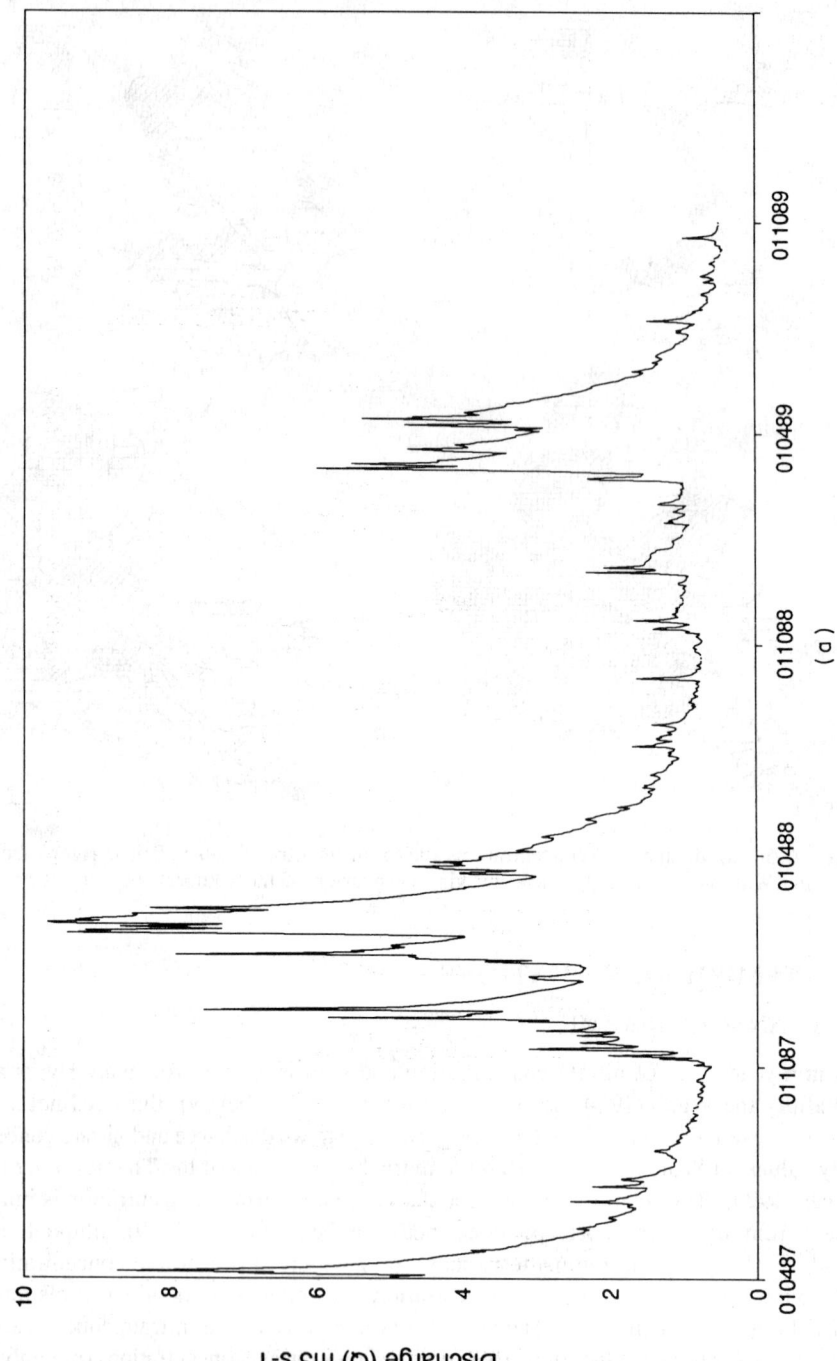

(a)

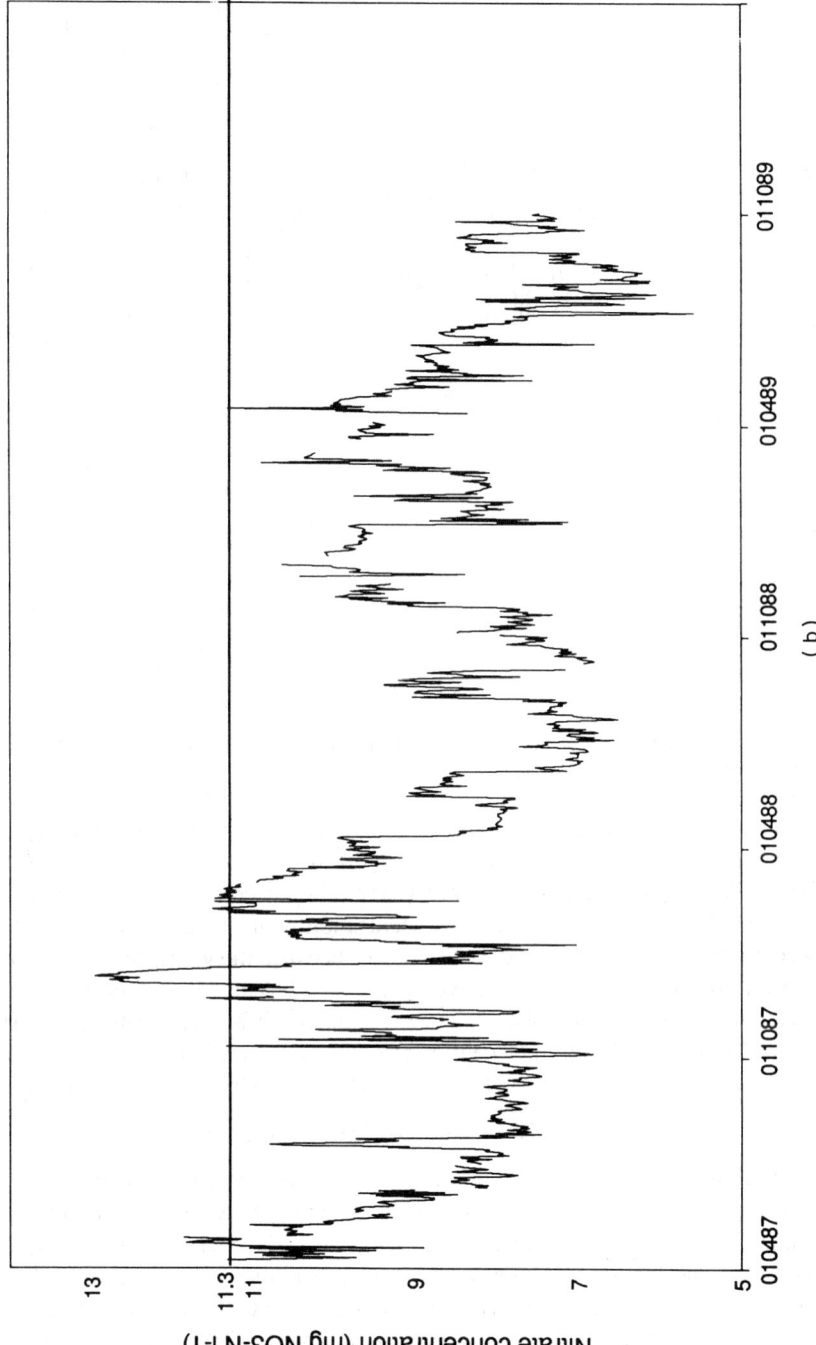

Figure 10.7 (a) Discharge and (b) nitrate concentrations at Worsham, 1987–9

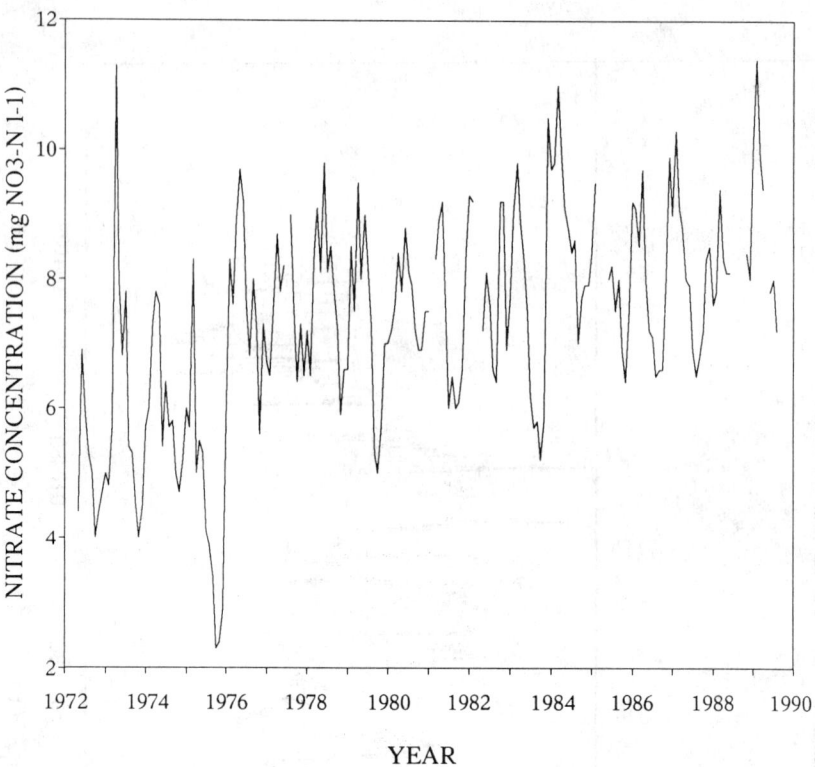

Figure 10.8 Monthly mean nitrate concentrations in the River Windrush 1973–90 (data provided by the former Thames Water Authority)

cyclical pattern, and to explain the regime. Casey and Clarke (1979) examined nitrate concentrations in weekly samples from the River Frome, Dorset, England, for an 11-year period (1965–75) inclusive; 51 samples per year). They used analysis of variance to see what proportions of the total variance in concentration could be attributed to differences between weeks (seasonal variation) and to differences between the years. Analysis of variance showed that differences between weeks (averaged over 11 years) accounted for 31% of the total variance, while 36% of the total variance was due to differences between years (each averaged over 51 weeks). Ninety per cent of the seasonal variation in weekly means (i.e. about 28% of the total variance in concentration) could be accounted for using a cosine index:

$$N_j = 2.752 + 0.47(\cos\ [2\pi(j-5)/51])$$

where N_j is the mean nitrate concentration in week j. Weekly means and the fitted cosine curve are shown in Figure 10.9; nitrate concentrations reach a peak in early February (week 5). Casey and Clarke then used multiple regression analysis in an attempt to combine the seasonal variation and long-term trends into a model of nitrate concentration for the River Frome. They assumed that the long-term trend was a continuous variable, quantifying the year variable exactly in terms of years (i) and fractions of years ($j/51$). The following equation explained 57.8% of the variance in nitrate concentration (individual samples):

$$N_{ij} = 2.12 + 0.109(i + j/51) + 0.485\cos\ [2\pi j - 5)/51]$$

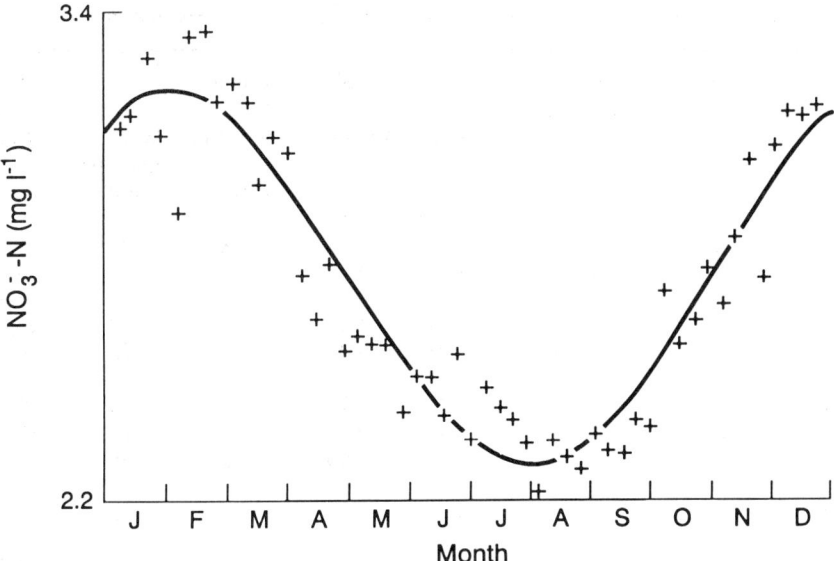

Figure 10.9 River Frome weekly nitrate concentration data and the fitted cosine curve (reproduced from Casey and Clarke, 1979, by permission of Blackwell Scientific Publications)

Because the two independent variables were not correlated, their joint effect in the multiple regression was roughly the sum of their individual effects. Adding discharge to the regression equation raised the level of explanation by only 4%. To test the hypothesis that the yearly cycle in nitrate concentrations could be explained by the variation in discharge alone, the seasonal cosine term was then excluded from the regression model; the level of explanation fell to 45%. This indicated to Casey and Clarke that although discharge was closely related to the seasonal variation in nitrate concentration when both were based on weekly data averaged over several years ($r = 0.86$), the individual discharges and nitrate concentrations were much less closely related, because both can vary considerably and independently over one week. Burt et al., (1988) mirrored Casey and Clarke's analysis for the Slapton Wood catchment. They also showed that seasonal variation in nitrate concentration could be satisfactorily described using a cosine index and, like Casey and Clarke, found (for weekly samples) that the correlation between nitrate and the cosine index ($r = 0.33$) was stronger than that between discharge and nitrate concentration ($r = 0.30$). Johnes (1990) found the converse when fitting cosine functions to the daily discharge and nitrate data for the River Windrush (Figure 10.8).

Walling and Webb (1984) fitted separate periodic functions, based on a Fourier series, to weekly sample data collected during 1980–83 from 15 monitoring stations in the Exe basin, south-west England, in order to isolate objectively and compare the nature of seasonal trends in nitrate concentration in different parts of the study basin. Results from fitting a first harmonic are given in Table 10.2 and a selection of the regimes is shown in Figure 10.10. Poor fits occurred at three sites: the Black Ball stream in the Exmoor upland where soil organic matter tends to be immobilised in peat soils so that nitrate levels in streams remain very low; and in the River Creedy and one other tributary where groundwater with a consistently high nitrate concentration mixes with quickflow of varying origin and concentration so that no clear cyclical variation results. The highest correlations and the most clearly defined annual regimes arise

Table 10.2   Results from fitting harmonics to River Exe data (adapted from Walling and Webb, 1984).

River	Drainage area (km)	Mean NO_3-N conc. (mg l^{-1})	Single harmonic				Two harmonics
			$R(\%)$	Amplitude	Week of max. con- centration	Week of min. con- centration	$R(\%)$
Black Ball	2.1	0.05	0.86*	0.01	8	34	1.77
Barle	128.0	0.87	18.44	0.17	49	23	25.20[b]
Exe (Pixton)	160.0	1.42	31.96	0.25	1	27	35.62[b]
Batherm	64.5	2.72	38.91	0.41	1	27	40.70[b]
Iron Mill	33.5	2.28	35.44	0.42	48	22	36.07
Exe (Stoodleigh)	422.0	1.50	34.73	0.28	3	29	39.73[b]
Lowman	49.1	4.39	15.95	0.45	1	27	17.91
Dart	46.0	2.55	42.87	1.06	52	26	44.73[b]
Exe (Thorverton)	601.0	2.34	9.34	0.21	6	32	13.15[b]
Culm (Woodmill)	226.0	4.99	16.59	0.54	1	27	20.43[b]
Culm (Rewe)	273.0	5.04	30.75	0.89	52	26	34.25
Jackmoor (Yendacott)	1.6	5.86	25.22	1.53	3	29	29.80[b]
Jackmoor (Pynes Cottage)	9.8	7.41	0.02[a]	0.02	27	1	0.50
Creedy	262.0	4.75	1.60[a]	1.40	46	20	2.51
Clyst	98.2	6.56	8.17	0.54	47	21	11.0[b]

[a] Single harmonic function is not statistically significant.
[b] Statistically significant improvement compared to single harmonic.

in tributaries such as the River Culm with more flow variability than those receiving substantial groundwater inputs.

Omitting the three sites where the harmonic function was not statistically significant, a clear trend of increasing amplitude with greater mean nitrate concentration was apparent. Addition of a second harmonic showed that for most stations the annual nitrate regime did not have a perfectly symmetrical sinusoidal form, but that the annual minimum and maximum occurred 4−6 weeks later and 2−3 weeks earlier than the timing suggested by the single harmonic. Betton (1990) extended this approach to some 900 stations in Great Britain. She found that second-order harmonics tended to fit best in lowland basins, especially in the drier east, where climate and geology cause the time of occurrence of runoff extremes to become later compared to the much wetter uplands to the north and west. January was the most likely month in which the second-order harmonic would possess a maximum and August the most likely month to possess a minimum and, as expected, eastern sites tended to have later peaks and more asymmetrical nitrate regimes (see also Betton, Webb and Walling, 1991).

Webb and Walling (1985) describe the annual nitrate regime of the River Dart, a tributary of the Exe. Hourly values of stream nitrate were obtained over an 8-year study period (1975−83). Mean daily concentrations showed an asymmetrical regime with a maximum in December and a minimum in early September. The autumn period was of particular note in that the range of daily mean concentrations was at its largest. Very high levels may occur

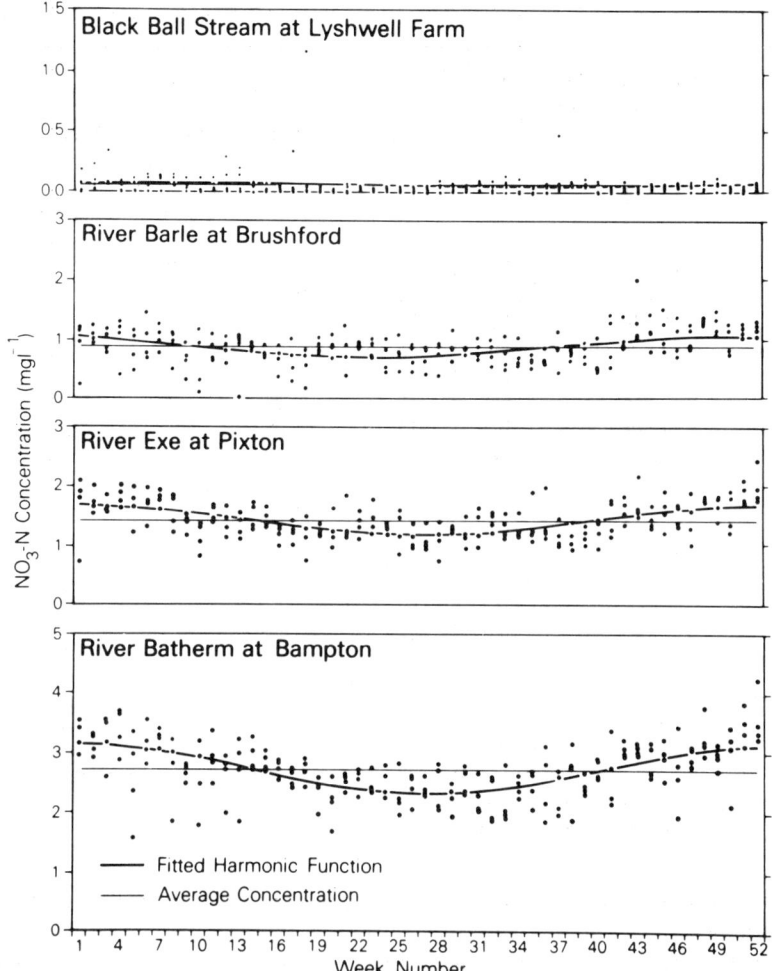

Figure 10.10 Selection of nitrate regimes for tributaries of the River Exe (from Walling and Webb, 1984, reproduced by kind permission of the authors)

if a wet autumn follows a dry summer. Sudden increases in stream nitrate are usually associated with storm events (see Section 10.3.2) and the term 'flushing effect' is sometimes used to describe the phenomenon (Walling and Foster, 1978). Similar results are shown in Figure 10.11 for the Slapton Wood catchment. Over the 20 years of observation, nitrate levels have been least variable in the spring and summer. In the autumn, positively skewed distributions confirm the occurrence of high nitrate concentrations in some years. Unlike concentration, the dominant control of nitrate load is discharge. Given that both concentration and discharge increase during the winter months, it is not surprising that there is such a clear annual regime for nitrate load. Total nitrate losses are strongly seasonal, with 80% of the load exported in December to February inclusive (Table 10.3).

Burt *et al.* (1988) describe monthly nitrate loads for the Slapton Wood stream for the period October 1974 to 1977, which included a major drought (May 1975 to August 1976). Throughout the drought, nitrate concentrations were little different from normal; nitrate loads

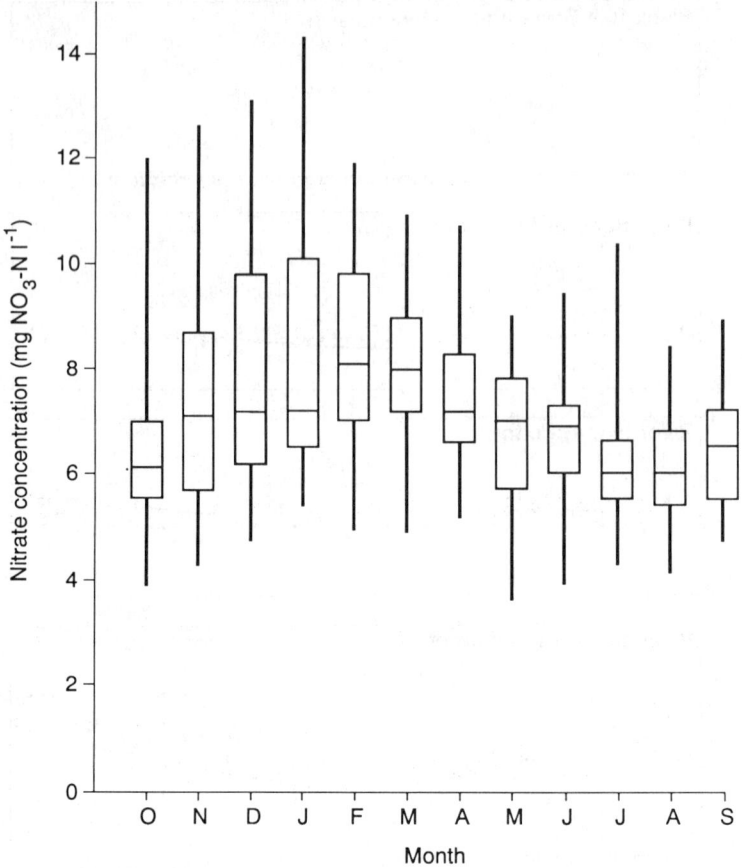

Figure 10.11 Box-whisker plots of monthly mean nitrate concentrations for the Slapton Wood catchment, 1970–90. Boxplots show the median, interquartile range and range for each month ($n = 20$)

were very low during the 1975/6 winter because stream discharge remained low through lack of soil moisture recharge. However, at the end of the drought, the very wet 1976/7 winter raised both discharge and nitrate concentrations to record levels, and consequently the nitrate loads were very large. Many (though not all) rivers in lowland England showed similar, dramatic responses in nitrate concentration and load after the 1976 drought (Betton, 1990). Recent droughts at the end of the 1980s have been followed by similar responses in stream nitrate (Haycock, 1991).

There are many reasons why the pattern of nitrate concentrations in so many rivers might display a marked seasonal regime. Enhanced leaching of nitrate in the autumn and winter provides the basic cause, though this in itself does not necessarily yield a smooth seasonal regime of the type described above (Figures 10.8–10.11). Even though there may be little inorganic nitrogen in the soil at harvest, rates of mineralisation are often high in the autumn when warm soils are rewetted (Section 12.1; see also Trudgill *et al.*, 1991). Micro-organisms are no longer isolated from decomposable substrate as larger pores in the soil begin to fill with water. Percolation also recommences as soils become wetter since the hydraulic conductivity of a soil depends in a highly non-linear manner on its water content. After an

Table 10.3 Nitrate losses from the Slapton Wood catchment, 1983—4

	Month	Rainfall (mm)	Runoff ($\times$ 1000 m³)	Mean $NO_3 - N$ concentration (mg l⁻¹)	Nitrate load (kg $NO_3 - N$)
1983	October	75	11.28	6.37	71.86
	November	69	11.83	7.00	82.25
	December	117	77.99	7.75	604.72
1984	January	243	172.93	7.79	1346.74
	February	70	97.79	9.15	894.58
	March	65	27.24	8.29	225.71
	April	8	23.18	7.05	163.44
	May	62	15.19	6.62	100.63
	June	9	7.90	6.19	48.88
	July	34	5.17	5.86	30.28
	August	54	4.96	6.01	29.83
	September	89	3.90	7.17	27.95
	Totals	895	459.36	* (flow-weighted mean = 7.90)	3626.90

extended drought, the amount of organic nitrogen mineralised may be great, since large amounts of labile organic matter may have built up when the soils dried out, with no opportunity in the intervening period for mineralisation, uptake or leaching to have taken place.

A smooth nitrate regime is best developed in catchments with deeper soils or aquifers, or both, so that some delay and attenuation occurs between leaching of nitrate from the soil in autumn and the time of peak nitrate concentration (and groundwater discharge) in midwinter. Deep soil and groundwater systems seem, therefore, to act as well-mixed reservoirs, providing a buffer between leaching from the near-surface soil and the stream (Figure 8.12). Catchments which display such regimes include the Slapton Wood catchment, where strong seasonality of subsurface runoff through the deep soil causes nitrate concentration to reach a peak in February, with 80% of the annual nitrate loss occurring in winter (Table 10.3) and groundwater-fed basins like the Windrush (Figure 10.8), Frome (Figure 10.9) and Leach (Section 12.4), where drainage through aquifers also produces a midwinter peak in nitrate concentration and load. Basins in generally drier areas such as the east of England are also included where soils require prolonged recharge in the autumn and early winter before drainage (and leaching) begins (Betton, Webb and Walling, 1991). One cause of a smooth annual regime may be uptake of nitrate by aquatic plants, causing a further significant decrease in stream nitrate concentration in the summer (see Chapter 12).

Some streams do *not* display clear annual nitrate regimes; this may be for a variety of reasons. Generally, sewage effluent forms a small part of the nitrate load in rural basins, but may constitute a large part of that load in urbanised basins, particularly in the summer, when non-point delivery of nitrate to surface waters is lowest. In addition, Walling and Webb (1985) demonstrate a peak in the nitrate concentration of sewage effluent in the summer. In soils where water is rapidly conducted into the stream (for example, in clay soils with underdrainage), peak nitrate concentrations occur early in autumn and decline steadily thereafter throughout the winter (Haigh and White, 1986); no clear annual regime for nitrate occurs in such basins either (Roberts and Marsh, 1987). In upland areas, little nitrate is leached from the permanently waterlogged, organic soils and no annual nitrate cycle is detected in flowing waters (Walling and Webb, 1985; Roberts, Hudson and Roberts, 1989). However, upland lakes very frequently show a winter nitrate peak and a summer trough. In temperate lakes, this pattern is particularly marked because nutrient supplies built up in the winter are depleted during the growth season, though inputs of combined nitrogen may occur throughout this period as a result of nitrogen fixation by heterocyctous cyanobacteria (Moss *et al.*, 1989).

Strong seasonality of the flow regime seems to aid the annual nitrate cycle; in basins with ample summer rainfall, especially those where baseflow is greatly diluted by surface runoff during storm events (e.g. River Creedy, Table 10.2), nitrate concentrations may vary more during storm events than from season to season. Nevertheless, the annual regime does seem to be a feature in many rivers of lowland England, even in larger basins of mixed geology (and therefore of spatially varying hydrology). However, it should be noted that the regime is often best described using monthly values; where sampling is frequent and averages are not used, a large scatter of nitrate concentrations may render the cycle less apparent. To an extent, therefore, the nitrate regime must be regarded as a smoothed seasonal characteristic which may not necessarily be detectable within a large scatter of individual sample values.

10.3.2 EPISODIC VARIATIONS IN FLOWING WATERS

The influence of hillslope hydrology on nitrate concentrations was discussed in some detail in Chapter 9. There it was concluded that the pathway by which water reaches the stream channel will largely determine its nitrate concentration. Significant fluctuations in stream nitrate concentration are thus likely during storm events when runoff from different source areas and generated by different runoff mechanisms mixes in the stream. The range of variation during storms may exceed that which occurs seasonally.

Intensive sampling during periods of storm runoff has not only highlighted the complexity of solute behaviour during storms at particular stations but also demonstrated considerable within- and between-catchment variability (Walling and Foster, 1975; Walling and Webb, 1984; Webb and Walling, 1985). This is illustrated by the changes in the nitrate concentration shown in Figure 10.12. An increase in nitrate concentration is typical of summer and autumn storms (Figure 10.12); often this peak concentration lags behind the discharge peak. Since there is little surface runoff at this time, 'old' soil water provides much of the storm runoff and no dilution effect occurs (Anderson and Burt, 1982). In any case, the build-up of dry deposition and dead plant material in a dry spell may mean that, unusually, overland flow will also have a high solute concentration. Delayed peaks in nitrate concentration during the hydrograph recession indicate subsurface soil horizons quickly acquires a high nitrate concentration as available nitrate is readily leached; this nitrate may have already been present in the soil before the storm, or resulted from mineralisation as the soil wetted up. Flushing events are particularly pronounced in the autumn storms after prolonged drought (Walling and Foster, 1978; Webb and Walling, 1985; Burt *et al.*, 1988; Haycock, 1991).

A rapid decrease in nitrate concentration is characteristic of flood events during winter and spring (Figure 10.12(b)). However, even in one small catchment there may be considerable variation in the timing and magnitude of the dilution, depending on the intensity of precipitation and antecedent climatic conditions (Webb and Walling, 1985). The dilution may 'lead' the flood peak where large quantities of dilute overland flow rapidly enter the stream; slower flow through the soil begins to raise nitrate concentrations as the peak discharge is reached (Anderson and Burt, 1982). The dilution may 'lag' behind the flood peak if soil nitrate is

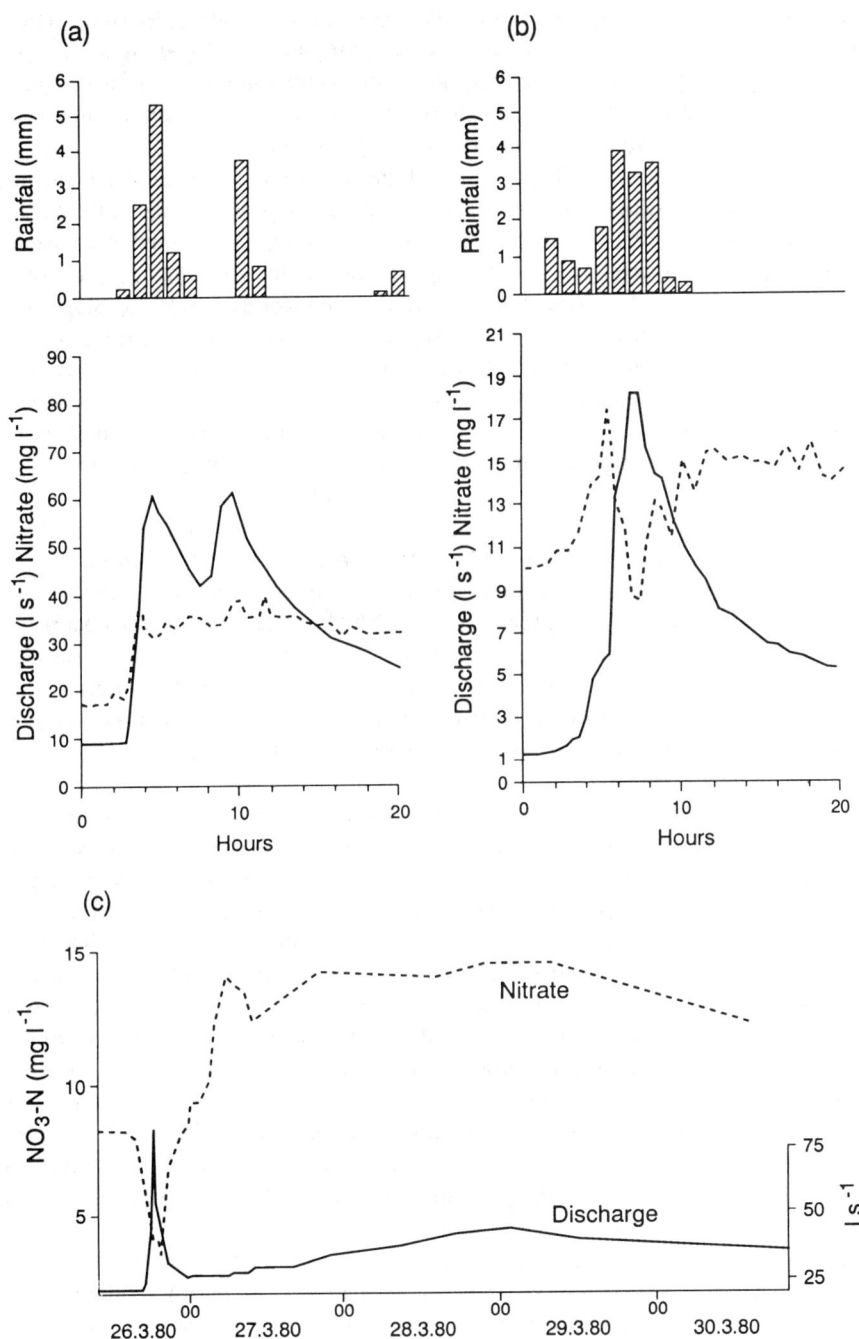

Figure 10.12 Nitrate concentrations during storm discharge. (a) and (b) Wytham stream, Oxford; (c) Slapton Wood stream, Devon

initially available but then becomes depleted so that later subsurface flow has a lower concentration than that leached early in the event (Walling and Foster, 1975). Summarising storm behaviour over an 8-year period for the River Dart, Webb and Walling (1985) showed that 'concentration' events were most likely in summer and autumn when subsurface flow dominates the storm runoff response. Dilution events were rare in summer but very much more common than concentration events from October to March.

In winter an important variant of the delayed flushing effect is seen in some catchments. A delayed hydrograph may occur several days after the initial peak in stream discharge. This delayed response is much more gradual than the first one and high flows may last for as long as two weeks. The hydrological mechanisms generating these delayed hydrographs have been fully described in Burt and Butcher (1985a,b). A typical event for the Slapton Wood catchment is shown in Figure 10.12(c). In some hydrographs (as here) there is a concurrent increase in discharge and nitrate concentration; in others the increase in nitrate concentration lags behind the discharge response. In either case the normal pattern is for concentrations to increase well beyond the value that existed before the storm so that the highest nitrate concentrations are associated with such delayed peaks in stream discharge. Since both discharge and nitrate concentration are high at such times, the total load removed is very large. In the 1983/4 winter, which was drier than normal, four such delayed hydrographs accounted for 70% of the total nitrate load in the Slapton Wood stream from December to February inclusive. This was 54% of the total nitrate load for the 1984 water year (Burt and Arkell, 1987). Haycock (1991) has demonstrated that delayed hydrographs which occur in the River Leach are also accompanied by large increases in nitrate concentration (see Section 12.4). At Slapton, throughflow in deep soil layers is responsible for the delayed discharge; in the Leach, drainage of a shallow aquifer generates the delayed response. In both cases, such delayed hydrographs are clearly the major mechanism of nitrate leaching both in terms of total nitrate load and of peak nitrate concentration.

The positive relationship between discharge and nitrate concentration, seen for many rivers (e.g. Figure 10.7), allows concentration to be adequately predicted from discharge using a simple regression equation (Section 10.1.5). However, as examples shown in Figure 10.12 indicate, the fact that nitrate concentrations may be markedly different at the same discharge on the rising and falling limbs of the flood hydrograph severely limits the predictive power of simple regression relations between discharge and concentration. This hysteresis phenomenon has often been investigated using plots of discharge against concentration for individual storm events. Simple clockwise and anticlockwise loops have been recognised and a wide range of causes have been proposed to account for this phenomenon (Walling and Webb, 1985). Anticlockwise loops are perhaps most common for nitrate given the important contribution of subsurface flow to the recession limb of the flood hydrograph. A systematic shift in the position of the hysteresis loop shows inter-storm variability: the shift may indicate progressive exhaustion of nitrate during a winter period (Burt et al., 1983) or the gradual development of the flushing effect in early winter.

In the long term, if linear trends and seasonal components are abstracted from the data. A fluctuating data series remains which may be, at one extreme, purely random, or, at the other, a smooth oscillatory movement. These oscillations are often related to episodes of disturbance within the catchment from which it gradually recovers. In itself, a major drought has surprisingly little effect on stream nitrate concentrations, though loads may be severely reduced through lack of discharge. However, after a major drought, considerable losses of nitrate can lead to very high concentrations of nitrate, often in excess of legal limits, causing difficulties for water supply companies. This was particularly the case after the 1975/6 drought in Britain (Burfield, 1977; Walling and Foster, 1978; Slack and Williams, 1985), but similar episodes have happened after more recent droughts (Figure 10.8). The large loss of nitrate may relate to lack of nitrate leaching during the drought, accumulations building up in the soil as a result. More likely, however, is that much nitrate is mineralised from stores of labile organic nitrogen, which have built up during the drought, when warm soils begin to wet up. High concentrations of nitrate are often observed in autumn, but this may be very protracted after a drought. Burt *et al.* (1988) showed that the Slapton Wood catchment exhibited a strong 'memory' with dry years followed by greater leaching in subsequent years, while wet years exhaust the system somewhat so that concentrations will be reduced in future years. This lag effect might relate both to mineralisation processes and to the time taken for subsurface water to move to the river.

A sudden and significant change in land use affecting a considerable area of the basin may also cause nitrate concentrations to increase markedly. This effect is distinct from gradual changes in land use which cause trends rather than sudden changes in nitrate concentrations (see Section 10.3.3). Hornung (1988) discusses non-agricultural sources of nitrate, emphasising clear-felling of forests and ploughing of moorland as important causes of major episodes of nitrate loss in the UK. Vitousek (1981) predicts excess mineralisation after clear-felling, the amounts varying with site fertility and felling practice. Whether the excess mineralisation leads to nitrate leaching depends on the supply of water. Several studies in the UK have shown greatly increased nitrate leaching following felling (Adamson *et al.*, 1987; Stevens and Hornung, 1988). Ploughing and drainage improvement schemes in the uplands may cause similar effects, depending on the degree of cultivation. Roberts, Hudson and Roberts (1989) showed that nitrate concentrations in an upland basin, normally very low, rose above EC limits for a brief period after tile-drainage and disc harrowing and remained higher than normal even after three years. Sustained losses in excess of 20 kg ha^{-1} a^{-1} nitrogen from partially improved upland pasture approach those for intensively farmed lowland basins; such losses could pose problems for eutrophication of upland reservoirs. The sudden change from grassland to arable which took place on the Chalk downlands of southern England during and after the Second World War might also be expected to produce a similar significant increase in nitrate levels in local rivers. However, the pulse of nitrate produced by ploughing is likely to be greatly attenuated by a deep aquifer like the Chalk so that the effect will be a gradual rather than a sudden increase in nitrate concentration (see Figure 8.9). In most cases, land use has intensified gradually within the basin; the result is a gradual upward trend in nitrate concentrations and loads. Such trends form the subject of the next section.

10.3.3 LONG-TERM TRENDS

The Royal Society (1983) notes that while UK rivers are among the cleanest in Europe, their nitrate levels have risen by 50–400% over the past 20 years. This has been particularly marked in the Midlands and in south-east England: a number of rivers in these areas are currently close to (and occasionally exceed) the World Health Organisation (WHO) recommendation of 11.3 mg $NO_3 - N$ l^{-1} (Figure 10.6). Publications on UK water quality trends date from the adoption of the WHO recommended limit in the early 1970s (e.g. Tomlinson, 1970). Recent publications from UNESCO (Meybeck, Chapman and Helman, 1989), José (1989) and Roberts and Marsh (1987) have provided detail of the global, national and European context of water quality change. All these authors stress the problems of comparability of data as a result of the use of different analytical methods. There can be difficulties of comparison if methods differ between sites or if different methods are used at the same site at different times. This needs to be taken into account in any analysis of such trends: the figures available should be used as indicators rather than absolute levels of water quality. The majority of publications focus on nitrate rather than total nitrogen or nitrogen speciation trends, largely as a result of the mobility of nitrate, concerns over health impacts, and the gradual increase in mandatory control of the nitrate content of drinking water and drinking-water sources.

Global freshwater quality has been reviewed in detail in a recent UNESCO report (Meybeck, Chapman and Helman, 1989), with an analysis presnted for nitrate essentially based on the GEMS/WATER database, and on data extracted from OECD reports (1985, 1987). A review is also presented of research findings for a range of world rivers. General trends in the nitrate content of a variety of national surface waters from this latter data bank are presented in Table 10.4 (from Meybeck, Chapman and Helman, 1989). This illustrates the variability of time trends at a global scale.

Table 10.4 Nitrate content of global waters (reproduced by permission from Meybeck, Chapman and Helman, 1989)

Country	River	1970	1975	1980	1985
Belgium	Meuse, Heer/Agimont	1.800[a]	7.800	2.180	3.120
	Meuse, Lanaye	3.900	9.400	2.520	2.790
	Escauf, Doel	3.000[a]	7.350	4.170	3.910
Canada	St Lawrence	0.193	0.230	0.160	0.210[b]
	Mackenzie	0.084	0.111	0.110	0.090[c]
	Fraser	0.049	0.300	0.060	0.120[c]
	Nelson	0.040	0.400	0.060	0.090[c]
France	Loire, Nantes	1.581[a]	1.445	1.987	
	Garonne, Bordeaux	1.152[a]	0.926	1.829	
Germany	Rhine, Bimmen L.	1.820	3.020	3.590	4.200
Italy	Po	0.946	1.350	1.630	3.280
	Tevere		1.500	1.370	
Japan	Ishikari	0.360		0.530[d]	
	Yodo			0.760[d]	
Netherlands	Meuse, Keizersv	3.070	3.690	3.770	4.280
	Meuse, Eijsden	2.450	2.510	2.780	2.920
	Ijssel, Kampen	2.760	3.460	4.270	4.330[e]
Sweden	Dalälven	0.120	0.107	0.136	0.106
USA	Delaware, Trenton		0.880[f]	1.080[f]	
	Mississippi, St Franc	0.360[f]	1.040[f]	1.300[f]	1.230[ef]

[a] 1971 data.
[b] 1984 data.
[c] 1983 data.
[d] 1979 data.
[e] 1983 data.
[f] Total concentrations.
Source: Adapted from OECD (1985, 1987).

Two useful reviews of water quality in the USA have been conducted by Smith and Alexander (1985) for 298 USA rivers based on the national USA river monitoring network (NASQC), and specifically for the Passiac River presented by Cirello *et al.* (1979). The latter also includes an assessment of ammonium concentrations over a 50-year period. The data presented by Smith and Alexander (1985) indicate a widespread increase in surface water nitrate concentrations throughout the USA, particularly in the drier eastern states, and in the Pacific coastal areas. The only rivers to show a decrease in nitrate concentration were those draining grassland catchments, again highlighting the importance of arable land and the ploughing up of pasture lands as a source of nitrate in surface waters. Cirello *et al.* (1979) concur with this trend; their survey of one 30-year record and five 13-year records for stations on the Passiac River indicates an increase in nitrate and ammonium concentrations at each site related to land-use intensification.

A review of the literature for European rivers reveals a general increase in nitrate concentrations over time. Wet mountainous regions (e.g. Scandinavia — Ryding and Forsberg, 1979; and the Alps — Meybeck, Chapman and Helman 1989) in general exhibit lower nitrate concentrations than lowland Europe. Distinct increases in nitrate concentrations are noted wherever there are significant agricultural inputs from a catchment. Examples of such trends published are presented in Figure 10.13. The lake data, although with significantly lower concentrations than those observed for river systems, nevertheless exhibit a distinct upward trend, with a doubling in nitrate concentration observed over the 24-year period of study. Data presented for the Rhine by Roberts and Marsh (1987) also indicate a doubling in nitrate concentration, but over a shorter time period of 20 years. However, they note that this increase has not been at a uniform rate throughout the Rhine drainage basin. Below Basle where the data for these analyses have been collected, the Rhine is dominated by sewage rather than agricultural sources. In the upper catchment regions dominated by viticulture and horticulture, the concentrations of nitrate are universally higher than those of the main Rhine below Basle, and have exhibited more rapid increases in nitrate concentration (Roberts and Marsh, 1987). This shows the importance of agriculture as a source of surface water nitrate. It also indicates the extent to which data for large rivers mask the true extent of inputs from tributaries and groundwater, and highlights the need for spatial analysis of both basin nutrient sources and water quality (e.g. Figures 10.2 and 12.8).

Further nitrate trends have been analysed for the River Elbe in Czechoslovakia (Paces, 1982), for the River Maas in The Netherlands (Roberts and Marsh, 1987) and for Finnish rivers (Kauppi, 1984) and French surface waters (Henin, 1986). All the data sets analysed show comparable and significant increases in nitrate loading over the past 20 years. The analysis by Paces (1982) covers the period 1877–1976 for the River Elbe, and shows a thirtyfold increase in ammonia and a twentyfold increase in nitrate concentrations. Analyses conducted by a number of authors (e.g. Kauppi, 1984; Roberts and Marsh, 1987) indicate a strong correlation with intensification of agricultural production over the same time period, and particularly with the increased use of nitrogen fertilisers throughout Europe.

A wide range of nitrate data are available for British rivers. Publications commence with that of Tomlinson (1970) for 18 rivers in England and Wales, leading on to the more recent reviews of José (1989) for the River Trent which also contains a valuable comparison with other British river trends, and by Betton (1990) for 743 sites (see also Betton, Webb and Walling, 1991). Other valuable analyses include those of Nicholson (1979), Marsh (1980), Rodda and Jones (1983), and Roberts and Marsh (1987) for the longer time trends available for British rivers. The longest records of water quality are available for the River Thames

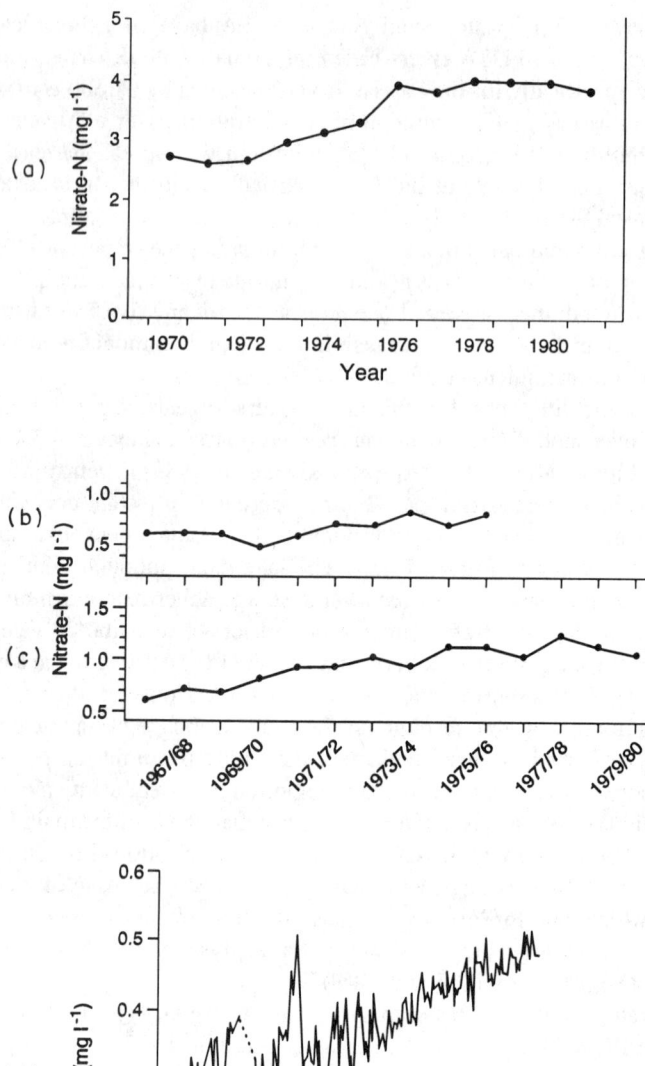

Figure 10.13 Long-term trends in nitrate concentration in (a) the River Rhine, (b) the River Skjern, (c) the River Karup and (d) Lake Geneva (compiled from Marsh, 1980; Dykseul, 1982; Meybeck, Chapman and Helman, 1989)

at Walton and the River Lee at Chingford, with nitrate concentration data for both rivers presented in the *Digest of Environmental Statistics* (DoE, 1978). The nitrate trend for the Thames has been reproduced by a number of authors (e.g. Onstad and Blake, 1980; Royal Society, 1983; Roberts and Marsh, 1987). The analysis conducted by Meybeck, Chapman and Helman (1989) on the Thames nitrate record indicates a slight but significant increase in nitrate concentrations in the period 1975–86, with a pronounced significant increase for the period 1928–78. Long-term trends for the River Stour at Langham, the River Tees at Broken Scar and the River Great Ouse at Bedford are illustrated with the Thames record in Figure 10.14.

Analyses of the Great Ouse record (Greene, 1978; Nicholson, 1979; Marsh 1980; Roberts and Marsh, 1987) and those of the Chelmer and Stour (Burfield, 1977; Greene, 1978; Nicholson, 1979; Slack and Williams, 1985) and Lee (DoE, 1978; Nicholson, 1979) indicate that for the south and east of England, particularly in East Anglia, there has been a significant and rapid increase in nitrate content since the 1960s, with many East Anglian water bodies now over the EC/WHO nitrate limit of 11.3 mg NO_3-N l^{-1}. Slack and Williams (1985) suggest that there has been no significant increase in the nitrate content of the River Chelmer in the period 1977–84, but do note that the 1984 figure was the highest recorded since 1977. However, the national review conducted by STACWQ (1983) on 25 British rivers monitored at water supply abstraction points and including the River Chelmer indicate that the rates of increase recorded for the period 1981–4 were greater than in any previous period. It should be noted that conclusions based on short records of only a few years may be suspect, since climatic variations may mask underlying trends related to land-use changes.

Analysis of the long-term nitrate levels in the River Windrush conducted by Johnes (1990) also confirms an upward trend in both concentration (Figure 10.7) and load. The mean annual nitrate concentration in this river has increased from 5.12 mg NO_3-N l^{-1} in the water year 1973/4 to 9.37 mg NO_3-N l^{-1} in 1989/90, with a mean annual of increase in nitrate concentration of 0.24 mg NO_3-N l^{-1} per annum. The trend of increasing nitrate concentration may be extrapolated to estimate future nitrate levels in the River Windrush,

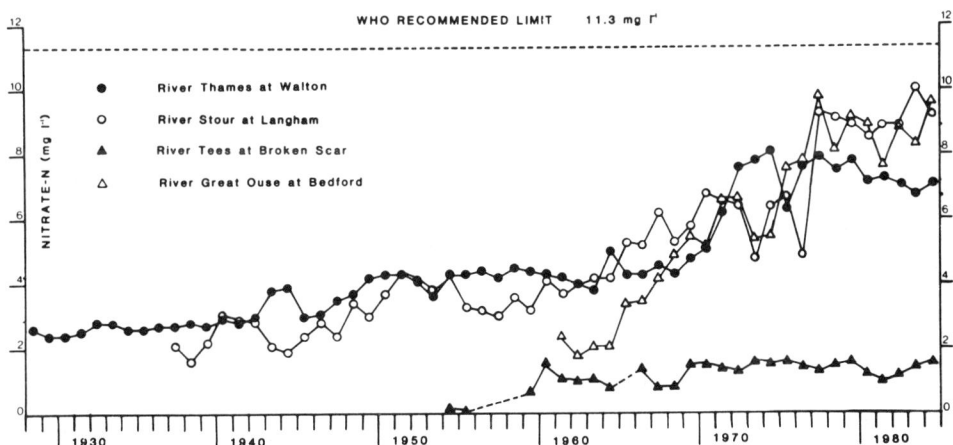

Figure 10.14 Long-term trends in nitrate concentration in the River Stour, River Tees, River Great Ouse and River Thames, UK (reproduced from Roberts and Marsh, 1989, by kind permission of the IAHS Press, Wallingford, UK)

Table 10.5 Estimated frequency of occurrence of nitrate concentrations exceeding the EC limit in the River Windrush

Year	Mean	Probability of exceedence (%)	Number of days that EC limit is exceeded
1985	8.3	2.27	8
1993	10.3	25.46	93
2001	12.3	74.54	272
2009	14.3	97.73	357

Note:
Assuming an increase of 0.24 mg l^{-1} a^{-1} from 1985. Mean $NO_3 - N$ concentration in 1985 = 8.3; standard deviation = 1.5. See also Figure 10.15.

and the frequency with which nitrate concentrations in the river will exceed the EC limit of 11.3 mg $NO_3 - N$ l^{-1}. The results of this analysis are presented in Table 10.5 and are illustrated in Figure 10.15. Clearly, if present upward trends in nitrate concentrations continue, then mean concentrations in the River Windrush will exceed the EC limit by the year 2000.

While the majority of the upward trend in nitrate concentrations in the River Windrush are due to changes in the availability of nitrogen sources in the catchment, a significant part of the long-term trend is also due to climatic variation in this basin, with a series of dry years being followed by a series of wetter ones in the period 1973 to 1990 (Johnes, 1990). Burt *et al.* (1988) found similar trends in the nitrate record for the Slapton Wood catchment (see section 10.4.2). Further alalysis of the Windrush record is presented in Section 10.4.3 with respect to changes in land use and farming practice in this predominantly rural catchment; extrapolation of land-use trends in this catchment suggests that mean annual nitrate concentrations will not exceed the EC limit until the year 2045.

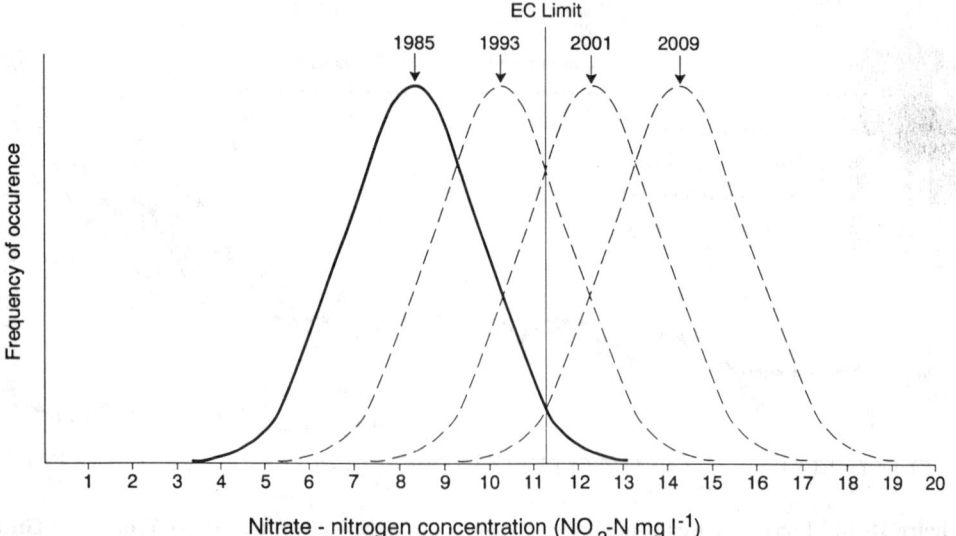

Figure 10.15 Estimated frequency of occurrence of nitrate concentrations exceeding the EC limit in the River Windrush

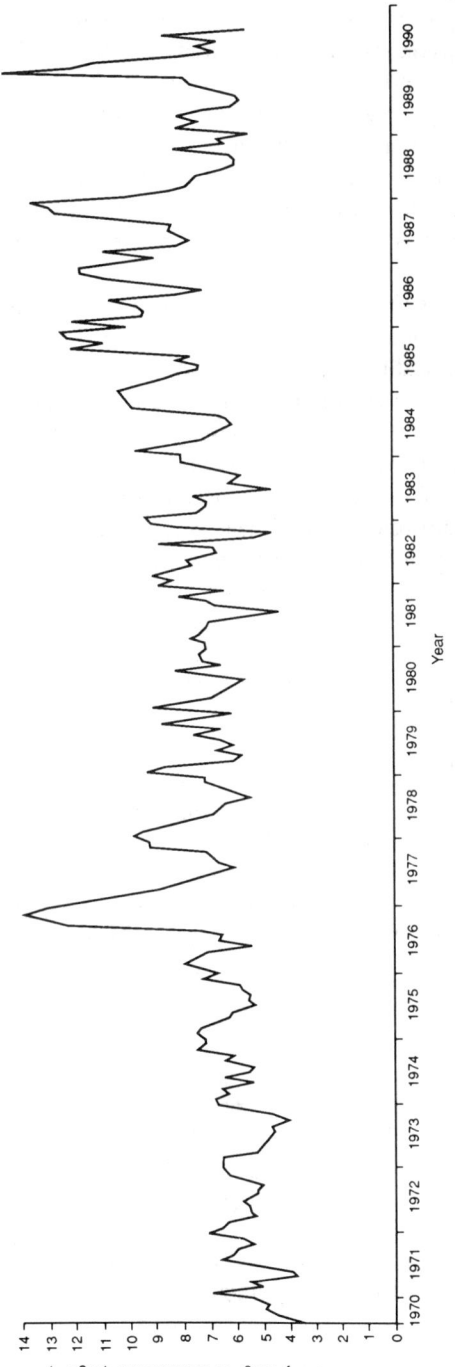

Figure 10.16 Long-term nitrate record for Slapton Wood, 1970–90

Long-term nitrate concentration data have been analysed for rivers in the south and west of England by Casey (1976) and Casey and Clarke (1979) for the River Frome, by Webb and Walling (1985) for the River Dart, and for the Slapton Wood stream by Burt *et al.* (1988; Figure 10.16). These analyses provide contradictory trends. The Frome and Slapton analyses indicate marked increases in nitrate concentrations for the periods of study. Casey and Clarke (1979) observe a 41% increase in nitrate concentration in the period 1965−72 for the Frome. Burt *et al.* (1988) also note a significant increase in nitrate concentrations, independent of climatic variation, which can be attributed to agricultural intensification (see Section 10.4.2). Webb and Walling (1985), however, state that no significant increase was observed for the River Dart for the period 1975−83; this may relate to the short period of study and to the unusual pattern of climatic variation during that time.

For the north-east of England, the reviews of nitrate and ammonia concentrations in the River Trent and its tributaries (Royal Society, 1983; José, 1989) also confirm an upward trend in nitrate concentration, but illustrate a decrease in ammonia concentration (see Figure 10.17). Two of the tributaries studied by José's analysis indicates increases in stream nitrate in all tributaries, but with a significant increase also observed for the rural streams. Overall, José calculates an annual increase of 455 tonnes per annum (0.43 kg ha^{-1} a^{-1}) in nitrogen loading on the River Trent. José concludes that the observed increases are comparable to those observed for European rivers. The decrease in ammonia concentration (Figure 10.17) is directly related to a reduction in the amount of effluent being discharged into the Trent.

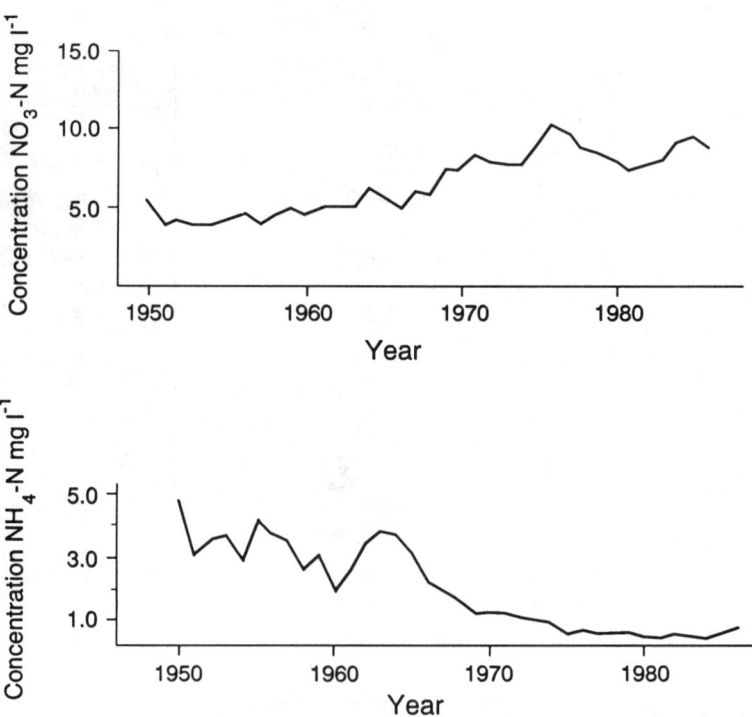

Figure 10.17 Long-term trends in nitrate and ammonia concentrations in the River Trent (reproduced by permission from José, 1989)

Analysis of the records for 12 Welsh rivers by Brooker and Johnson (1984) over the period 1969−79 also indicated no change in nitrate concentrations, which the authors contrast with the published records for English rivers. This can perhaps be attributed to the limited distribution of arable land in the catchments of the 12 rivers studied, and the relative stability of their human population levels. Few data have been analysed for Irish rivers, but those presented for six rivers draining into Lough Neagh from 1969 to 1979 by Smith (1977) and Smith *et al.* (1982), exhibit an increase in nitrate concentrations of 0.1 mg $NO_3 - N \, l^{-1}$ per annum over a 10-year period. These results are attributed to agricultural and sewage loading increases (José, 1989). Data presented by Toner and Lennox (1980) for Irish rivers over the period 1972−8 show a twofold increase over this period.

Such results clearly illustrate a general increase in nitrate concentrations in surface waters throughout Britain, Europe and the USA, with the only exceptions arising in predominantly grassland watersheds. Globally, nitrate concentrations are highest in the surface waters of industrialised nations, with no clear patterns apparent for developing countries at present. Current intensification of agricultural enterprises, together with very large population increases, suggest that careful monitoring is warranted for such countries.

10.4 MODELLING NITRATE IN RIVERS

10.4.1 CHOICE OF MODEL

Anderson and Burt (1985) have reviewed the types of model structure which are classically identified in hydrology. The same divisions may be adopted here: empirical, conceptual and physically based. Addiscott and Wagenet (1985) used a similar framework to review models of solute leaching in soils. Cutting across this choice is the decision about how to include spatial variations in the model. The most common simplification made in catchment modelling is to use a *lumped* or spatially averaged approach; here the catchment system, its inputs and response are represented mathematically using only the dimensions of depth and time (Blackie and Eeles, 1985). *Distributed* models are capable of forecasting the spatial pattern of hydrological conditions within a catchment as well as simple outflows and bulk storage volumes.

Among their other advantages, distributed models permit forecasting of the delivery of pollutants, such as nitrate, to be coupled to the movement of water through the catchment system and any changes taking place within a catchment (Beven, 1985). In most cases, nitrate modelling has involved the use of lumped models. Most progress in nitrate studies has been made in the modelling of nitrate leaching in soil profiles (Addiscott and Wagenet, 1985). With the exception of empirical models, far fewer attempts have been made to model nitrate transport at the catchment scale. The relatively complicated physically based models serve primarily as research tools, while the less demanding conceptual models offer the hope of being more widely applicable, particularly in the field of catchment management.

Empirical models
These contain no physically based transfer functions to relate input to output. Empirical models include those which establish a statistical correspondence between input and output and those which use simple coefficients to quantify outputs from different parts of the system. Such models may be highly successful despite their lack of physical basis, but it may be of dubious worth to try to extrapolate beyond the range of actual experience. The more successful

empirical models appear to be those which predict nitrogen loss to aquatic environments using coefficients for nitrogen loss derived from independent, process-based studies (Johnes, 1990; Section 10.4.3). Because of their minimal data requirements, empirical models have proved useful for management purposes.

Smith and Stewart (1989) note that two types of statistical model have been developed to forecast futured trends in nitrate concentration. The first type has simply aimed at determining the underlying trends after removing variation due to flow and seasonality. Examples of this type of approach include the studies of Warn and Page (1984) on the River Welland, and of Warn (1984) on the River Stour. The second type of model seeks to derive causal relationships between catchment condition and nitrate concentration in the river. Onstad and Blake (1980) derived a relationship between nitrate available for transport from the catchment and nitrate levels in the River Thames using time-series analysis over a 50-year period. They showed that 78% of the variation in annual mean nitrate concentration could be explained by changes in agricultural inputs. The model was then used to estimate future nitrate levels in the Thames under a variety of plausible agricultural trends to the year 2000. Their analysis showed that, even with the most optimistic assumptions, a strategy to deal with nitrate in potable water would be needed if the EC limit of 11.3 mg $NO_3 - N\ l^{-1}$ remained unchanged. Smith and Stewart (1989) developed a model based on multiple regression which explains 95% of the variation in nitrate loading on the major rivers in the Lough Neagh catchment for the years 1971–87. The model relates loading of nitrate to fertiliser use, previous summer rainfall, summer temperature, and flow in the previous winter. The model indicated that there was an increase in nitrate loading associated with fertiliser use, and that the equivalent of 13% of nitrogen fertiliser that is lost as leachate comprises 50% of the river loadings. Neither human population nor land-use change were significant contributors of nitrate in this catchment. Combining maximum fertiliser use and the 'worst-case' climatic scenario, the model estimates nitrate concentrations will increase by two to three times, though remaining well below EC limits. Other uses of multiple regression to model nitrate concentrations in rivers include Casey and Clarke (1979; Section 10.3.1) and Burt et al. (1988; Section 10.4.2).

Recently, there has been much interest in the use of export coefficient models, originally developed to predict nutrient loading on lakes (Vollenweider, 1968; Jorgensen, 1980. USEPA, 1980; Beaulac and Reckhow, 1982) to predict nitrogen loading on riverine systems (Johnes and O'Sullivan, 1989; Johnes, 1990; Johnes and Burt, 1990). Nitrogen losses from the catchment to the river are computed using a range of export coefficients derived from the literature for each nitrogen source within the catchment area. This approach has a key advantage over other models in that it can predict the total nitrogen load delivered to a water body as well as estimating the mean annual nitrate concentration in a river at any given point along its course. Johnes (1990) has recently developed this approach to include separate coefficients for each category of land use, livestock and anthropogenic input of nitrogen to a catchment in order to take into account spatial variations in fertiliser use, nitrogen fixation rates, manure application, types and numbers of livestock, and human population size. Export coefficient models are empirical in that the export coefficients have no physical basis, although they are derived from actual experiments. These models are amenable to management application because of their proven accuracy in predicting instream nitrogen concentrations in relation to land use. Export coefficient models are cheap and quick to apply, allowing a distributed approach to be taken which can be used to identify the spatial distribution of major nutrient export zones within a catchment. A full description of this type of model is given in Section 10.4.3.

Conceptual models

These occupy an intermediate position between the physically based models and empirical black-box models. Such models are formulated on the basis of a simple arrangement of a relatively small number of components, each of which is a simplified representation of one process element in the system being modelled (Anderson and Burt, 1985). Addiscott and Wagenet (1985) use the term 'functional' for models that adopt simplified treatments of solute and water flow; by not incorporating fundamental process mechanisms, they thereby require fewer input data and computer expertise for their use. A number of such models have been developed to model nitrate leaching in soil profiles, often dividing the soil into horizontal layers and using simple functional models to describe nitrate transport between the layers (e.g. Burns, 1974; Addiscott, 1977). Such models are often mathematically very simple and their requirements for input data are modest, making them suitable for management purposes. In addition, they offer the attractive possibility of relating small-scale observations of soil water and nitrate leaching to larger-scale observations of nitrate loss from a catchment; however, as yet, such models have not been aggregated to the catchment scale.

Current interest in the effects of acid precipitation and the buffering capacity of the catchment has highlighted interest in the links between storm runoff mechanisms and solute leaching. Models have been developed which reproduce the major trends in chemical and hydrological behaviour in a catchment (Christophersen, Seip and Wright, 1982; Cosby *et al.*, 1985). On the whole, such models have tended to take a fairly simple view of catchment hydrology, employing two or three reservoirs to model runoff components, in order to emphasise the chemical reactions occurring, notably anion retention, cation exchange and weathering of minerals. Since such models contain all the principal processes controlling anion and cation behaviour, nitrate leaching could also be simulated (Whitehead, 1988).

Whitehead (1988) describes a semi-distributed (lumped subcatchments linked by a river network) model of the Thames which included a lumped daily hydrological model for each of 17 tributary subcatchments, a soil zone and aquifer model for calculating the nitrate concentrations of surface runoff and groundwater given a particular land use and fertiliser application rate, and an integrated mass balance model of flow and nitrate transport along 22 reaches of the main river. The model contains some elements which are physically based, and yet others are much simplified. This pragmatic approach is necessary to give some chance that the model may be applied rather than remain a research tool.

Physically based models

These are based on complex physical theory, incorporating the most fundamental mechanisms of the process as presently understood. For leaching, this implies the use of equations derived from Darcy's law for water flow and the expression of resulting solute transport as the combination of mechanisms of mass flow and dispersion−diffusion (Addiscott and Wagenet, 1985). Such models make large requirements in terms of computing and data, are very costly to develop and operate, and may be difficult or impossible to calibrate because of our inability to collect sufficient field data. Where objectives or resources are limited, a simpler type of model may be more cost-effective and no less accurate (Anderson and Burt, 1985). Nevertheless, by offering a physically based approach which is necessarily distributed in space, such models have much to offer in studying nitrate loss from catchments. Process-based models are currently being developed to simulate cycling of carbon and nitrogen in soil and to link this to crop growth and leaching (Addiscott and Whitmore, 1987). Future developments will aim at catchment-scale applications: to extend existing distributed hydrological models such as SHE and IHDM to include nitrate, and to attach physically based soil nitrogen models to lumped hydrological reservoir models. As always, it will be necessary to sacrifice some complexity in order to concentrate upon what the modellers consider to be the salient features of the system.

10.4.2 MODELLING NITRATE IN THE SLAPTON CATCHMENT

Regression type models of water quality data have been frequently used, both to assist planning in the water industry (e.g. Warn 1984; Warn and Page, 1984) and to derive causal relationships between nitrogen inputs and changes in nitrate concentration in rivers (e.g. Onstad and Blake, 1980; Smith and Stevens, 1989). In general, ordinary regression models with time as an independent variable are not suitable for describing time series for two reasons. First, the observations making up the time series are usually independent. Second, the use of regression models to predict values of the dependent variable outside the fitting region can be extremely dangerous (Miller and Wichern, 1977). However, for short series (less than 50 values), and where there is no significant autocorrelation in the dependent variable, then regression analysis may be used with caution. Accordingly, Burt et al. (1988) applied regression and correlation techniques to the annual mean nitrate series for the Slapton Wood catchment (Figure 10.16), 1970−85. Here, we briefly update that analysis to 1990 (20 annual means). It is interesting to note that the resulting regression model remains remarkably stable when the series is extended, suggesting that the use of regression techniques is, in this case, justified.

Table 10.6 gives the annual nitrate, rainfall and runoff series, and a summary of relevant regression results. Lack of significant autocorrelation between residuals in the regression between year and annual mean nitrate ($r = 0.27$) shows that the use of a regression model is permissible in this case. Bivariate correlations show that nitrate is most strongly related to year, though correlations with rainfall and runoff are also significant at the 0.05% level. The stepwise multiple regression results indicate that 48.3% of of the total variation in nitrate is explained by a combination of year and rainfall. Analysis of this result using partial correlation (Johnston, 1980) shows that year alone accounts for 25.3% of the explained variance (52.3% of the total explained), rainfall alone accounts for 8.2% (17.1% of the total) and that year and rainfall together jointly account for 14.8% of the explained variance (30.6% of the total) in that later years have generally been wetter. Thus, notwithstanding the influence of hydrology on nitrate loss (there is more leaching and so a higher concentration in a wet year), the strong upward trend in nitrate concentration over this 20-year period is confirmed. The partial regression coefficient given in step 2 of the multiple regression (0.126) is typical for English rivers (Betton, Webb and Walling, 1991), implying an approximate increase of 1 mg $NO_3 - N$ l^{-1} every 8 years if the trend remains linear.

Finally, one notable point made by Burt et al. (1988) was that runoff in previous years is an important control of nitrate concentration. This is confirmed in an analysis of the 20-year record: a 64% level of explanation is provided by a stepwise regression involving year, last year's rainfall and rainfall two years ago (both rainfall variables were log. transformed). The form of the regression equation shows that a dry year will lead to high nitrate concentrations in subsequent years, and vice versa. These results show that antecedent conditions are important in regulating the supply of nitrate, implying that the soil system has a strong 'memory' with respect to nitrate production and leaching.

Table 10.6 Results of a regression analysis of the 20-year series of annual mean nitrate concentrations for the Slapton Wood catchment

(a) Annual series

Water year	Nitrate (mg NO$_3$ − N l^{-1})	Rainfall (mm)	Runoff (mm)
1971	5.21	847	401
1972	5.80	928	434
1973	5.47	963	382
1974	5.81	1090	485
1975	6.49	921	538
1976	6.66	576	168
1977	9.68	1212	689
1978	7.71	970	569
1979	6.88	1063	580
1980	7.05	1089	557
1981	6.64	1162	601
1982	7.69	1119	607
1983	6.64	1108	654
1984	6.98	895	488
1985	8.32	1135	630
1986	10.06	1110	607
1987	9.20	997	546
1988	8.81	1161	647
1989	6.46	892	436
1990	8.44	1055	564

(b) Correlation matrix

	Nitrate	Rainfall	Runoff
Rainfall	0.480		
Runoff	0.591	0.917	
Year	0.633	0.330	0.439

(c) Stepwise multiple regression including rainfall in current year

Step	1	2
Constant	−288.6	−244.6
Year	0.149	0.126
Rainfall		0.0029
R^2	40.1	48.4

(d) Stepwise multiple regression including rainfall in previous years

Step	1	2	3
Constant	−288.6	−312.5	−345.1
Year	0.149	0.173	0.201
log (Rf−1)		−7.6	−8.9
log (Rf−2)			−6.5
R^2	40.1	54.5	64.0

10.4.3 EXPORT COEFFICIENT MODELLING IN THE WINDRUSH CATCHMENT

Traditionally, the control of nitrate pollution in surface waters has relied on the evaluation of pollution control strategies at the plot or small catchment scale, using detailed process-based models. For larger drainage basins, such models can be expensive to construct and difficult to calibrate, depending on their data requirements. In order to control nitrogen loading on surface waters a manageable, relatively inexpensive model is desirable, providing it is accurate and capable of prediction. Export coefficient modelling, originally developed in North America for prediction of nutrient loading on lakes (e.g. Vollenweider, 1968; Jorgensen, 1980; USEPA, 1980; Beaulac and Reckhow, 1982; Delwich and Haith, 1983) has the potential to meet this need. Johnes (1990) has evaluated the export coefficient modelling approach (as outlined by Jorgensen, 1980) as a management tool for appraisal of nitrate control strategies for large river systems. Export coefficient modelling is based on a distributed approach to basin study, predicting the nitrogen export to a surface water body from each nitrogen source within its catchment (Johnes, 1990). This allows account to be taken of the spatial distribution of nutrient export zones within the catchment, in conjunction with spatial and temporal trends in surface water nitrate concentrations. The catchment of the River Windrush was selected as a suitable drainage basin for such an evaluation, being large enough (363 km^2) to assess the capabilities of this approach for catchment management.

Field investigations established current land use and farming practice for the catchment. This involved a detailed questionnaire survey of land use and agricultural practice on each farm and large estate in the catchment. Information derived from this survey was then used to establish fertiliser application (rate and timing) for each land-use type, to determine livestock levels in the catchment and to detail the use of animal wastes on the land. A parallel programme of water sampling and analysis was also conducted from which total nitrogen and nitrate loads in the Windrush were calculated.

Past land use and livestock levels were established for the period 1925−85, based on parish summaries of the annual MAFF agricultural returns. Long-term trends in nitrogen loads in the River Windrush were obtained for the period 1973−90 from Water Authority records (see Figure 10.8) and for earlier years by extrapolation from the River Thames record (Figure 10.14). A literature survey was conducted to establish past fertiliser practice, the human population of the catchment, the amount of nitrogen in human wastes before and after treatment, the nitrogen content in animal wastes, and the nitrogen input from atmospheric sources. Export coefficients were derived from published data for the percentage nitrogen loss from each nutrient source in the catchment to the drainage system. The sources of aquatic nitrogen loading have been detailed in full in Chapter 5. Examples of the export coefficients derived from the literature survey are given in Table 10.7 for land use types, and for livestock and human wastes.

The export coefficient model was contructed using the land-use and nitrogen input data, together with the export coefficients, for the year 1989/90. Initial model predictions were calibrated using data from the water quality sampling programme for the water year 1989/90 (cf. James, 1980; Jorgensen, 1980; Figures 10.3 and 10.7). Sensitivity analysis was conducted on the model to determine the key parameters controlling model response for this basin, and the dimensionless response of the model itself. The results of the basin sensitivity analysis were subsequently used to formulate a basin-specific management strategy to reduce nitrogen loading on the River Windrush (see below). The results of the dimensionless model sensitivity analysis were used to adjust the export coefficients of the controlling model parameters within

Table 10.7 Export coefficients for (a) land use and (b) livestock and human waste

(a)

Land use	N export (% total input)
Cereals	12
Root crops	20
Oilseed rape	30
Temporary grass	5
Permanent grass	5
Rough grazing	2
Woodland	13 kg ha^{-1} a^{-1}

(b)

Animal	N content per head (kg a^{-1})	Net loss (%)
Cow	70.2	16.15
Pig	18.75	14.45
Sheep	8.9	17.0
Horse	76.8	16.15
Poultry	0.3	15.3

the known range of each export coefficient to produce a final model calibration. The observed and predicted loads in this model calibration are presented in Table 10.8. The model was then used to provide predictions for the period 1925–90 for the Windrush based on the past land use, agricultural practice and human population levels. These predictions were then compared to long-term water quality data for the River Windrush (Figure 10.18). Clearly, the model predictions for 1925–72 are less accurate than those for the 1973–90 period. Nevertheless, the results illustrate the sensitivity of this catchment to change, and the suitability of past records of agriculture, population and rainfall for such a reconstruction. The model was therefore used with some confidence to evaluate pollution control strategies for the Windrush catchment.

Six potential management strategies were evaluated for the Windrush catchment. Option 1 was a basin-specific strategy, involving conversion of all oilseed rape in the catchment to permanent grass with a fertiliser application rate of 150 kg N ha^{-1} a^{-1}, restriction of

Table 10.8 Observed and predicted loads of N: calibration results

	Tonnes a^{-1}	kg ha^{-1} a^{-1}
Observed load	1014.37	34.27
Predicted load		
Initial calibration	1048.87	35.43
% error	(3.29)	
Final calibration	1025.17	34.63
% error	(1.06)	

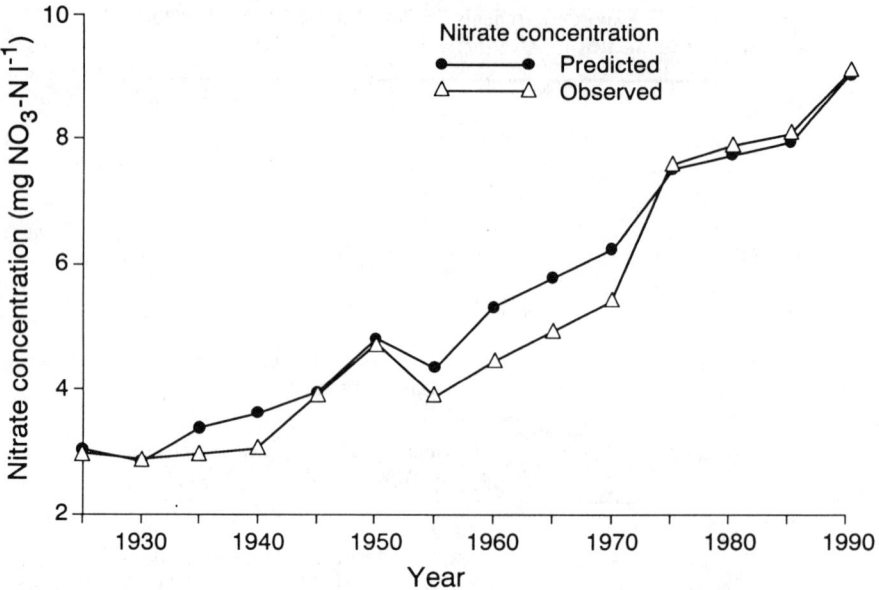

Figure 10.18 Observed and predicted trends in nitrate concentration in the River Windrush 1925−90

fertiliser application rates to all cereal crops, conversion of all temporary grass to permanent grass with no further ploughing up of permanent grass, a reduction of cattle numbers by 25%, and connection of all households to a sewerage system. Option 2 involves a 20% decrease in fertiliser application rates, advocated by Addiscott and Powlson (1989) as a means of maintaining river nitrate levels at below the EC limit. Options 3 to 5 derive from the proposals laid out in the Nitrate Sensitive Areas scheme (MAFF, 1989; Section 12.3); option 3 relates to the basic NSA scheme, option 4 to the premium NSA scheme (conversion of all arable land to permanent unfertilised, ungrazed grass), and option 5 to the premium NSA scheme (conversion of all arable land to permanent grass with <150 kg N ha^{-1} a^{-1} applied). Option 6 requires the creation of a riparian strip 50 m wide either side of the surface drainage network, with unfertilised grazed grass (Section 12.4).

All the options were evaluated using the 1990 figures for land use, livestock numbers and human population for the catchment. The results of this evaluation are presented in Figure 10.19 as a percentage change in nitrate concentration from the 1989/90 level. Each of the management strategies proposed has the potential to maintain the Windrush water supply at a level below the EC limit for nitrate. The question of which management strategy to adopt is dependent upon the degree of reduction in concentration below the EC limit, and to reduce the number of incidences when the nitrate concentration exceeds this limit, then the logical choice of management strategy is option 6. This requires the least change in land use, while achieving the objectives of the management strategy. All other options would require substantial change in the agricultural economy of this region and substantial financial compensation to be paid.

Clearly, such a modelling strategy provides considerable scope for developing an accurate approximation of the Best Practicable Environmental Option (BPEO, Royal Commission on Environmental Pollution, 1976, 1988) for the control of nitrate pollution in surface waters.

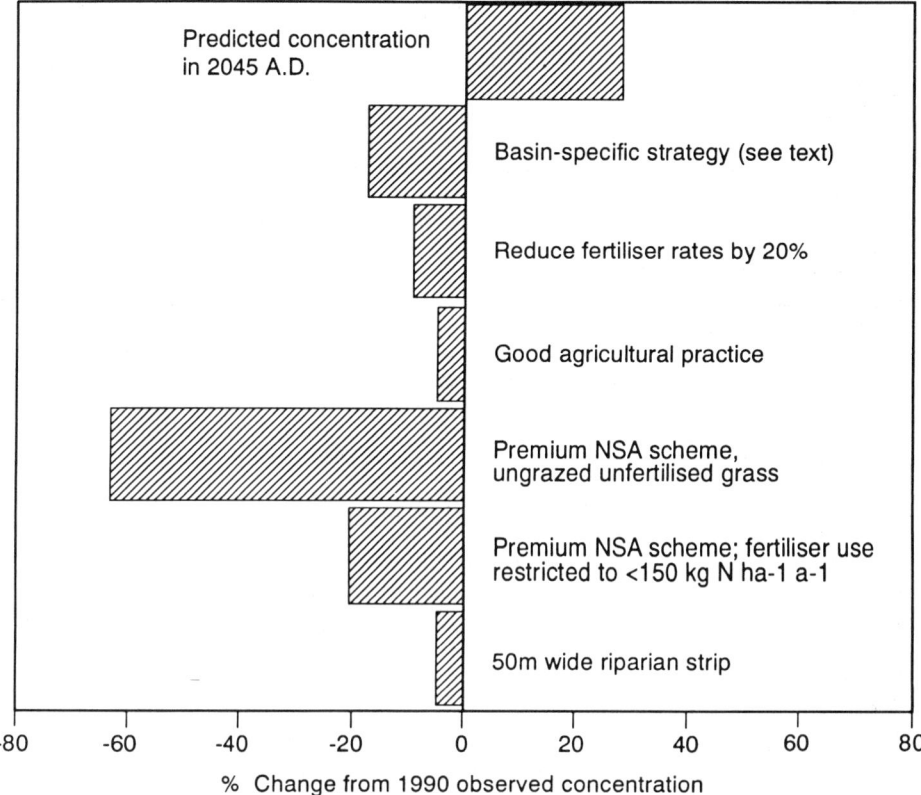

Figure 10.19 Model predictions of changes in nitrate concentration in the River Windrush for selected catchment management strategies

The simplicity of the model format, combined with its proven accuracy and mobility make it an attractive tool. However, while the approach allows a detailed scientific appraisal to be made of past and current land-use impacts upon an aquatic environment, the BPEOs formulated will not necessarily result in a change in policy. Such change appears particularly unlikely when considering the past history of non-plural policy formulation dominated by political and economic interest groups in Great Britain. However, the provisions of the Water Act 1989 for legislative control of pollution in vulnerable zones may dictate a shift in the balance in interests in policy formulation (see Chapter 15). The success or failure of the Nitrate Sensitive Areas scheme will provide a useful indication of any further change of policy in the near future. The options for control of nitrate pollution, and the constraints on the implementation of legislative controls within the framework of such control, are discussed in greater detail in the following chapters.

REFERENCES

Adamson, J.K., Hornung, M., Pyatt, D.G. and Anderson, A.R. (1987) Changes in solute chemistry of drainage waters following the clearfelling of Sitka spruce plantation. *Forestry*, **60**, 165−77.
Addiscott, T.M. (1977) A simple computer model for leaching in structured soils. *J. Soil Sci.*, **28** 554−63.

Addiscott, T.M. and Powlson, D. (1989) Laying the ground rules for nitrate. *New Scientist*, 29 April, 28–29.

Addiscott, T.M. and Wagenet, R.J. (1985) Concepts of solute leaching in soils: a review of modelling approaches. *J. Soil Sci.*, **36** 411–24.

Addiscott, T.M. and Whitmore, A.P. (1987) Computer simulation of changes in soil mineral nitrogen and crop nitrogen during autumn, winter and spring. *J. Agr. Sci., Cambridge*, **109**, 141–57.

Anderson, M.G. and Burt, T.P. (1982) The contribution of throughflow to storm runoff: an evaluation of a chemical mixing model. *Earth Surf. Proc.*, **7**, 6, 565–74.

Anderson, M.G. and Burt, T.P. (1985) *Hydrological Forecasting*, Wiley, Chichester.

Batley, G.E. and Gardiner, D. (1977) Sampling and storage of natural waters for trace metal analysis. *Water Research*, **11**, 9, 745–56.

Beaulac, M.N. and Reckhow, K.M. (1982) An examination of land use–nutrient export relationships. *Water Resources Bulletin*, **18**, 1013–24.

Berg, E.L. (ed.) (1982) *Handbook for Sampling and Sample Preservation of Water and Wastewater*, US National Technical Information Service, Report PB 83 124503, Springfield, Virginia.

Betton, C. (1990) *Nitrate levels in British streams and rivers*, Unpublished PhD thesis, University of Exeter.

Betton, C., Webb, B.W. and Walling, D.E. (1991) Recent trends in $NO_3 - N$ concentration and loads in British rivers. IAHS 203, 169–80.

Beven, K. (1985) Distributed models. In Anderson, M.G. and Burt, T.P. (eds), *Hydrological Forecasting*, Wiley, Chichester, pp. 405–35.

Blackie, J.R. and Eeles, C.W.O. (1985) Lumped catchment models. In Anderson, M.G. and Burt, T.P. (eds), *Hydrological Forecasting*, Wiley, Chichester, pp. 311–45.

Brady, N.C. (1984) *The Nature and Properties of Soils*, 9th edition, Collier-Macmillan, London.

Brooker, M.P. and Johnson, P.C. (1984) The behaviour of phosphate, nitrate, chloride and hardness in twelve Welsh rivers. *Water Research*, **18**, 9, 1155–64.

Burfield, I. (1977) Public health aspects of nitrates in Essex water supplies. *Public Health Eng.*, **5**, 5, 116–24.

Burns, I.G. (1974) A model for predicting the redistribution of salts applied to fallow soils after excess rainfall or evaporation. *J. Soil Sci.*, **25** 165–78.

Burt, T.P., Butcher, D.P., Coles, N. and Thomas, A.D. (1983) The natural history of Slapton Ley nature reserve: XV Hydrological processes in the Slapton Wood Catchment. *Field Studies*, **5**, 731–52.

Burt, T.P. and Butcher, D.P. (1985a) Topographic controls of soil moisture distributions. *J. Soil Sci.*, **36**, 469–86.

Burt, T.P. and Butcher, D.P. (1985b) Stimulation from simulation? A teaching model of hillslope hydrology for use on microcomputers. *J. Geog. Higher Educ.*, **10**, 23–9.

Burt, T.P. and Arkell, B.P. (1986) Variable source areas of stream discharge and their relationship to point and non-point sources of nitrate pollution. IAHS 157, 155–64.

Burt, T.P. and Arkell, B.P. (1987) Temporal and spatial patterns of nitrate losses from an agricultural catchment. *Soil Use and Management*, **3**, 4, 138–42.

Burt, T.P., Arkell, B.P., Trudgill, S.T. and Walling, D.E. (1988) Stream nitrate levels in a small catchment in South West England over a period of 15 years. *Hydrol. Processes*, **2**, 267–84.

Casey, H. and Clarke, R.T. (1979) Statistical analysis of the nitrate concentrations from the River Frome (Dorset) for the period 1965–76. *Freshwater Biol.*, **9**, 91–7.

Christophersen, N., Seip, H.M. and Wright, R.F. (1982) A model for streamwater chemistry at Birkenes, Norway. *Water Resources Res.*, **18**, 4, 977–96.

Cirello, J., Rapaport, R.A., Strom, P.F., Matulewich, V.A., Morris, M.L., Goetz, S. and Finstein, M.S. (1979). The question of nitrification in the Passiac River, New Jersey: an analysis of historical data and experimental investigation. *Water Res.*, **13**, 525–37.

Cosby, B.J., Hornberger, G.M., Galloway, J.N. and Wright, R.F. (1985) Modelling the effects of acid deposition: assessment of a lumped parameter model of soilwater and streamwater chemistry. *Water Resources Res.*, **21**, 1, 51–63.

Delwich, L.L.D. and Haith, D.A. (1983) Loading functions for predicting nutrient losses from complex watersheds. *Water Resources Bull.*, **198**, 753–62.

Department of the Environment (1978) *Digest of Environmental Statistics*, HMSO, London.

Dykseul, A. (1982) *The water quality of the river Rhine in the Netherlands over the period 1970–1981*,

Govt. Inst. Waste Water Treat, Report 82-061, 43−6.

ECETOC (1988) *Nitrate and Drinking Water*, European chemistry Industry Ecology and Toxicology Centre (ECETOC) Technical Report 27.

Elwood, J.W., Newbold, J.D., O'Neill, R.V. and Van Winkle, W. (1983) Resource spiralling: an operational paradigm for analyzing lotic ecosystems. In Fontaine, T.D. and Bartell, S.M. (eds), *Dynamics of lotic ecosystems*, Ann Arbor Science, Michigan, pp. 3−27.

Falkenmark, M. and Chapman, T. (1989) *Comparative Hydrology: an ecological approach to land and water resources*, UNESCO, Paris.

Ferguson, R.I. (1986) River loads underestimated by rating curves. *Water Resources Data*, **22**, 1, 74−6.

German Chemists Association (1981) Preservation of water samples. Report by the Working Party on 'Stabilization of Samples', Hydrochemistry Team of the German Chemists Association. *Water Research*, **15**, 233−41.

Giesy, J.P. and Briese, L.A. (1978) Particulate formation due to freezing humic waters. *Water Resources Res.*, **14**, 3, 542−4.

Greene, L.A. (1978) Nitrates in water supply abstractions in the Anglian region: current trends and remedies under investigation. *Water Poll. Cont.*, **77**, 4, 478−91.

Hagebro, C., Bang, S. and Somer, E. (1983) Nitrate load/discharge relationships and nitrate load trends in Danish Rivers. IAHS 141, 377−86.

Haigh, R.A. and White, R.E. (1986) Nitrogen leaching from a small, under-drained, grassland, clay catchment. *Soil Use and Management*, **2**, 65−70.

Haycock, N.E. (1991) *Riparian land as buffer zones for agricultural runoff*, Submitted DPhil thesis, University of Oxford.

Heathwaite, A.L., Burt, T.P. and Trudgill, S.T. (1990) The effect of land use on nitrogen, phosphorus and suspended sediment delivery to streams in a small catchment in Southwest England. In Thornes, J.B. (ed.), *Vegetation and Erosion: processes and environments*, Wiley, Chichester, pp. 161−78.

Heathwaite, A.L. and Johnes, P.J. (in press) Sample collection and preservation for nitrogen and phosphorus analysis in freshwaters. *Water Res.*

Hornung, M. (1988) Modelling non-agricultural sources of nitrate. In *MAFF Report on Nitrate Modelling*, MAFF, London.

Hunt, D.T.E. and Wilson, A.L. (1986) *The Chemical Analysis of Water*, 2nd edition, The Royal Society of Chemistry, London.

Jacobs, T.C. and Gilliam, J.W. (1985) Riparian losses of nitrate from agricultural drainage waters. *J. Environ Quality*, **14**, 4, 472−8.

James, A. (1980) Water quality modelling. In A.M. Gower (ed.), *Agriculture and Water Quality*, Wiley, Chichester.

Jenkins, D. (1968) The differentiation, analysis and preservation of nitrogen and phosphorus forms in natural waters. *Advances in Chemistry*, **73**, 265−80.

Johnes, P.J. (1990) *An investigation of the effects of land use upon water quality in the Windrush catchment*, Unpublished DPhil thesis, University of Oxford.

Johnes, P.J. and Burt, T.P. (1990) Modelling the impact of agriculture upon water quality in the Windrush catchment — an export coefficient approach in a representative basin. *Proc. Conf. Hydrol. Res. Basins & Environment, Wageningen*, **44**, 245−52.

Johnes, P.J. and Burt, T.P. (1991) Water quality trends in the Windrush catchment: nitrogen speciation and sediment interactions. IAHS 203, 349 57.

Johnes, P.J. and O'Sullivan, P.E. (1989) Nitrogen and phosphorus losses from the catchment of Slapton Ley, Devon — an export coefficient approach. *Field Studies*, **7**, 285−309.

Johnston, R.J. (1980) *Multivariate Statistical Analysis in Geography*, Longman, Harlow.

Jorgensen, S.E. (1980) *Lake Management*, Pergamon Press, Oxford.

José, P. (1989) Long-term nitrate trends in the River Trent and four major tributaries. *Reg. Rivers*, **4**, 43−57.

Kauppi, L. (1984) 1: Contribution of agricultural loading to the deterioration of surface waters in Finland, 2: Nitrate in runoff and river waters in Finland in the 1960s and 1970s, *Public Water Res. Inst. & Nat. Board Waters, Finland*, **57**, 24−40.

Kesner, B.T. and Meentemeyer, V. (1989) A regional analysis of total nitrogen in an agricultural landscape. *Landscape Ecol.*, **2**, 3, 151−63.

Kovacs (1989) Measurement and estimation of hydrological processes. In Falkenmark, M. and Chapman,

T. (eds), *Comparative Hydrology: an ecological approach to land and water resources*, UNESCO, Paris, pp. 75–104.

Labadz, J.C. (1988) *Runoff and sediment production in blanket peat moorland: studies in the Southern Pennines*, Unpublished PhD thesis, Huddersfield Polytechnic.

Lowrance, R.R., Todd, R.L. and Asmussen, L.E. (1984) Nutrient cycling in an agricultural watershed. *J. Environ. Qual.*, **13** 1, 22–32.

Marsh, T.J. (1980) Towards a nitrate balance for England and Wales. *Water Services*, October, 601–6.

Meybeck, M., Chapman, D., and Helman, P. (1989) *Global Freshwater Quality: a first assessment*, Global Environment Monitoring System/UNEP/WHO.

Miller, R.B. and Wichern, D.W. (1977) *Intermediate Business Statistics* Holt, Reinhart and Winston, New York.

Moss, B., Booker, I., Balls, H. and Manson, K. (1989) Phytoplankton distribution in a temperate floodplain lake and river system. I. Hydrology, nutrient sources and phytoplankton biomass. *J. Plankton Res.*, **11**, 4, 813–38.

Newbold, J.D., Elwood, J.W., O'Neill, R.V. and Sheldon, A.L. (1981) Measuring nutrient spiralling in streams. *Can. J. Fish Aquat. Sci.*, **38**, 860–63.

Nicholson, N.J. (1979) A review of the nitrate problem. *Chem. Ind.*, 189–95.

Odum, E.P. (1989) Input management of production systems. *Science*, **243**, 177–82.

OECD (1985) *Environmental Data Compendium 1985*, OECD, Paris.

OECD (1987) *Environmental Data Compendium 1987*, OECD, Paris.

OECD (1990) *Environmental Data Compendium 1990*, OECD, Paris.

O'Neill, P. (1985) *Environmental Chemistry*, Allen & Unwin, London.

Ongley, E.D. (1987) Scale effects in fluvial sediment-associated chemical data. *Hydrol. Proc.*, **1**, 171–9.

Onstad, C.A. and Blake, J. (1980) Thames basin and agricultural relations. *Proc. Symp. Watershed Management*, ASCE/Boise, pp. 961–73.

Paces, T. (1982) Long-term changes in concentration and fluxes of nitrogen species in the Elbe River basin. In *Impact of Agricultural Activities on Groundwater*, Int. Symp. IAH Memoires, Prague, **16**, 2, pp. 299–315.

Pinay, G., Decamps, H., Chauvet, E. and Fustec, E. (1990) Functions of ecotones in fluvial systems. In Naiman, R.J. and Decamps, H. (eds), *The Ecology and Management of Aquatic–Terrestrial Econtones*, Parthenon, Paris, pp. 141–70.

Roberts, G. and Marsh, T. (1987) The effects of agricultural practices on the nitrate concentrations in the surface water domestic supply sources of Western Europe. IAHS 164, 365–80.

Roberts, A.M., Hudson, G. and Roberts, G. (1989) A comparison of nutrient losses following grassland inprovement using two different techniques in an upland area in Mid-Wales. *Soil Use & Management*, **5**, 174–9.

Rodda, J.C., and Jones, G.N. (1983) Preliminary esturies and coastal waters around Gret Britain derived from the harmonized monitoring scheme. *J. Inst. Wat. Eng. Sci.*, 529–39.

Royal Commission on Environmental Pollution (1976) *Air Pollution Control: an Integrated Approach*, 5th Report, HMSO, London.

Royal Commission on Environmental Pollution (1988) *Best Practicable Environmental Option*, 12th Report, HMSO, London.

Royal Society (1983) *The Nitrogen Cycle of the United Kingdom*, Royal Society, London.

Ryding S-O and Forsberg, C. (1979). Nitrogen, phosphorus and organic matter in running waters: studies from six drainage basins. *Vatten*, **1**, 46–58.

Slack, J.G. and Williams, D.N. (1985) Long term trends in Essex river water nitrates and hardness. *Aqua*, **2**, 77–8.

Smith, R.A. and Alexander R.B. (1985) Trends in concentrations of dissolved solids, suspended sediments, phosphorus and inorganic nitrogen at US Geological Survey National Streams Quality Accounting Network Stations (NASQAN). *National Water Summary 1984 — Hydrologic perspectives*, US Geological Survery, 66–73.

Smith, R.V. (1977) Domestic and agricultural contributions to the inputs of phosphorus and nitrogen to Lough Neagh. *Water Res.*, **11** 453–9.

Smith, R.V., Stevens, R.J., Foy, R.M. and Gibson, C.E. (1982) Upward trends in nitrate concentrations in rivers discharging into Lough Neagh for the period 1969–79. *Water Res.*, **16**, 183–8.

Smith, R.V. and Stewart, D.A. (1989) A regression model for nitrate leaching in Northern Ireland. *Soil Use & Management*, **5**, 71−6.

STACWQ (1983) *4th Biennial report of the Standing Advisory Committee on Water Quality*, Department of Environment/National Water Council, Standing Technical Committee Report 37, HMSO, London.

Standing Committee of Analysts (1980) *General Principles of Sampling and Accuracy of Results*, HMSO, London, pp. 15−48.

Stevens, P.A. and Hornung, M. (1988) Nitrate leaching from a felled Sitka spruce plantation in Beddgelert Forest, North Wales. *Soil Use and Management*, **4**, 1, 3−8.

Stevens, R.J. and Stewart, B.M. (1982) Concentration, fractionation and characterisation of soluble organic phosphorus in river water entering Lough Neagh. *Water Res.*, **16** 1507−19.

Suess, M.J., (ed.) (1982) *Examination of Water for Pollution Control*, Volume 1: *Sampling, Data Analysis and Laboratory Equipment*, WHO, Copenhagen.

Tomlinson, T.E. (1970) Trends in nitrate concentration in English rivers in relation to fertiliser use. *Water Treat Exam.*, **19**, 277−93.

Toner, P.F. and Lennox, L.J. (1980) Nitrate content in Irish rivers, *Irish J. Environ. Sci.*, **1**, 75−6.

Trudgill, S.T., Burt, T.P., Heathwaite, A.L. and Arkell, B.P. (1991) Soil nitrate sources and nitrate leaching losses, Slapton, South Devon. *Soil Use & Management*, **5**, 200−6.

United States Environmental Protection Agency (1980) *Modelling phosphorus loading and lake response under uncertainty: a manual and compilation of export coefficients*, Report EPA-440/5-80-011, Environmental Protection Agency, Washington, DC.

Vannote, R.L., Minshall, G.W., Cummins, K.W., Sedell, J.R. and Cushing, C.E. (1980) The river continuum concept. *Can. J. Fish Aquat. Sci.*, **37**, 130−37.

Vitousek, P.M. (1981) Clearcutting and the nitrogen cycle. In Clark, F.E. and Rosswall, T. (eds), *Terrestrial Nitrogen Cycles*, Ecol. Bull., Stockholm, **33**, 631−42.

Vollenweider, R.A. (1968) *Scientific fundamentals of stream and lake eutrophication, with particular reference to nitrogen and phosphorus*, OECD Technical Report No. DAF/DST/88.

Walling, D.E. (1988) Measuring sediment yield from river basins. In Lal, R. (ed.), *Soil Erosion Research Methods*, Soil and Water Conservation Society, Ankeny, Iowa, pp. 39−73.

Walling, D.E. and Foster, I.D.L. (1975) Variations in the natural chemical concentration of river water during flood flows, and the lag effect: some further comments. *J. Hydrol.*, **26**, 237−44.

Walling, D.E. and Foster, I.D.L. (1978) The effects of the 1976 drought and autumn rainfall on stream solute levels. *Earth Surf. Proc.*, **3**, 393−406.

Walling, D.E. and Webb, B.W. (1978) Mapping solute loading in an area of Devon, England. *Earth Surf. Proc.*, **3**, 85−99.

Walling, D.E. and Webb, B.W. (1981a) The reliability of suspended sediment load data. IAHS 133, 177−84.

Walling, D.E. and Webb, B.W. (1981b) Water Quality. In Lewin, J. (ed.), *British Rivers*, Allen and Unwin, London, pp. 126−69.

Walling, D.E. and Webb, B.W. (1982) The design of sampling programmes for studying catchment nutrient dynamics. *Proc. Symp. Hydrol. Res. Basins*, Berne, 747−58.

Walling, D.E. and Webb, B.W. (1984) Local variation of nitrate levels in the Exe basin, Devon, England. *Beitrage zur Hydrologie*, **10**, 71−100.

Walling, D.E. and Webb, B.W. (1985) Solutes in river systems. In Trudgill, S.T. (ed.), *Solute Processes*, Wiley, Chichester.

Warn, A.E. (1984) Calculating future levels of nitrate in rivers and pumped storage reservoirs. *Water Sci. Tech.*, **16**, 635−42.

Warn, A.E. and Page, C. (1984) Estimating the effect of water quality on surface water supplies. *Water Res.*, **18**, 167−72.

Webb, B.W. and Walling, D.E. (1985) Nitrate behaviour in streamflow from a grassland catchment in Devon, UK. *Water Res.*, **19**, 8, 1005−16.

Whitehead, P.G. (1988) Modelling nitrate in surface water systems. In *MAFF Report on Nitrate Modelling*, MAFF, London.

World Health Organisation (1970) *European Standards for Drinking Water*, 2nd edition, WHO, Geneva.

Part III

MANAGEMENT STRATEGIES

11 Changes in Agricultural Practice

R.J. PARKINSON

Department of Agriculture, University of Plymouth

11.1 INTRODUCTION

During the last four decades agricultural productivity in the developed world has increased considerably. Scientific developments, technological innovation and more sophisticated husbandry methods have all contributed to increased yields of both arable and grass crops. For example, average wheat yields in the UK increased from less than 3 to nearly 7 t ha^{-1} in the period 1950 to 1985 (Marks and Britton, 1989). Similar trends have been observed throughout Europe and the United States. During the same period, maize yields in the USA have risen from 2.5 to 6.5 t ha^{-1} (Hauck, 1990). Increased fertiliser use has been acknowledged as a major contributory factor in these yield increases, with the growth in the use of nitrogen being the most significant.

It is now recognised that increased agricultural productivity has major environmental consequences of which nitrate leaching is only one. During the 1980s several articles of legislation (see Chapters 1 and 14) were passed which impinge upon agricultural practice, in an attempt to reduce the quantity of nitrate leaching from farmland. Although these processes of loss are still imperfectly understood, it was felt by many national and international authorities that steps must be taken to reduce the transfer of nitrate from soil to aquifers and watercourses. This legislation, and the increasing willingness of the farming community to use agrochemicals more efficiently, has led to a number of changes in agricultural practice which can be considered to be a part of a catchment-wide control strategy to reduce nitrate loadings in surface and groundwater. In this chapter, changes are considered in relation to fertiliser use, cropping methods, both arable and grassland, and finally livestock waste-handling systems. It must be emphasised that fertiliser is only one source of N that crops utilise; the productivity of most agricultural soils depends heavily on mineralisation of organic N (see Chapter 3). As non-fertiliser sources can contribute up to 70% of the total N uptake by crop plants (Christian, Cress and Dowdell, 1985), other soil and crop management practices have a central role to play in nitrate control strategies.

11.2 CROPPING SYSTEMS

11.2.1 FERTILISER PRACTICE

Worldwide, fertiliser N use continues to increase at a steady rate, estimated by Jenkinson (1990) to be roughly 5% per annum over the decade 1980−90. The pattern of use within individual countries, however, varies widely according to a variety of factors, which include population growth rate, the recent development of agricultural technology and the extent of concern over the purity of potable water. Hauck (1990) notes that in developing countries,

Nitrate: Processes, Patterns and Management. Edited by T.P. Burt, A.L. Heathwaite and S.T. Trudgill
© 1993 John Wiley & Sons Ltd

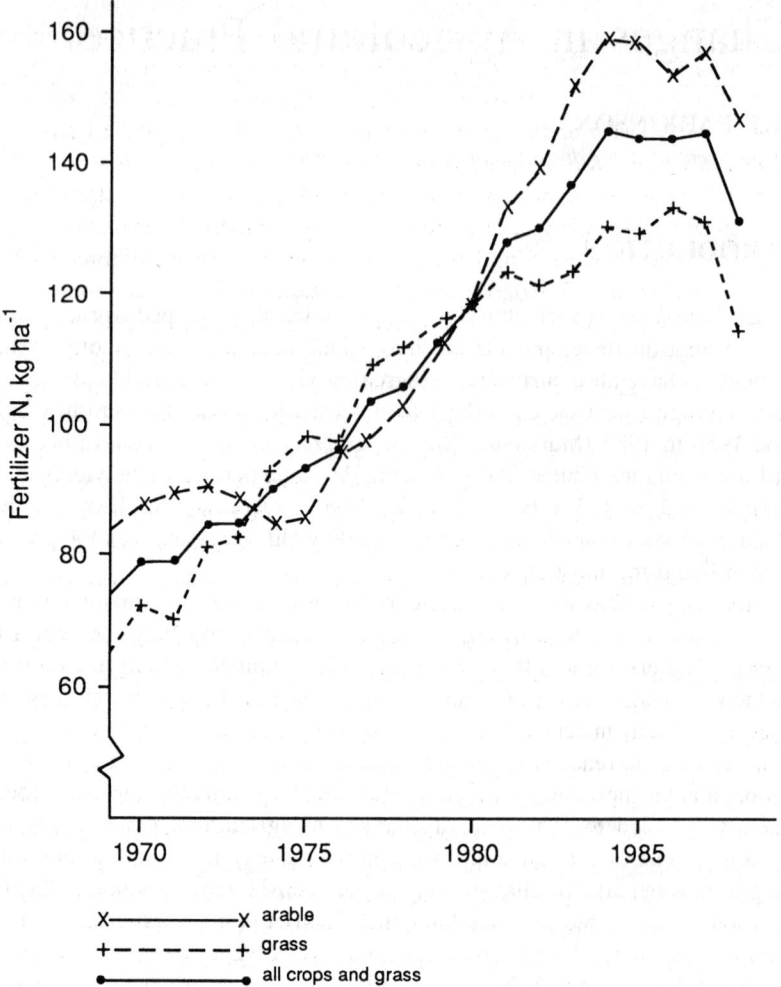

Figure 11.1 Use of N on arable and grass since 1969 (after Chalmers, Kershaw and Leech, 1990)

the consumption of N fertilisers is expected to grow by 6% per annum over the next ten years, while the growth rate in developed countries is expected to be less than 1% per annum.

Statistics for the UK indicate that the 'exponential' trend of fertiliser N use recorded in the 1960s and 1970s had reached a plateau in the mid- 1980s (Figure 11.1) as a variety of pressures were brought to bear on the agriculture industry (Church and Lewis, 1977; Chalmers, Kershaw and Leech, 1990). Taking the 40-year period from 1945 to 1985, fertiliser N use rose by 900%. It is interesting to note that during this same period, use of other fertilisers. also increased. Phosphorus fertiliser use rose relatively steadily, by 50%, while potassium fertiliser use increased by 500%. The changing price structure in the fertiliser manufacturing industries over recent decades has led to a consistent fall in the relative price of fertiliser N. In consequence, the development of soil and crop management strategies to use fertiliser N efficiently, and hence minimise nitrate losses, have tended to be neglected.

Since 1985 there has been a marked change in fertiliser use by farmers in Western Europe; both the timing and rate of application have changed in response primarily to recommendations from the agricultural advisory bodies. In the UK, fertiliser application rates to cereal crops fell during the latter half of the 1980s (Sylvester-Bradley et al., 1987; Chalmers, Kershaw and Leech, 1990). Average applications to winter wheat fell from 192 to 181 kg N ha^{-1} while winter barley rates were reduced from 150 to 142 kg N ha^{-1}. These reductions may not seem large, but they are important in that the trends of the last three decades now seem to have been reversed. Advisory bodies are continuing to modify their recommendations in the light of recent research work stimulated mainly by the increasing environmental concerns over inefficient fertiliser use. Standard rates of application for individual cropping situations are being replaced by site-specific advice based in many cases on the determination of soil mineral nitrate-N early in the spring. Such a system has been in operation in the Netherlands since 1986 (Neeteson, Dilz and Wijnen, 1989) and has since been implemented on a trial basis in the predominantly arable cropping areas of Germany and the UK. In Denmark, N fertiliser recommendations are based on nitrate concentrations measured in soil samples taken in early spring at fixed locations on a 7 km grid square basis across the entire country (Østergaard, 1989). Prediction methods for fertiliser N requirements will, in the future, assume a much greater importance as the agriculture industry is encouraged to match nutrient supply to crop plant requirements, taking into account soil reserves of these nutrients at the onset of the spring growth period.

Timing of application has been shown to exert a crucial influence on the fate of applied N. In the 1970s and early 1980s fertiliser recommendations for cereal crops commonly included autumn nitrogen, at rates up to 50 kg N ha^{-1}, particularly when bulky residues were to be incorporated. In these situations, the natural leakage of N from soils, when the summer soil water deficit has been satisfied and deep drainage out of the profile recommences, is supplemented by N mineralised due to cultivations and a significant proportion of any fertiliser added in the autumn (see Chapter 3). Powlson et al. (1986) demonstrated that the recovery of autumn-applied fertiliser N by winter wheat is low, varying between 11% and 42%, depending upon rainfall receipt. A recent survey of fertiliser practice (Chalmers, Kershaw and Leech, 1990) concluded that in 1989 only 18% of the UK's winter wheat crops received an autumn dressing, a marked decline from the 1985 level of 56%. Currently, it is recommended that only crops which show consistent yield benefits, such as winter oilseed rape, should receive autumn N (MAFF, 1988).

The form and method of application of fertiliser N has changed in recent years, but primarily in response to changing price per unit of N, rather than considerations of the likely efficiency of use. Bremmer (1990) notes that urea (NH_2CONH_2) has become the most important nitrogen fertiliser in world agriculture, due primarily to its increasingly competitive unit cost of production. The susceptibility of different forms of N to leaching bears little relationship to current agricultural practice. For example, the use of nitrification inhibitors, such as nitrapyrin and dicyandiamide means of delaying the production of nitrate from urea- and ammonium-based fertilisers continues to be restricted to specific situations, mainly horticultural, due to the lack of consistent benefits derived from the use of these products (see, for example, Vaughan, 1985). Nitrification inhibitors have traditionally been employed to reduce the rate of ammonia conversion to nitrate in an attempt to provide a more uniform supply of nitrate to crops, rather than to prevent leaching of excess nitrate. Current practice is concentrating on the reduction of inputs rather than a modification of the soil chemical reactions by the use of microbial inhibitors. Methods of application continue to be dominated by surface application of solid 'prilled' N, with alternative, liquid-based, forms such as aqueous ammonia, which often are placed subsurface, being used only infrequently.

It is important to note that the wider environmental impact of any changes in nitrate fertiliser use by the agriculture industry will be felt slowly (see also Chapter 12). Addiscott (1988) reported losses of nitrate from the Rothamsted drain gauges over a 38-year period at the turn of the century, without addition of fertiliser or manure, and concluded that the half-life, or time taken for the nitrate concentration to fall to half of the original value measured when cropping ceased, was 41 years. The manipulation of the fertiliser regime of agricultural soils to minimise nitrate leaching must be viewed as a long-term process.

11.2.2 CULTIVATION AND DRAINAGE

Research carried out over the past two decades has clearly demonstrated the role of routine cultivations in stimulating N mineralisation and hence creating the potential for leaching loss. In particular, the switch from spring to winter cereals has led to more cultivation activity, both primary and secondary, being carried out in the late summer/early autumn period. The inevitable consequence is enhanced nitrate leaching in autumn and winter (see Chapters 3 and 9). Increased autumn workloads for cereal growers led to the development and adoption of time-saving reduced and even zero tillage methods. Many experiments have been carried out into the consequence of these contrasting cultivation methods on the soil environment and nutrient losses (for example Dowdell, Colbourn and Cannel 1987). The consensus is that leaching losses are greater with more extensive soil disturbance. But for the recent upsurge in environmental concerns relating to other agricultural practices, such as straw burning and the widespread use of herbicides to control annual weeds, it is likely that soil management recommendations for nitrate control would have led to the wider adoption of minimal and zero tillage systems. However, as a result of strong public pressure, straw burning has been banned in many European countries, notably Germany, Denmark and the Netherlands, and under the provisions of the Environmental Protection Act (1990), will be prohibited in England and Wales from 1 January 1993. Without the ability to control weeds and pests, and to dispose of bulky crop residues by burning, the majority of cereal growers will return to a cultivation system based on soil inversion, hence stimulating the mineralisation of organic matter. Whether more nitrate is leached as a consequence will depend upon several factors, most importantly the recent cropping history (see Section 11.2.3).

There are certain situations where ploughing does not increase the magnitude of nitrate leaching in comparison with other cultivation methods. Richter *et al.* (1989) investigated the rate of nitrogen mineralisation in arable soils which were deep ploughed. Ploughing up 100 mm of subsoil was found to slow the quantity of nitrate mineralised by 50%, and hence reduced the leaching potential. Deep ploughing was, therefore, recommended as a strategy to control nitrate losses, for example when ploughing out grassland, when large leaching losses can occur.

Modification of subsoil physical conditions and water regimes by underdrainage, subsoiling and soil loosening can have a considerable effect on the soil nitrogen balance. Harris *et al.* (1984) and Dowdell, Colbourn and Cannel (1987) compared nitrate-N losses between drained and undrained plots growing winter wheat in a clay soil, and observed that losses from drained plots were typically an order of magnitude greater from the undrained plots (40−55 compared to less than 5 kg $NO_3 - N$ ha^{-1}). These losses were correlated with larger volumes of water moving vertically through the soil to drain depth. Drainage of grassland also can result in large increases in nitrate leaching. Losses from both arable and grassland systems are discussed in more detail in Chapters 3 and 9, and the implications for surface waters are discussed in Chapters 5 and 10. During the early 1970s, underdrainage in England and Wales was being installed at a rate of 100 000 ha per year, aided by generous Ministry of Agriculture grants, which at that time were allocated on the basis of 60% of the total cost of installation. Since that time, grant aid has been progressively reduced, and now stands at 15%. This reduction, and falling farm incomes, have resulted in a considerable decline in the rate of land drainage, down to approximately 20 000 ha per year in the late 1980s. It may reasonably be postulated that this trend has certainly not contributed to any increase in the total quantity of N that has been lost from agricultural land during the 1980s, and may therefore be considered an unintentional nitrate control mechanism.

Subsoiling and soil loosening can have similar although less significant effects on nitrate losses. For example, Parkinson, Twomlow and Reid (1988) described a soil-loosening experiment in which the pathways of water movement were modified by a soil-loosening treatment (carried out to a depth of 400 mm) over an existing drainage system. During the autumn and winter immediately following the loosening operation, nitrate-N losses were significantly greater from the loosened soil, with concentrations in the loosened subsoil typically reaching $40-50$ mg NO_3-N kg^{-1} in comparison with $20-30$ mg NO_3-N kg^{-1} for the unloosened control site. With the exception of the deep ploughing and mixing situation (Richter *et al.*, 1989) disturbance of the subsoil will lead to enhanced nitrate losses. Strategies for minimising such losses should therefore aim to avoid major soil disturbance.

11.2.3 ARABLE CROP MANAGEMENT

There are several other crop husbandry practices, in addition to changing fertiliser regimes and cultivation practice (discussed in Sections 11.2.1 and 11.2.2, respectively), that have a significant impact on the risk of nitrate leaching from arable soils. In particular, a sound understanding of both the timing of establishment and sequence of successive crops has been shown to be central to any nitrogen management strategy (Pedersen, 1990; MAFF/WOAD, 1991).

In order to minimise the risk of leaching during the critical autumn period, maintenance of crop cover is crucial (Addiscott, 1988). Results from Danish experimental investigations, quoted by Pedersen (1990), confirm the importance of autumn crop cover. Nitrate-N losses from plots after harvesting spring barley were greatly reduced when grass was undersown (Figure 11.2), demonstrating that the presence of a well-established crop cover during the autumn can 'mop up' excess nitrate that has been mineralised during the summer months and hence is available for crop uptake as the soil rewets in the autumn and temperatures are still sufficiently high to allow plant growth. Results from recent research conducted in England tell a similar story: the presence of a growing crop can decrease leaching by 30 kg N ha^{-1} (Christian *et al.*, 1990).

While the beneficial rôle of undersowing in cereals is well known, the advantages of earlier sowing of cereal crops in terms of reducing N losses are less certain. The optimum sowing date for winter cereals varies depending upon climatic conditions. In southern Britain, the optimum dates in terms of yield potential are mid-September for barley and mid-October for wheat (Toosey, 1989). However, in order to remove sufficient nitrate from the mineralised pool, and hence act as effective 'catch crops', earlier sowing in necessary. Such a practice may well lead to yield reductions due to increased disease and, ironically, may result in additional use of autumn pesticides and fungicides in an attempt to control diseases and the

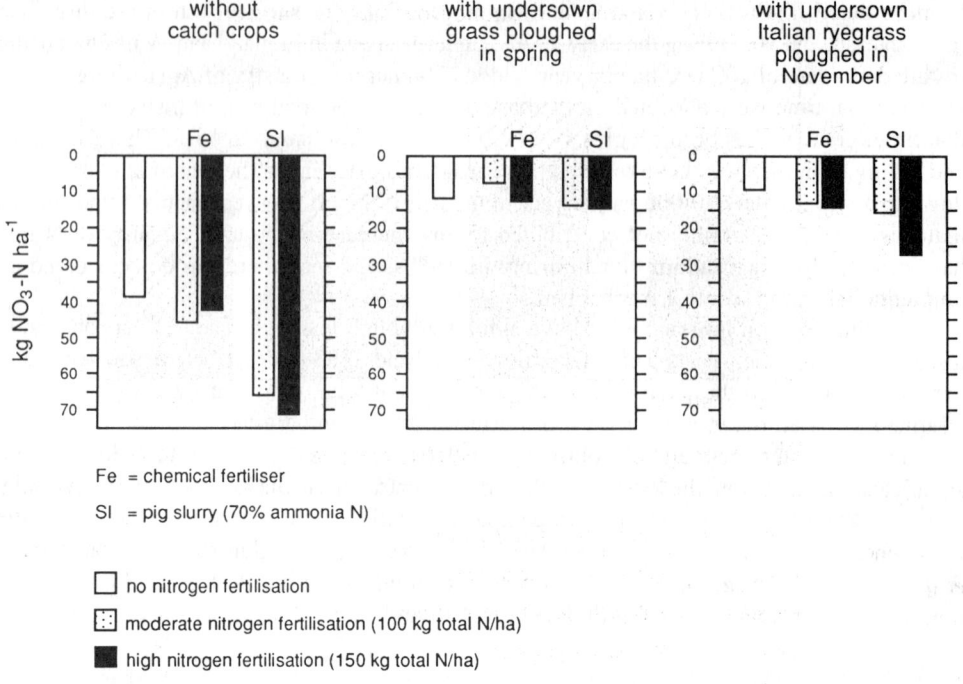

Figure 11.2 Patterns of nitrate leaching after harvesting spring barley undersown with Italian ryegrass (after Pedersen, 1990)

insects that transmit them. Pedersen (1990) quotes further examples from Denmark that indicate the varying efficiency of winter cereals and other crops in removing N from the available pool (Figure 11.3). Highest concentrations of mineral N in the soil profile were measured in early winter (when crop growth can be assumed to have ceased) under a wheat crop established after peas, a leguminous crop. Cultivation prior to sowing wheat in the continuous cereal treatment (column 1) stimulated mineralisation processes; in consequence, more residual mineral N was found under this treatment than in the fallow soil. Danish legislation to control N leaching stipulates the establishment by 1 September for autumn-sown crops unless straw is incorporated. Vinten, Howard and Redman (1991) confirm that winter cereal crops are of little use in controlling N losses by leaching but also noted that, in the case of crops grown in Central Scotland, winter cover crops may be of limited effectiveness.

Horticultural crops pose more of a problem for nitrate control in restricted areas of productive soils. The low percentage ground cover and the fact that many such crops show yield benefits from high levels of mineral N up to harvest, which may include the vulnerable autumn and winter months. Wehrman and Scharpf (1989) measured reductions in soil profile nitrate concentrations (N_{min}) from December to March, in an unspecified year, after a variety of crops (Table 11.1(a)). The high values after the vegetables and resulting serious losses have lead to the development in Lower Saxony of N_{min} sampling and analysis programmes to determine nitrate status while the vegetable crop is growing. Such a strategy can lead to major reductions in actual use of N (Table 11.1(b)).

The incorporation of bulky crop residues with a wide carbon : nitrogen ratio has been

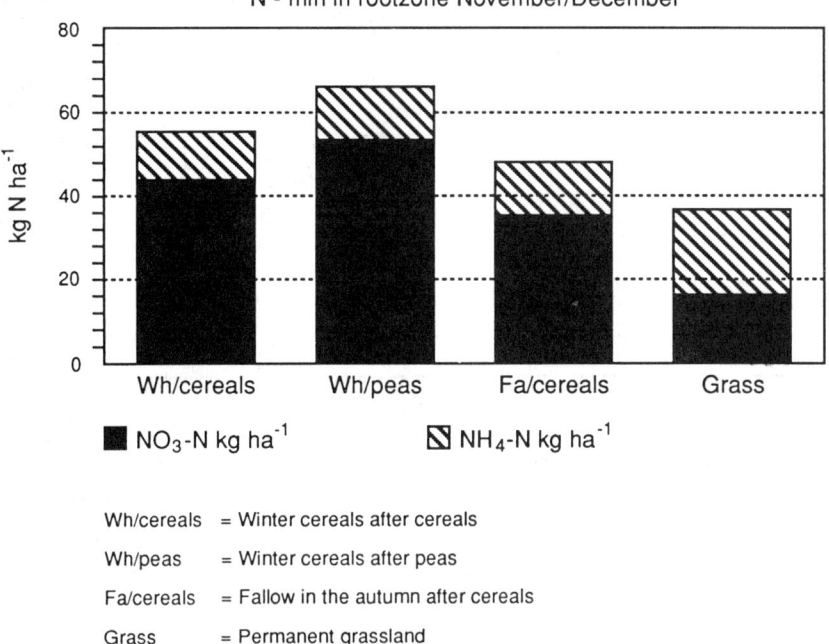

Figure 11.3 Mineral N in the rootzone (0–100 cm) in November/December as affected by current and previous crop (after Pedersen, 1990)

advocated as a means of reducing leaching losses from arable land as a result of bacterial immobilisation of mineral N post-incorporation (Powlson *et al.*, 1985). Jarvis *et al.* (1989a) observed short-term reductions in nitrate leaching of 25–50% from arable soils, but only in the absence of autumn N. Christian *et al.* (1990) reported a reduction in leaching losses of 10 kg $NO_3 - N$ ha^{-1} when straw was incorporated in comparison with burning. Such losses are considerably less than those previously predicted, which may be due to the slow decomposition rate of lignin (Powlson, 1990). Immobilisation of 5–10 kg N ha^{-1} per tonne of straw incorporated are now seen to be excessive. In response to restrictions and, in some cases, the banning of straw burning, many farmers, particularly those with all-arable enterprises, will, of necessity, resort to incorporation. The long-term benefits of incorporation include changes in soil physical conditions, such as increased aggregate stability and reduced susceptibility to compaction and erosion. However, the limited experimental evidence that is available indicates that once the soil organic matter and micro-organism activity has equilibrated at a new, higher level under an incorporation regime, nitrate losses can equal or exceed those observed when straw disposal is carried out by burning or baling (Powlson, Brookes and Christensen, 1987).

11.2.4 GRASSLAND MANAGEMENT

The dynamics of N use in grassland systems differ from arable in one essential aspect, as pointed out by Jenkinson (1990): major losses of N occur after crops are consumed by humans or animals, whereas the main losses in pastoral systems occur before the product (meat, milk,

Table 11.1 (a) Changes in nitrate content between December and March after selected arable crops, 0–900 mm soil depth

Crop	December (kg N ha^{-1})	March (kg N ha^{-1})	Difference (kg N ha^{-1})
Carrot	139	31	− 108
Carrot	91	24	− 67
Table beet	161	31	− 130
Lupine-leek	197	73	− 124
Celery	191	113	− 78
Winter wheat	20	29	+ 9
Winter barley	25	36	+ 11
Sugar beet	25	35	+ 10

Table 11.1 (b) N fertilisation of vegetable crops in practice compared to recommendations at the beginning of winter (after Wehrmann and Scharpf, 1989)

Crop rotation	Total fertilisation (kg N ha^{-1})			N_{min} Nov./Dec. (kg N ha^{-1}) (0–90 cm)
	According to N_{min}-method	in practice	Difference	
Cauliflower	384	813	+429	564
Cauliflower	70	267	+197	198
Kohlrabi leek	150	432	+282	248
Celery	62	96	+ 34	156
Savoy	232	307	+ 75	71

wool, etc.) is obtained. In contrast with arable sytems, grasslands are characterised by much higher rates of nitrogen turnover and, in the case of young swards, by a positive nitrogen balance (Jenkinson, 1988). Potentially these swards are likely, therefore, to become major sources of significant nitrate losses once the sward has matured, and particularly if the sward is ploughed out. Average use of N on grassland in the UK has remained consistently below that of arable land since 1980 (see Figure 11.1) but these average figures hide a wide range of fertiliser application rates according to grassland utilisation. Table 11.2 provides a breakdown of grassland-utilisation categories. Within the 'intensively grazed' and 'cut for silage' categories, use of nitrogen fertiliser can be well in excess of the average for the class. Jarvis et al. (1989a) noted that in 1983, 17% of intensively managed grassland received in excess of 300 kg N ha^{-1}, and 5% received more than 400 kg. The utilisation and management of grass must be a central component of any N control strategy, particularly in catchments where grass crops are common. While cut swards may contribute little to leaching losses at fertiliser rates up to 250 kg ha^{-1} per annum, it is now well established that grazed swards can cause significant N control problems (White, 1989).

Grazing practices introduce a serious complication to any N control strategy. For example, in the case of cattle, Jarvis, Hatch and Roberts (1989b) estimated returns of 320 kg N ha^{-1} per annum from young steers on a grass sward receiving 420 kg N ha^{-1}. The degree of spatial and temporal variability associated with N returns from grazed land can be very large,

Table 11.2 Fertiliser N use according to grassland utilisation (after Chalmers, Kershaw and Leech, 1990)

Total N	1983	1984	1985	1986	1987	1988
			Overall kg ha^{-1}			
Grazed intensively, not mown	166	154	171	180	169	148
Grazed intensively, mown	175	186	189	184	211	178
Cut for silage	198	219	197	207	195	190
Cut for hay	88	88	80	83	83	78
Other grazings, not mown	76	83	79	79	76	64
All grass	125	132	131	135	133	116
P_2O_5						
Grazed intensively, not mown	25	20	23	23	22	23
Grazed intensively, mown	30	30	30	29	32	26
Cut for silage	38	39	33	34	32	32
Cut for hay	27	22	20	16	20	21
Other grazings, not mown	22	21	19	16	16	16
All grass	26	25	24	22	23	21
K_2O						
Grazed intensively, not mown	22	22	24	26	23	24
Grazed intensively, mown	44	53	52	53	62	48
Cut for silage	57	73	64	69	66	63
Cut for hay	25	24	23	21	25	24
Other grazings, not mown	17	19	17	16	16	14
All grass	28	33	32	33	33	30

as demonstrated by nitrate-N concentrations, which under urine patches can range from 300−600 kg N ha^{-1} (Ball and Ryden, 1984). Under such conditions, gaseous losses of nitrogen as nitrous oxide and ammonia can total more than leaching losses (Ryden, 1984). Nevertheless, leaching losses of nitrate-N from grass systems must not be underestimated. Ryden, Ball and Garwood (1984) demonstrated that leaching losses from a medium-textured soil developed over chalk was more than five times higher from a grass sward that was grazed as opposed to cut. In addition, the recorded losses exceeded those normally observed from arable soils in south-east England.

Available evidence indicates that attempting to reduce the incidence of denitrification in grazed grass situations, for example, by increasing the efficiency of drainage systems, will lead to greater leaching losses of N due to increased mineralisation (Scholefield, Garwood and Titchen, 1988). Results from the long-term drainage trial at North Wyke, Devon, conducted by the Institute of Grassland and Environmental Research (Garwood, 1988), clearly demonstrates the extent to which pasture management, through fertiliser application, drainage or reseeding, can modify N leaching (Table 11.3). Drainage significantly increased nitrate losses, for example, by 180% and 290% for the permanent grass, low and high N treatments, respectively. In this experiment, reseeding has increased the efficiency of N use, such that losses from the new sward are approxiately half that from the permanent pasture. The results quoted in Table 11.3 are averages from several years of observations. Ploughing out grass

Table 11.3 Annual N-balance for undrained and drained permanent grass and reseeded perennial ryegrass swards, kg N ha^{-1} (after Garwood, 1988)

	Permanent grass				Reseed	
	Undrained	Drained	Undrained	Drained	Undrained	Drained
N-input	200	200	400	400	400	400
Animal output						
(cattle LWG[a])	24	27	28	30	27	34
Leached	20	56	48	187	24	74
Denitrified	90	56	111	84	115	81
Ammonia loss	30	30	60	69	101	113
Storage in soil						
organic matter	72	64	48	45	137	128
Total N						
accounted for	236	233	295	415	404	430

[a] LWG = Liveweight gain.

and establishing a new crop, whether arable or grass, invariably leads to high losses of nitrate in the year immediately succeeding pasture inversion. This is particularly true in the case of grass/clover or other legume-based systems, which form the cornerstone of many low-input and organic farming systems. EFRC (1990) reported nitrate concentrations during the winter after ploughing out legume-rich swards which peaked at 120, 200 and 290 kg $NO_3 - N$ ha^{-1} for three sites in southern England. It is important to stress that over the course of a six-year rotation, total leaching losses are much smaller in low-input and particularly organic systems.

11.3 MANURE AND SLURRY MANAGEMENT

11.3.1 TIMING OF MANURE AND SLURRY APPLICATION

Animal wastes contain extremely variable quantities of available nitrogen, either as ammonium or nitrate, depending on collection and storage conditions. Smith, Unwin and Williams (1985) reviewed a number of experiments conducted in the UK, and concluded that, on average, 30% of the total N in slurries applied to land in spring was available in the season of application, but noted that autumn applications were less efficient. Archer (1991) argued that there is a strong case for banning the autumn/early winter application of livestock wastes with a high available nitrogen content to arable land. Such wastes include poultry manure and pig slurry, but not cow slurry (Table 11.4). The risk of nitrate leaching from cow slurry is lower than pig slurry due to a higher dry matter and lower relative availability of N (Archer, 1991).

The dominant factor influencing the timing of manure and slurry application to land is the storage capacity of the livestock enterprise in question. The rapid expansion of intensive dairying and pig production systems in recent years was not matched by the development of adequate slurry handling facilities; the increased number of farm pollution incidents recorded during the 1980s in the UK illustrate this fact (NRA/MAFF, 1990). This lack of sufficient storage capacity forced farmers to spread slurry during the winter months, but rarely in the critical autumn period, as slurry lagoons were invariably empty at the end of the summer.

Table 11.4 Typical total and available nitrogen content of livestock wastes (after Archer, 1991)

	Dry matter (%)	Total nitrogen (fresh basis)	N available (%)
Farmyard manure — pig	25	6.0 kg t^{-1}	25
— cattle	25	6.0 kg t^{-1}	25
Pig slurry — undiluted	10	6.0 kg m^{-3}	65
— separated solid	15	6.0 kg m^{-3}	40
— separated liquid	4	6.0 kg m^{-3}	85
Cow slurry — undiluted	10	5.0 kg m^{-3}	30
Dairy dirty water	0.1 – 1	0.02 – 2 kg m^{-3}	85
Broiler manure	70	24 kg t^{-1}	65
Battery manure	70	42 kg t^{-1}	65

In the UK, new slurry lagoons must now have a 4-month storage capacity (MAFF/WOAD, 1991); in Denmark, the storage requirement is 9 months (Hansen, 1989). Larger stores will obviate the need for spreading during the winter months when the crop is not growing, hence reducing the nitrate leaching risk.

11.3.2 METHODS OF MANURE AND SLURRY APPLICATION

Methods of handling and spreading farm waste play a relatively small role in any strategy to control nitrate leaching, but experience from the Low Countries, particularly the Netherlands, indicates that where direct runoff in ditches is a potential threat, alternative methods of disposal other than surface application must be considered. Tine injection of liquid wastes beneath the soil surface has been employed in the Netherlands as a means of reducing surface runoff and volatilisation losses of N (Korevaar and den Boer, 1990). In addition to increasing the efficiency of N use, injection can cut down the risk of surface runoff. Steenvoorden (1989) demonstrated that, provided the technique is carried out with care, injection can result in lower leaching losses in the autumn/winter period following slurry application due to the enhanced take-up of N. Injection of slurry containing up to 100 kg N ha^{-1} did not lead to higher leaching losses. In Figure 11.4 the extent of leaching from slurry applied by different means is compared with surface spread artificial N. Only when total available mineral N exceeds 800 kg ha^{-1} per annum does injection lead to high leaching losses. In this experiment, injection was carried out into at a depth of 150 mm into a growing crop in the spring months. Clearly, this process is not suitable for the disposal of liquid wastes during the autumn/winter period, or at greater depths over permeable subsoils or intensive underdrainage systems.

In the UK, current advice within the *Code of Good Agricultural Practice for the Protection of Water* (MAFF/WOAD, 1991) allows for the surface application of dirty water from livestock units (less than 3% dry matter, but very variable available N content — see Table 11.4) at any time of the year provided the conditions are suitable. Separation of slurry (typically, 4–7% dry matter) into dirty water and a solid for handling separately during the spring and

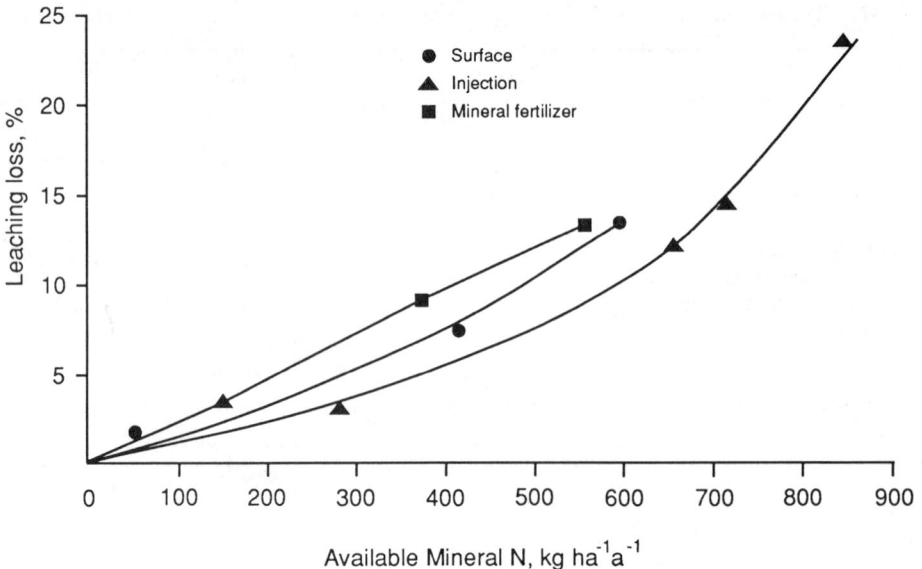

Figure 11.4 Nitrate leaching as a percentage of available N (mineral N from fertilisers and slurry) as influenced by available inorganic N from slurry and fertiliser (after Steenvoorden, 1989)

summer offers farmers with a limited liquid waste storage capacity a solution to the winter storage problem without the need to invest in a significantly enlarged slurry store. The implications of this increasingly common practice are yet to be evaluated.

11.4 IMPACT OF RECENT LEGISLATION

11.4.1 RECENT LEGISLATION

Several important items of legislation that relate to water quality and agricultural practice have been introduced in recent years within the European Community. These are introduced in Chapter 1 and described in detail in Chapter 14. Successive measures that are now having or will soon have an impact on agricultural practice include (1) Directive 80/778/EEC, relating to the quality of water for human consumption (CEC, 1980), which came into effect in 1985 and (2) Draft Directive 4136/89 relating to the protection of fresh and marine waters against pollution by nitrates from diffuse sources (European Commission, 1989). The 1980 directive relates directly to water quality, and influences agricultural practice only indirectly. The draft directive on nitrate from diffuse sources will, if enforced in its present form, have rather more serious and direct consequences for the agriculture industry. Although this directive is only in draft form at present, several of the recommendations in the *Code of Good Agricultural Practice* adopted by MAFF/WOAD (1991) have been written with this potential directive in mind.

   Individual European countries have progressed with 'farming-orientated' nitrate control measures at varying rates according to the extent of the problem and the political climate in each country concerned. In Denmark, where the regulations are most strict, the following measures were in place in 1988 (Hansen, 1989):

(1) Farms with more than 30 livestock units (LSUs) must have a waste storage capacity of 9 months;
(2) Slurry must not be applied to uncropped soil between harvest and 1 November;
(3) Maximum stocking density must not exceed specified limits (for example, 2.3 LSU ha^{-1} for cattle);
(4) Slurry must be incorporated into the soil within 12 hours of application;
(5) All farms must prepare crop rotation and fertiliser requirement programmes;
(6) Sixty-five per cent of the area of each farm must be cropped during the autumn (as from 1990).

In the Netherlands, nitrate control has assumed less importance than the restriction of phosphorus losses from agricultural systems, due to the more important role that P plays in controlling the eutrophication of surface waters (see Chapter 5). Regulations in place in 1987 prohibit the land spreading of slurry in the months of October and November, and during January and February when the fields are snow-covered (Steenvoorden, 1989).

Three recent articles of legislation have affected, and will continue to affect, agricultural practice in many parts of the UK in future years. Section 112 of the Water Act 1989 made provision for the establishment of Nitrate Sensitive Areas (see Chapters 1 and 12 for a fuller discussion). Section 92 of the Water Resources Act 1991 reinforced these provisions (NRA, 1992). This latter article of legislation makes provision for the implementation of a *Code of Good Agricultural Practice*, with the purpose of (1) giving practical guidance to persons engaged in agriculture with respect to activities that may affect controlled waters and (2) promoting desirable practices for avoiding or minimising the pollution of any such waters. This code, which has major implications for the agriculture industry, is discussed in detail in Section 11.4.2. Contravention of the code will not itself give rise to criminal or civil liability, but failure to comply with it could be taken into account in any legal proceedings following a pollution incident (MAFF/WOAD, 1991). The second recent article of legislation that has a direct bearing on the nitrate problem is the Environmental Protection Act 1990, which under Section 152 empowers the Minister for Agriculture to prohibit or restrict the burning of crop residues on agricultural land by persons engaged in agriculture. This ban was imposed after disposal of residues from the 1992 harvest, in England and Wales only. The consequences of the adoption of straw incorporation on a much wider scale than hitherto can only, in the short term, aid nitrogen immobilisation and reduce leaching losses (see Section 11.2.3).

11.4.2 GOOD AGRICULTURAL PRACTICE

The guidelines which form the basis of the *Code of Good Agricultural Practice for the Protection of Water* (MAFF/WOAD, 1991) concentrate heavily on the control of pollution from slurry-handling systems, parlour washings, silage effluent and surface runoff, which in 1989 comprised 73% of farm pollution incidents (NRA/MAFF, 1990). The final chapter of the *Code* deals specifically with aspects of nitrate control, with the emphasis being placed on methods by which losses can be reduced at little or no cost to the farmer. It is stressed that in some catchments where high nitrate levels are already a problem, measures beyond those laid down in the *Code* will need to implemented, particularly when the proposed directive on nitrates from diffuse sources comes into force. Many of the recommendations embodied in the *Code* have been founded on the extensive research work conducted primarily in Europe and the USA over the last 20 years, much of which has been reported in this chapter and elsewhere in this volume. Adherence to aspects of this *Code* which relate to N control, the major points of which are summarised below, will lead to major changes in agricultural practice (MAFF/WOAD, 1991). A more detailed summary is given in the Appendix to this chapter.

(1) *Ploughing up of grass*. This should be avoided wherever possible. Reseeding should be done with the minimum of cultivation and with full crop cover ensured by early October. Arable crops preceding grass should be established as early as practicable, especially autumn cereals.

(2) *Organic manures*. The maximum recommended application rate is 250 kg N ha^{-1} per annum determined on a LSU basis. Liquid wastes, such as cow and pig slurry, should not be applied to arable land between harvest and 1 November.

(3) *Inorganic N fertilisers*. The economic optimum should be assessed for each field, taking into account crop requirement and soil N supply. MAFF recommendations should be followed (MAFF, 1988). Autumn application of N should only be made where there is good evidence of uptake. There is no such evidence for autumn cereals. Single applications are limited 120 kg N ha^{-1}.

(4) *Crop cover*. Autumn sown crops should be established as soon as is agronomically possible, and ideally by early September. Bare soil over the winter period should be avoided.

(5) *Crop residues*. Low N residues, such as cereal straw, should be incorporated into the soil in the autumn, but high N residues, such as vegetable waste, should not be incorporated until just before sowing the next crop.

(6) *Autumn cultivations*. These should be delayed as long as possible without delaying the establishment of the next crop. Residues of late harvested crops, such as maize and sugar beet, can be left undisturbed until just before sowing the following spring crop.

(7) *Grassland management*. Reduction in the grazing intensity, particularly in the autumn, will reduce the amount of nitrate leached.

It is difficult to evaluate the likely consequences of adoption of this *Code of Good Agricultural Practice* for water quality on a catchment basis. Although the code is a statutory instrument under the Water Resources Act 1991, there is no legal requirement to adhere to it, although, as mentioned above, failure to do so may be taken into account in a legal action where a pollution event has occurred (MAFF/WOAD, 1991). In the short term it is likely that there will only be a patchy adoption of the measures listed above. Further discussion on this issue can be found in Chapters 12 and 15.

11.4.3 THE NITRATE SENSITIVE AREAS SCHEME

The MAFF introduced the Nitrate Sensitive Areas (NSA) Scheme in 1989 in areas where nitrate concentrations in water sources exceed, or are at risk of exceeding, the limit of 11.3 mg NO_3-N l^{-1} specified by the European Community Drinking Water Directive 80/778. The aim of the scheme is to control the entry of nitrate from agricultural land into water sources (MAFF, 1989). Further discussion of catchment-scale implications and the legal aspects of the scheme can be found in Chapters 1, 12 and 14, respectively. Only the proposed changes in agricultural practice specified in the scheme are discussed here.

The NSA scheme is structured on three tiers:

(1) *The Advisory Campaign*, targeted at both the NSAs proper and catchments where only the intensive advisory campaign would operate. Advice is given free by ADAS

representatives on how to adopt good agricultural practices which will reduce the amount of nitrate leached from their land. Most of this advice is now embodied in the *Code of Good Agricultural Practice*, outlined above.

(2) *The Basic Rate Payment Scheme* involves substantial restrictions in agricultural practice which go far beyond those identified in the *Code of Good Agricultural Practice*. Payment under the scheme, at a rate of £40 ha^{-1}, would require acceptance of a variety of measures on all the farm NSA land. The more important conditions which are more stringent than the *Code* are given below. Full details are supplied in MAFF (1989).

    (a) Fertiliser use — apply less than the economic optimum amount of N fertiliser on specified crops, for example, winter wheat: 25 kg N ha^{-1} below optimum, winter oilseed rape: 50 kg N ha^{-1} below optimum.

    (b) Crop cover — autumn sow crops should be drilled by 15 October if harvest of the previous crop allows, but September drilling is preferred. If the previous crop is removed after 15 October, it is recommended that a spring crop is drilled in November. Any cover crop should not be removed before 1 February on sandy soils or 1 December on loamy soils.

    (c) Organic manures — annual applications are limited to 175 kg N ha^{-1}. No slurry or poultry manure should be applied between 1 September and 1 November of grassland or between 1 July and 1 November on annual by cropped fields.

    (d) Ploughing up grass — latest sowing date for the following crop is 1 October.

    (e) Others — right of access to fields to carry out leaching measurements must be provided. Hedgerows, woodlands and ponds should be maintained.

(3) *The Premium Rate Payment Scheme* is only available in addition to the basic scheme, and can apply to all or part of a farm. The scheme calls for a complete change in farming system, usually from arable to permanent pasture or woodland. Compensation may be given up to £210 ha^{-1} in addition to any payments under the Basic Rate Payment Scheme. The level of compensation depends upon the degree of change. For example, conversion of arable land to permanent, ungrazed grass, sown to a conservation seeds mixture, excluding clover and receiving no artificial or organic nitrogen will qualify for the highest rate of payment. If grazing or fertiliser application up to 150 kg N ha^{-1} is allowed, then payments are reduced accordingly.

11.4.4 USE OF SEWAGE SLUDGE ON AGRICULTURAL LAND

Sewage sludge contains valuable plant nutrients, including N in various forms, which can substitute for fertiliser nutrients. Liquid digested sewage sludge typically contains 0.2% N, of which approximately half is in an available form (Water Research Centre, 1985). The total percentage figure compares directly with FYM, although sludge has a higher N availablility. The total production of sewage sludge in the UK exceeds 1 million t dry solids, of which approximately 40% is dumped or recycled to agricultural land. In the light of the ban on sea dumping of sludge in 1998, this figure may well increase. However, contamination problems caused by heavy metals and pathogens have lead to less frequent use of this N source than might otherwise have been the case. The Sludge (Use in Agriculture) Regulations 1989 (Statutory Instruments, 1989) and the *Code of Practice for Agricultural Use of Sewage Sludge* (DOE, 1989) define the conditions under which sewage sludge can be used by the agriculture industry. The latter reference and the more recent *Code of Good Agricultural Practice for the Protection of Water* both stress the beneficial aspects of this nutrient source and place restrictions on its use in vulnerable situations, such as when runoff or rapid sludge movement to land drains may occur (see Appendix).

11.5 FUTURE CHANGES

Many agricultural practices, such as those described in Section 11.4.2, have a direct effect on nitrate losses from soils. The measures included in the UK *Code of Good Agricultural Practice* and the more restricive NSA regulations are designed to reduce nitrate leaching losses, and are based on experimentally derived relationships between agricultural practice and N loss. These measures are only voluntary in the UK at present, and monitoring is underway to establish whether significant reductions in leaching losses can be achieved at catchment scale as a result of field and farm-scale changes in practice. In other European countries changes have been imposed upon the agricultural community. If the draft directive on nitrates from diffuse sources is adopted, then further restrictions in farming activity will occur. The implementation of a more comprehensive Set-Aside scheme throughout the EC in 1992/93 will also have a major, but as yet unquantified, impact on nitrate losses from farmland. The management rules for Set-Aside land stipulate that uncropped land should not be left bare over winter. The use of either natural regeneration or planted cover is allowed (MAFF, 1992). In addition, use of fertilisers and manures on land taken out of production is severely restricted. For the cropping year 1992/93, the Set-Aside area will be equivalent to 15% of the total eligible for Area payments, which applies to land cropped with either cereals, oilseeds or protein crops. These changes in land management practices will have a major impact on nutrient losses from arable land.

It is difficult to speculate as to whether the changes embodied in the various Codes of Practice and EC Directives will have immediate effects on nitrate losses from soils. In the case of non-point source aquifer contamination, where there is a long time lag in the soil/aquifer leaching system, it will be many years before the full impact of these changes is felt (see Chapter 8). Direct runoff from point sources should be reduced in the future if the recommendations are adopted by the industry. More research and advisory effort must be put into developing techniques which will minimise the quantity of nitrate available for leaching each autumn and into determining the amount of mineral N available in the spring, so that fertiliser inputs can be adjusted accordingly.

APPENDIX

The Code of Good Agricultural Practice for the Protection of Water (MAFF/WOAD, 1991) is a practical guide to help farmers and growers avoid causing water pollution. As a Statutory Code under Section 97 of the Water Resources Act 1991, it will not be an offence to ignore the guidelines laid down in the *Code*, but failure to adhere to the advice could be taken into account in any legal action. The *Code* contains chapters on a variety of agricultural materials whose mishandling may affect water quality, including slurry, farm yard manure, dirty water, silage effluent, fertilisers, fuel oil, sheep dip and pesticides. The main points in the *Code* are listed below. It is stressed that in some catchments action beyond the good agricultural practice described will be needed to keep water below the levels specified in EC Drinking Water Directive.

(1) *Ploughing up of grass.* This should be avoided if possible. If permanent grass or short-term leys need reseeding, it should be done with the minimum of cultivation and in a way which ensures full crop cover by early October. Arable crops following grass should be established as early as practicable, especially autumn cereals.

(2) *Organic manures.* The maximum recommended application rate is 250 kg N ha^{-1} per annum, determined on a LSU basis. Total N content of manures, slurries and sewage sludges is calculated from standard tables giving %N content and average daily waste production by farm animals. Liquid wastes, such as cow and pig slurry, should not be applied to arable land between harvest and 1 November. There are no time limitations on the spreading of solid manures and sludges, where the concentration of readily available forms of N are low.

(3) *Inorganic N fertilisers.* The economic optimum should be assessed for each field, taking into account crop requirement and soil N supply. MAFF recommendations should be followed (MAFF, 1988). Autumn application of N should only be made where there is good evidence of uptake. This may include winter oilseed rape but not autumn cereals. Applications to grass from mid-season onwards should be reduced if growth is checked (for example, by drought). Spring seedbed applications should be limited to that which will be taken up by the crop soon after application. Records of the amount and dates of fertiliser and organic manure applications should be kept in order to facilitate nutrient loading calculations. Single application limits are 120 kg N ha^{-1}.

(5) *Crop cover.* Autumn-sown crops should be established as soon as is agronomically possible, and ideally by early September. Sowing crops after mid-October will have little effect on winter nitrate leaching. Where possible, select varieties that are suitable for early sowing. Bare soil over the winter period should be avoided.

(5) *Crop residues.* Low N residues, such as cereal straw, should be well mixed into the soil in the autumn to encourage N immobilisation. High N residues, such as vegetable waste and most other non-cereal residues, should not be incorporated until just before sowing the next crop.

(6) *Autumn cultivations.* These should be delayed as long as possible without delaying the establishment of the next crop. This particularly applies to establishment after leguminous crops, such as vining peas. Residues of late-harvested crops, such as maize and sugar beet, can be left undisturbed until just before sowing the following spring-crop.

(7) *Grassland management.* Risk of nitrate loss from intensively grazed grassland is high in the autumn. Reduction in the grazing intensity will reduce the amount of nitrate leached.

(8) *Irrigation.* A reliable irrigation scheduling system should be used to prevent excessive water applications which may lead to free water draining out of the profile, and consequent nitrate leaching.

REFERENCES

Addiscott, T.M. (1988) Long-term leakage of nitrate from bare unmanured soil. *Soil Use and Management*, **4**, 91–5.

Archer, J.R. (1991) Management of livestock wastes for nitrate control. In *Pollution — the New Standards*, RASE/ADAS Conference, 6 February 1991.

Ball, P.R. and Ryden, J.C. (1984) Nitrogen relationships in intensively managed temperate grasslands. *Plant and Soil*, **76**, 23−33.

Bremner, J.M. (1990) Problems in the use of urea as a nitrogen fertilizer. *Soil Use and Management*, **6**, 70−71.

Chalmers, A., Kershaw, C. and Leech, P. (1990) Fertiliser use in farm crops in Great Britain: results from the survey of fertiliser practice, 1969−88. *Outlook on Agriculture*, **19**, 269−78.

Christian, D.G., Crees, R. and Dowdell, R.J. (1985) Yield and uptake of fertilizer nitrogen by direct-drilled winter barley growing on a chalk soil. *Soil Use and Management*, **1**, 74−9.

Christian, D.C. Goss, M.J., Howse, K.R., Powlson, D.S. and Pepper, T.J. (1990) Leaching of nitrate through soil. *Institute of Arable Crops Research Report for 1989*, 67−8.

Church, B.M. and Lewis, D.A. (1977) Fertiliser use on farm crops in England and Wales. Information from the survey of fertiliser practice 1942−72. *Outlook on Agriculture*, **9**, 186−93.

Council of European Communities (1980) Directive relating to the quality of water for human consumption, Directive 80/778/EEC, Brussels.

DOE (1989) *Code of Practice for Agricultural Use of Sewage Sludge*, HMSO, London.

Dowdell, R.J., Colbourn, P. and Cannel, R.Q. (1987) A study of mole drainage with simplified cultivation for autumn sown crops on a clay soil. 5. Losses of nitrate-N in surface run-off and drain water. *Soil and Tillage Research*, **9**, 317−31.

EFRC (1990) *Nitrogen mineralisation in organic ley/arable farming systems*, Research Note No. 7. Elm Farm research Centre, Newbury.

European Commission (1989) The protection of fresh and marine waters against pollution by nitrates from diffuse sources, Doc No. 4136/89 (Com (88) 708 final).

Garwood, E.A. (1988) Water deficiency and excess in grassland: the implications for grass production and for the efficiency of use of N. In Wilkins, R.J. (ed.), *Proceedings of an IGAP Symposium*, 27 March 1987, pp. 24−41.

Hansen, J.F. (1989) Nitrogen balance in agriculture in Denmark and ways of reducing the loss of nitrogen. In Germon J.C. (ed.), *Management Systems to Reduce Impact of Nitrates*, Elsevier, London, pp. 1−14.

Harris, G.L., Goss, M.J., Dowdell, R.J., Howse, K.R. and Morgan, P. (1984) A study of mole drainage with simplified cultivation for autumn-sown crops on a clay soil. 2. Soil water regimes, water balances and nutrient losses in drain water, 1978−80. *Journal of Agricultural Science, Cambridge*, **102**, 561−81.

Hauck, R.D. (1990) Agronomic and public aspects of soil nitrogen research. *Soil Use and Management*, **6**, 66−70.

Jarvis, S.C., Barraclough, D., Unwin, R.J. and Royle, S.M. (1989a) Nitrate leaching from grazed grassland and after straw incorporation in arable soils. In Germon, J.C. (ed.), *Management Systems to Reduce Impact of Nitrates*, Elsevier, London, pp. 110−23.

Jarvis, S.C., Hatch, D.J. and Roberts, D.H. (1989b) The effect of grassland management on nitrogen losses from grazed swards though ammonia volatilization; the relationship to excretal returns from cattle. *Journal of Agricultural Science, Cambridge*, **112**, 205−16.

Jenkinson, D.S. (1988) Soil organic matter and its dynamics. In Wild, A. (ed.), *Russell's Soil Conditions and Plant Growth*, Longmans, London, pp. 564−607.

Jenkinson, D.S. (1990) An introduction to the global nitrogen cycle. *Soil Use and Management*, **6**, 56−60.

Korevaar, H. and den Boer, D.J. (1990) Practical measures to reduce nutrient losses from grassland systems. *Proceedings of the Fertilizer Society*, **301**, 1−34.

MAFF (1988) *Fertilizer Recommendations for Agricultural and Horticultural Crops*, HMSO, London.

MAFF (1989) *The Nitrate Sensitive Areas Scheme*, HMSO, London.

MAFF (1992) *Arable Area Payments*, MAFF Explanatory booklet, London.

MAFF/WOAD (1991) *The Code of Good Agricultural Practice for the Protection of Water*, MAFF, London.

Marks, H.F. and Britton, D.K. (1989) *One Hundred Years of British Food and Farming: a Statistical Survey*, Taylor and Francis, London.

NRA/MAFF (1990) *Water Pollution from Farm Waste: 1989 Survey of Reported Water Pollution Incidents*, NRA South West, Exeter.

NRA (1992) *The Influence of Agriculture on the Quality of Natural Waters in England and Wales*.

NRA Water Quality Series No. 6, Bristol.

Neeteson, J.J., Dilz, K. and Wiljnen, G. (1989) N-fertilizer recommendations for arable crops. In Germon, J.C. (ed.), *Management Systems to Reduce Impact of Nitrates*, Elsevier, London, pp. 253—63.

Østergaard, H.S. (1989) Analytical methods for optimization of nitrogen fertilization in agriculture. In Germon, J.C. (ed.), *Management Systems to Reduce Impact of Nitrates*, Elsevier, London, pp. 224—35.

Parkinson, R.J., Twomlow, S.T. and Reid, I. (1988) The hydrological response of a silty clay loam following drainage treatment. *Agricultural Water Management*, **14**, 125—36.

Pedersen, C.A. (1990) Practical measures to reduce nutrient losses from arable land (annual crops). *Proceedings of the Fertilizer Society*, No. **300**, 1—24.

Powlson, D.S. (1990) Straw incorporation — nitrate implications. In *Burning Ban — The Final Straw*, RASE/ADAS Agricultural Conference, 28 November 1990.

Powlson, D.S., Jenkinson, D.S., Pruden, G. and Johnston, A.E. (1985) The effect of straw incorporation on the uptake of nitrogen by winter wheat. *Journal of the Science of Food and Agriculture*, **136**, 26—30.

Powlson, D.S., Hart, P.B.S., Pruden, G. and Jenkinson, D.S. (1986) Recovery of [15]N- labelled fertilizer applied in autumn to winter wheat at a few sites in eastern England. *Journal of Agricultural Science*, **107**, 611—20.

Powlson, D.S., Brookes, P.C. and Christensen, B.T. (1987) Measurement of soil microbial biomass provides an early indication of changes in total soil organic matter due to straw incorporation. *Soil Biology and Biochemistry*, **19**, 159—64.

Richter, G.M., Hoffman, A., Nieder, R. and Richter, J. (1989) Nitrogen mineralization in loamy arable soils after increasing the ploughing depth and ploughing grasslands. *Soil Use and Management*, **5**, 169—73.

Ryden, J.C. (1984) The flow of nitrogen in grassland. *Proceedings of the Fertilizer Society*, **229**, 1—44.

Ryden, J.C., Ball, P.R. and Garwood, E.A. (1984) Nitrate leaching from grassland. *Nature*, **311**, 50—53.

Scholefield, D., Garwood, E.A. and Titchen, N.M. (1988) The potential of management practices for reducing losses of nitrogen grazed pastures. In Jenkinson, D.S. and Smith, K.A. (eds), *Nitrogen Efficiency in Agricultural Soils*, Elsevier, London, pp. 220—31.

Smith, K.A., Unwin, R.J. and Williams, J.H. (1985) Experiments on the fertilizer value of animal waste slurries. In Williams, J.H., Guidi, G. and L'Hermite, P. (eds) *Long Term Effects of Sewage Sludge and Farm Slurries Applications*, Elsevier, London, pp. 124—35.

Steenvoorden, J.H.A.M. (1989) Agricultural practices to reduce nitrogen losses via leaching and surface runoff. In Germon, J.C. (ed.) *Management Systems to Reduce Impact of Nitrates*, Elsevier, London, pp. 72—81.

Sylvester-Bradley, R., Addiscott, T.M., Vaidyanathan, L.V., Murray, A.W.A. and Whitmore, A.P. (1987) Nitrogen advice for cereals: present realities and future possibilities. *Proceedings of the Fertilizer Society*, **263**, 1—36.

Toosey, R.D. (1989) Arable crops. In Halley, R.J. and Soffe, R.J. (eds), *The Agricultural Notebook*, Butterworths, London, pp. 90—176.

Vaughan, J. (1985) Effects of source and amount of fertilizer nitrogen and nitrification inhibitors on the yield and nitrate concentration of glasshouse lettuce. *Soil Use and Management*, **1**, 80—81.

Vinten, A.J.A., Howard, R.S. and Redman, M.H. (1991) Measurement of nitrate losses from arable plots under different nitrogen input regimes. *Soil Use and Management*, **7**, 3—13.

Water Research Centre (1985) *The Agricultural Value of Sewage Sludge. A Farmers' Guide*, WRc, Medmenham.

Wehrmann, J. and Scharpf, H.-C. (1989) Reduction on nitrate leaching in a vegetable farm — fertilization, crop rotation, plant residues. In Germon, J.C. (ed.), *Management Systems to Reduce Impact of Nitrates*, Elsevier, London, pp. 147—56.

White, R.E. (1989) Leaching. In Wilson, J.R. (ed.), *Advances in Nitrogen Cycling in Agricultural Ecosystems*, CAB International, Wallingford, pp. 193—211.

12 Controlling Losses of Nitrate by Changing Land Use

T.P. BURT

School of Geography, University of Oxford

and

N.E. HAYCOCK

Department of Agricultural Water Management, Silsoe College

12.1 FARMING AND NITRATE POLLUTION

12.1.1 NITRATE FERTILISER AND NITRATE LEACHING

Modern agriculture is now recognised by both farmers and environmentalists as a significant source of water pollution. Sediment from eroded soil, pesticides and fertilisers, both organic and inorganic, can adversely affect the quality of surface and ground waters. These diffuse or 'non-point' pollutants enter rivers and aquifers over wide areas of a drainage basin (in contrast to 'point' sources such as the effluent from a sewage-treatment works). The environmental effects of agricultural pollutants may be of little consequence to the farmer, at least in the short term, but this may not be so for other members of the community. A strong concern has been expressed about increases in the nitrate concentration of rivers and aquifers over recent years. In the United Kingdom, publications such as those by the Department of the Environment (1986, 1988) and the Royal Society (1983) reflect this concern by both government and non-governmental agencies.

The popular misconception is that the nitrate problem is caused by farmers applying too much nitrate fertiliser to crops so that the surplus left after harvest is leached away in the following winter. This is too simplistic. Nevertheless, there is now little doubt that runoff from agricultural land is responsible for the high concentrations of nitrate in fresh waters noted in recent years and that progressive intensification of agricultural practices, with increasing reliance on the use of nitrogen fertiliser, has contributed to this problem. As demonstrated in Chapter 3, a complex and prolonged cycling of applied nitrogen through various compartments of the soil nitrogen cycle is involved, so that it is not necessarily the same molecules of nitrogen applied as fertiliser which are lost by leaching. In one experiment, use of ^{15}N-labelled fertiliser showed that very little labelled inorganic nitrogen remained in the soil at harvest; about 70% of the labelled nitrogen was in the above-ground parts of the crop, with most of the remainder in the soil organic matter (Powlson *et al.*, 1986a).

Organic matter in the soil is of varying complexity and stability; nitrogen immobilised in organic matter derived from the microbial biomass may be mineralised most rapidly (Skjemstad, Vallis and Myers, 1988). As noted below, soils which have regularly received nitrogen fertiliser may contain much easily mineralised organic matter. This nitrogen can be rapidly mineralised when conditions are appropriate (warm, moderately wet soil) and may be liable to leaching if there is no actively growing crop. Thus, where the amount of fertiliser

Nitrate: Processes, Patterns and Management. Edited by T.P. Burt, A.L. Heathwaite and S.T. Trudgill
© 1993 John Wiley & Sons Ltd

applied is large, the loss of nitrate can be equivalent to a significant proportion of the fertiliser nitrogen added. The word 'equivalent' indicates that most of the nitrate leached does not come directly from the nitrate applied in the previous season. Most of the applied nitrogen which is not taken up by the crop, or that which is returned to the soil as plant residues or as dung and urine, will become incorporated into the soil organic matter and microbial biomass, a proportion of this organic matter being mineralised each year (Severn-Trent Water, 1988).

12.1.2   TIME OF APPLICATION OF NITRATE FERTILISER

A number of studies have provided annual nitrogen budgets for plots or lysimeters. These are most helpful in demonstrating the relation between inputs, outputs and the storage of nitrogen in the soil, but cannot in themselves provide information on the cycling of nitrogen within the soil. Powlson *et al.* (1986a) provide a nitrogen cycle for the Broadbalk plot 8 at Rothamsted (UK) on which winter wheat has been grown for many years, receiving 144 kg N ha^{-1} every spring (Figure 12.1). Inputs of nitrogen in fertiliser, from wet and dry deposition, and from biological fixation are balanced by losses in crop uptake, denitrification and leaching; the longevity of the experiment ensures a steady state. Fertiliser is applied in the spring and leaching losses are low; nevertheless, only about 72% of the total nitrogen added is recovered in the crop. An equivalent of 28% of the total input of N is lost from the system, mainly by denitrification. The level of inorganic nitrogen present is highly variable; 30–60 kg N ha^{-1} is typical in autumn, falling to 2–16 kg N ha^{-1} in spring when the fertiliser is applied. For a nearby plot where additional fertiliser (50 kg N ha^{-1}) is added in autumn, losses increase considerably. Recovery of the autumn-applied fertiliser nitrogen is low (11–40%). Between 40% and 80% of the fertiliser nitrogen is lost mainly (if not entirely) by leaching (Powlson *et al.*, 1986b). Autumn-applied nitrogen increases the grain yield slightly, but from an environmental point of view, an application of nitrogen in early spring is safer than an equivalent application in autumn as nitrate is exposed to leaching for a shorter period and is thus less likely to enter rivers or aquifers. Many studies have shown that nitrate losses from grassland are small when fertiliser applications are modest (<250 kg ha^{-1} a^{-1}). However, even for grassland, large applictions of nitrogen fertiliser can lead to large leaching losses, especially if the sward is grazed (Ryden, Ball and Garwood, 1984).

12.1.3   FERTILISER NITRATE AND SOIL MICROSTRUCTURE

Arable soils often contain more inorganic nitrogen in autumn than is required by overwintering crops. Some of this nitrogen will be from fertiliser not taken up by the previous crop, but most will come from mineralisation of soil organic matter. In either case, the more inorganic nitrogen remaining, the greater is the potential for loss by leaching (Powlson *et al.*, 1986b). Even though there may be little inorganic nitrogen in the soil at harvest, rates of mineralisation are often high in autumn when warm soils begin to wet up. Substrate may be inaccessible to soil micro-organisms during dry periods; large (>10 μm) soil pores are dry while those below 0.48 μm are small enough to exclude micro-organisms (McGill and Myers, 1987). As soil water content increases, microbial activity increases as micro-organisms are no longer isolated from decomposable substrate. Powlson *et al.*, (1986a) note that soils with a long history of receiving nitragen fertiliser, and greater returns of crop residue, contain more mineralisable nitrogen.

Crops given too much nitrogen return more organic matter to the soil as roots, root exudates

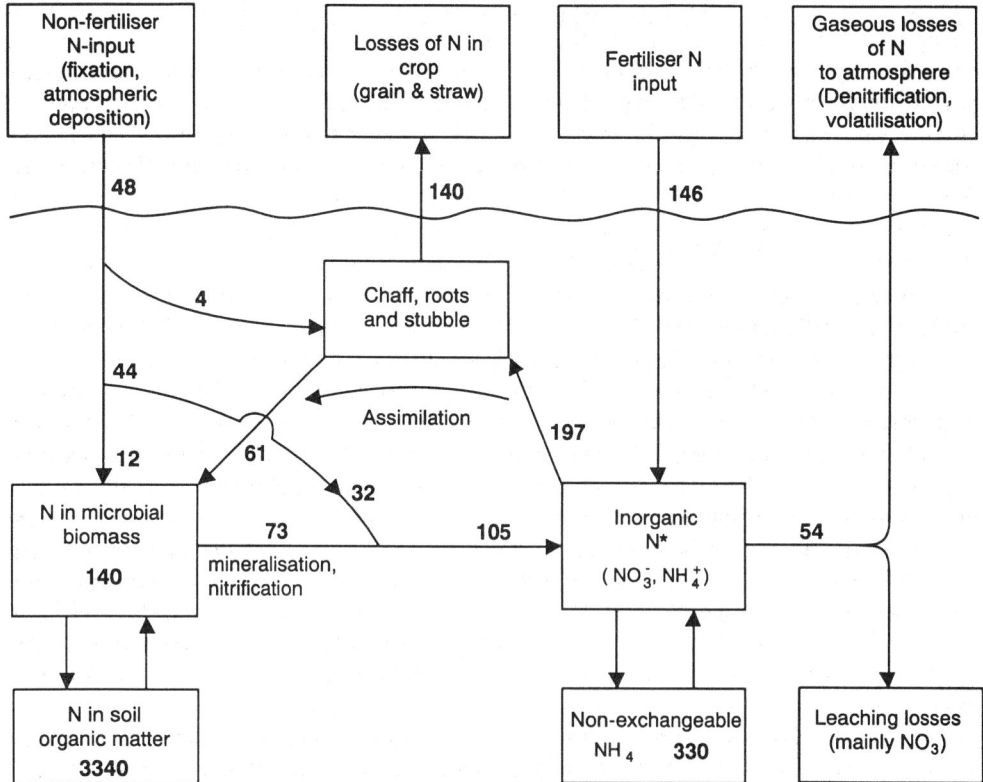

Figure 12.1 The nitrogen cycle for a plot under continuous winter wheat receiving approximately 144 kg ha^{-1} N every year (adapted from Powlson *et al.*, 1986a). Figures in kg ha^{-1} in boxes; in kg ha^{-1} a^{-1} otherwise. (Since the level of inorganic nitrogen present is highly variable, no figure is given; as a rough guide, 30−60 kg N ha^{-1} would be typical for this soil in autumn falling to 2−16 kg N ha^{-1} in spring before fertiliser is applied)

and stubble than crops given little nitrogen, so more organic nitrogen accumulates in the soil. Mineralisation of nitrogen in plots given high rates of fertiliser will therefore be correspondingly large. Total soil nitrogen increases as fertiliser application rises, mainly as organic matter. Small amounts of additional soil nitrogen, because of adding fertiliser over many years, therefore have a large effect on mineralisation rates (Powlson *et al.*, 1986b). Furthermore, it seems likely that the most recent additions of soil organic matter are the most labile and thus most easily mineralised.

Addiscott and Powlson (1989) suggest that fertilisers could indirectly affect the leaching of nitrate by increasing the amount of readily decomposable organic matter. The amount of labile organic matter is small compared to the main bulk of largely inert organic matter, but it could have a disproportionately large effect on mineralisation and on subsequent leaching of nitrate. It may be that there is never 100% recovery in the crop of even small amounts of nitrogen fertiliser because of competition for nitrate by the soil microbial population responding to an input of carbon to the soil via root exudates (Johnston and Jenkinson, 1990). The biomass so formed is, however, readily mineralised though not necessarily at a time convenient for crop uptake.

12.1.4 LEACHING LOSSES IN RELATION TO APPLICATION RATES

A number of studies have shown that nitrate losses increase significantly where the amount of nitrogen fertiliser added is large. However, there is no simple relation between nitrogen application and leaching loss: too many factors serve to complicte the situation — crop type and cover, method of cultivation (including fertiliser treatment), soil type, climatic conditions, and so on. Nitrate losses may remain small where application rates are appropriate for crop requirements, especially for grassland. However, the amounts of nitrogen lost by leaching substantially increase as the amount of added nitrogen is increased. Where nitrogen in soil is limiting, plants typically respond to increasing amounts of applied nitrogen with increased growth and/or nitrogen uptake. As nitrogen becomes less limiting, each additional increment of nitrogen applied is less efficient at higher yields; excessive amounts of nitrogen can result in a decrease in yield (Figure 12.2). However, as Hauck (1990) notes, the point of interest illustrated by Figure 12.2 is the amount of leachable nitrogen (as nitrate) in the soil profile in relation to application rate. At rates that produce little or no additional yield, soil nitrate increases rapidly. Ample field data are available to suggest that somewhere near (but below) maximum yield, the probability of accumulating leachable nitrogen increases markedly (Hauck, 1990; van der Meer, 1990). Johnston and Jenkinson (1990, table 5) show that the recovery of fertiliser nitrogen applied to winter wheat on the Broadbalk plots falls at higher application rates: apparent percentage recovery is 73% for 96 kg N ha^{-1} applied, 64% for 144 kg N ha^{-1}, 58% for 192 and 240 kg N ha^{-1} and only 52% for 288 kg N ha^{-1}. Much of the larger dressings of nitrogen may well remain as nitrate in the soil at harvest and be at risk to loss by leaching.

   Barraclough, Hyden and Davies (1983) presented results from a 3-year study of nitrate leaching from isolated 0.4 ha grassland plots fertilised with 250, 500 and 900 kg N ha^{-1} a^{-1}; cumulative leaching over the 3 years was equivalent to 1.5%, 5.4% and 16.7% of the fertiliser applied, respectively. For grassland, Garwood, Salette and Lemaire (1980) observed that on increasing the fertiliser nitrogen application from 250 to 500 kg N ha^{-1}, the amount lost through leaching increased from 8 to 142 kg ha^{-1}. Ryden, Ball and Garwood (1984) argue that, as the management of grassland is intensified, the amounts of nitrate leached are likely to equal or exceed the range observed for continual arable production (40–100 kg N ha^{-1} a^{-1}), the losses being greatly accentuated by ruminant production.

12.1.5 CULTIVATION PRACTICES AND NITRATE LOSSES

The way in which the land and specific crops are managed may greatly influence the loss of nitrate by leaching, even if the amount of fertiliser added is correct. This matter is fully reviewed in Chapter 11. The effect of cultivation is particularly important: the break-up of soil aggregates exposes substrates less accessible to microbial metabolism. Destruction of macro-aggregates may provide a short-term increase in soil fertility through the aerobic degradation of the labile organic matter. Reorganisation of soil architecture also improves substrate/organism contact, and often infiltration and aeration ((Skjemstad, Vallis and Myers, 1988). Cultivation in the autumn is known to release much nitrate, often in excess of the requirements of the newly sown crop. McEwen et al. (1989) showed that cultivated fallow produced far more nitrate in the soil in autumn than a whole range of crops. The results of Dowdell, Crees and Cannell (1983) imply that contrasting methods of cultivation might be used to limit the large nitrate concentrations present in ploughed soil.

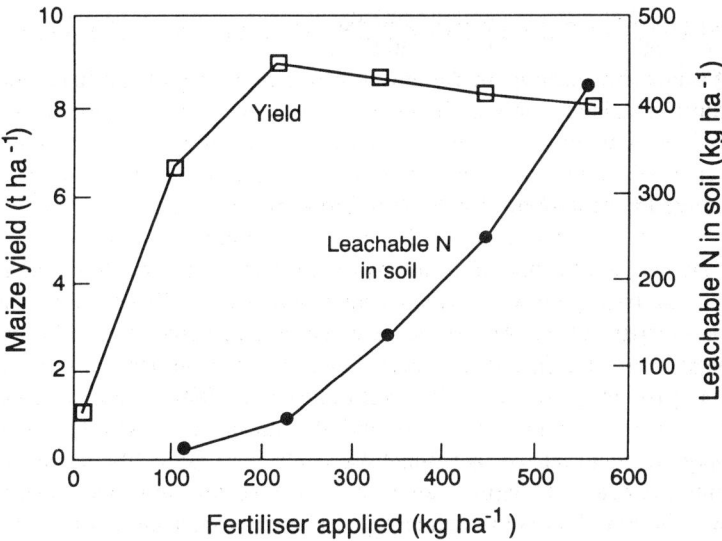

Figure 12.2 The relationship between fertiliser application, crop yield (maize) and leachable N in the soil (based on Hauck, 1990)

Soil drainage is also a significant control of the way in which nitrogen is lost from the soil. Harris *et al.* (1984) showed that nitrate losses on undrained plots were relatively low, equivalent to 5% of fertiliser added. On drained plots, losses were much higher with over 90% of runoff and nitrate lost through mole drains, equivalent to 34% of fertiliser input. Colbourn and Harper (1987) showed that drainage limited denitrification at this site to about 65% of losses from undrained soil. Gambrell (1975) demonstrated that land drains in the USA controlled the oxidation processes in the overlying soil, giving preference to the degradation of soil organic matter and an increase in the nitrate concentration of drainage water.

One further way in which researchers have sought to establish links between land use and nitrate loss is through the study of spatial patterns of river water quality. A full review of this work is given in Chapter 10. A detailed sampling framework is needed to indicate clear links between land use and stream nitrate concentrations. Burt and Arkell (1987) mapped nitrate concentrations in the Slapton Wood catchment (Figure 9.7). Using the results of dilution gauging experiments, they were able to identify source areas for nitrate loss within the basin. The results suggest a general association between land use and the loss of nitrate.

Given the lack of a simple relation between fertiliser application and nitrate leaching, much attention has been devoted to the development of models which can be used to predict nitrogen fertiliser requirements. The nitrogen requirement of a given crop varies considerably, from field to field and from season to season. It depends on the amount of nitrogen that the crop must absorb to permit maximum growth, on cropping history, soil type, rooting depth and on the residual inorganic nitrogen present in the rooting zone, which, in turn, depends on past weather conditions (Royal Society, 1983). At present, physically based models for the uptake of nitrogen and for the various transformations undergone by nitrogen in the soil are still in the prototype stage (e.g. Whitmore and Addiscott, 1987). Currently, therefore, leaching losses from fields are often estimated using the results of field experiments such as those described above. This creates a dilemma. Policy decisions are most likely to require surveys of large areas; in the absence of suitable process models, best estimates of nitrate leaching may have to be based on relatively crude field evidence (see Chapter 10). Even when theoretical models become available, their application to large drainage basins may well remain practicably (and practically) impossible.

12.2 CATCHMENT-WIDE RESTRICTIONS

After several years of application of its Common Agricultural Policy, with emphasis mainly on increasing production, the philosophy of the European Community (EC) has started to change (Nychas, 1990). One aspect of this is that the environmental impact of agriculture has been considered for the first time. The EC has accepted the need for specific schemes, including compulsory measures, to avoid deterioration of the environment. The need to fund compensation to offset the effects of 'natural' handicaps on agriculture has also been recognised. Specific schemes have been established which provide aid for farmers in designated environmentally sensitive areas; farmers are compensated for loss of income caused by adopting practices compatible with environmental protection. Similar measures, voluntary and compulsory, have been more widely established to limit the effect of agriculture on the environment. Some of these are relevant to the nitrate issue (Nychas, 1990; see also Chapter 15, this volume).

12.2.1 MODIFICATION OF CROPPING PRACTICE

This aspect has been fully discussed in Chapter 11. Part of the Nitrate Sensitive Area scheme (see Section 12.3.2 below) involves an advisory campaign to promote good agricultural practice relating to nitrate leaching. Farmers conforming with such 'no regrets' advice would not qualify for financial compensation.

12.2.2 SET-ASIDE

Set-aside is a voluntary scheme designed to reduce surpluses of arable crops (MAFF, 1988). In return for taking at least 20% of their land out of production farmers receive compensation; the scheme covers all main arable crops except potatoes and fodder crops. Farmers may put land set aside to fallow (land must not be left bare), plant it with trees or use it for permitted non-agricultural purposes (e.g. horse training). Any agricultural production is prohibited on set-aside land. Although the scheme was introduced to decrease production, its indirect effects on fertiliser use are obvious. However, there was no attempt to target set-aside to specific areas which are sensitive to nitrate loss, nor to limit fertiliser use on fields not included within the scheme. Also, by including a rotational fallow option, the scheme may have less impact as periodic ploughing would release substantial amounts of nitrate. In the event, the amount of land set-aside in the UK — about 650 000 ha — has been somewhat below expectation and real potential (Nychas, 1990) and the effect on nitrate loss in any one basin is therefore likely to have been minimal. The Farm Woodland and Farm Diversification schemes have also effectively set land aside from cultivation; again, however, their impact on overall leaching losses is thought to have been small.

   Johnes (1990) used an export coefficient model to predict nitrate concentrations in the river Windrush, a tributary of the Thames, for a variety of possible changes in land use (Table

Table 12.1 Model predictions of nitrate concentrations in the river Windrush for selected catchment management strategies (reproduced by permission from Johnes, 1990)

Land use	Mean annual NO_3-N concentration $(mg\ l^{-1})$
Current (1990) land use	9.14
(1) NSA compensation scheme (convert all arable to permanent unfertilised, ungrazed grass)	3.31
(2) NSA compensation scheme (as (1) but grazed)	5.45
(3) NSA compensation scheme (as (1) but with < kg $ha^{-1}\ a^{-1}$ fertiliser)	7.27
(4) Reduce all fertiliser applications by 20%	8.25
(5) 'Extensification' scheme: Reseed all oilseed rape to grassland (<150 kg N $ha^{-1}\ a^{-1}$ fertiliser) Restrict cereals to <180 kg N $ha^{-1}\ a^{-1}$ fertiliser No direct grazing of fodder crops Reduce cattle numbers by 25% Convert temporary to permanent grass (<150 kg N $ha^{-1}\ a^{-1}$ fertiliser)	7.62
(6) NSA Good Agricultureal Practice	8.83
(7) Setting aside an area of land equivalent area to a 50 m riparian zone in the Windrush Valley — unfertilised, ungrazed	7.74
(8) As (7) but grazed	8.92

12.1). Further details of the model are given in Chapter 10. Using strategies suggested in the Nitrate Sensitive Area (NSA) scheme (see Section 12.3.3), Johnes (1990) showed that setting aside the entire Windrush catchment (350 km^2) to permanent unfertilised grassland would bring considerable reductions in nitrate loss (Table 12.1(1)). Of course, such a change in land use would have significant repercussions on the economic viability of farms. Nevertheless, the results do imply that set-aside schemes could bring major benefits in water quality. Setting aside land to grazed, permanent grassland produced less dramatic results, though still a reduction of 40% compared to current nitrate levels (Table 12.1(2)). Limitations on fertiliser use bring smaller but still quite significant benefits (Table 12.1(3) and (4)).

12.2.3 EXTENSIFICATION

This EC scheme was also introduced as a mechanism to reduce surpluses. Farmers have to reduce output of a surplus product by at least 20%, by means other than set-aside, without other surplus production capacity being increased. Farmers may chose between a 'quantitative' method, based on actual reduction in quantitative terms, and a 'production' method option, based on the adoption of less intensive production practices for the product in question (Nychas, 1990). Again, the potential effect on fertiliser use is clear, especially for the latter option.

12.2.4 PROTECTION OF DRINKING-WATER RESOURCES — COMPARISON OF OPTIONS

In upland Britain the traditional method of controlling water quality was to buy up the entire catchment and then restrict land use within that area. Such a policy would be difficult in lowland areas with high-quality agricultural land and, in any case, it is unlikely that water supply companies could afford to buy such land. However, if a protection policy were applied in such areas individual farmers could be compensated for being prevented from farming at the economic optimum within good farming practice (DOE, 1986).

In the Hatton catchment study (Severn-Trent Water, 1988) a computer simulation model (Oakes, 1987) was used to assess options for the control of nitrate in water supply. Two different approaches which could be adopted to limit nitrate leaching from land were considered. The end result, to limit the total leaching within the catchment area, is the same for both approaches but it is achieved in two different ways: local protection zones (see Section 12.3) and catchment-wide restrictions. A water-treatment option was also considered. The catchment-wide scheme, essentially one of extensification, was applied over the whole catchment area supplying the borehole. The object of this approach was to spread the burden of the effect over all the relevant land and thereby minimise the impact on individual farmers. The types of farming change which were considered were a 20% reduction of fertiliser application to cereals, replacement of oilseed rape, sugar beet and potatoes by peas and beans, and the substitution of an area of cereals by grassland without increasing the number of livestock, so providing more extensive grassland management.

Various combinations of these changes were investigated using the model to identify those which achieve the required reduction in nitrate leaching (Severn-Trent Water, 1988, para. 3.5). Assessment included a financial comparison of the costs which would be incurred to implement each of the options. Estimated leaching losses assumed in the model are shown in Table 12.2; studies such as that of Foster, Cripps and Smith-Carington (1982) provide the basis for these estimates. Finally, it should be noted that the conclusions of the Hatton catchment study only relate to water supply in one specific area and do not take into account wider issues of agricultural policy — for example, who meets the cost of compensation?

The Hatton study concluded that, unless land use in the catchment changed, groundwater would exceed the EC standard for nitrate in drinking water early in the next century. Action by Severn-Trent to blend supplies could defer the impact, but eventually it would be necessary

Table 12.2 Leaching losses assumed in the Hatton catchment modelling study (reproduced by permission from Severn-Trent Water, 1988)

Land use	Equivalent N leached
Winter cereals	40% N applied
Peas	50% kg N ha^{-1}
Other arable	50% N applied
Cut grass	10% N applied
Grazed grass	15% N applied
Ploughed grass (3-year ley)	280 kg N ha^{-1}
Woodland/urban	10 mg NO3 l^{-1}
Grass reseeding (over 1 year)	150 kg N ha^{-1}

to install treatment plants to remove nitrate. Chemical treatment could be avoided if land use in the catchment is changed to reduce nitrate leaching. It was found that catchment-wide restrictions would be far more expensive to implement and police and thus, although they have a lower financial impact on individual farmers, the total cost can be greater. Catchment-wide restrictions would reduce the productivity and profitability of a large number of farmers, whereas local protection zones would mean a severe curtailment of farming activities for just a small number (Severn-Trent Water, 1988, para. 5.12).

In a similar study (DOE, 1986) it was also concluded that protection of an entire catchment area is impractical, and that it is preferable to enforce graded levels of restriction on non-water supply activities in annular zones around each borehole source. Restrictions on fertiliser use was the only agricultural option considered, though changes in land use could be used to effect the same net loss of nitrate. Results are shown in Figure 12.3. The Lincolnshire Limestone responds quickly to reductions in nitrate loss because of the rapid transport of percolating water through fissures in the unsaturated zone. By contrast, the Cambridgeshire Chalk is a deep aquifer dominated by slow percolation through this fine-grained porous rock. Thus groundwater responds extremely slowly to changes in agricultural practice; even if the entire catchment area was put down to non-fertilised grass now, it could take several decades before the nitrate concentration in the borehole falls below the EC limit. This does not rule out land-use change as an option on the Chalk, but means that water treatment may also be needed in the short term. In addition, the long-term nature of such a land-use policy on aquifers like the Chalk must be borne in mind. Further modelling studies (DOE, 1988) also found in favour of local protection zones, which showed greater savings in national resource terms than catchment-wide protection.

All three of the modelling studies reviewed above apply only to groundwater. Trudgill and Burt (1993) have considered the cost of protection of the Slapton Ley Nature Reserve, the largest freshwater body in south-west England. Assuming that nitrate loss equates with fertiliser application and that there is a need to return to pre-1970 nitrate concentrations in the lake (when much less eutrophication was evident), they calculate that the cost would be about £1.5 million per year for the entire catchment area (46 km^2). Johnes (1990) showed that limitations on fertiliser application might reduce nitrate losses in the Windrush catchment by between 10% and 20% (Tables 12.1(3) and (4)). Using her model to identify crops and cropping practices most susceptible to nitrate leaching, Johnes devised her own 'extensification' scheme for the Windrush (Table 12.1(5)); these measures together would reduce nitrate concentrations in the Windrush by about 17%. 'Good farming practice' as defined in the NSA scheme brings only small benefits, a reduction of only 3% in nitrate levels (Table 12.1(6)). However, as a no-cost option, even such small gains may be useful in holding nitrate losses at current levels and helping to prevent further increases.

12.2.5 ORGANIC FARMING

A proposal for an EC regulation on organic production of agricultural products is currently under discussion for eventual adoption. Its basic principle is that products composed wholly or in part of synthetic chemicals may not be used as fertilisers. Though no particular scheme of aid is proposed to promote this type of farming, its growth is expected due to recent consumer interest (Nychas, 1990).

Organic farming systems rely upon crop rotations, animal manures, crop residues, legumes and green manures to maintain soil productivity and tilth, to supply plant nutrients, and to

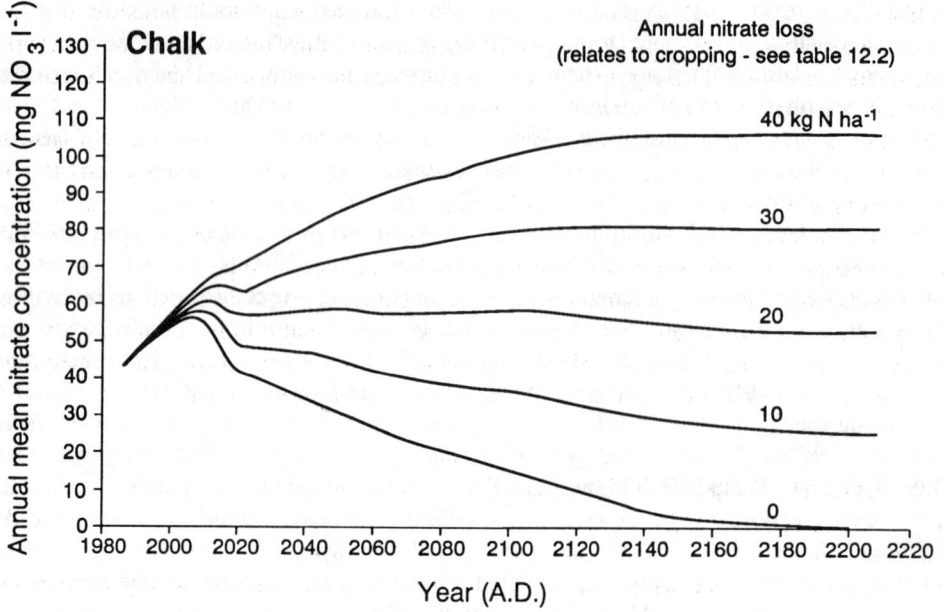

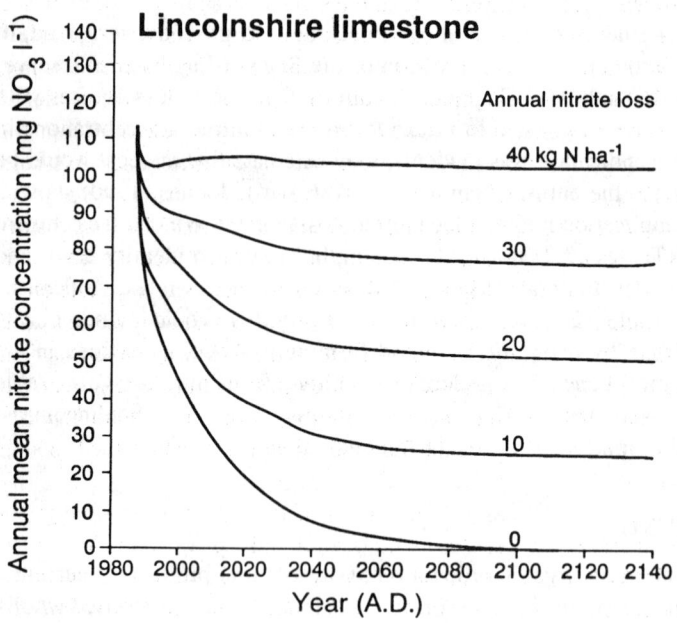

Figure 12.3   Predicted nitrate concentrations in discharge from two UK aquifers following reductions in fertiliser application assuming constant leaching losses (reproduced from DoE, 1986, with the permission of the Controller of Her Majesty's Stationery Office)

control insects, weeds and other pests (USDA, 1980; Widdowson, 1987). Since all intensive agriculture seems to carry with it the risk of significant nitrate loss, low- or medium-input systems such as organic farming, in which the use of inorganic fertilisers is low or non-existent, may leak less nitrogen than intensive systems.

In an orgainc farming system the main cause of nitrate leaching is the ploughing-up of grass/clover leys. Davies and Barraclough (1988) showed that the average annual loss of nitrate from an organic farm in southern England was 19.7 kg ha^{-1}, rather less than an all arable conventional enterprise of the same area (15−65 kg ha^{-1} a^{-1}) and roughly the same as a conventional mixed farm (50% dairy/beef; 50% arable), assuming 5 kg ha^{-1} and 40 kg ha^{-1} losses from the grass and arable sectors, respectively. Any change in management that reduced the loss of nitrogen after ploughing would have a major effect on nitrate loss and probably would increase cereal yields. There may well be potential for the increased use of a winter catch crop or cover crop followed by a spring-sown cereal.

12.3 LOCAL PROTECTION ZONES

12.3.1 MODELLING STUDIES

The local protection zone option assumes that a small area close to and usually surrounding an individual borehole will be changed to a land use which reduces nitrate leaching to a minimum. The size of area is selected to achieve a balance between the reduced leaching from severe restrictions within the local protection zone and the greater nitrate leaching from 'normal' farming in the rest of the catchment, resulting in the same overall reduction in nitrate leaching as would be achieved with catchment-wide restrictions. The burden of the effect is felt by fewer farmers who, within the local protection zone, must cease their normal farming (Severn-Trent Water, 1988). In the Hatton catchment study, the types of land use considered for this option were: conversion to permanent pasture cut for hay; broadleaved forestry (evergreens would alter the evaporation loss and so reduce recharge); grass and woodland for recreation (e.g. camping, horse training); and natural reversion to scrub.

The reduction in nitrate concentration which may be achieved in any particular case results from a combination of the size of the zone and the reduction in nitrate leaching from that which would otherwise occur (DOE, 1986). Figure 12.4 shows the percentage reduction in nitrate concentrations which would result from a given area of protection and a given leaching rate. To achieve an overall reduction in nitrate loss of 20%, any of the following would suffice; a protection zone equal to 80% of the catchment within which the leaching rate is reduced to 75%; a protection zone equal to 40% of the catchment within which the leaching rate is reduced to 50%; or a protection zone covering just 20% of the catchment in which the leaching rate is reduced to zero (DOE, 1986).

The last of these accords most closely with the definition of local protection zone given above. The results give the reduction in nitrate loss which would take place in the long term (cf. Figure 12.3), and are relative to concentrations which would ultimately be reached in the absence of land-use controls. The model predicts the final equilibrium concentration and does not have the capability to describe the length of any delay between the imposition of land-use controls and the attainment of stable, lower nitrate levels. Thus, the effects on nitrate concentrations of implementing a local protection zone policy may be experienced the same year (unusually) or up to several decades later, depending on local hydrogeological conditions (see Chapter 8).

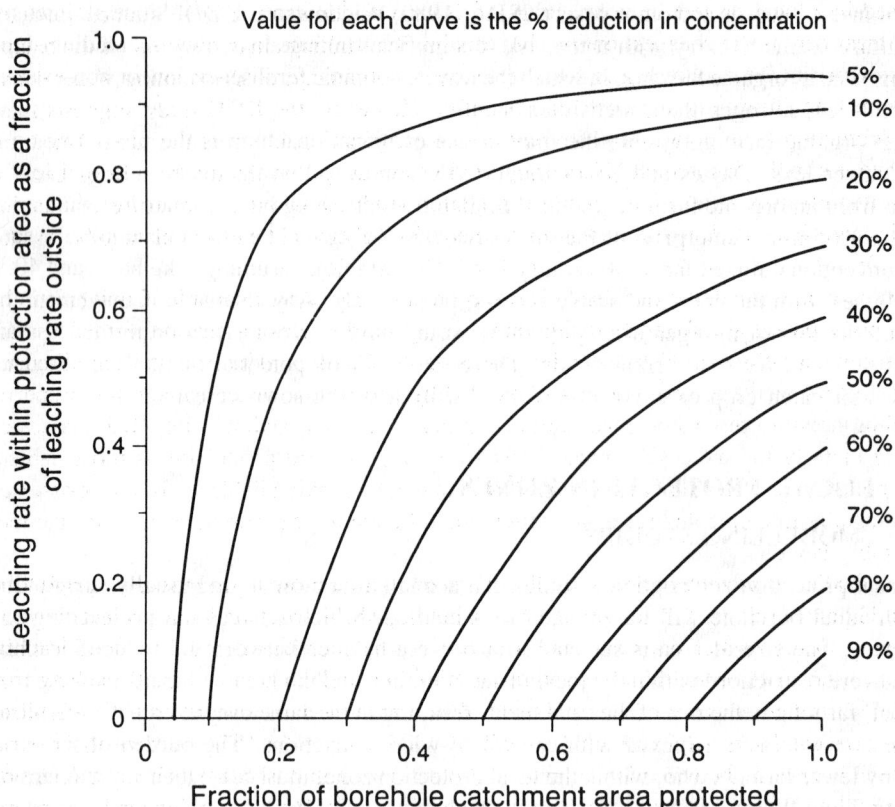

Figure 12.4 Reductions in nitrate concentration in relation to the fraction of a borehole catchment area which is protected by land-use control and the leaching rate within the protection zone (reproduced from DoE, 1986, with the permission of the Controller of Her Majesty's Stationery Office)

The Hatton catchment study (Severn-Trent Water, 1988) compared catchment-wide restriction and local protection zones. The study concluded that on practical grounds the local protection zone option has a number of attractions. Because the areas involved are smaller and the land-use changes more significant, monitoring is easier and there would be greater confidence in their effectiveness. However, the study did not consider the wider issues which use of this option raises: the economic and social impact on those living and working in the protection zones, the implications for local employment and for suppliers of farm inputs, the time taken for farmers to develop new business skills, the question of who pays the compensation, and the additional environmental benefits which alternative land uses might bring.

The DOE(1988) study concluded that, in national resource terms, localised controls often showed greater savings than catchment-wide protection. Their costings took into account the fact that additional production of agricultural commodities does not necessarily yield benefits to the nation of the same order as the private gains to individual farmers. In particular, the DOE study assumed continuing high levels of exchequer support for cereals, including long-term intervention and large export refunds under CAP. Their results imply that it would be in the national interest to compensate farmers for income lost in complying with the restrictions imposed in a local protection zone. Both the Hatton catchment and DOE studies show that water treatment is the cheapest option, but this involves taking a narrow view of the balance between on-site farm income and costs to the water company, so discounting wider social, economic and environmental costs and benefits. However, the DOE study suggests that a local protection zone option might repay closer examination in particular areas (moderate effective rainfall, fissured aquifers, no nearby source of water for blending, areas of predominantly mixed farming), while the Hatton study shows that a mixture of blending supplies and land-use change would avoid the need for a chemical treatment plant to be installed in the future.

Very few other modelling studies of local protection zones are available. Cook (1991) has developed a potential risk model using map overlays of the relevant factors deemed to affect nitrate concentrations at abstraction boreholes in the Chalk aquifer of east Kent, England. These factors were: effective rainfall, soil liability to nitrate leaching, aquifer properties including depth to the water table, and proximity to the abstraction point. Risk classes are used to identify the most vulnerable areas where agricultural production is most liable to cause pollution of water supplies. Cook points out that by ranking liable areas, a risk-assessed, incrementally implemented set-aside policy could be employed, commencing with the most vulnerable land.

Most studies of protection zones are limited to a consideration of groundwater supply from boreholes and springs, Johnes (1990) has modelled the impact of local protection zones bordering surface water courses. She used her export coefficient model to predict nitrate concentrations in the river Windrush for a situation in which all land within 50 m of the river channel was set aside as permanent grassland. Where the grassland is ingrazed and unfertilised, nitrate concentrations are estimated to fall by 15% (Table 12.1(7)). Where the riparian grassland is grazed, the reduction amounts only to 2.5% (Table 12.1(8)). Johnes' calculations do not include any assessment of the nutrient retention capacity of riparian land and merely imply a reduction in nitrate leaching through changes in land use. The subject of nutrient retention zones is further explored in Section 12.4 below. For aquifers, the use of buffer zones is clearly impossible since leachate percolates directly down to the water table. In groundwater-fed catchments, preventative options remain the only alternative to water treatment as a means of controlling nitrate concentrations in the water supply.

12.3.2 FIELD EVIDENCE

DOE (1988) gives two examples where detailed practical options have been put into operation. Knight and Tuckwell (1988) describe operations undertaken by Wessex Water to control agricultural use of water supply catchments for the city of Bath in southern England. Bath obtains most of its supply from springs in the surrounding hills; these feed two reservoirs, at Monkswood and Batheaston, which together provide 80% of the supply. Within the catchment, areas of 'greater pollution risk' were identified. These are sites where the springs are shallow and where rapid recharge is likely as thin limestone beds with freely draining soils overlie impermeable clays. A study in 1984 indicated that if the increase in nitrate in the Monkswood reservoir continued at the current rate, nitrate concentrations could exceed the EC limit within 20 years. The case for Batheaston reservoir was less clear. The reservoirs could be replaced by other, lower-nitrate supplies at a cost of £2.14 million or by ion-exchange treatment facilities installed at a cost of £1.41 million.

Land in the catchment is principally used for mixed farming; in 1986 one-quarter of the catchment was occupied by cereals and about half by permanent pasture, the rest being short leys, fodder crops and rough grazing. Two farms on the plateau above the spring line occupy land which recharges several important groups of springs. One farm was already owned by Wessex Water; in 1985 the farm became vacant, giving them the opportunity to amend agricultural practices to reduce nitrate leaching and to seek to implement these practices throughout the catchment by negotiating agreements with other farmers.

In 1987 the water authority purchased the second farm. Guidelines on cropping and fertiliser practice were drawn up; these included a preference for autumn-sown cereals, maintenance of current areas of rough grazing and woodland, a preference for sheep rather than cattle, and guidelines on the amount and timing of fertiliser application. In the areas of 'greater risk' more stringent restrictions were necessary: permanent grass is to replace cereals and should only be used for hay, with a maximum annual nitrogen application of 80 kg N ha^{-1}; there will be restrictions on grazing animals over-wintering in valley bottoms and no manure or slurry spreading will be allowed; ploughing and reseeding is to be limited to 10% per year of any area.

Through a series of tenancy agreements and negotiations with other landowners, Wessex Water were able to protect 43% of the Monkswood catchment with cropping and fertiliser agreements, including 40% of the area of greater risk, and 73% of the Batheaston catchment, including 90% of the area of greater risk. Where tenancy agreements were changed, any loss of a tenant's income was compensated by increasing the area of land or reducing rents. Other landowners applied the guidelines in return for grazing leases on Wessex land. The 1987 value of the current agricultural restrictions was estimated at £200 000. Even when the cost of purchasing the remaining land was taken into account, the total cost of controlling agricultural activity in the two catchments was estimated at £1.15 million, still well below the cost of water treatment or replacement of the sources. The upward trend in winter nitrate concentrations at Batheaston has been reversed since 1985. Shallow springs have also shown decreasing concentrations of nitrate since 1985; deep springs have as yet shown no change (Knight and Tuckwell, 1988). The indications are that the measures taken have avoided the need for additional water treatment or additional supply.

DOE (1988) also gives details of the control of agricultural practices in water supply catchments of the Eastbourne Water Company in south-east England. Three farms occupy the catchment which lies on the Chalk aquifer. All three farms are under the control of the Company, directly in one case, and through agreement with the local Borough Council which manages the other two. Though guidelines have existed since 1952, control of fertiliser use dates from 1971. An agreement restricted applications in an inner zone to 63 kg N ha^{-1} and in an outer zone to 75 kg Nha^{-1}; in 1983 the limit was raised to 94 kg N ha^{-1} for the whole area. It has not been possible to match the pattern of fertiliser use to the nitrate concentration in the supply; given the deep unsaturated zone of the Chalk aquifer, no rapid response would be expected (Section 8.2). After rising steadily for three decades (with the exception of a sharp fall in 1972), nitrate concentrations fell in 1982 and have remained steady since then. DOE (1988) concluded that, although the reduction in nitrate levels cannot be attributed directly to the protection policy, continued control is helping to keep nitrate levels well below the EC limit.

12.3.3   THE UK'S NITRATE SENSITIVE AREA AND NITRATE ADVISORY AREA SCHEMES

In November 1988 the UK government first announced its intention to introduce agricultural measures aimed at reducing the level of nitrate in water. These made clear that, wherever possible, agricultural restrictions should be introduced on a voluntary basis with compulsory measures as a fallback, and that farmers should be compensated where they were subject to substantial restrictions beyond the degree which could be regarded as good agricultural practice. The government designated a pilot sample of 19 sites at which a range of measures could be tested; all 19 sites selected were groundwater sources, springs or boreholes. The aim of the scheme, which commenced in the autumn of 1990, is to control the entry of nitrate from agricultural land into groundwater. Under the scheme two sets of areas were designated. In the ten Nitrate Sensitive Areas (NSAs) voluntary but substantial agricultural restrictions involving financial compensation have been introduced. At the other set of nine Nitrate Advisory Areas (NAAs) a campaign will promote good farming practices aimed at reducing nitrate leaching but without compensation.

Among the voluntary measures relating to good agricultural practice are: avoidance of autumn fertiliser, slurry or manure spreading; strict observance of recommended maximum rates of fertiliser and manure application; planting of a winter cover crop in the autumn before spring-sown crops; avoidance of grassland ploughing, particularly in autumn; and avoidance of late-season fertiliser applications. These measures are discussed more fully in Chapter 11.

The NSA scheme is itself divided into two. There is a Basic Scheme, which is designed to reduce nitrate leaching within broadly existing agricultural practices, and a Premium Scheme, involving more fundamental changes in agricultural practices. Full details are given in the Statutory Instruments issued by the UK government in its Nitrate Sensitive Areas (Designation) Order SI 1990/1013 (see also Section 14.8). In summary, measures in the Basic Scheme include detailed requirements concerning:

- Limits on levels of organic and inorganic fertiliser at or below the economic optimum and constraints on the timing of applications;
- Requirements to sow a crop or cover crop to avoid bare land in the autumn;
- Limitation on grassland ploughing to include only leys in arable rotation;
- Retention of hedgerows and woodland (or replaced by similar features);
- Requirement to produce a manure plan for permanently housed pig and poultry units which shows that the unit has or will have sufficient storage handling and transport to meet the spreading/timing requirements of the Scheme.

There are four options in the Premium Scheme:

(1) Conversion of arable land to grassland, unfertilised and ungrazed;
(2) Conversion of arable land to grassland, unfertilised but with grazing allowed;
(3) Conversion of arable land to grassland with limited fertiliser use and optional grazing;
(4) Conversion of arable land to grassland with woodland.

Rates of compensation payment range from £55 to £95 per hectare per annum in the Basic Scheme and from £90 to £380 in the Premium Scheme (depending on the option selected and the percentage of a holding which is subject to a premium scheme agreement).

Johnes (1990) applied the recommendations contained in the NSA scheme to her export coefficient model to predict changes in nitrate concentrations in the river Windrush, assuming that such measures were adopted on a catchment-wide basis (Table 12.1); her results have already been discussed earlier in this chapter. While the NSA scheme is designed to give local protection only, Johnes' results do indicate (like the WRC model — Oakes, 1987) that such measures will bring significant benefits. However, like the other models discussed, Johnes' model is an equilibrium approach and gives no indication of how long the new situation would take to happen. In the fissured limestones of the Windrush, this might take only a few years, but in a deep aquifer like the Chalk, perhaps several decades.

12.3.4 BUFFER ZONES BETWEEN FARMLAND AND SURFACE WATER COURSES

The notion of creating buffer zones between sources of pollution and surface water courses is not a new one. They are used in many European countries to safeguard surface water from contamination by point sources of pollution. A number of countries, notably Germany, use artificial wetlands below sewage works to provide 'tertiary' treatment before effluent is discharged into rivers. In agricultural situations the use of buffer strips at the bottom of fields to trap eroded soil is long established, especially in the USA (Dillaha, 1989; Heede, 1990). Recent studies have begun to focus on the role of streamside land in regulating nutrient fluxes between agricultural land and surface waters. This subject is discussed fully in the next section. Unlike the local protection zones described above, such buffer zones interact with agricultural drainage *after* leaching has occurred rather than seeking to prevent nitrate loss at source.

12.4 NUTRIENT-RETENTION ZONES

12.4.1 THE ECOLOGICAL IMPORTANCE OF TERRESTRIAL–AQUATIC ECOTONES

Ecotones (or boundary zones) are important regulators of the flux of energy and material across landscapes. *The Ecology and Management of Aquatic–Terrestrial Ecotones* (Naiman and Décamps, 1990) focuses on the dynamic nature and management potential of the ecotone occurring at the interface of terrestrial and aquatic ecosystems. This section is concerned with two questions related to this topic. Can a terrestrial–aquatic ecotone function effectively as a nutrient-retention zone for non-point source nitrate pollution? Further, can nearstream land be suitably managed so as to mitigate the worst excesses of nitrate leaching from agricultural land before nitrate-rich water drains into a river or lake?

Décamps and Naiman (1990) note that, until recently, ecological research focused on the functioning of individual ecosystems, avoiding consideration of the boundaries, (or ecotones) between them. They identify two reasons why the ecotone concept has recently been studied in more detail. First, freshwater ecologists have begun to realise that the well-being of freshwaters relates not just to point sources of pollution but in many basins depends to a large extent on diffuse non-point pollution from the terrestrial environment (Hynes, 1975). Second, growing interest in disturbance and patch dynamics has led to numerous studies orientated towards interactions and exchanges across heterogenous landscapes, influences of heterogeneity on biotic and abiotic processes, and management of that heterogeneity (Forman and Godron, 1986). An ecotone is now regarded as a dynamic rather than a static zone, possessing properties of its own, and is now considered as integral parts of the landscape, deriving its properties from its position in the landscape (Risser, 1990). Thus, an ecotone is now regarded as a functional entity whose attributes and interconnections depend on its transitional position between adjacent ecological systems. The ecotone so defined is not only a boundary but also an area of interaction with two or more adjacent systems (Pinay *et al.*, 1990).

The *terrestrial–aquatic ecotone* occupies the zone between hillslope and river channel.

Usually this coincides with the floodplain. Here water tables remain high for much of the time. Fluxes of water and nutrients will usually be from the terrestrial to the aquatic system, although seepage of river water into the channel bank or overbank flooding may reverse the normal direction of flow. Given their position, nearstream ecotones can potentially function as natural sinks for sediment and nutrients emanating from farmland. This, in turn, raises the possibility of actively establishing and maintaining buffer zones at the bottom of agricultural fields adjacent to streams to trap both soil particles and nutrients so as to protect bodies of water from possible nutrient enrichment (Odum, 1990). The term *riparian* may be used to describe this ecotone, although traditionally this has included only a very narrow strip bordering the channel. Our concept of a riparian ecotone would encompass a much broader zone, in some cases the entire floodplain (if undrained), though the zone might be very narrow where steeply sloping fields come right to the edge of the channel.

Pinay and Décamps (1988) proposed a conceptual model to explain nitrogen fluxes in the soils of riparian zones in relation to varying degrees of waterlogging and nitrate input. They argued that denitrification rates observed in floodplain sediments would be sufficient to remove all nitrate from groundwater flowing under a riparian wood, given a floodplain width of 30 m. A number of studies have shown the influence of riparian woods on the retention of nitrate in groundwater (e.g. Brinson, Bradshaw and Holmes, 1983; Lowrance *et al.*, 1984). Peterjohn and Correll (1984) demonstrated that about 19 m of flow under a riparian wood can be sufficient to reduce 90% of the nitrate load in the groundwater. Similarly, Jacobs and Gilliam (1985) have shown that nitrate in the groundwater was eliminated after about 30 m of flow. However, none of these studies involved intensive field monitoring, in space or time, and details of ecotone functions remain uncertain.

Although the gradient of vegetation change within the terrestrial−aquatic ecotone may be gradual, subsurface processes across this continuum show distinct zones of nutrient retention (Figure 12.5). The terrestrial−aquatic ecotone is itself enclosed by boundaries, where nutrient absorption is optimised. These boundaries can best be described as the 'nutrient absorption interfaces' of the terrestrial−aquatic ecotone. One boundary comprises the stream channel, its bed and banks. This zone is traditionally defined as the riparian zone (Tansley, 1911). Nutrients are absorbed from the stream water, and as such, process activity is orientated parallel to the direction of stream flow. The absorption of nitrate within the *aquatic−terrestrial interface* is enhanced by the presence of vegetation (macrophytes) and underflow (water movement through sediments on the channel bed; Munn and Meyer, 1988) and the extent of the hyporheic zone (Duff and Triska, 1990). Pinay *et al.* (1990) show that unidirectional water flow promotes mixing of materials from upstream towards downstream, giving a dependence of downstream sites on upstream zones. These linkages are underlined in unifying ideas in aquatic ecology such as the river continuum (Vannote *et al.*, 1980) and nutrient spiralling (Webster, 1975; Newbold *et al.*, 1981) concepts. During their displacement downstream, nutrients are absorbed, utilised and released by organisms; their downstream transport is coupled with transformations from organic to mineral matter. Riparian zones may play a major role in the retention of nutrients during their transport downstream. strong interactions between streams and the riparian ecotone can entail substantial exchanges of matter and energy within the floodplain (Pinay *et al.*, 1990). However, nitrate absorption within this boundary is a seasonal process with high stream discharge degrading the channel sediment system in the winter months.

The other boundary of the terrestrial−aquatic ecotone is found at the point where the hillslope joins the floodplain. This *terrestrial−aquatic interface* (Figure 12.5) exists where saturated, anoxic soils, rich in carbon, are exposed to nitrate-rich groundwater. Rates of denitrification

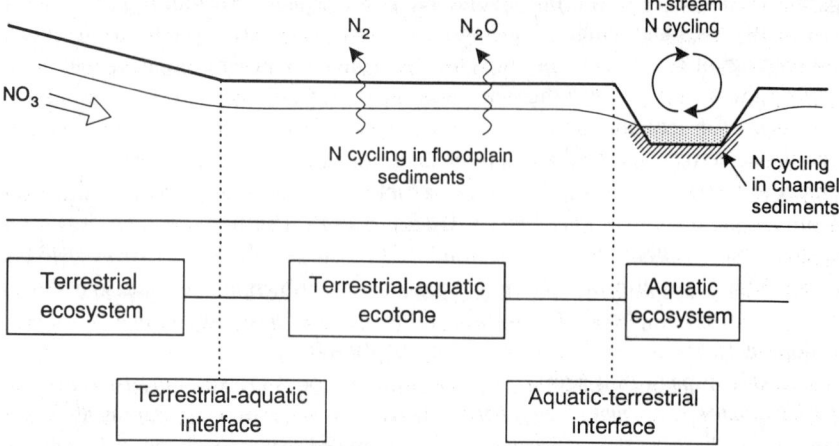

Figure 12.5 A schematic diagram of the terrestrial—aquatic ecotone

are high within this zone since the nutrients required by the denitrifying bacteria are abundant. Denitrification does not occur further across the floodplain due to the lack of nitrate in the soil water, all of which has been removed at the terrestrial—aquatic interface (Cooper, 1990; Haycock, 1991). Nutrient absorption processes at the terrestrial—aquatic interface essentially operate at right-angles to the channel, although the topography of the floodplain may mean that there is a significant down-valley component to the subsurface flow as it moves from the hillslolpe to the channel. The nutrient-retention processes operating within this interface are not exposed to the perturbations that the sediments in the aquatic—terrestrial interface experience. Jacobs and Gilliam (1985) noted that a 1 m width of floodplain sediment could absorb the same quantity of nitrate as a 10 m length of channel. Haycock's (1991) study of the river Leach in England (see next section) showed that, during the winter when nitrate concentrations in streamflow are at their highest, the nitrate absorption potential of 1 m of floodplain sediment was equivalent to 230 m of channel sediment.

The type of vegetation found on the floodplain has traditionally been considered to be of primary importance in controlling the efficiency of nitrate absorption processes within the terrestrial—aquatic interface. Several studies have argued that the presence of trees is crucial (Yates and Sheridan, 1983; Lowrance et al., 1983; Peterjohn and Correll, 1984). However, other studies have contended that the role of surface vegetation is secondary compared to other features of the floodplain, notably the presence of saturated conditions together with a carbon-rich sediment (Pinay and Décamps, 1988; Cooper, 1990; Haycock, 1991). Trees may enhance the potential denitrifying capacity of the terrestrial—aquatic interface because they contribute significant amounts of carbon to the soil (Cuffney, 1988). Even so, the type of vegetation cover may be irrelevant as long as the floodplain surface is vegetated. It is the maintenance of a subsurface environment within which denitrification rates can be optimised which seems to be more essential.

12.4.2 ECOTONE FUNCTIONS OF THE LEACH FLOODPLAIN: A FIELD STUDY

Haycock (1991) monitored the passage of nitrate-rich groundwater across an undrained floodplain in order to evaluate the nitrate-retention capacity of the terrestrial—aquatic interface.

The study was conducted in central England on the floodplain of the River Leach, a headwater tributary of the Thames. This is a groundwater-fed basin, the aquifer being the Cotswold limestone (see Section 8.2). The area supports intensive mixed farming. The results presented here relate mainly to the 1989/90 winter. Winter is the time of highest nitrate concentration in stream water and thus the period when the nutrient retention capacity of riparian ecotones may be crucially important. The mean monthly nitrate concentration of the Leach at the study site is 11.3 mg NO_3-N l^{-1}, exactly equal to the EC limit. There is strong seasonality in stream nitrate levels with a maximum in February (17.3 mg NO_3-N l^{-1}) and a minimum in August (5.5 mg NO_3-N l^{-1}).

Two floodplain sections of the River Leach were selected. Because the floodplain is undrained, effluent groundwater from the aquifer cannot short-circuit the floodplain via tile drains or open ditches. Moreover, the water table will be maintained at a higher level in undrained soils, which may enhance denitrification. Selection of two sites allowed comparison between a grazed pasture and a plantation of poplar trees (*Populus italica*). As noted above, other studies have concentrated on the role of nutrient retention by riparian woodland. However, comments here are confined to the grassland site since no differences could be detected between the two. Slopes above the floodplain have freely draining soils and are underlain by fissured limestone. The floodplain lies on impermeable clay. Its soil appears to be a mixture of colluvium and alluvium and has a saturated hydraulic conductivity of 0.012 mm s^{-1} (0.432 m h^{-1}). This indicates a permeable soil, well capable of conducting large volumes of subsurface flow. At each site, grids of unlined boreholes (depths from 0.7 to 3.1 m) were augered on the floodplain and on the adjoining slope. These holes allowed water table elevation to be measured and water samples to be taken. The grassland site has a grid of 24 boreholes across an area measuring 24 m by 16 m, with six holes upslope and four holes upvalley.

Stage and nitrate concentrations for the river Leach are shown in Figure 12.6 for the period May 1989 to December 1990. Also shown are nitrate concentrations for a nearby spring which flows directly into the stream, and the difference in nitrate concentration between spring and stream water over this period. Both summers were times of significant drought. Stream nitrate concentrations were significantly below those of the spring throughout both summers, presumably because instream absorption processes were active. Nitrate concentrations in the spring water fell gradually throughout both summers, though remaining above the EC limit of 11.3 mg NO_3-N l^{-1} until late in the autumn. Modest rainfall in October and November 1989 served to reduce the soil moisture deficit, initiating recharge of the aquifer, and stream discharge began to rise slowly. Heavy rainfall in December (1989) and January and February (1990) caused sharp rises in streamflow and concomitant increases in nitrate concentrations of both spring and stream. A similar response was noted at the end of 1990.

Figure 12.7(a) shows a topographical survey of the grassland site. The present stream channel does not occupy the lowest part of the floodplain. In winter, when the water table is close to the soil surface, the flow direction of water draining from the hillslope is not straight across the floodplain to stream. Instead, flow lines curve downvalley, and the water eventually drains into the stream at some point downstream. During the dry summer of 1989 the floodplain was very dry; few boreholes contained water, except those closest to the stream. Flow direction was influent from channel to floodplain, rather than vice versa. From mid-November onwards, sufficient recharge had taken place that drainage from the aquifer through the floodplain soil to the stream was re-established.

Figure 12.7(b) uses box-whisker plots to show mean water table elevation, averaged by row, for the grassed floodplain from November 1989 to February 1990. Each 'box' shows

(a)   Daily mean stage height at the catchment outlet

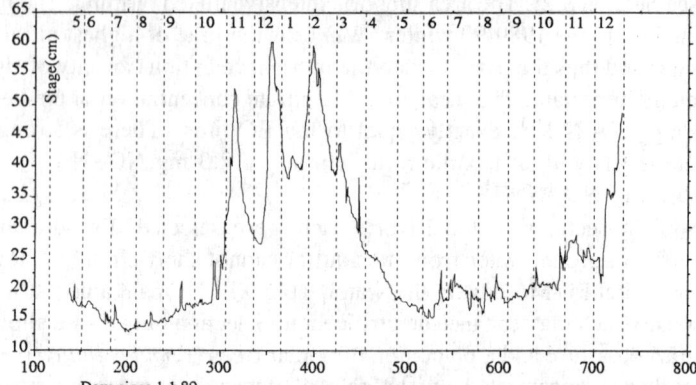

(b)   Daily nitrate-N concentration at the catchment outlet

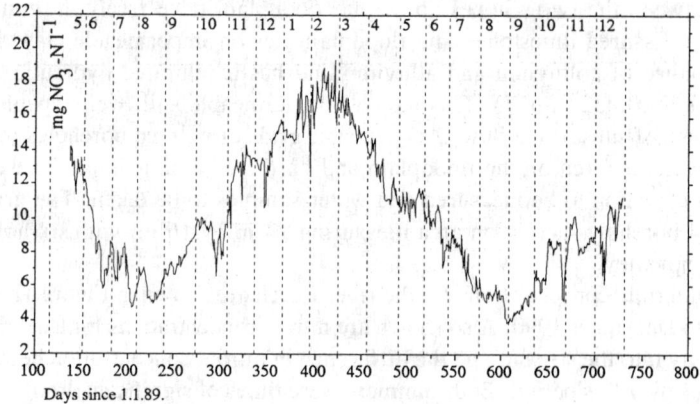

(c)   Daily nitrate concentration of Twin Head Spring

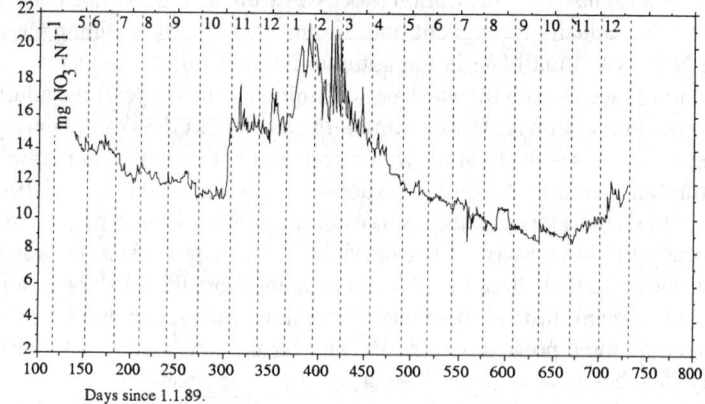

Figure 12.6 River stage and stream nitrate concentrations for the River Leach, May 1989 to December 1990. Also shown are nitrate concentrations for a local spring (reproduced by permission from Haycock, 1991)

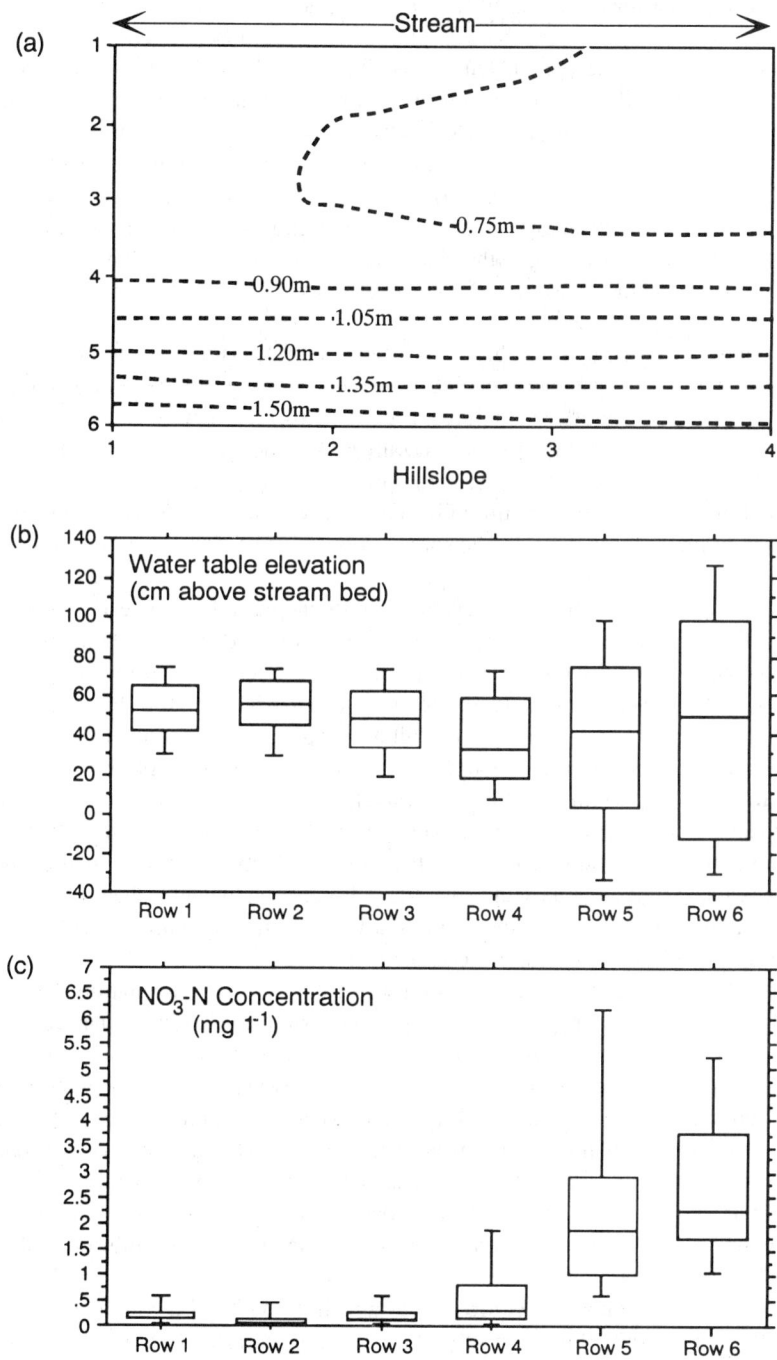

Figure 12.7 (a) Topographical survey of a section of floodplain on the River Leach; (b) water table elevations; (c) nitrate concentrations across the floodplain (reproduced by permission from Haycock, 1991)

the median level and interquartile range, while the 'whiskers' show the 10% and 90% deciles. Row 4 is consistently the site of lowest water level; in winter, water flows continuously towards this point from the hillslope, driven by a high hydraulic gradient. Given the relatively high hydraulic conductivity of the soil, calculations show that the amounts of subsurface flow through the floodplain soil, when the water table is high, are significant and quite sufficient to account for increases in stream discharge along the stream section. Figure 12.7(b) shows that, for the instrumented plot, only rows 1 and 2 contribute directly to the channel; otherwise, the zone around row 4 represents the focus for drainage, from where water moves downvalley. This situation is not unusual: many floodplains have complex topography within which the direction of flow for water draining to the channel may well be tortuous. The important point is that floodplains are characterised (at least seasonally) by high water tables and low hydraulic gradients.

Results plotted in Figure 12.7(c) show nitrate concentrations across the grassed section, again averaged by row. These cover the 1989/90 winter period when the nitrate concentration of water draining from the aquifer is high (cf. Figure 12.6). The pattern is one of consistently low nitrate concentrations on the floodplain with high concentrations only found in those boreholes within the hillslope. These results show that there is a sharp loss of nitrate as groundwater drains into the floodplain soil. We believe that this loss is caused mainly by denitrification, although there may also be some assimilation of nitrogen by the soil microbial biomass.

Denitrification can be rapid where anaerobic conditions prevail and where a carbon source is available to fuel the process, and proceeds rapidly in undrained soils (Colbourn and Harper, 1987). No measurements of denitrification rates have yet been made at the site, but Ambus and Christensen (1990) showed that denitrification in a riparian meadow can rapidly remove nitrate from agricultural drainage. While some dilution may take place, our calculations suggest that flow volumes from aquifer to floodplain are too large for a simple mixing of aquifer and floodplain water to explain the nitrate profile shown in Figure 12.7(c). We interpret the reduction in nitrate concentration to be an active denitrification loss as groundwater enters the floodplain soil. Other results show that the nitrate-retention capacity of the floodplain remains large, even at times of maximum subsurface flow. Since winter is the time of maximum nitrate concentration in river water, the fact that a floodplain can retain nutrients at this time is of some importance to water supply companies.

The results for the floodplain section can be set in context using information from a spatial water sampling programme (Figure 12.8). The protection afforded by a riparian ecotone is made clear by examining point source inputs where no such buffer zone exists between source and stream. With the exception of a site immediately downstream of a sewage works, all stream sites sampled throughout the catchment had lower mean nitrate concentrations than the springs. This may happen for at least two reasons. The stream itself has a capacity for nutrient retention, although this function is seasonal and the release of inorganic nitrogen may occur in winter when macrophytes decay. Also, the mitigating effects of nitrate retention by the floodplain, as discussed above, may help reduce the nitrate concentration of the stream water, especially in winter.

Further field measurements are required to establish the rates of denitrification operating within these floodplain soils; this is the subject of ongoing research. Nevertheless, our results to date already demonstrate what other researchers have suggested from more cursory field studies: that riparian ecotones are capable of functioning effectively as nutrient-retention zones. The presence of a saturated, carbon-rich soil seems crucial in encouraging a high rate of denitrification. The role of surface vegetation remains unclear.

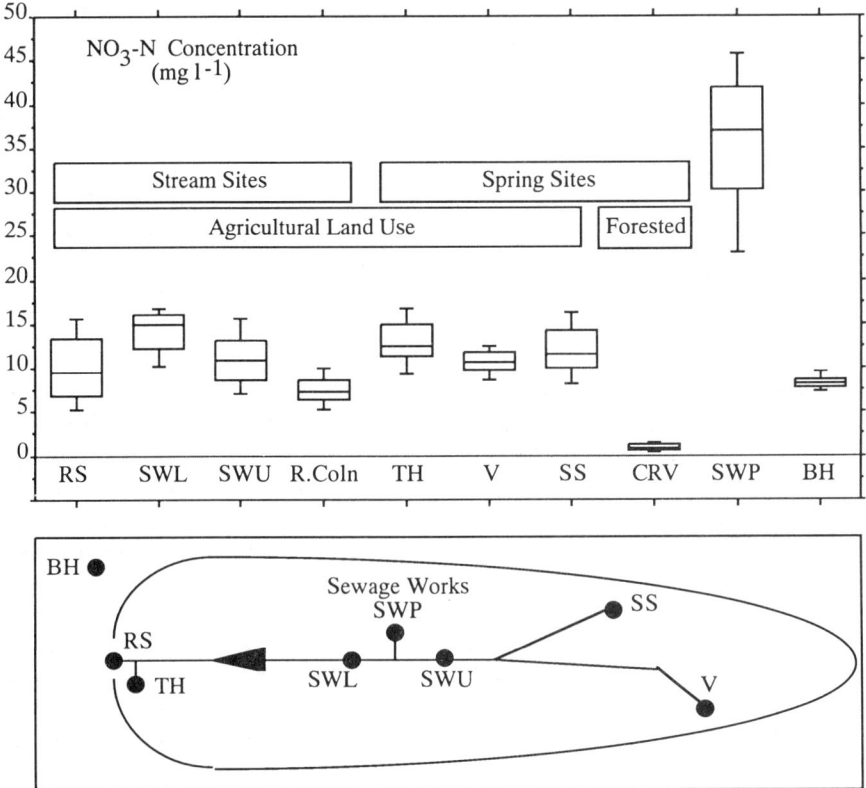

Figure 12.8 Nitrate concentrations at sampling sites on springs and streams (mapped schematically) within the catchment of the river Leach and at other nearby locations in the Cotswolds (reproduced by permission from Haycock, 1991). RS: research site (see Figure 12.6); SWL: below Northleach sewage works; SWU: above Northleach sewage works; R.Coln: river Coln near CRV; TH: Twin Head spring (see Figure 12.6); V: Hampnett village spring; SS: Seven Springs, Northleach; CRV: spring at Chedworth Roman Villa below ancient woodland; SWP: sewage works outfall; BH: borehole in Inferior Oolite (all other springs are in the Great Oolite)

12.4.3 SETTING ASIDE LAND FOR NUTRIENT-RETENTION ZONES

The Leach floodplain, where present, is narrow and generally aquifers discharge directly into the stream via springs so that there is little opportunity for the mitigating function of the floodplain to operate. For this reason, spring and stream concentrations are very similar during the winter period. In other words, the Leach is too well coupled to its aquifer (Figure 12.6). Further downstream, where floodplains are wider, underdrainage has often been installed so that aquifer drainage will by-pass the floodplain, better drainage allowing arable crops to be cultivated rather than the traditional meadow or woodland usage of such land. Ambus and Christensen (1990) showed, if drains are allowed to spill their water on to the floodplain so that water must flow over or through the floodplain soils to reach the stream, that denitrification is rapid and the quality of agricultural drainage water is greatly and rapidly improved. Our results also show that nitrate retention may take place within a very small distance, (just a few metres). As mentioned earlier, denitrifying bacteria operate best at the junction of anaerobic/aerobic zones where both carbon and nitrate are abundant. The Leach is not a special case: floodplain soils are typically carbon-rich and waterlogged. Since the grassland and poplar sites function in a similar way, we cannot determine whether woodland provided a more effective nutrient buffer than grassland, since subsurface processes appear to operate independently of surface vegetation type.

It follows that nitrate losses may be reduced by creating a nutrient-retention zone between the farmland and the river. Where floodplains have been underdrained, one option might be to block drains and so force water to move slowly through floodplain soils to the stream. Such strategic removal of nearstream land from cultivation might require compensation to be paid to farmers for loss of production, but might well prove a more cost-effective remedy than alternative measures. In addition, re-establishment of water meadows, whether lightly grazed or ungrazed, would bring the benefits of enhanced conservation and amenity value. Our results imply that, where point sources enter the stream directly, construction of a narrow seepage zone between spring and stream could be of great benefit. Wetlands have been used as a means of tertiary treatment of polluted effluents for many years. Wetlands constructed around springs might serve a similar purpose in dealing with the worst effects of agriculturally contaminated water — and possibly prove less costly than other means of dealing with nitrate pollution.

A policy of setting aside agricultural land in sensitive locations is not restricted to the floodplains of large alluvial rivers but also has relevance for headwater catchments. This is where relatively unpolluted water — to dilute pollution entering the river further downstream — is commonly thought to originate. Our results show that headwater streams in agricultural areas may also be highly polluted, often above legal limits. The concept of setting aside land from agricultural use in order to protect groundwater sources has already been implemented in the UK (Section 12.3.3). An extension of this scheme to include protection of surface waters may be in order (and indeed seems likely, given the terms of EC Directive 91/676 — see Chapter 14). The use of undrained floodplains as barriers to nitrate movement could allow modern farming and water supply to co-exist in the same basin. Such nutrient-retention zones might also trap eroded soil and associated pollutants, and would additionally provided habitats of high conservation value. Burt, Heathwaite and Trudgill (1993) stress that particular locations within a basin may be especially sensitive to land-use change, and that the effect of changes in small basins may well prove significant at the larger scale. The suggestion here to 'set aside' nearstream land might solve some of the pollution problems associated with modern agriculture and make rivers less sensitive to land-use changes within their catchment area.

REFERENCES

Addiscott, T. and Powlson, D.S. (1989) Laying the ground rules for nitrate. *New Scientist*, 29 April, 28—9.

Ambus, P. and Christensen, A. (1990) Cleaning of agricultural drainage by denitrification in a riparian meadow. Paper presented at the Sixth Workshop on Nitrogen in Soils, The Queen's University of Belfast, December.

Barraclough, D., Hyden, M.J. and Davies, G.P. (1983) Fate of fertiliser nitrogen applied to grassland. I. Field leaching results. *Journal of Soil Science*, **34**, 483—98.

Brinson, M.M., Bradshaw, H.D. and Holmes, R.N. (1983) Significance of floodplain sediments in nutrient exchange between a stream and its floodplain. In Fontaine, T.D. and Bartel, S.M. (eds), *Dynamics of Lotic Ecosystems*, Ann Arbor Science, Michigan, pp. 199—220.

Burt, T.P. and Arkell, B.P. (1987) Temporal and spatial patterns of nitrate losses from an agricultural catchment. *Soil Use and Management*, **3**, 138−43.

Burt, T.P., Heathwaite, A.L. and Trudgill, S.T. (1993) Catchment sensitivity to land use controls, In Thomas, D.S.G. and Allison, R.J. (eds), *Landscape Sensitivity*, Wiley, Chichester.

Colbourn, P. and Harper, I.W. (1987) Denitrification in drained and undrained arable clay soil. *Journal of Soil Science*, **38**, 531−40.

Cook, H.F. (1991) Nitrate protection zones: targetting and land use over an aquifer. *Land Use Policy*, **8**, 16−28.

Cooper, A.B. (1990) Nitrate depletion in the riparian zone and stream channel of a small headwater catchment. *Hydrobiologia*, **202**, 13−26.

Cuffney, T.F. (1988) Input, movement and exchange of organic matter within a subtropical coastal backwater river floodplain system. *Freshwater Biology*, **19**, 305−20.

Davies, G.P. and Barraclough, D. (1988) Nitrate leaching at Rushall Farm, Wiltshire, 1985−8. Unpublished Report, ICI Jealotts Hill Research Station, Bracknell, UK.

Décamps, H. and Naiman, R.J. (1990) Towards an ecotone perspective. In Naiman, R.J. and Décamps, H. (eds), *The Ecology and Management of Aquatic−Terrestrial Ecotones*, Parthenon, Paris, pp. 1−6.

DOE (1986) *Nitrate in water*. Department of the Environment Pollution Paper 26, HMSO, London.

DOE (1988) *The Nitrate Issue*, Department of the Environment, HMSO, London.

Dillaha, T.A. (1989) Water quality impacts of vegetative filter strips. Paper Number 89-2043, ASAE-CSAE Meeting Presentation, St Joseph, Michigan, pp. 1−9.

Dowdell, R.J., Crees, R. and Cannell, R.Q. (1983) A field study of effects of contrasting methods of cultivation on soil nitrate content during autumn, winter and spring. *Journal of Soil Science*, **34**, 367−80.

Duff, J.H. and Triska, F.J. (1990) Denitrification in sediments from the hyporheic zone adjacent to a small forested stream. *Canadian Journal of Fisheries and Aquatic Sciences*, **47**, 1140−47.

Forman, R.T.T. and Godron, M. (1986) *Landscape Ecology*, Wiley, New York.

Foster, S.S.D., Cripps, A.C. and Smith-Carington, A. (1982) Nitrate leaching to groundwater. *Philosophical Transactions of the Royal Society of London*, **296**, 477−89.

Gambrell, R.P. (1975) Denitrification in subsoils of the North Carolina coastal plains as affected by soil drainage. *Journal of Environmental Quality*, **4**, 311−16.

Garwood, E.A., Salette, J. and Lemaire, G. (1980) The influence of water supply to grass on the response to fertiliser nitrogen and nitrogen recovery. In Prins, W.H. and Arnold, G.H. (eds), *Role of Nitrogen in Intensive Grassland Management*, Pudoc Wageningen, Netherlands, pp. 59−65.

Harris, G.L., Goss, M.J., Dowdell, R.J., Howse, K.R. and Morgan, P. (1984) A study of mole drainage with simplified cultivation for autumn-sown crops on a clay soil. 2. Soil water regimes, water balances and nutrient loss in drain water, 1978−80. *Journal of Agricultural Science, Cambridge*, **102**, 561−81.

Hauck, R.D. (1990) Agronomic and public aspects of soil nitrogen research. *Soil Use and Management*, **6**, 66−70.

Haycock, N.E. (1991). *Riparian land as buffer zones in agricultural catchments*, Unpublished DPhil thesis, Oxford University.

Heede, M. (1990) *Vegetation strips control erosion in watersheds*, USDA Forest Service, Rocky Mountain Forest & Range Experiment Station, Report RM 499.

Hynes, H.B.N. (1975) The stream and its valley. *Verhandlungen Internationale Vereiningung Limnologie*, **19**, 1−15.

Jacobs, T.C. and Gilliam, J.W. (1985) Riparian losses of nitrate from agricultural drainage waters. *Journal of Environmental Quality*, **14**, 472−8.

Johnes, P.J. (1990). *An investigation of the effects of land use upon water quality in the Windrush catchment*, Unpublished DPhil thesis, Oxford University.

Johnson, A.E. and Jenkinson, D.S. (1990) The nitrogen cycle in UK arable agriculture. *Proceedings of the Fertiliser Society*, **286**, 1−24.

Knight, M.S. and Tuckwell, J.B. (1988) Controlling nitrate leaching in water supply catchments. *Journal of the Institution of Water and Environmental Management*, **2**, 248−52.

Lowrance, R.R., Todd, J., Fail, J., Hendrickson, O., Leonard, R. and Asmussen, L. (1984) Riparian forests as nutrient filters in agricultural watersheds. *BioScience*, **34**, 374−7.

MAFF (1988) *Set Aside*, Ministry of Agriculture, Fisheries and Food, London.

McEwan, J., Darby, R.J., Hewitt, M.V. and Yeoman, D.P. (1989) Effects of field beans, fallow, lupins, oats, oilseed rape, peas, ryegrass, sunflowers and wheat on nitrogen residues in the soil and on the growth of a subsequent wheat crop. *Journal of Agricultural Science, Cambridge*, **115**, 209−19.

McGill, W.B. and Myers, R.J.K. (1987) Controls on dynamics of soil and fertiliser nitrogen. In Follett, R.F., Stewart, J.W.B. and Cole, C.V. (eds), *Soil Fertility and Organic Matter as Critical Components of Production Systems*, Soil. Sci. Soc. A. & A. Soc. Agron., Madison, pp. 73−99.

Munn, N.L. and Meyer, J.L. (1988) Rapid flow through the sediment of a headwater stream in the southern Appalachians. *Freshwater Biology*, **20**, 235−40.

Naiman, R.J. and Décamps, H. (1990) *The Ecology and Management of Aquatic−Terrestrial Ecotones*, Parthenon, Paris.

Newbold, J.D., Elwood, J.W., O'Neill, R.V and Van Winkle, W. (1981) Measuring nutrient spiralling in streams. *Canadian Journal of Fisheries and Aquatic Sciences*, **38**, 860−63.

Nychas, A. (1990) An EEC perspective on fertiliser use. *Chemistry and Industry*, 17 December, 828−31.

Oakes, D.B. (1987) *Prediction of groundwater nitrate concentrations*, Water Research Centre Publication PRU 1566-M, Medmenham.

Odum, W.E. (1990) Internal processes influencing the maintenance of ecotones: do they exist? In Naiman, R.J. and Décamps, H. (eds), *The Ecology and Management of Aquatic−Terrestrial Ecotones*, Parthenon, Paris, pp. 91−101.

Peterjohn, W.T. and Correll, D.L. (1984) Nutrient dynamics in an agricultural watershed: observations on the role of a riparian forest. *Ecology*, **65**, 1466−75.

Pinay, G. and Décamps, H. (1988) The role of riparian woods in regulating nitrogen fluxes between the alluvial aquifer and surface water: a conceptual model. *Regulated Rivers*, **2**, 507−16.

Pinay, G., Décamps, H., Chauvet, E. and Fustec, E. (1990) Functions of ecotones in fluvial systems. In Naiman, R.J. and Décamps, H. (eds), *The Ecology and Management of Aquatic−Terrestrial Ecotones*, Parthenon, Paris, pp. 141−69.

Powlson, D.S., Pruden, G., Johnson, A.E. and Jenkinson, D.S. (1986a) The nitrogen cycle in the Broadbalk Wheat Experiment: recovery and losses of ^{15}N-labelled fertiliser applied in spring and inputs of nitrogen from the atmosphere. *Journal of Agricultural Science, Cambridge*. **107**, 591−609.

Powlson, D.S., Hart, P.B.S., Pruden, G. and Jenkinson, D.S. (1986b) Recovery of ^{15}N-labelled fertiliser applied in autumn to winter wheat at four sites in eastern England. *Journal of Agricultural Science, Cambridge*, **107**, 611−20.

Risser, P.G. (1990) The ecological importance of land−water ecotones. In Naiman, R.J. and Décamps, H. (eds), *The Ecology and Management of Aquatic−Terrestrial Ecotones*, Parthenon, Paris, pp. 7−21.

Royal Society (1983) *The Nitrogen Cycle of the United Kingdom*, The Royal Society, London.

Ryden, J.C., Ball, P.R. and Garwood, E.A. (1984) Nitrate leaching from grassland. *Nature*, **311**, 50−54.

Severn-Trent Water (1988) *The Hatton Catchment Study*, Severn-Trent Water, Birmingham.

Skjemstad, J.O., Vallis, I. and Myers, R.J.K. (1988) Decomposition of soil organic nitrogen. In Wilson, J.R. (ed.), *Advances in Nitrogen Cycling in Agricultural Ecosystems*, CAB International, Wallingford, pp. 134−44.

Tansley, A.G. (1911) *Types of British Vegetation*, Cambridge University Press, Cambridge.

Trudgill, S.T. and Burt, T.P. (1993) Occasional Publication, Field Studies Council, Preston Montford, Shrewsbury, UK, in press.

USDA (1980) *Report and Recommendations on Organic Farming*, United States Department of Agriculture, Washington, DC.

Vannote, R.L., Minshall, G.W., Cummins, J.R., Sedell, J.R. and Cushing, C.E. (1980) The river continuum concept. *Canadian Journal of Fisheries and Aquatic Sciences*, **37**, 130−37.

van der Meer H.G. (1990) Nitrogen budgets. Paper presented at the Sixth Workshop on Nitrogen in Soils, The Queen's University of Belfast, December.

Webster, J.R. (1975) Analysis of potassium and calcium dynamics in stream ecosystems on three southern Appalachian watersheds of contrasting vegetation. Dissertation. University of Georgia, Athens, Georgia.

Whitmore, A.P. and Addiscott, T.M. (1987) Applictions of computer modelling to predict mineral nitrogen in soil and nitrogen in crops. *Soil Use and Management*, **3**, 38−42.

Widdowson, R.W. (1987) *Towards Holistic Agriculture: a Scientific Approach*, Pergamon, Oxford.
Yates, P. and Sheridan, J.M. (1983) Estimating the effectiveness of vegetated floodplains/wetlands as nitrate, nitrite and orthophosphorus filters. *Agriculture, Ecosystems and Environment*, **9**, 303–14.

13 Treatment Processes for Nitrate Removal from Water Supplies

T. HALL
WRc Swindon

and

B.T. CROLL
Anglian Water Services Limited

13.1 INTRODUCTION

Historically, water treatment for the production of potable supplies was concerned primarily with disinfection and removal of suspended or colloidal material contributing to visible turbidity or colour. The earliest treatment plants were based on the use of slow sand filtration through beds of fine sand to remove turbidity of mineral or algal origin. Such processes were labour intensive (for cleaning accumulated material from the beds) and required large areas of land. However, the ability to produce a high-quality water has led to their continued use, and many slow sand filter plants are still in operation worldwide. In particular, the London area is served by several very large slow sand filter works.

In order to reduce the labour and land requirements for water treatment, more rapid processes were developed. The new treatment plants were still based on filtration through beds of sand but also used chemical coagulation to improve the removal of material through the coarser sand needed to establish the higher rates of filtration. Such plants often include a clarification stage based on settlement or, more recently, flotation, to reduce solids loadings to the filters. The removal of turbidity and colour from the water not only improves its appearance but also helps to increase the efficiency of disinfection necessary to prevent the spread of waterborne disease.

The most widely used and well-established disinfectant for water treatment is chlorine, although concern over the production of chlorination by-products has resulted in a trend towards other disinfectants (for example, ozone or chlorine dioxide). Modern treatment works can involve many process stages, including initial screening, chemical coagulation, clarification, sand filtration, ozonation and chlorination. Concern over organic compounds in water supplies, particularly pesticides, is leading to the widespread inclusion of adsorption processes into the treatment stream, using granular activated carbon. Chlorination by-products and pesticides are among the group of health-related parameters of newly developed concern within the water supply industry. Nitrate is also included within this group.

High-nitrate concentrations in water supplies can lead to a rise in bottle-fed babies of methaemoglobinaema (see Chapters 1 and 15), a condition in which the oxygen-carrying capacity of the blood is impaired. There is also a reported link between high nitrate in drinking

Nitrate: Processes, Patterns and Management. Edited by T.P. Burt, A.L. Heathwaite and S.T. Trudgill
© 1993 John Wiley & Sons Ltd

water and stomach cancer, but the evidence for this is inconclusive (DoE, 1986). The medical implications of high-nitrate concentrations in water supplies have resulted in the EC Drinking Water Directive (EEC, 1980) stipulating a Maximum Admissible Concentration (MAC) of 50 mg $NO_3^-.l^{-1}$ and a Guide Level (GL) of 25 mg $NO_3^-.l^{-1}$. The EC Directive standards are now incorporated into UK legislation by the Water Supply (Water Quality) Regulations 1989 (Statutory Instrument, 1989).

Nitrate in water supplies can be controlled at source through the construction of protection zones or other limitations on nitrate application (see Chapters 11 and 12), but, in addition to the impact on the farming industry of such approaches, the benefits for some aquifers may not become apparent for many years because of the long retention times involved (see Chapter 8). However, protection zones have been established and are under investigation as a long-term solution to the problem of high-nitrate concentrations in some areas.

Another option for control is the use of blending of local low-nitrate sources with the high-nitrate supplies. The low-nitrate sources may have other quality problems, particularly high iron and manganese, but these are cheaper and easier to remove than nitrate. It is also necessary to take into account other chemical changes in the water when introducing blending, which may, for example, lead to corrosion problems because of the alteration in the ionic composition of the water. Blending offers an easy-to-operate solution to nitrate control in water supplies, and is in widespread use. The main drawback is the fact that for some areas the nearest low-nitrate source will be long distances away, and costs for installation of pipes and pumping would be excessive. In such cases it would be necessary to install treatment to remove the nitrate from the supply before distribution. Nitrate is not removed by conventional water-treatment techniques, and a range of nitrate removal process types have been developed in response to rising nitrate concentrations in water sources.

The two types of process currently favoured for nitrate removal from public water supplies are ion exchange and biological denitrification. Membrane processes (reverse osmosis or electrodialysis) can remove nitrate, but are likely to be expensive for all but the very smallest supplies, where they might offer advantages in terms of ease of operation. They may also be chosen for sites where disposal of ion-exchange waste would be expensive.

In the UK, process development started in the early 1970s. Pilot plant trials of biological denitrification led to the development of a biological fluidised bed process (WRc, 1979) and experimental work with ion exchange (Gauntlett, 1975) identified the need for nitrate-selective resins. Similar development work occurred throughout Europe and the USA, leading to the wide range of processes now available, particularly with regard to biological denitrification. Installation of plants commenced in the late 1980s in the UK and other European countries, notably France (Philpot and Larminat, 1988; Richard, 1989) and, to a limited extent, in the USA (Lauch and Guter, 1986).

13.2 ION EXCHANGE

13.2.1 GENERAL PRINCIPLES

In ion exchange, the water to be treated is passed through beds of ion-exchange resin beads, which absorb anions, including nitrate, in exchange for another anion, normally chloride. When the capacity for exchange becomes exhausted the resin is regenerated, normally with sodium chloride solution, to return it to the chloride form ready for the next operational run. The spent regenerant, high in nitrate and chloride, can represent up to 1% of the volume

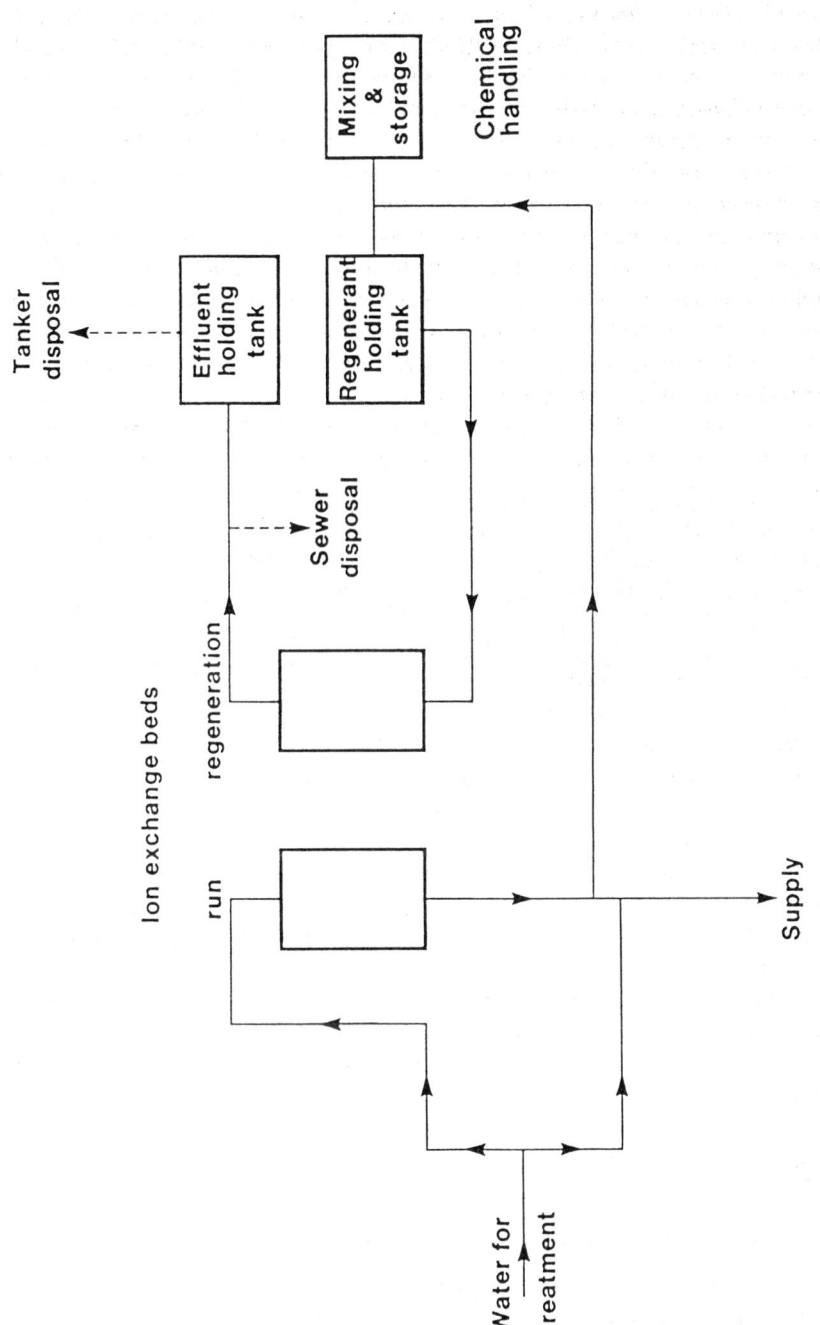

Figure 13.1 Ion-exchange plant schematic

treated; disposal would be by discharge to sewer or tankering to a large river or the sea. The high chloride concentration makes it unsuitable for disposal to agricultural land.

The need for waste disposal can have a major influence on process operating costs. It is possible to reduce the disposal volume using membrane processes, particularly electrodialysis, to treat the waste, but this would also be expensive and the costs must be balanced against those for direct disposal without waste treatment. The use of biological denitrification to treat ion exchange waste has been investigated (van der Hoek, van der Ven and Klapwijk (1988)); removal of nitrate from the waste could make disposal easier and can also allow limited re-use of the regenerant, thus reducing waste volumes.

A layout for a typical ion exchange plant is shown in Figure 13.1. A minimum of two beds would normally be used, operated out of phase with each other to even out the treated water quality variations which occur during a single run. An example of these quality variations for a nitrate selective resin, as discussed below, is shown in Figure 13.2. Counterflow regeneration (regenerant flow in the opposite direction to the treated water flow) is normally used for nitrate removal ion exchange.

The anion exchange resins used for the production of demineralised water (conventional resins) are capable of removing nitrate but suffer from the following disadvantages:

(1)   Increased chloride in the treated water as a result of regeneration with sodium chloride, coupled with reduced bicarbonate, a situation which could lead to the corrosion of brass fittings in the distribution system.

(2)   Low nitrate removal capacity for water high in sulphate because of the higher selectivity for sulphate over nitrate shown by the conventional resins; lower nitrate capacity results in greater salt requirements and larger volumes of waste regenerant for disposal.

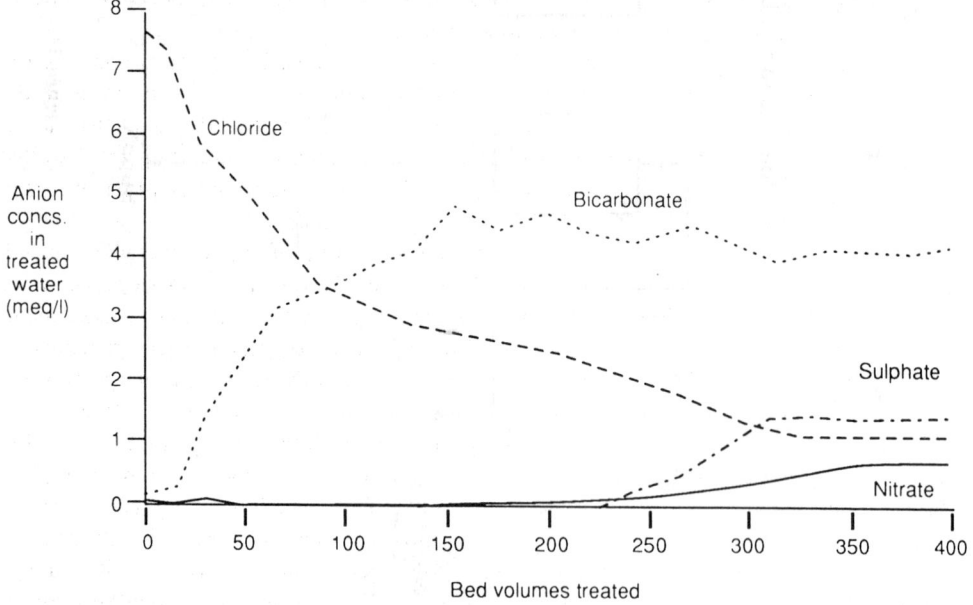

Figure 13.2   Treated water quality variation from a single ion exchange bed over one run (using nitrate selective resin)

(3)   Displacement of nitrate from the resin by sulphate (because of the sulphate selectivity) if the run is continued past the point at which nitrate leaving the bed is equal to the inlet concentration.

Nitrate selective ion exchange resins are now available, which have a higher selectivity for nitrate over sulphate. These have a lower total capacity for ion exchange but a higher nitrate removal capacity for high sulphate waters. The lower total capacity results in lower chloride concentrations in the treated water, and the nitrate selectivity prevents the displacement of nitrate by sulphate. The higher nitrate capacity can result in lower costs for regenerant chemicals and waste disposal when treating high-sulphate waters. These advantages of nitrate-selective resins are illustrated in Figure 13.3.

The potential problem of corrosion of brass fittings (particularly dezincification (Croll, 1991) occurs as a result of the increased ratio of chloride to bicarbonate (alkalinity) in the treated water, rather than from the higher chloride concentration itself. In most water supply applications the increase in corrosion potential will be marginal. However, in some cases the corrosion will be unacceptable and measures to reduce the chloride to bicarbonate ratio will be necessary. One way of alleviating this problem is to use bicarbonate during regeneration

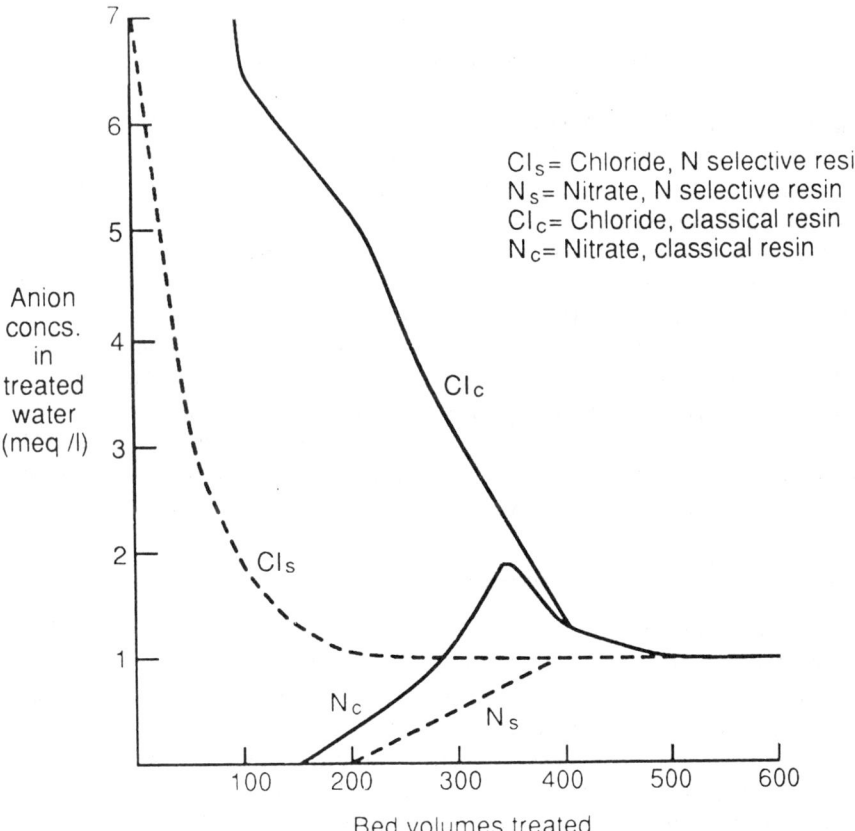

Cl_s = Chloride, N selective resin
N_s = Nitrate, N selective resin
Cl_c = Chloride, classical resin
N_c = Nitrate, classical resin

Figure 13.3   Advantages of nitrate selective resins over conventional anion exchange resins

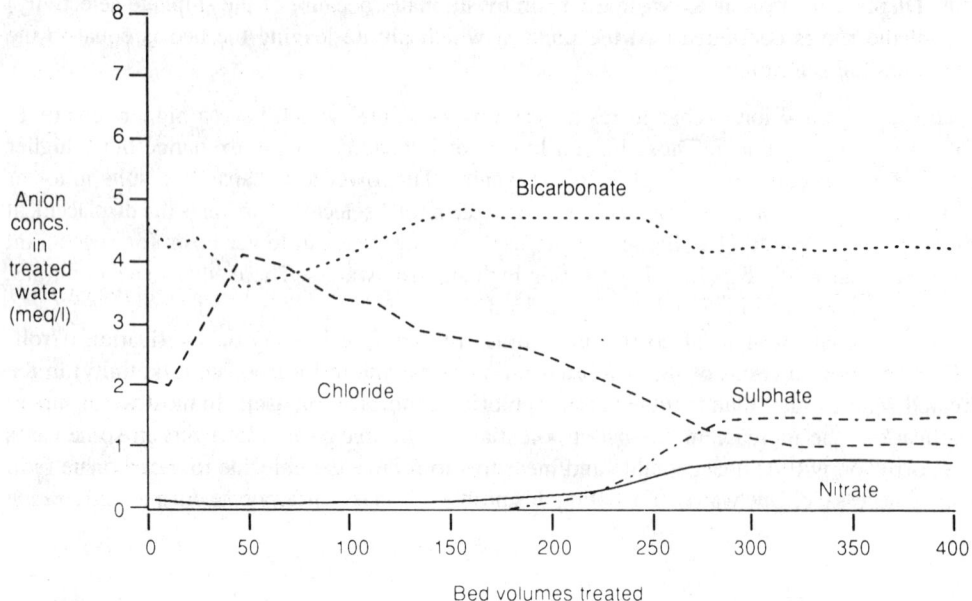

Figure 13.4 The effects of bicarbonate conditioning on treated water quality

to condition the resin after normal counterflow regeneration with sodium chloride. This involves passing a small volume of sodium bicarbonate solution upwards through the bed, such that resin in the lower part of the bed is in the bicarbonate rather than chloride form. The initial high chloride peak passing down the bed is partly exchanged for bicarbonate, reducing the early chloride:bicarbonate ratio in the treated water. An example of this is illustrated in Figure 13.4. Sodium bicarbonate is expensive compared with sodium chloride; the minimum amount needed to achieve a satisfactory chloride: bicarbonate ratio would be used, and in most circumstances the effects on total operating costs would not be large.

One further problem observed in the past has been the deposition of calcium carbonate within the bed (Harries, 1981) probably as a result of the formation of localised conditions of high pH and bicarbonate concentration, which increase the headloss through the bed. Until more information is available on this problem, current advice includes the provision for a facility for an acid wash and/or full bed backwash in any water-treatment plant installed.

13.2.2 PLANT DESIGN AND OPERATION

The design of ion exchange nitrate removal plants is discussed in detail in a WRc report (WRc, 1989). The plant size is based on hydraulic considerations and the degree of nitrate removal required. Ion exchange can achieve nitrate removal of 90% or more, and would be used to treat as small a proportion of the total flow needed to give the necessary overall nitrate removal after blending the treated water with the remainder of the flow. Treatment rates are typically 25 $m^3.m^{-3}$ bed per hour (i.e. 25 bed volumes per hour), which, for most waters, should give operating run lengths of up to 24 hours between regenerations. Regeneration can be initiated on the basis of monitoring the treated water nitrate concentration, although for many groundwater sources, the feed nitrate may be constant enough to allow run length to be based on volume treated, once the nitrate breakthrough curve has been established from initial runs. Small-scale pilot plant trials can give useful information for proposed full-scale plant design and operation, for example, with regard to selection of suitable resins and regeneration conditions.

The factors to be considered when choosing the optimum operating conditions for an ion exchange plant are (1) nitrate removal:salt use ratio and (2) nitrate removal:waste volume ratio. These two factors will vary in importance from site to site. On a site where disposal of the waste regenerant is relatively cheap (for example, to a sewer) the process can be optimised on the salt use, thus minimising chemical costs. If, however, the waste regenerant had to be tankered to a separate disposal site, the transport costs could outweight the chemical costs and the waste volume would need to be minimised. This can be done by:

(1) Using an increased concentration of sodium chloride brine for regeneration (although there is a limit to the maximum concentration that can be used for a particular resin because of the osmotic effects which can physically disrupt the resin structure);
(2) Minimising the volume of wasted rinse water (the water used to displace brine from the resin bed, and the first few bed volumes of treated water which may still contain traces of residual brine).

If no rinse water was needed, then the same regeneration level (in grams NaCl per litre of resin) would be the optimum for both waste volume and salt use. However, as this is not the case, and a fixed volume of rinse, independent of regeneration level, is needed, then the optimum regeneration level for the waste volume can be higher than the optimum volume for salt use. This is illustrated in Table 13.1, which shows examples of typical values for nitrate capacity for a nitrate-selective resin at four regeneration levels using a brine concentration of 5% w/v, with different amounts of rinse water (2 ot 4 bed volumes).

Table 13.1 shows how the optimum regeneration level tends toward the higher values as more rinse water is wasted, and to a fairly low value (100 g NaCl per litre resin) if salt use is the most important factor. The table also illustrates the trend of average chloride to bicarbonate ratio in the treated water with increasing regeneration levels. In areas where dezincification could be a problem the average chloride to bicarbonate ratio of the treated water could be important to minimise blending requirements, and so a higher regeneration level would be used.

Table 13.1 Optimisation of ion exchange

Regenerant level (g NaCl per litre resin)	Bed volumes of regenerant (BV)	NO_3 capacity (meq l^{-1} resin)	NO_3: Salt ratio (meq g^{-1} NaCl)	NO_3: Waste volume (2 BVs rinse) (meq l^{-1})	NO_3: Waste volume (4 BVs rinse) (meq l^{-1})	Average treated water Cl/HCO_3 (meq/meq)
50	1	100	2.0	33.3	20.0	0.9
100	2	250	2.5[a]	62.5	41.7	0.85
150	3	330	2.2	66.0[a]	47.1	0.8
200	4	380	1.9	63.0	47.5[a]	0.75[a]

[a] Optimum values.
Salt concentration = 5% w/v.

13.3 BIOLOGICAL DENITRIFICATION

13.3.1 GENERAL PRINCIPLES

Biological denitrification occurs naturally in sediments of rivers, lakes and reservoirs (see Chapter 5) where, in the absence of dissolved oxygen (anoxic conditions), certain widely occurring bacterial species can use nitrate for respiration, converting it to nitrogen gas:

$$NO_3^- \rightarrow NO_2^- \rightarrow NO \rightarrow N_2O \rightarrow N_2 \text{ (gas)}$$

To use this natural reaction in a water-treatment process requires the establishment of high bacterial concentrations, achieved by providing a food/energy source for bacterial growth and a large surface area for bacterial attachment in biological reactors. A wide range of biological denitrification processes have been developed throughout Europe. They vary in the type of food/energy source used and the nature of the support bed. The principal variations are:

(1) Processes using heterotrophic bacteria which obtain the energy and carbon required for growth from organic compounds — the carbon sources normally used for denitrification are methanol, ethanol or acetic acid;

(2) Processes using autotrophic bacteria which obtain energy from oxidation of inorganic chemicals (hydrogen (Gahrs, Rutten and Schnoor, 1989) or sulphur (Schippers *et al.*, 1987)) and carbon from CO_2/bicarbonate;

(3) Support beds of gravel, plastic media or fluidised sand.

In operating the denitrification process it is important to minimise the amount of food/energy source entering the water supply system in order to prevent any problems of bacterial growth in the distribution pipework. In heterotrophic processes this is done by controlling the carbon source dose in response to the dissolved oxygen and nitrate concentrations in the feedwater. The autotrophic processes offer advantages in this respect. Excess hydrogen can be easily removed after denitrification by aeration of the water (Gahrs, Rutten and Schnoor, 1989). Autotrophic denitrification by sulphur-oxidising bacteria uses beds of sulphur granules upon which the bacteria grow; the energy source (sulphur) is utilised directly by the bacteria, and does not enter the water being treated. This process also offers simplicity of operation, which could be of benefit for remote rural sites. A disadvantage of the autotrophic processes arises from the relatively low reaction rates compared with heterotrophic denitrification. This results in the need for larger autotrophic plants for the same degree of treatment and, therefore, higher capital costs. Furthermore, nitrate reduction using sulphur increases the sulphate concentration in the water:

$$5S + 6 NO_3^- + 2H_2O \rightarrow 3N_2 + 5 SO_4^{2-} + 4H^+$$

For some waters, this could result in the EC Directive MAC for sulphate ($250 \text{ mg.SO}_4^{2-} \text{ l}^{-1}$) being exceeded.

The range of heterotrophic processes developed differ with respect to the carbon source and biomass support material. However, the choice of carbon source is primarily economic and/or political rather than technical, and each of the process types is likely to operate satisfactorily with methanol, ethanol or acetic acid as the carbon source. In the UK, methanol is the cheapest carbon source, and the use of ethanol or acetic acid would increase chemical costs by factors of about 2 or 3, respectively.

The biomass support material is in either fixed beds or fluidised beds. The former are

377

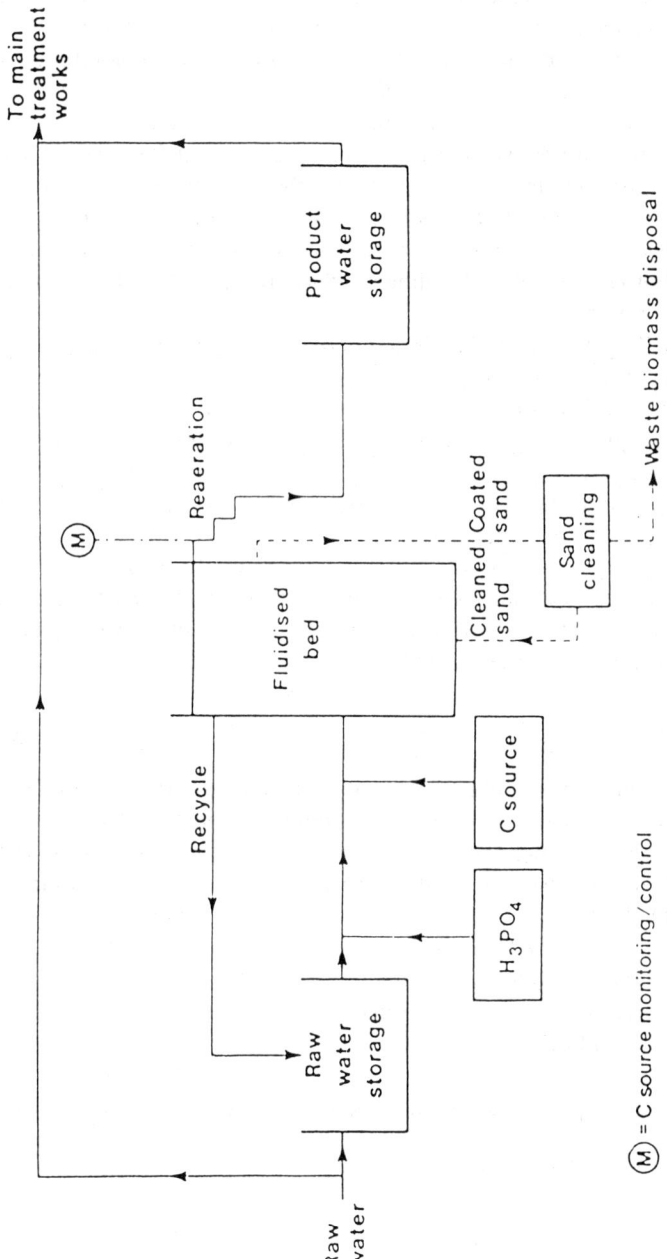

Figure 13.5 Biological fluidised bed schematic

periodically taken out of operation and backwashed to remove excess biomass accumulated as a result of bacterial growth. Biological fluidised bed (BFB) processes are described in more detail below.

'*In-situ*' biological denitrification techniques have been developed (Braester and Martinell, 1988; Mercado, Libhaber and Soares, 1988) which rely on bacterial growth within the aquifer, initiated by injection of a carbon source into the water within the aquifer at strategic points around an abstraction well. *In-situ* denitrification offers potential advantages with regard to capital costs because plant and equipment requirements are minimal. However, its application is likely to be limited to relatively homogeneous aquifers and the area of land required may be substantial, depending upon the properties of the aquifer and the level of denitrification required. The risks associated with failure of *in-situ* systems are much greater than those for treatment plant, in that it may be necessary to abandon boreholes because of aquifer blockage (Kruihof *et al.*, 1985). Treatment plants are, therefore, likely to be the preferred option for most circumstances.

Work in the UK by WRc, Anglian Water and the Department of the Environment has led to the development of a heterotrophic biological fluidised bed (BFB) process (Croll *et al.*, 1985) (Figure 13.5). The use of fluidised sand provides a large surface area for bacterial growth, and consequently high biomass concentrations without the problems of clogging of the bed. The sand is fluidised by upflow of the water being treated. The carbon source (methanol, ethanol or acetic acid) is dosed to the influent water; phosphate dosing may also be necessary to stimulate bacterial growth. Excess biomass is removed, without shutting down the process, by taking sand from the bed through a pump to strip off the attached bacterial film, returning the cleaned sand to the bed and transferring the waste biomass to a suitable point for thickening and storage before disposal. Waste volumes are very small, and the sludge is suitable for disposal to agricultural land.

13.3.2   DESIGN AND OPERATION OF BFB DENITRIFICATION PLANTS

Details of BFB denitrification plant design and operation are given in a WRc report (WRc, 1989). The process uses fine sand (0.3–0.5 mm) as a biomass support, in beds fluidised by upflow of the treated water at rates of 20 m per hour. Removal of dissolved oxygen (DO) occurs at the bottom of the bed, establishing anoxic conditions for nitrate removal in the upper regions of the bed.

The depth of fluidised bed required is a function of the dissolved oxygen and nitrate concentrations in the feedwater, the average biomass concentration in the bed and the rates of DO and nitrate removal at the minimum operating temperature. Other design figures that can be used are as follows:

Average biomass (volatile solids, VS) concentration = 15 kg VS/m^{-3} bed.

Removal rates at 2°C using methanol:

$$DO\ removal\quad = 18\ g\ DO/(kg\ VS.h^{-1})$$
$$Nitrate\ removal = 5.5\ g\ NO_3^-N/(kg\ VS.h^{-1})$$

Using these figures, the required fluidised bed depth can be calculated from the raw water DO and nitrate concentration, the total depth required being the sum of the requirements for DO and nitrate. The approximate requirement for DO removal is 0.75 m depth per 10 mg

Table 13.2 Carbon source requirements

Carbon source	For DO removal (mg/mg DO)	For nitrate removal (mg/mg NO$_3^-$N)
Methanol	1.0	2.5
Ethanol	0.5	2.0
Acetic acid	1.2	3.5

DO l^{-1} feedwater, and for nitrate removal is 0.25 m per mg NO$_3^-$N l^{-1} removed. As with ion exchange, the plant would be used to treat only a proportion of the total works output to achieve the desired overall nitrate removal. Carbon source requirements for dissolved oxygen and nitrate removal are shown in Table 13.2.

Operating experience with all types of biological denitrification process has shown that, at times, elevated nitrate concentrations can occur in the denitrified water, to concentrations well in excess of the EC Directive MAC of 0.1 mg NO$_2^-$ l^{-1} (0.03 mg NO$_2^-$N l^{-1}). Nitrite is easily removed using chlorine, but the chlorine demand is high (5 mg Cl$_2$ per mg NO$_3^-$N) and the nitrite is oxidised back to nitrate, impairing the efficiency of the denitrification process. Biological denitrification processes are normally operated with an underdosing of the carbon source, to leave a small residual nitrate concentration in the treated water, so that virtually complete removal of the carbon source is achieved. Any carbon source remaining in the treated water could lead to problems of bacterial growth in the distribution system. Operation with an underdosing of the carbon source prevents bacterial growth in distribution, but has a tendency to result in nitrite production.

Pilot plant trials have been carried out to investigate the ways of operating the process to maintain nitrite residual concentrations at acceptable levels. This work involved overdosing of the carbon source (methanol) and removal of excess methanol through the development of biological activity in post-denitrification conventional water treatment. The pilot plant was

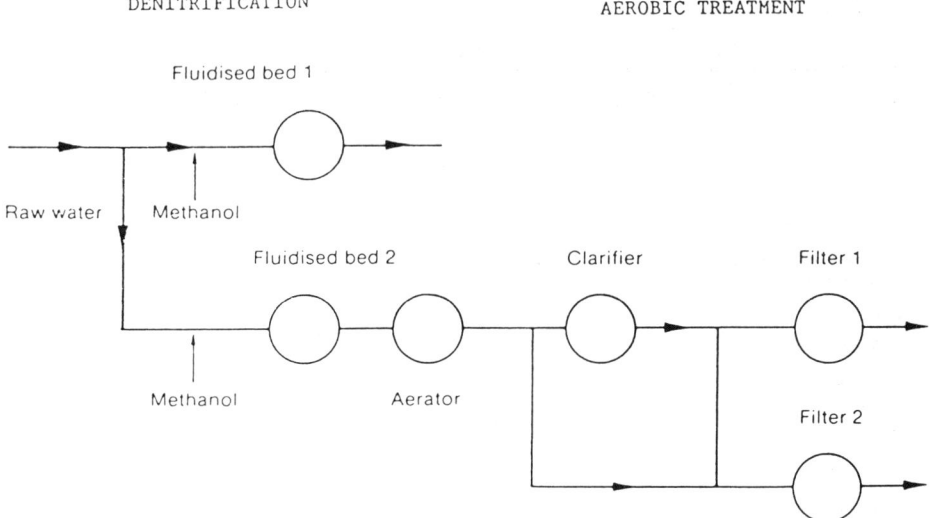

Figure 13.6 BFB pilot plant for investigating control of nitrite residual concentration

designed to compare underdosing and overdosing of carbon source in two fluidised beds operated in parallel. Denitrified water from the overdosed bed was aerated before further treatment using pilot scale floc blanket clarification and/or rapid gravity filtration (both process types being commonly used in conventional water treatment). A schematic of the pilot plant is shown in Figure 13.6.

Overdosing of methanol was successful in controlling nitrite concentrations. This is illustrated in Table 13.3, which shows profile concentrations of nitrate, nitrite and methanol up the fluidised beds. When the methanol is used up in an underdosed bed, high nitrite concentrations remain in the treated water leaving the bed. In the overdosed bed, however, the remaining nitrite is removed, using the excess methanol, once nitrate removal is complete. In principle, the same effect could be achieved by dosing the exact amount of methanol needed for complete denitrification, but problems would be encountered in controlling the dose, which would need to be adjusted as the dissolved oxygen and nitrate concentrations in the feedwater varied.

The results obtained indicated that a methanol excess of up to 5 mg l^{-1} can be removed by conventional treatment processes subsequent to BFB denitrification. This is likely to result from the accumulation of bacteria (carried over from the BFB) within the floc blanket or

Table 13.3 Example of profile results from BFB pilot plant

Height above base of BFB (m)	Underdosed bed (mg l^{-1})			Overdosed bed (mg l^{-1})		
	NO_3N	NO_2N	Methanol	NO_3N	NO_2N	Methanol
0	13	0.12	33	13	0.12	43
1	11	0.37	22	11	0.28	40
2	7	0.49	10	4	0.38	19
3	5	0.48	5	<0.1	0.11	9
4	3	0.60	1	<0.1	<0.1	6
4.5	3	0.42	<0.5	<0.1	<0.001	6

Table 13.4 Methanol removal in conventional treatment processes

Treatment	Mean methanol (mg l^{-1}) in:	
	Feed	Treated
Rapid gravity filtration (coarse sand)	1.7	<0.5
Rapid gravity filtration (fine sand)	1.8	1.0
Chemical coagulation and rapid gravity filtration (coarse sand)	4.8	<0.5
Chemical coagulation and floc blanket clarification	4.8	0.7
Chemical coagulation, floc blanket clarification and rapid gravity filtration (coarse sand)	4.8	<0.5
Chemical coagulation using ferric sulphate		

Coarse sand = 8/16 mesh grade
Fine sand = 16/30 mesh grade
Water temperature 16°C

filter. Regular sampling of the plant over the whole period of operation indicated that good removal could be obtained at water temperatures of below 10°C. Examples of the results are shown in Table 13.4.

In summary, the pilot plant work indicated that nitrite production could be avoided by using a slight overdose of carbon source (<5 mg l^{-1}) which could be reduced to <0.5 mg l^{-1} by the development of biological activity in conventional water treatment processes.

13.4 MEMBRANE PROCESSES

In reverse osmosis (RO), the hydraulic pressure exerted on one side of a semi-permeable membrane forces water across the membrane, leaving salts, including nitrate, behind (WRc, 1989). This results in the production of a treated water stream and a concentrate waste stream, the relative sizes of each being governed by the pressure differential across the membrane. The principle is illustrated in Figure 13.7. As well as the use of high pressure RO (typically, 15−20 bar) for direct removal of nitrate, it has been proposed that lower pressure (e.g. 5 bar) 'nanofiltration' membranes could be used as a pre-treatment to ion-exchange nitrate removal using conventional resins. These membranes remove sulphate and other divalent ions (e.g. Ca^{2+}) but allow the passage of monovalent ions such as nitrate. Removal of sulphate before ion exchange will increase the nitrate removal capacity of conventional resins (see Section 13.2.1). The economics of such an approach would need to be evaulated for specific circumstances.

In electrodialysis, the membranes allow ions to pass but not water. The driving force is an electrical current which carries the ions through the membranes. By using stacks of alternate anion and cation permeable membranes, it is possible to produce a treated and concentrate stream. Electrodialysis reversal (EDR) is a modification of the process in which the current is reversed on a regular basis to prevent problems of calcium scale formation. The principle of electrodialysis is illustrated in Figure 13.8. Nitrate-selective membranes are now available which make the process more cost-effective for nitrate removal, because the proportion of nitrate removed relative to other ions is increased, compared with conventional EDR. This reduces the overall power requirements per unit nitrate removed.

For both reverse osmosis and electrodialysis the concentrate flows would typically be between 5% and 20% of the throughput; waste volumes are therefore large compared with ion exchange, but the quality of the membrane process waste is higher, and disposal may not present a serious problem. Operating costs for membrane processes are high as a result of power consumption. The economies of scale for membrane processes are low compared with ion exchange or biological denitrification, giving relatively high capital costs for larger-membrane plants.

13.5 PROCESS SELECTION

The main factors influencing the process selection for nitrate removal from public water supplies are:

(1) The need for filtration after biological denitrification, to remove carry-over from the bacterial bed; and
(2) The disposal of ion exchange waste regenerant.

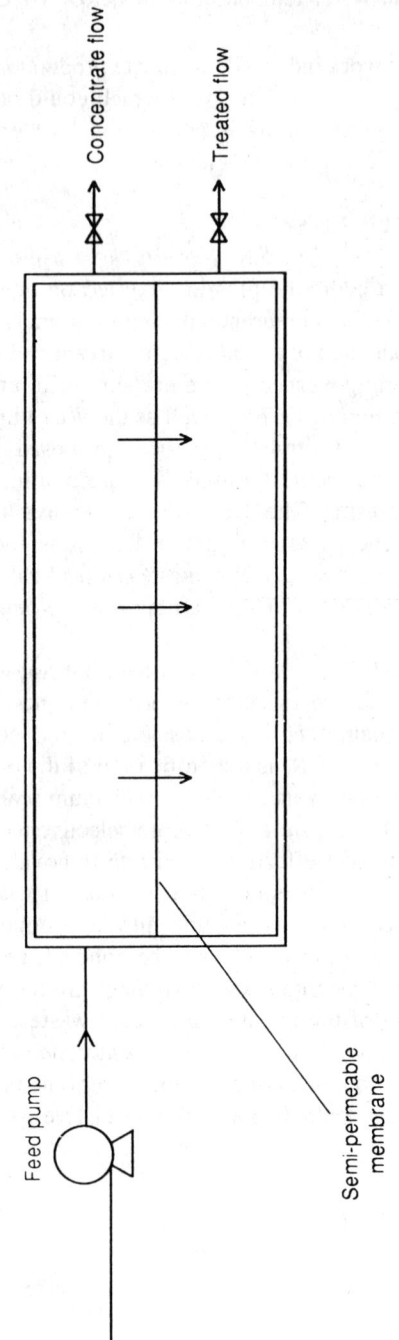

Figure 13.7 Principle of reverse osmosis

Feed
A
C
A
C
A
C
ANODE
CATHODE
Anode (waste) stream
Cathode (waste) stream
Treated (diluate) flow
Concentrate (waste) flow
A = Anion permeable membrane
C = Cation permeable membrane
⊕ = Cation (e.g. Ca^{2+}, Mg^{2+}, Na^{+})
⊖ = Anion (e.g. NO_3^{-}, SO_4^{2-}, Cl^{-}, HCO_3^{-})

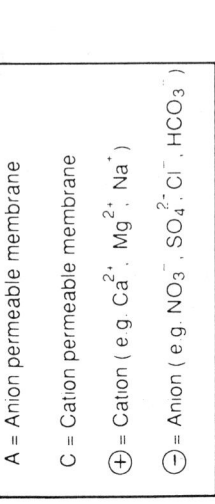

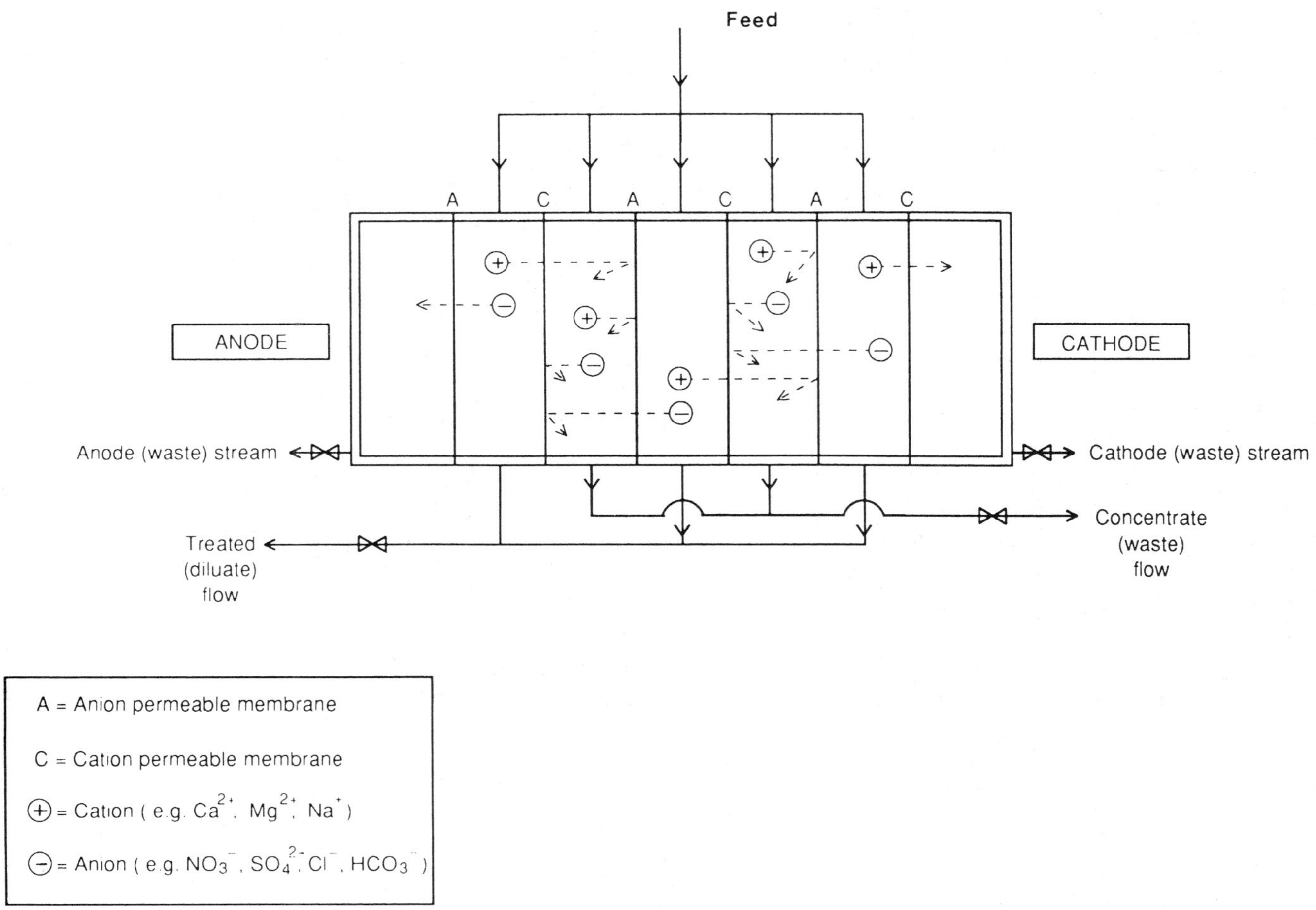

Figure 13.8 Principle of electrodialysis

For most public supplies derived from surface water sources (see Chapter 5) (which are normally relatively large compared with groundwater supplies and are already treated by filtration) with continuously high nitrate concentrations, biological denitrification is likely to be more cost-effective than ion exchange. For groundwater sources (see Chapter 8) (without existing filters), ion exchange would normally be the cheaper process. However, for large groundwater sources (above, say, 20 000 m^3 per day), or where waste disposal by tankering was necessary over distances in excess of 20 km, biological denitrification might offer cost advantages. For very small supplies, the higher costs of membrane processes might be acceptable on the grounds of easier operation or where tankering of ion exchange waste was necessary.

Where nitrate concentrations are intermittently high, the slow start-up characteristics of biological processes may preclude their use in favour of ion-exchange or membrane processes which can be started up rapidly.

The costs for nitrate removal processes are reviewed by Booker, Hall and Hyde (1989), and are discussed in more detail in a WRc report (WRc, 1989). Generally, operating costs plus capital repayment for ion exchange or biological denitrification would be roughly in the range $5-15$ p m^{-3} distributed (1991 prices), depending on plant size, nitrate concentration and waste disposal options, among other factors.

13.6 OVERVIEW

The action being taken to maintain nitrate concentrations in water supplies below regulatory levels includes the control of agricultural activity, blending and water treatment. In the long term, the control of agricultural activity through the implementation of nitrate protection zones (Chapters 5, 10 and 12) is likely to offer the most favourable solution to the problem. However, in some areas the full benefits from protection zones may not be derived for many years, and short-term control measures are needed. Blending with local low-nitrate sources is being used, but the availability of low-nitrate sources will limit the potential of this option because of the high costs for pumping over long distances. At many sites, nitrate removal treatment processes will be the only viable control option, at least in the short term and perhaps, in some areas, in the long term.

Treatment by ion exchange offers advantages with respect to operation and costs for smaller groundwater sources. However, the environmental effects of the disposal of ion-exchange waste regenerant, high in nitrate and chloride, are of increasing concern to the water supply industry. As a result, there is likely to be a trend towards the more widespread use of biological denitrification, either to treat the water directly or to treat the ion-exchange waste to reduce its impact on the environment.

REFERENCES

Booker, N.A., Hall, T. and Hyde, R.A. (1989) Removal of industrial and agricultural contaminants from groundwater. *Proceedings of the Conference: World Water '89*, Institution of Civil Engineers, London, November 1989, Thomas Telford, London, pp. 109−15.
Braester, C. and Martinell, R. The Vyredox and Nitredox methods of *in-situ* treatment of groundwater. *Water Science and Technology,* **20**, No. 3, 149−63.

Croll, B.T. (1991) Nitrate removal using ion exchange: brass corrosion considerations. *Institution of Water and Environmental Management Yearbook*, pp. 39–46.

Croll, B.T., Green, L.A., Hall, T., Whitford, C.J. and Zabel, T.F. (1985) Biological fluidised bed denitrification for potable water. In Tebbutt, T.H.Y. (ed.), *Advances in Water Engineering*, Elsevier Science, New York, pp. 180–87.

DoE (1986) *Nitrate in Water*, A report by the nitrate co-ordination group, Pollution Paper No. 26, HMSO, London.

EEC (1980) EC Directive relating to the quality of water intended for human consumption. *Official Journal of the European Community*, No. L229, 80/778/EEC, available in DoE Circular No. 20/82, HMSO, London, 1982.

Gahrs, J.H., Rutten, P. and Schnoor, G. (1989) Drinking water treatment using hydrogen. *Water and Sewage International*, 1, No. 1, 35–9.

Gauntlett, R. (1975) Nitrate removal from water by ion exchange. *Water Treatment and Examination*, 24, 172–93.

Harries, R.C. (1981) *Studies of the removal of nitrate from potable surface water by ion exchange*, Unpublished PhD Thesis, Hatfield Polytechnic.

Kruihof, J.C., Van Paassen, J.A.M., Hijnen, W.A.M., Dierx, H.A.L. and Van Bennekom, C.A. (1985) Experience with nitrate removal in the Eastern Netherlands. Paper presented at the Conference 'Nitrates Dans les Eaux', Paris, 1985.

Lauch, P.R. and Guter, G.A. (1986) Ion exchange for the removal of nitrate from well water. *Journal American Water Works Association*, May, 83–8.

Mercado, A., Libhaber, M. and Soares, M.I.M. (1988) *In-situ* biological groundwater denitrification: concepts and preliminary field tests. *Water Science and Technology*, 20, No. 3, 197–209.

Philpot, J.M. and Larminat, G. (1988) Nitrate removal by ion exchange: the Ecodenit process, an industrial scale facility at Binic (France). *Water Supply*, 6 (Brussels), 45–50.

Richard, Y.R. (1989) Operating experience of full-scale biological and ion exchange plants in France. *Journal Institution of Water and Environmental Management*, 3, No. 2, 154–67.

Schippers, J.C., Kruihof, J.C., Mulder, F.G. and Van Lieshout, J.W. (1987) Removal of nitrate by slow sulphur/limestone filtration. *Aqua*, No. 5, 274–80.

Statutory Instrument (1989) No: 1147, The Water Supply (Water Quality) Regulations 1989.

Van der Hoek, J.P., Van der Ven, P.J.M. and Klapwijk, A. (1988) Combined ion exchange/biological dentrification for nitrate removal from groundwater under different process conditions. *Water Research*, 22, (6), 679–84.

WRc (1979) *Biological removal of nitrate from river water*, Technical Report TR98.

WRc (1989) *Nitrate removal from drinking water: a technical and economic review of treatment processes*, Report 856-S.

14 Nitrate and the Law

S. BALL

Department of Law, University of Sheffield

14.1 INTRODUCTION

The law on nitrate in water must be understood in the context of the general law on water pollution. This is because the development of legislation specifically directed at the problems posed by nitrate is as yet only in its infancy. It must also be understood in the context of controls on the quality of the public water supply, since much of the impetus for the control of nitrate in surface and groundwater has come from the need to control levels of nitrate in drinking water.

This chapter will therefore outline the general law on water pollution and water supply, including the standards set for nitrate in drinking water, before explaining in greater detail the provisions on nitrate sensitive areas and the EC Directive on Nitrates in Water 91/676. It should however, as always with matters of environmental law, be borne in mind that the law is liable to change and that the exact rules set out may well alter in the next few years.

14.2 DOMESTIC AND EC LAW

14.2.1 DOMESTIC LAW

In the environmental field domestic law mainly results from legislation (i.e. Acts passed by Parliament). However, because of the impossibility of providing for the detail of the law in the restricted time available in Parliament, it is common for Acts to include within them powers for specified individuals or bodies to make delegated legislation. This is normally a member of the elected government, such as the Secretary of State for the Environment or the Minister of Agriculture, Fisheries and Food.

14.2.2 EC LAW

EC law is of enormous importance in relation to environmental protection (see Haigh, 1989). In signing the various treaties establishing the EC, the UK has agreed to transfer some of its law-making powers to the EC institutions, albeit within limited spheres. EC law comes in a number of types, but the main one of relevance to environmental law is the Directive. A Directive is addressed to Member States, and imposes on them a duty to implement its terms, usually within a specified time limit. They will be in breach of EC law if they fail to implement the Directive fully. This may result in infringement proceedings being brought before the European Court of Justice, which, in turn, may declare whether the Directive is being properly implemented. However, one weakness of EC law is that there is no effective

Nitrate: Processes, Patterns and Management. Edited by T.P. Burt, A.L. Heathwaite and S.T. Trudgill
© 1993 John Wiley & Sons Ltd

remedy for non-implementation of a Directive, except pressure from the EC institutions, the other Member States and the public.

Directives are not normally directly effective within Member States in the sense that they do not give rise to rights and obligations which can be enforced by individuals and companies before national courts. As a result, the law which is enforced before UK courts is domestic law: if EC law is different it acts as a reminder that the domestic law is likely to be changed to comply at some stage.

14.3 THE LEGISLATION ON WATER POLLUTION

The major part of the law on the quality of drinking water and on water pollution is in the Water Resources Act 1991, the Water Industry Act 1991 and delegated legislation made under those Acts (see Howarth, 1990a; Bell and Ball, 1991). They replace earlier Water Acts and the Control of Pollution Act 1974. However, in terms of content these Acts are the same as the Water Act 1989, which they replaced from 1 December 1991.

The shape of much of this law is now dictated by a string of EC Directives on water quality. In line with the aims of the EC, these normally have the twin objectives of seeking to ensure uniformity of laws and practices between all Member States and improvement of environmental standards throughout the EC. The Directives tend to follow two basic models: (1) those which impose standards on emissions, which are mainly used for the prevention or restriction of discharges of dangerous and toxic substances to surface and groundwaters; and (2) those which specify quality objectives for certain types of waters, mainly by reference to the use which is to be made of those waters (Somsen, 1990b). It is this second category which is of greatest relevance in relation to nitrate (see the Directives on Drinking Water and on Surface Water for Drinking referred to below).

Owing to the limited space available, the law in Great Britain only will be considered in this chapter. However, it can be reflected that the relevant EC standards are equally applicable in the other EC Member States.

14.4 THE STRUCTURE OF THE WATER INDUSTRY

14.4.1 WATER ACT 1989

The water industry of England and Wales was extensively reorganised by the Water Act 1989 from 1 September 1989. Prior to then, the ten Regional Water Authorities carried out water management on an integrated basis in hydrologically defined areas. The main reason for the reorganisation was the political one of privatising the operational side of the industry, but the objective of splitting the operational and regulatory sides of the industry was also achieved, thus retreating from a position where the Regional Water Authorities, as a result of their wide range of functions, were both 'poachers' and 'gamekeepers' at the same time.

Operational activities relating to the provision of the public water supply and sewerage services (including the treatment of sewage) are now carried out by the ten privatised Water and Sewerage Undertakers. There are also 29 Statutory Water Companies, which are responsible solely for water supply in their designated areas. Regulatory functions in relation to water supply and sewerage provision are the responsibility of the Office of Water Services (OFWAT).

14.4.2 THE NATIONAL RIVERS AUTHORITY

The 1989 Act created the National Rivers Authority as an independent non-departmental public agency. It has taken over most of the regulatory powers of the Regional Water Authorities and has the primary responsibility for dealing with pollution of inland, underground and coastal waters, although Her Majesty's Inspectorate of Pollution now has responsibility for discharges of certain dangerous substances under the Environmental Protection Act 1990. Its other functions relate to flood defence and land drainage, the maintenance of water resources and the operation of the system of licensing of abstractions of water, salmon and freshwater fisheries, navigation, and conservancy and harbour authority activities. In addition, it has a number of environmental duties relating to recreation and the promotion of conservation of nature (see Howarth, 1990b).

One of the strengths of the National Rivers Authority is its national status, and it is striving to build up a uniform system of pollution control throughout the country. This is to be achieved by the establishment and enforcement of uniform policies and practices on discharge consents and water quality.

14.4.3 SCOTLAND

The water industry is organised differently in Scotland. Regional and Island Councils are responsible for the provision of the water supply and for sewerage services. The control of water pollution is the responsibility of River Purification Boards, seven independent catchment area bodies with members appointed by the Regional and District Councils and the Secretary of State. As a further complication, the legislation is different too: the Control of Pollution Act 1974 remains in force in Scotland, although it has been amended by the Water Act 1989, Schedules 22 and 23. The result is that the substance of the law is similar in Scotland, though not always exactly the same. Any significant differences will be noted.

14.5 WATER SUPPLY

A major reason for protecting inland waters against nitrate is the public health reason of protecting the public water supply from pollution. British law is effectively dictated by EC standards in this respect.

14.5.1 DRINKING WATER DIRECTIVE

The EC Directive on Drinking Water 80/778 lays down 62 parameters relating to the quality of water provided for human consumption. The standards required are divided into two categories: Maximum Admissible Concentrations (or in some cases Minimum Required Concentrations) and Guide Levels. For Maximum Admissible Concentrations Member States must set values which are at least as strict as those in the Directive and must ensure that those values are met in practice. Guide Levels represent a concentration which the EC considers it desirable that waters should meet, but they are not legally binding. For nitrate, the Maximum Admissible Concentration is 50 mg l^{-1}, with a Guide Level of 25 mg l^{-1}. This compares with the recommended (i.e. non-binding) level of 100 mg l^{-1} established by the World Health Organisation.

Article 9 of the Directive allows States to grant derogations from its requirements where compliance is not possible due to the nature and structure of the ground in the supply area. The wording envisages derogations being granted for natural problems only and does not appear to cover the situation where the failure to reach the required standards is artificial (for example, due to nitrate leaching). The UK government originally granted a number of derogations on precisely this ground, but withdrew those that were not based on natural causes in April 1988 (Bates, 1989). Derogations may also be made on the grounds of adverse meteorological conditions, such as drought. Any derogation must not constitute a public health hazard.

Compliance with the Directive should have been achieved by 1985, but there is also a power in Article 20 for the Commission to authorise a delay in reaching the required standards if special difficulties are being experienced. No such authorisation appears to have been given and infringement proceedings against the UK are currently before the European Court of Justice for non-compliance with the standards for nitrate in respect of 28 zones in England (Case 337/89, see Somsen, 1990a). It is understood that the UK government is arguing that compliance will be achieved by 1995, but any argument that the deadline was too short is likely to be rejected on the grounds that the Member States agreed it in the first place.

One other Directive requires a mention. As part of the strategy for achieving drinking water standards, the Directive on Surface Waters for Drinking 75/440 defines classes of water (A1, A2 and A3), together with 46 relevant parameters with which surface sources of drinking water must comply in order to fall within each class. Waters below Class A3 should not normally be used for drinking. The standard for nitrate is set at a maximum of 50 mg l^{-1} in each class.

14.5.2 IMPLEMENTATION IN THE UK

These Directives have been implemented as follows. Under the Water Industry Act, Section 67 (in Scotland the Water (Scotland) Act 1980, Section 76A), domestic water must be 'wholesome'. This is officially defined for the first time in the Water Supply (Water Quality) Regulations SI 1989/1147 (SI 1990/119 in Scotland). These lay down a large number of specific criteria with which water must comply. In each case the limits have been set so as to conform to the Drinking Water Directive. The Regulations also provide for a system of public registers of water quality, and a Drinking Water Inspectorate has been established within the Department of the Environment with responsibility for monitoring water quality. The requirements of the Surface Waters for Drinking Directive have been implemented by the Surface Waters (Classification) Regulations SI 1989/1148, which call the three classes DW1, DW2 and DW3.

14.6 GENERAL CONTROLS ON WATER POLLUTION

14.6.1 GENERAL OFFENCE OF WATER POLLUTION

It is a criminal offence to 'cause or knowingly permit any poisonous, noxious or polluting matter or any solid waste to enter controlled waters' (Water Resources Act 1991, Section 85). 'Controlled waters', defined in Section 104, include virtually all inland and coastal waters. It is of special relevance that they include groundwaters (i.e. any waters contained in underground strata, or in wells or boreholes). This offence is very widely drafted and will cover such things as agricultural runoff that pollutes surface or groundwaters and accidental escapes of polluting materials. It is not necessary to show that there was any negligence involved, merely that a particular activity caused the entry. Almost the only effective defence is to have a consent from the National Rivers Authority.

There used to be a defence to this general offence if it was committed by a farmer who was acting in accordance with 'good agricultural practice', which was defined in a Code of Guidance issued by the Minister of Agriculture, Fisheries and Food. This defence was repealed by the Water Act 1989, although the Code of Guidance remains as a non-binding document (MAFF and WOAD, 1991). Breach of it gives rise to no criminal offence, but adherence will be taken into account in the decision whether to prosecute and in fixing any penalty.

On conviction in a Magistrates' Court there is a maximum fine of £20 000 and/or 3 months in prison. On conviction in the Crown Court there can be an unlimited fine and/or a 2-year jail sentence. The National Rivers Authority has a discretion over whether to prosecute for any breach. Although prosecution has in the past only been used as a last resort, or for serious incidents, current policy is clearly to prosecute more readily, and it is significant that the largest category each year relates to agricultural pollution incidents (National Rivers Authority, 1992).

14.6.2 CONSENTS FOR DISCHARGES

A consent is required from the National Rivers Authority before any trade or sewage effluent may be discharged into controlled waters. This covers active discharges through pipes or channels. 'Trade effluent' is defined very widely to include effluent from agricultural, fish farming or research establishments, so a consent is required for nearly all industrial or agricultural discharges.

The National Rivers Authority may attach a wide range of conditions to a consent, setting out what may be discharged and under what circumstances. It is normal to attach absolute numerical limits, with the result that *any* discharge in excess of the terms of a consent is in breach of it. It is a criminal offence, with the same potential penalties as for the general offence, to discharge trade or sewage effluent without a consent, or in breach of any conditions attached to one.

A fee is paid for the application, and from 1 July 1991, an annual charge has been payable by dischargers to reflect the cost to the National Rivers Authority of monitoring the discharge. There is a little-used right of appeal to the Secretary of State against refusal of a consent or against any conditions imposed. Applications for consent are publicised and there is a public register of all applications, consents, water quality objectives and samples taken by the National Rivers Authority.

Once granted, a consent may be varied or revoked by the National Rivers Authority without compensation, illustrating the position that no-one has a right to pollute. It is currently reviewing existing consents. However, a variation or revocation may not normally be made within 2 years of the previous variation.

14.6.3 SEWAGE WORKS

Discharges from sewage works are subject to the same requirements. However, compliance is normally only required on a 95 percentile rate. This is a hangover from the days when sewage works were operated by the same bodies which regulated water pollution, and the National Rivers Authority has proposed that the position is changed so that they are brought into line with other dischargers (National Rivers Authority, 1990). (A discharge *to* the sewers requires a trade effluent consent from the sewerage undertaker.)

14.6.4 WATER QUALITY OBJECTIVES

Consents have traditionally been set on an individualised basis by reference to the effects of the discharge on the receiving waters. Particular emphasis has always been placed on Biochemical Oxygen Demand. This involves consideration of the use to which water is to be put. The Water Act 1989 introduced a statutory system of water quality classifications, with specific standards set for various uses. Water quality objectives for individual stretches of controlled waters will then be set by the Secretary of State (after a public procedure). These will replace the non-statutory objectives currently in operation and will be of enormous importance in setting discharge consents and in making other decisions about water quality. The National Rivers Authority will be under a duty to exercise its powers to achieve these statutory objectives at all times, so far as it is practicable to do so.

14.6.5 WATER POLLUTION AND COMMON LAW

At common law surface and groundwaters are not owned by anyone, but any owner of a river bank, or of land through which water percolates, or of a property right such as a fishery, has a right to receive water in its ordinary state. Any interference with the quality of this water may amount to a nuisance, giving the owner a right to claim damages or to seek an injunction to halt the pollution. These rights act as a significant complement to the regulatory and criminal laws on water pollution in relation to toxic substances, but an action for nitrate pollution is unlikely because of the difficulty of showing any damage to the relevant owner.

14.6.6 LIMITATIONS OF GENERAL CONTROLS

The emphasis in water pollution control has tended to be on discharge consents and dealing with accidents. One limitation is that, while the system is generally adequate for controlling emissions from definite sources, such as pipes, it is less successful at controlling pollution from diffuse sources. As a result, the consent system is increasingly being supplemented by a range of preventative powers, which aim to prevent pollution arising at source.

14.7 PREVENTATIVE CONTROLS

Under Section 161 of the Water Resources Act 1991 the National Rivers Authority has widely drafted powers to prevent pollution incidents where there is a threat to controlled waters, and to carry out remedial and clean-up operations. It can recover the cost from the person who caused the pollution. This section is mainly used for one-off acts of pollution, but there is no reason why it could not be used for continuing leaching that was causing pollution.

Section 92 empowers the Secretary of State to make Regulations concerning precautions to be taken to prevent any poisonous, noxious or polluting matter from entering controlled waters. The Control of Pollution (Silage, Slurry and Agricultural Fuel Oil) Regulations SI 1991/324 have been made setting minimum standards for silage operations and for slurry and agricultural fuel oil stores from 1 September 1991. It is a criminal offence to breach the fairly specific standards laid down in the Regulations and the National Rivers Authority is given clean-up powers (Ball, 1991).

Section 93 empowers the Secretary of State to designate water protection zones. These would effectively establish a system of local law with regard to water pollution within the zone. Similar powers were available in the Control of Pollution Act 1974, but no such zones have ever been designated. It was originally intended that water protection zones should be used to control nitrate, but the publicity engendered by the nitrate problem during the passage of the Water Act 1989 was such that a special section was inserted on nitrate sensitive areas. Accordingly, a Water Protection Zone cannot now include restrictions on nitrate.

Finally, it should be noted that a waste-disposal licence is required from the local authority for the disposal of waste to land. Although there are exceptions for agricultural wastes disposed on the farm of origin, such activities as the spreading of sewage sludge may require such a licence, and the potential for water pollution is a ground for refusal or the application of conditions. This system is soon to change, so that a waste management licence is required under the Environmental Protection Act 1990.

14.8 NITRATE SENSITIVE AREAS

14.8.1 DEFINITION

Specific powers in relation to nitrate are set out in Sections 94 and 95 of the Water Resources Act 1991 (Section 31B of the Control of Pollution Act 1974 in Scotland). This allows for the designation of nitrate sensitive areas, in which the entry of nitrate into controlled waters may be prevented or controlled. The section only applies where the entry takes place as a result of agriculture: the entry of nitrate from other sources can only be controlled by the other mechanisms already mentioned. However, since controlled waters are defined in Section 104 so as to include groundwaters, this section potentially covers all leaching of nitrate from agricultural activities.

These powers are complementary to the more general legal controls and to other methods of reducing nitrate in drinking water, such as the blending of different supplies. But they are important because they reflect a policy of tackling the problem at source rather than once it has arisen. It is clear that such preventative powers will be used more commonly in the future.

14.8.2 DESIGNATION

Designation of a nitrate sensitive area is effected by the relevant Secretary of State making a statutory Order to that effect. (In England the Secretary of State must act jointly with the Minister of Agriculture, Fisheries and Food.) Because of the possible expenditure implications, the consent of the Treasury is also required. An Order can only be made if requested by the National Rivers Authority which, in making the request, must identify controlled waters at risk from nitrate pollution and the agricultural land from which the entry of nitrate into those waters is likely. The National Rivers Authority should only apply for an Order if it considers that its other more general powers are inadequate to control nitrate pollution. This is the limit of the Authority's formal involvement in nitrate sensitive areas, since the special regulatory powers available within them are exercised by the relevant Agricultural Department (i.e. the Ministry of Agriculture, Fisheries and Food, or the Agriculture Department of the Welsh or Scottish Office).

14.8.3 VOLUNTARY CONTROLS

There are effectively two types of Order available, respectively imposing voluntary and mandatory controls. Under the voluntary scheme the relevant Agriculture Minister is empowered to enter into agreements with owners or tenants of agricultural land in which obligations with respect to the management of the land are agreed in return for compensation. There is no compulsion involved: the farmer has a choice whether to enter an agreement or not. Under the mandatory scheme controls over activities may be imposed on farmers in return for compensation.

In keeping with its clear preference for such methods in dealing with agricultural problems, the Conservative government made public its intention to use only the voluntary controls initially, and for the mandatory powers to be held in reserve as back-up powers (MAFF and WOAD, 1989). In the Nitrate Sensitive Areas (Designation) Order SI 1990/1013, ten areas have been designated which are subject to the voluntary scheme (Forster, 1990). This is a pilot scheme to assess the effectiveness of the controls, hence areas with a range of agricultural practices and hydrogeological conditions have been selected. The ten areas, which are fairly limited in size, are Sleaford (Lincolnshire), Branston Booths (Lincolnshire), Ogbourne St George (Wiltshire), Old Chalford (Oxfordshire), Egford (Somerset), Broughton (Nottinghamshire), Wildmoor (Hereford and Worcester), Wellings (Staffordshire and Shropshire), Tom Hill (Staffordshire) and Kilham (North Humberside).

The Order lists two sets of standard terms which a farmer must agree to in order to qualify for the payments offered — respectively, a Basic Scheme Agreement and a Premium Scheme Agreement.

14.8.4 BASIC SCHEME AGREEMENT

To qualify for this, the farmer must agree to some very detailed restrictions on farming activities. These involve changes in farming practice often going well beyond good agricultural practice, but which reduce the prospect of nitrate leaching, such as the reduction of uses of fertiliser in autumn and winter, the avoidance of grassland ploughing and the maintenance of winter cover. Specific restrictions relate to:

- The amounts of inorganic (i.e. chemical) nitrogen fertiliser that may be applied.
- The amounts of organic nitrogen fertiliser that may be applied.
- The timing of any applications of fertilisers, especially with a view to limiting or preventing the application of fertiliser in the autumn and winter.
- The storage of slurry or liquid sewage sludge.
- The cultivation of grassland in autumn and winter.
- The removal of hedgerows or woodland.
- The use of irrigation.
- The conversion of grassland to arable land.

In addition, the agreement requires that winter cover is retained either by an autumn-sown crop or by sowing a cover crop. No inorganic nitrogen fertiliser may be applied to a cover crop and there are further restrictions on its removal.

Farmers producing pigs or poultry which are permanently housed must also submit a plan with their application indicating how they propose to store, handle and dispose of slurry or manure to avoid the entry of nitrate into controlled waters. Once an agreement is made they should adhere to this plan.

14.8.5   PREMIUM SCHEME AGREEMENT

A farmer may only agree a Premium Scheme Agreement if already in the Basic Scheme. The essential objective is to remove land from arable use. Thus a farmer must agree to stop arable use before 1 October in any relevant year and replace it with grassland which is maintained without cultivation (apart from reseeding, which requires the permission of the relevant Agriculture Department) for the period of the agreement.

There are four options available to the farmer, which attract varying rates of compensation. In each case, any grass cuttings must be removed. Irrigation is not permitted, except under option D. The options are:

(A)   Conversion to unfertilised grassland with no grazing.
(B)   Conversion to unfertilised grassland with grazing that is limited to the number of stock which may be fed from the grass alone.
(C)   Conversion to grassland used for grazing, but with limited applications of fertiliser permitted.
(D)   Conversion to unfertilised grassland with woodland, but with no grazing. At least 50% of the woodland must be broadleaved, and alder is not permitted.

14.8.6   COMPENSATION AND MONITORING

There are standard amounts of compensation set out in Schedule 4 to the Order. These vary between the designated areas, ranging from £55 to £95 per hectare per year for a Basic Scheme Agreement, and from £90 to £380 per hectare per year for a Premium Scheme Agreement, with different payments applying according to the option chosen. These rates will be reviewed in 1993. There are special payments for pig and poultry producers who build a slurry store or transport manure in order to comply with the agreement. A farmer must normally agree in respect of the whole of the farm which is within the designated area. For the areas already designated, applications should have been made by May 1991 and the government has reported that agreements have been made on 87% of the land area within the nitrate sensitive areas. Payments will not duplicate those made under other schemes in which similar restrictions may be agreed. For example, in designated environmentally sensitive areas farmers may receive annual payments for agreeing to certain specified farming practices in the interests of landscape and wildlife protection (see the Agriculture Act 1986, Section 18).

An agreement is a contract between the farmer and the relevant Agriculture Department. Unlike the general provisions on water pollution, it is not a criminal offence to fail to comply with the terms of an agreement. There are provisions through which compliance with the agreement may be monitored, and the farmer must keep records of fertiliser application. The agreement may be terminated, or payments withheld or recovered, if a farmer fails without reasonable excuse to comply with its terms. An agreement will bind those deriving title from the original farmer, such as new owners.

There is also an advisory campaign within designated areas to promote good agricultural practice relating to nitrate leaching; compliance with this advice does not give rise to compensation.

14.8.7 MANDATORY CONTROLS

No mandatory Orders have yet been made, but under Section 94 restrictions similar to those above may be imposed on farmers in designated areas. In addition, a mandatory Order may require positive obligations to be undertaken, such as the construction of containment walls around slurry stores. The section provides for the introduction of systems requiring consent for specified activities from the Minister responsible for the designation (not the National Rivers Authority), and also for the creation of criminal offences of ignoring the provisions of the Order, with the same maximum penalties as for the offences under Section 85. The Order may provide for compensation to be paid to anyone affected by the obligations, and the government is clearly committed to paying compensation wherever restrictions go beyond good agricultural practice.

The procedures for making a mandatory Order are set out in Schedule 12 and are more complex than those for making a voluntary Order. The National Rivers Authority must apply for an Order by submitting a draft to the relevant Minister. Precise publicity requirements are laid down, including a duty to notify any local authority and water undertaker within the designated area, and to notify any owner or occupier appearing to the relevant Minister to be likely to be affected. The relevant Minister has a power (not a duty) to hold a public inquiry before making the Order and may modify it in the light of representations received.

14.9 EC DIRECTIVE ON NITRATES 91/676

14.9.1 THE PASSAGE OF THE DIRECTIVE

The Directive on the protection of waters against pollution caused by nitrates from agricultural sources was formally issued by the EC Council in December 1991. It goes quite a bit further than current British measures (Forster, 1989; Somsen 1990a) by laying down some fairly specific requirements and restrictions, and may well necessitate the use of stronger powers to combat nitrate pollution in Britain, including the use of mandatory Orders. However, it should be noted that the final form of the Directive does not go anywhere near as far as earlier drafts in laying down detailed requirements. It also, unlike earlier drafts, only applies to nitrate pollution caused by agricultural sources and not where it has resulted from other sources.

14.9.2 VULNERABLE ZONES

The Directive has two main objectives: to avoid a concentration of nitrate in surface and groundwaters, and to avoid eutrophication of surface, estuarial, coastal and marine waters. To achieve these objectives, Member States are first required to identify waters which could be affected by nitrate pollution. These are defined in Annex I as:

(1) Surface freshwaters, especially those intended for abstraction for drinking water, which either contain, or could contain if protective action is not taken, more than the concentration of nitrate laid down in Directive 75/440 (i.e. 50 mg l^{-1}).

(2) Freshwaters, estuaries and coastal waters which are eutrophic, or may become eutrophic in the near future, if protective action is not taken.

(3) Groundwaters which contain more than 50 mg l^{-1} nitrate, or could contain more than that amount if protective action is not taken.

There is a degree of flexibility involved here. The application of the criteria is a matter for the Member States. Not all waters that fit the description are to be covered: Member States must also take account of other factors set out in Annex IB, such as the physical and environmental characteristics of the waters and the land. This rather vague formulation is a recipe for future argument as to the scope of the Directive, particularly in areas where waters are eutrophic but nitrate pollution is not the sole factor.

Having identified waters which may be affected by nitrate pollution, Member States must then designate 'vulnerable zones'. These are defined as areas which drain into the polluted (or potentially polluted) waters described above and which contribute to the nitrate pollution in them. Again the vagueness of this last phrase is likely to give rise to debate. Member States must identify these vulnerable zones by 19 December 1993. They must also review the designations at least every 4 years, and add extra zones if appropriate.

14.9.3   RESTRICTIONS WITHIN VULNERABLE ZONES

Within the vulnerable zones, Member States are required, within 2 years of the designation of the zone, to establish action programmes, which must then be implemented within a further 4 years. The action programmes must include the measures laid down in Annex III, though these are not as detailed as in earlier drafts of the Directive. They mainly concern general requirements, such as that rules shall be formulated about periods when the application of certain types of fertiliser to land is prohibited and the capacity of storage vessels for livestock manure. They must also include limitations on the application of fertilisers to land, taking into account the soil, climate and land-use characteristics of the vulnerable zone and the balance between nitrogen supply and foreseeable nitrogen requirements of the crops. One matter of detailed control is that the measures adopted should ensure that the amount of livestock manure applied to land each year should not exceed 210 kg of nitrogen per hectare in the first 4 years and 170 kg thereafter. However, Member States may permit higher levels if they inform the Commission and the derogation can be justified objectively on the basis of such things as a long growing season, high rainfall, or the growing of crops with a high nitrogen uptake.

There is a further duty to take additional measures if it is clear that the action programmes are not meeting the objectives of the Directive, though this is limited by a duty to take into account the cost-effectiveness of any proposed additional measures.

14.9.4   RESTRICTIONS OUTSIDE THE VULNERABLE ZONES

Outside the vulnerable zones, Member States are required to draw up codes of good agricultural practice with the aim of reducing nitrate pollution. Compliance by farmers with these codes is to be entirely voluntary, but general requirements as to their scope are laid down in Annex II of the Directive. The existing British codes will have to be amended to comply with these requirements.

14.9.5   MONITORING IMPLEMENTATION OF THE DIRECTIVE

Member States are required to set up monitoring programmes to assess the effectiveness of the action programmes, which must, in any event, be reviewed at least every 4 years. They are also required to establish programmes for monitoring nitrate levels. In both cases guidelines may be drawn up by a committee of national experts. Finally, Member States are to send implementation reports to the Commission every 4 years from the notification of the Directive.

14.9.6 CONSULTATION PAPER ON DESIGNATING RELEVANT WATERS

In March 1992 the government issued a consultation paper on the proposed criteria and procedures for identifying the relevant waters (Scottish Office Environment Department, 1992). For criterion (1), monitoring of water for compliance with Directive 75/440 has now commenced. For criterion (2), it is recommended that different factors be considered for estuaries and coastal waters on the one hand and freshwaters on the other. For estuaries and coastal waters, the consultation paper recommends that the following factors should be taken into account in deciding whether any stretch of water is likely to suffer from eutrophication: enhanced winter nitrate concentrations; unusual algal blooms; long duration of algal blooms; oxygen deficiency; reductions in fauna; changes in macrophyte growth; occurrence and magnitude of paralytic shellfish poisoning; and formation of algal scums.

For freshwaters, the consultation paper also recommends that simple numeric chemical criteria are not appropriate, but that a range of factors should be considered before deciding on the facts whether any stretch of water is likely to suffer from eutrophication. For still freshwaters, the following symptoms are recommended: phosphorus concentrations; chlorophyll-a concentrations; water clarity; dissolved oxygen; effects on fauna; effects on macroflora; effects on microflora. For running freshwaters, the following symptoms are recommended: phosphorus concentration; algal biomass; water-retention time; dissolved oxygen; effects on fauna; effects on macroflora; effects on microflora.

As far as groundwaters are concerned (criterion (3) of the Directive), it is suggested that all groundwater sources with nitrate levels in excess of 50 mg l^{-1} or a positive linear trend should first be identified. All sources which are not projected to have a concentration in excess of 50 mg l^{-1} by 2010 should then be rejected, as should those where sources other than agriculture are implicated. Those sources where the expected equilibrium nitrate concentration is 45 mg l^{-1} or less can also be rejected, leaving candidate areas where more refined tests can be undertaken.

Further consultation will follow in due course on the mechanisms to be used to designate vulnerable zones under the Directive and on where they should be located.

14.10 URBAN WASTE WATER TREATMENT DIRECTIVE 91/271

The Urban Waste Water Treatment Directive, which was agreed by the EC Council on 18 March 1991, is a far-reaching Directive which will have an important impact on discharges of effluent from sewage works. However, once again the original proposals put forward by the Commission and supported by the European Parliament were weakened significantly by the Council, which put deadlines back and increased thresholds for action (Collins, 1991).

The Directive lays down minimum standards for sewerage systems and sewage treatment. These vary according to the size of the urban area concerned and the nature of the receiving area. Thus, secondary (i.e. biological) treatment is required for discharges from municipalities with a population equivalent of 15 000 or more by 31 December 2000, with smaller municipalities required to comply, in most cases, by the end of 2005.

In relation to nitrate, the importance of the Directive is strengthened by the fact that additional treatment and higher standards will be required for discharges to sensitive areas. The criteria for identifying sensitive areas are laid down in Annex IIA of the Directive and include the following:

(1) Surface freshwaters intended for abstraction for drinking water which could contain more than the concentration of nitrate laid down in Directive 75/440 if protective action is not taken (i.e. 50 mg l^{-1}); and

(2) Freshwaters, estuaries and coastal waters which are eutrophic, or may become eutrophic in the near future if protective action is not taken.

These requirements are almost identical to those under the Nitrates Directive and the consultation paper mentioned above (Scottish Office Environment Department, 1992) treats them in the same manner. It should also be noted that under the Directive exceptions may be made for less sensitive areas (although primary treatment is always required). Member States are required to identify sensitive and less sensitive areas by the end of 1993.

APPENDIX: DOMESTIC AND EC LEGISLATION

Control of Pollution Act 1974
Water (Scotland) Act 1980
Agriculture Act 1986
Water Act 1989
Environmental Protection Act 1990
Water Resources Act 1991
Water Industry Act 1991

Water Supply (Water Quality) Regulations SI 1989/1147
Water Supply (Water Quality) (Scotland) Regulations SI 1990/119
Surface Waters (Classification) Regulations SI 1989/1148
Nitrate Sensitive Areas (Designation) Order SI 1990/1013
Control of Pollution (Silage, Slurry and Agricultural Fuel Oil) Regulations SI 1991/324

Directive on Surface Waters for Drinking 75/440
Directive on Drinking Water 80/778
Directive on Nitrates 91/676 (OJ 1991 L 375/3)
Directive on Urban Waste Water Treatment 91/271 (OJ 1991 L 135/40)

REFERENCES

Ball, S. (1991) Current survey: control of pollution, *Water Law*, 2, 70–2.
Bates, J. (1989) Nitrates and the law, *Land Management and Environmental Law Report*, 1, 6–7.
Bell, S. and Ball, S. (1991) *Environmental Law*, Blackstone Press, London.
Collins, K. (1991) The Directive on municipal waste water treatment, *Water Law*, 2, 116.
Forster, M. (1989) EEC proposals on pollution of water by nitrates, *Land Management and Environmental Law Report*, 1, 7.
Forster, M. (1990) Nitrate sensitive areas — too voluntary a settlement, *Land Management and Environmental Law Report*, 2, 48–51.
Haigh, N. (1989) *EEC Environmental Policy and Britain*, 2nd revised edition, Longman, Harlow.
Howarth, W. (1990a) *Water Pollution Law*, Shaw and Sons, London (first published 1988, with 1990 supplement).

Howarth, W. (1990b) *The Law of the National Rivers Authority*, Centre for Law in Rural Areas, University College of Wales, Aberystwyth.

Ministry of Agriculture, Fisheries and Food and Welsh Office Agriculture Department (1989) *Nitrate Sensitive Areas Scheme: A Consultation Document*.

Ministry of Agriculture, Fisheries and Food and Welsh Office Agriculture Department (1991) *Code of Good Agricultural Practice for the Protection of Water*.

National Rivers Authority (1990) *Discharge Consent and Compliance Policy: A Blueprint for the Future*.

National Rivers Authority (1992) *Water Pollution Incidents in England and Wales — 1990*.

Scottish Office Environment Department (1992) *Criteria and Procedures for Identifying Sensitive and Less Sensitive Areas (Urban Waste Water Treatment Directive) and 'Polluted Waters' (Nitrates Directive)*.

Somsen, H. (1990a) Current survey: water supply and resources, *Water Law*, **1**, 5—6.

Somsen, H. (1990b) EC water directives, *Water Law*, **1**, 93—8.

Part IV

OVERVIEW

15 The Politics of Nitrate in the UK

T. O'RIORDAN and **G. BENTHAM**

School of Environmental Sciences, University of East Anglia

15.1 INTRODUCTION

The nitrate issue falls into a class of environmental disputes which muddle the relationships between scientific judgement and public perception, and which reveal tensions within society over how far humanity should interfere with natural processes in the name of progress. The politics of nitrate, therefore, illustrates an important trend in modern times, namely the struggle by a minority to take what they regard as the misuse of a substance as a symbol, or metaphor, of the lack of environmental accountability in contemporary democracies. In this case 'big' government backed by publicly funded research and advisory services and 'big' business (the extremely profitable agrochemical industry) are assumed effectively to dictate the pattern of events against what is regarded as the public interest. As in all politics, the politics of nitrate is about rearranging effective power.

In such disputes science, as manifested in independent assessments of causes and effects of environmental change, can become a weapon in the political fray. Antagonists on all sides seek to call on environmental science to buttress their case and confound their enemy. In so doing, scientific judgement becomes part of the political struggle to be used at the command of the power brokers. This is not a happy picture for those who like to see science as both objective and committed to a sincere search for truth. Contemporary environmental politics is changing the character of scientific authority and political accountability during a period when large corporations are under scrutiny for their social conscience and environmental performance.

Nitrogen is vital for plant growth, for the rich green hue in leaves, and for the size and protein content of cereals. Nitrogen fixation from the atmosphere is limited. Its transfer to plants requires the activity of bacteria in plants and soils, at rates which effectively control natural plant productivity. Organic matter recycling and fixation by leguminous plants can increase this process, but the demand for even greater productivity from poorer soils has led to the high profitability of the artificial fertiliser industry which now operates successfully in all corners of the globe. Various studies of nitrogen uptake suggest that the efficiency of plant use of synthetic fertiliser is about six times that of extraction from manure or rotting biomass (Smil, 1991, p. 425). It is the combination of an artificially accelerated natural process and the scope for high and guaranteed profit that has led to the rapid growth of the fertiliser industry since 1945. World production now exceeds 90 million tonnes annually, with over one billion tonnes produced since 1975.

According to a recent report (FAO, 1988, p. 5) the application of nitrogenous fertilisers averages 550 kg ha^{-1} in the Netherlands and 350 kg ha^{-1} in the UK. There is plenty of evidence that these levels of application leave significant residues in the soil, possibly as much

Nitrate: Processes, Patterns and Management. Edited by T.P. Burt, A.L. Heathwaite and S.T. Trudgill
© 1993 John Wiley & Sons Ltd

as 250 kg ha^{-1} over a 7–10 year poeriod (Goodman and Redclift, 1991, p. 211). Smil (1991, p. 427) estimates on a crude guess, that only about 70% of nitrogen applied as fertiliser actually is taken up by the intended crop. The rest is either stored, eroded or denitrified. This total could amount to some 10–12 million tonnes annually. Throughout the cultivated world nitrogenous fertiliser application averages out at 50 kg ha^{-1}. Rates of change are universally positive, but slowing in the USA and in the more intensively cultivated parts of Europe as diminishing returns of yield increase are recorded for a wide range of crops and soils. In the developing world, with its increasing dependence on high-yield crops and cultivation methods, the growth of fertiliser application continues virtually without interruption.

The release of nitrate into water and the air is by no means confined to artificial chemical application. As previous chapters of this book have shown, nitrate enters the environment through the disturbance of soils by draining and ploughing, and through the release of ammonium and nitrate from organic manures. In many areas where agriculture is intensifying, soil disturbance is a primary source of nitrogen release, for the process of mineralisation takes many years to complete.

Nitrate in the environment evokes hope and anger. Applied sensibly at modest scales to appropriately designed plants, it is a true blessing for a beleaguered farmer with insufficient organic residue and infertile soil. To rid the world of nitrate-based fertiliser would be a grievous error, particularly for the poorer farmer of the overpopulated developing world. But in a planet of glut and famine, where nitrate fertiliser is used in superabundance for output that is already grossly in surplus, and subsidised for already wealthy farmers, this use of the chemical becomes a cause for anger and frustration. It is this symbol of a mismanaged agricultural economy, protected by specialised, self-serving cliques, that helps to generate much of the anger over nitrate in the environment.

In the section that follows, the arguments for and against the effects of nitrate on human health, in various forms and in various concentrations, are carefully examined (see also Chapter 1). The other aspect for environmental concern is eutrophication, or the excessive nutrient enrichment of waterbodies, notably rivers, lakes and the coastal zone. Here the scientific arguments are more clear-cut, though by no means unambiguously proven. Eutrophication results in a simplification of aquatic ecosystems with a decline in the higher-order plants (macrophytes) and a proliferation of algae (phytoplankton). Certain species are unsightly, others are toxic to fish and especially shellfish, and others still, may be toxic to mammals, including human pets. There are the blue-green species which have caused much adverse publicity in Europe recently because of their appearance in water-supply reservoirs and other lakes where people congregate for environmentally clean enjoyment. These points are expanded in Chapters 5 and 7 in this volume.

15.2 HEALTH EFFECTS OF NITRATE

One of the great triumphs of Victorian environmental engineering was the development in Britain of safe public water supplies. For more than a hundred years the British public has come to expect the water from their taps will be wholesome and do no damage to their health. It is easy to imagine the shock and alarm that would be caused if they were told that some of this water was capable of causing life-threatening disease in young babies and cancer in adults. This, of course, is precisely the accusation that has been made against high concentrations of nitrate in public water supplies which have been alleged to cause *methaemoglobinaemia* ('blue baby syndrome') and stomach cancer. Environmental groups have not been slow to realise that the issue of possible health impacts provides them with a powerful weapon in their campaigns against nitrate pollution. Others have recognised an opportunity to make profits and, in less than a decade, bottled water in Britain has been transformed from an eccentricity into a multi-million pound business. There can be no doubt about the impact on public opinion of the vigorous shroud-waving that has been part of the nitrate debate, but what is the evidence that the alleged health threats are real?

15.2.1 METHAEMOGLOBINAEMIA

Methaemoglobinaemia is a condition that mostly affects babies less than six months old (House of Lords, 1989). It is caused by the reaction of nitrite in the bloodstream with haemoglobin to form methaemoglobin. This normally accounts for less than 2% of total haemoglobin but when levels rise above 10% the flow of oxygen to body tissue is restricted and the child develops a characteristic blue tinge which gives the condition its common name of 'blue baby syndrome'. Death can result, once somewhere in the range of 45–65% of haemoglobin has been converted. The link with nitrate in water is that the low gastric acidity of infants favours bacterial growth that encourages the reduction of nitrate to nitrite. This, together with the high fluid intake of infants and their poorly developed capacity to reconvert methaemoglobin to haemoglobin, explains the concentration of risk in younger infants. There is also evidence that infants with gastro-intestinal infections are at greater risk and therefore bacteriologically contaminated water supplies may pose particular problems.

Methaemoglobinaemia is now a very rare condition in developed countries. In Britain the Department of Health is aware of 14 confirmed reports of cases directly linked to nitrate in water in the last 35 years. Several of these were associated with shallow private water supplies where there was evidence of significant bacteriological contamination as well as high nitrate levels. There has been one death reported which occurred in 1950 and the last reported case was in 1972 (House of Lords, 1989). However, since methaemoglobinaemia is not a notifiable disease it is not possible to be certain about the true incidence.

The origin of the officially designated limits on nitrate in drinking water can be traced back to epidemiological evidence on methaemoglobinaemia risk (Comly, 1945). A limit of 45 mg l^{-1} was established for drinking-water supplies in the USA in 1962 and this was adopted by the World Health Organisation. In 1970 the WHO relaxed this standard for Europe by recommending a limit of 50 mg l^{-1}. This formed the basis for the Community Standard of 50 mg l^{-1} in the 1980 Drinking Water Directive (House of Lords, 1989). However, there has been a great deal of controversy about this limit with most reported cases of the disease in Europe being associated with water nitrate concentrations of more than 100 mg l^{-1}, often from bacteriologically contaminated private sources. Therefore, there is little evidence of substantial risk from public water supplies in Britain, even where nitrate levels exceed the 50 mg l^{-1} limit. Private supplies can pose greater problems and here there may be a need for supplies of bottled water to be made available.

On balance, there is little evidence to support the view that nitrate pollution of water supplies in Britain poses a significant risk of methaemoglobinaemia. There has been a virtual absence of cases during a period of rising nitrate concentration in surface and groundwaters and there are doubts about the degree of risk that exists at current levels of exposure. This raises serious doubts about whether the substantial costs of reducing nitrate pollution can be justified on the grounds of avoiding this health risk, although such expenditure might be justified on other grounds.

15.2.2 STOMACH CANCER

Potentially much more serious is the putative risk of stomach cancer. Although the incidence of this cancer has been declining it remains a common disease that is usually fatal, and in England and Wales it remains responsible for several thousand deaths each year. One of the main reasons for suspecting a cancer risk arises from the role ascribed to nitrate in a model of stomach cancer causation proposed by Correa *et al.* (1975). Bacteria in the human gastro-intestinal tract are capable of reducing nitrate to nitrite and this can react with the products of digestion to produce N-nitroso compounds. Evidence from animal experiments provides strong evidence that these are potent carcinogens that might be capable of initiating stomach cancers in humans (Preussman and Stewart, 1984).

The results of some early epidemiological studies supported the possibility of a link between stomach cancer and nitrate exposure. Hartman (1983) compared national stomach cancer mortality rates for 12 countries with estimates of per capita nitrate intake and found a significant positive association. Studies in Chile have shown an association between geographical variations in stomach cancer and exposure to nitrate fertiliser (Zaldivar and Robinson, 1973; Armijo and Coulson, 1975; Zaldivar, 1977). In Colombia two communities with high rates of stomach cancer were shown to have a greater dependence on local well water with high nitrate concentration as compared with two communities with low rates of stomach cancer (Cuello *et al.*, 1976). In England, Hill, Hawksworth and Tatersall (1973) were able to demonstrate the existence of an excess of stomach cancers in the town of Worksop, which has a high concentration of nitrate in its public water supply.

Subsequent investigations have mostly failed to confirm the positive association found in these earlier studies (Forman, 1989). Davies (1980) was able to show that the apparent excess of stomach cancer in the town of Worksop was the result of its distinctive occupational structure. Once adjustments had been made for social class and the high proportion of coal-miners in the area no excess was apparent in the high-nitrate area. Fraser and Chilvers (1981) compared regional data on stomach cancer mortality and fertiliser use in rural areas of England and Wales and found a clear negative association. They also analysed stomach cancer death rates and nitrate levels in water for smaller rural districts in two regions for the periods 1969−73 and 1974−8. In the Anglian Water Authority area they found a positive association for males but not females. In the Yorkshire Water Authority area no association was apparent for males while for females there was a positive correlation with nitrate for one time period but not for the other. Their conclusion was that the results were inconclusive. This was not the case for a much larger study by Beresford (1985), who examined stomach cancer mortality rates for 1969−73 and nitrate concentrations in public water supplies in 253 urban areas in Great Britain. After adjusting for social factors she found significant negative associations between mortality and nitrate levels for both males and females. Her conclusion was that this lack of evidence of a health risk from nitrate in drinking water was an important finding for the British water industry.

A rather different research design was adopted in an important study by Forman, Al-Dabbagh and Doll (1985). Their approach was based on the measurement of concentrations of nitrate and nitrite in samples of saliva which should give a good indication of the amounts that will be conveyed to the stomach. Samples were taken from populations in two regions of Britain with high rates of stomach cancer (Wales and the North-east) and two with low rates (Oxford and the South-east). Information was also collected on the social class and diet of the respondents to the survey. One of the main findings of the study was that nitrate and nitrite concentrations in the population were significantly higher in the regions with a low risk of stomach cancer. That is, the results were the opposite of what would be expected if high body burdens of nitrate and nitrite were important risk factors for stomach cancer. The results of the comparisons between social classes were also of considerable interest. Stomach cancer mortality displays a pronounced social class gradient with higher risks in lower social classes. However, in the low risk regions Forman, Al-Dabbagh and Doll (1985) found higher levels of nitrate and nitrite in social classes I and II. This means that body burdens of the suspected carcinogens were highest in the groups likely to have the lowest incidence of the disease by virtue of their region of residence and social class.

As well as providing striking evidence that is counter to the hypothesis that nitrates are a significant factor in stomach cancer risk, the study by Forman, Al-Dabbagh and Doll also presents useful data on the contribution of different sources of total nitrate intake. They show that in both the high- and low-risk regions the contribution from drinking water was considerably lower than from food, with the largest source being from vegetables. It is interesting that the estimated intake of nitrate from vegetable sources was significantly higher in the low-risk regions. This raises the possibility that some aspect of a diet rich in vegetables (for example, vitamin C) might inhibit carcinogenesis (Bartsch, Ohshima and Pignatelli, 1988). If this is the case it could be that in the parts of the country with a high concentration of nitrate in water, any cancer risk from this source is being more than offset by possible beneficial effects of a diet rich in vegetables. However, the possibility remains that individuals with unusually low consumption of vegetables could be at risk from nitrate in drinking water in areas where concentrations are high. If such individuals formed a minority of the population in such areas any excess risk among this group might be masked by the lower risks in the majority. Being based on estimates of risks and risk factors in populations, it is a feature of all the studies reviewed above that they are likely to be insensitive to such effects. What is needed is studies based on data for individuals.

Unfortunately, there are relatively few studies that have attempted to relate individual levels of nitrate exposure to individual risk of stomach cancer. A study by Britton et al. (1990) goes some way towards meeting this requirement. It linked individual data on housing tenure (as an index of socio-economic circumstances) from the 1971 Census with mortality records for 1971−81 and with data on nitrate in water for 1969−73 for the local authority of residence of each person in the study. Analysis of data for the whole country showed no evidence of a positive association between nitrate levels and stomach cancer mortality. However, when the analysis was restricted to the south and east regions having the highest nitrate concentrations the results were different. For local authority tenants (but not owner−occupiers or private renters) there was evidence of excess stomach cancer mortality in the highest nitrate group. The authors concluded that high levels of water nitrate are only a risk factor for stomach cancer when present with socio-economic, perhaps specifically dietary, disadvantage.

Other studies using data on individuals have not found similar positive associations. In Canada a study by Risch et al. (1985) compared the recent diets of patients with stomach cancer with those of a control group of healthy individuals. They found a significant negative association between estimates of food nitrate intake (mostly from vegetables) and cancer risk. However, intake of exogenous nitrite, largely from preserved meat, showed a positive association with cancer risk. An important limitation of this study was that it did not consider waterborne nitrate. Another problem is that the study collected data on recent diet whereas the long latency period for stomach cancer indicates that diet 20 or more years ago is probably more important.

Al-Dabbagh *et al.* (1985) investigated a group of 1300 male employees from a fertiliser manufacturer in the United Kingdom. These workers were exposed to high levels of nitrate from the inhalation and ingestion of dust. If nitrate is a significant risk factor for stomach cancer an enhanced incidence of the disease would be expected in this highly exposed group. The high levels of exposure to nitrate were confirmed by measurement of salivary nitrate but there was no evidence of an excess of stomach cancer in the exposed group compared to the rates that were typical of the locality.

15.2.3 CONCLUSIONS ON NITRATE AND CANCER

Although results have been mixed, the epidemiological evidence does not provide any strong support for the hypothesis that the concentrations of nitrate in drinking water in Britain poses a significant risk of stomach cancer. Given the well-established trend of a risk in nitrate concentrations in water supplies, an increase in stomach cancer rates might be expected, but the actual trend has been downwards. Neither is there any support for the hypothesis from the geography of the disease. The highest rates of stomach cancer are in the north and west of the country, whereas it is in the south and east that nitrate concentrations are highest. This lack of support for the hypothesis is also true of comparisons between different social groups. There is evidence that nitrate intake is greater in social classes I, II and III whereas mortality from stomach cancer is highest in classes IV and V.

Of course, simple comparisons of time trends, geography and social patterns are always prey to the confounding effects of other variables. However, it is worth noting that in most of the studies that have attempted to control for confounding factors, little evidence has been found of a positive association between nitrate intake and stomach cancer risk. Notwithstanding these conclusions, there remains a need for a careful monitoring of the situation. Given the long latency of the disease and the fact that nitrate concentrations in surface and groundwater are continuing to rise there is always the possibility that effects will only be apparent some time in the future.

There is also a need for further investigation of the possibility that some groups might be at greater risk than the majority of the population, perhaps because of poor-quality diet. What is needed is further research on populations with long-term exposure to high nitrate concentrations preferably incorporating data for individuals on nitrate intake, health outcomes and possible confounding factors. However, on the basis of the information that is currently available the most plausible conclusion is that observed levels of nitrate in drinking water are unlikely to be associated with an elevated risk of stomach cancer.

15.3 SCIENCE, POLITICS AND THE PRECAUTIONARY PRINCIPLE

The connection between algae, nitrate and health effects is by no means fully understood (see Chapter 1). In the absence of definitive scientific statements, groups which wish to act in the interests of *precaution* call for a ban on the use of the substance in certain vulnerable areas, for example on the catchments around sensitive lakes, rivers and groundwater recharge zones. The concept of precaution is still evolving in environmental science. In its literal form, it means acting in advance of scientific certainty in circumstances where it is deemed prudent on the evidence available to take avoidance action. Action is adopted even if it incurs costs for certain groups of producers and consumers, and even when the benefits of putative environmental gain cannot be fully justified.

At one level, then, precaution implies a 'sub-optimal' economic assessment of the gains and losses of acting early, where the sub-optimality can be justified on the grounds of prudence, or good housekeeping. But at quite another level, precaution is a deeply political concept. It is being used to justify expensive intervention in the name of 'environmental common sense'. This, in effect, means that certain interests who formerly had relatively little influence over the course of events now have considerable weight. These interests include non-establishment (but not necessarily misguided) scientists whose styles of analysis and reasoning are unusual (but by no means 'wrong'). They also include what are known as 'cause' environmental pressure groups, such as Greenpeace and Friends of the Earth, both now very well organised and committed groups with money, support and good connections, who seek to create the right political conditions to give them leverage in crucial decisions. This is because the higher level of the precautionary principle is steadily shifting the burden of proof, away from the victim or possible victim, who are often too poor and ill-organised to prosecute their case, towards those whose actions or proposed actions could lead to unacceptable environmental alteration. This shift of proof means that the axiom of innocent until proven guilty is becoming reversed to 'guilty until proven innocent'.

This is the case for nitrate, as it is for a whole host of other chemicals, particularly the heavy metals and the bioaccumulative compounds such as the organophosphates and the chlorinated hydrocarbons (PCBs, dioxin). This presents a real dilemma for science. Science proceeds by axiom, searching for the truth by trial and error, by eliminating what can be shown to be false, and by accepting that uncertainty is an essential component of any conclusion. Thus, when debating environmental health risk it is never possible to be sure, only to be as certain as the evidence available at the time suggests.

What we are beginning to witness is a change in the concept of 'expertise' and 'science'. The notion of expert authority is being reconstructed to include scientifically educated community leaders and advisers, and the informed members of civic-minded pressure groups. The context, therefore, in which 'science' has to operate is being extended to include the hopes, fears, prejudices and trusting beliefs of an ever-widening net of activated citizens.

In this changing world of shifting alliances it is increasingly difficult to de-politicise the context of scientific advice. The link to gastric cancer may be all but non-existent, but cancer is a highly emotive disease: a mere whiff of it in a causal charge activates precaution. Similarly, the connection between nitrate and algal blooms is by no means made: but in the public mind, at a scale too well established to be politically ignored, nitrate is the culprit, and farmers are the pawns in a chain of forces that lead to the multinational agribusinesses and pliant governments. The 'public' face of science is strongly coloured by events, by symbols and by images of power and powerlessness. This is why millions of European Currency Units will be spent on reducing nitrogen inputs into the southern and eastern North Sea, and why European Community nationals have a legal right to drink water at 50 mg l^{-1} almost regardless of cost.

The politics of nitrate in the UK tell us a lot about the frills of environmental concern in a world where 3000 children die each day from avoidable, but human-caused environmentally induced disease, and where, in many parts of eastern Europe, the average length of life is shortening by as much as two weeks every year because of acute toxification of the air and water from grossly mismanaged and underfunded waste-treatment works and industrial installations. Such is the culture of modern environmentalism (see Timberlake and Thomas, 1991).

15.4 RESTRAINING APPLICATIONS OF ARTIFICIAL FERTILISERS

In many parts of agriculturally rich Europe the nitrate concentrations of groundwater are rising. In southern England, over 20% of public water supplies exceed the 50 mg l^{-1} limit and a number breach the 100 mg l^{-1} concentration, regarded by the World Health Organisation as unacceptable (Department of the Environment, 1986). Furthermore models of a number of groundwater-fed streams indicate that nitrate levels in the aquifer are so high, that even with 50% reduction of current use, the concentrations would continue to rise for another 50 years (Chapter 8).

One important issue here is the degree to which the British government is prepared to comply with what is regarded officially as an unsupportable argument on scientific grounds but which meets the needs of a collective entity called the European Community. Put in its stark form, the political reality is that, since other Community countries are spending money on agreed pollution-control measures, so the UK should pull its weight regardless of the scientific niceties. Throughout the 1980s Britain was dubbed 'the dirty man of Europe' for its intransigence over reduction of SO_2, NO_x and nitrate in the face of a European panic over acid rain and eutrophication. This story has been well analysed by Rose (1990). Despite all its attempts to argue that its position was scientifically justifiable, Britain was simply treated by the Community as an environmental 'dog in the manger'. Once Mrs Thatcher had delivered her Royal Society speech on the importance of environmental sustainability in September 1989, Britain was in a politically defenceless position in Europe. Compliance with community environmental directives was therefore inevitable (see, *inter alia*, Porritt and Winner, 1989; McCormick, 1991).

One of these was the Drinking Water Directive of 1980, which requires that sources of water for public consumption should have no more than a limit of 50 mg l^{-1}. For the three years 1985—8, Britain avoided its responsibilities in this regard by the application of derogations, namely a legal disregard of a directive on the grounds of special circumstances. Table 15.1 provides data on derogations that applied during this period. By 1988 the UK had abandoned this position, partly in the face of legal challenge from the European Commission, and looked at ways of blending supplies from lower nitrate areas, and denitrifying water supplies. The Department of the Environment (DoE) Nitrate Co-ordination Group estimated that to reduce all water supplies to below 50 mg l^{-1} would require immediate capital expenditure of £50 million: £45 million in 10 years and £100 in the 10-year period following that. This would mean an increase of 10—15% on water bills, compared with only 3% to meet the 80 mg l^{-1} level, which, on health grounds, is arguably tolerable.

At first, the government line was to identify 'good agricultural practice' for fertiliser application and deploy the official Agricultural Development and Advisory Service (ADAS) of the Ministry of Agriculture, Fisheries and Food (MAFF), as the agency to limit 'excessive'

Table 15.1 Derogations for nitrate in drinking water in England, 1985

Water Authority area	Number of supplies	Population affected
Anglian	26	500 000
Severn-Trent	24	372 000
Thames	1	48 000
Yorkshire	1	1 000
Total	52	921 000

applications (see Chapter 11). An unused section of the Control of Pollution Act 1974 permitted the government to allow the regional water authorities to designate water-protection zones in which water authorities could restrict fertiliser use, without compensation, so long as good agricultural practice could not be proven (Chapters 12 and 14). The key question was therefore: should some farmers be ordered to reduce their profitability just because of an accident of geology? Put another way: should a polluter pay when neighbours on more favourable or tolerable soils do not? There is precedent for this in the Clean Air Act 1956, which establishes smoke-abatement zones. Anyone who wants to burn a coal fire in such areas must use more expensive smokeless fuels even if their neighbours outside the zone can burn the cheaper, smoky, fuels.

By 1983 the nitrate issue had become a political issue. The drought of 1976 had concentrated the substance to an unusual degree, mostly because sewage works discharges were proportionately more important at a time of low (summer) water flow and because of the large amounts of leaching which followed the end of the drought (Chapter 10). In 1983 it was revealed that 10 water sources had been temporarily or permanently closed due to nitrate pollution. In 1985 the Directive on Drinking Water Quality came into effect and the government announced that it would allow water companies to spend up to £50 million in blending supplies, but would also seek the derogations as listed in Table 15.1. By the late 1980s three developments threatened this position. These were the government's determination to privatise the water industry, legal challenges to the derogations and EC concern regarding agricultural over-production (Hill, Aaronovitch and Baldock, 1989, p. 233).

The privatisation process resulted in the creation of an independent regulatory agency, the National Rivers Authority, charged with the job of establishing water quality targets for all watercourses and for meeting the requirements of EC water quality directives. Meanwhile, Friends of the Earth (FoE) mounted a campaign to force the government to meet the 50 mg 1^{-1} figure for nitrate concentrations. By 1988, with a new Environment Secretary in place, the government had changed its views and accepted the 50 mg 1^{-1} limit with a phased period of compliance. The fact that FoE had threatened to take the government to the European Court of Justice over the issue was relevant. But more significant was the decision, emanating from 10 Downing Street, that Britain should comply with existing EC directives for the government was embarking on larger battles over monetary and political union.

A third factor was the changing political attitude to agricultural surpluses. This was coupled to a logical interest in redirecting agriculture towards more environmentally friendly practices, a move that had followed a natural evolution from 1985 onwards (see Lowe et al., 1986). During this period the main UK agricultural lobby, the National Farmers Union (NFU), tinkered with the issue for nitrate reduction as a means of responding to over-production. By 1990 a working party of NFU actually recommended this as a viable option compared with other more draconian means of taking land out of production, or reducing support prices for cereals and livestock.

Hill and his colleagues (1989) suggest two prime reasons for this period of political inaction between 1985 and 1989. One was the addition of scientists on official committees which consisted of 'balanced' representation to protect the *status quo*. This process of using power to keep a potentially threatening issue from being determined at all, is known as 'non-decision making'. The other was the unwillingness of the Water Authorities to take action on a differential and localised basis where only a few farmers would be singled out as 'villains', all of whom were following official guidance for ADAS.

In their analysis, Hill *et al.* (1989) argue that the influence of the agricultural lobby supported by the agribusiness industry and MAFF generally would be sufficient to maintain the *status quo* on fertiliser application. That grouping was sufficiently powerful to insist on compensation for any profit foregone as a payment for restricting fertiliser use. MAFF backed this position while DoE argued for the 'polluter pays' principle. This latter position was seen to be untenable given the geography of nitrate pollution.

The break point came with the overwhelming pressure for the EC to reduce production surpluses: one place to start was in the over-fertilised cereal garden of eastern and southern England. One particular context was the alarm over blue-green algal blooms in the same areas during the dry summer of 1989. Another was the incidence of the diatom *Navicula* in some south-western rivers, a proliferation that adversely affected salmon and trout spawning. These outbreaks resulted in MPs being pressured by angry constituents, and demands on the privatised water companies to invest in costly protective treatment works. Of interest here is the reply by a DoE minister to a concerned MP in the constituency of one of the largest inland reservoirs, Rutland. The Minister said that his medical adviser had been 'unable to give definitive advice because of the unreliability of the analytical information available and the absence of any reliable toxicological data for algal toxins'. When science cannot deliver, prejudice, passion and power plays are encouraged to reign.

So by mid-1989 nitrate pollution was firmly on the political agenda, irrespective of the scientific position on health effects. The formation of the NRA led to a much more open and critical appraisal of farm-pollution incidents. The leniency which characterised the era when the Water Authorities refused to prosecute because of their own failure to control their own sewage-treatment works was drawing to an end. Nevertheless, the NRA was at first unwilling to be too tough on farmers. That attitude changed by 1990 when new EC regulations on farm wastes were translated into UK practice and grant aid schemes. In addition, evidence from the NRA's published reports showed that farm-pollution incidents were growing. By 1988 almost half of all serious water-pollution incidents were due to agriculture (NRA, 1989). By 1990 new regulations on slurry disposal and silage-effluent controls were implemented in a tough manner, regardless of the considerable costs to small livestock farmers. Between 1988 and 1989 the number of prosecutions against farmers rose, with almost a third of the offences being reported as serious (NRA, 1990).

15.5 POLICY OPTIONS FOR NITRATE REDUCTION

The major shift in policy was the introduction of Nitrate Sensitive Areas (NSAs) in July 1990 (see Figure 15.1). These are essentially the water-protection zones of the 1974 Act translated into the Water Act of 1989 (see Chapters 12 and 14). The key difference was the willingness of the government to compensate farmers for not applying fertilisers up to the level that would be regarded as good agricultural practice. Ten pilot areas were designated under the scheme. Under one provision, known as the 'Basic Scheme', farmers are paid between £55 and £95 per ha for not applying fertilisers above the 'economic optimum' of 165 kg ha^{-1} a^{-1} for winter wheat, barley, oilseed rape and forage catch crops. Within this package were controls on manure and sewage sludge spreading. Additional payments of £200–£380 per ha are available to farmers participating in what is known as the 'Premium Scheme', where cereal production is diverted to grasses and non-leguminous species. Table 15.2 provides the details.

By 1991 it was evident that the primary purpose of the NSA scheme had failed. While

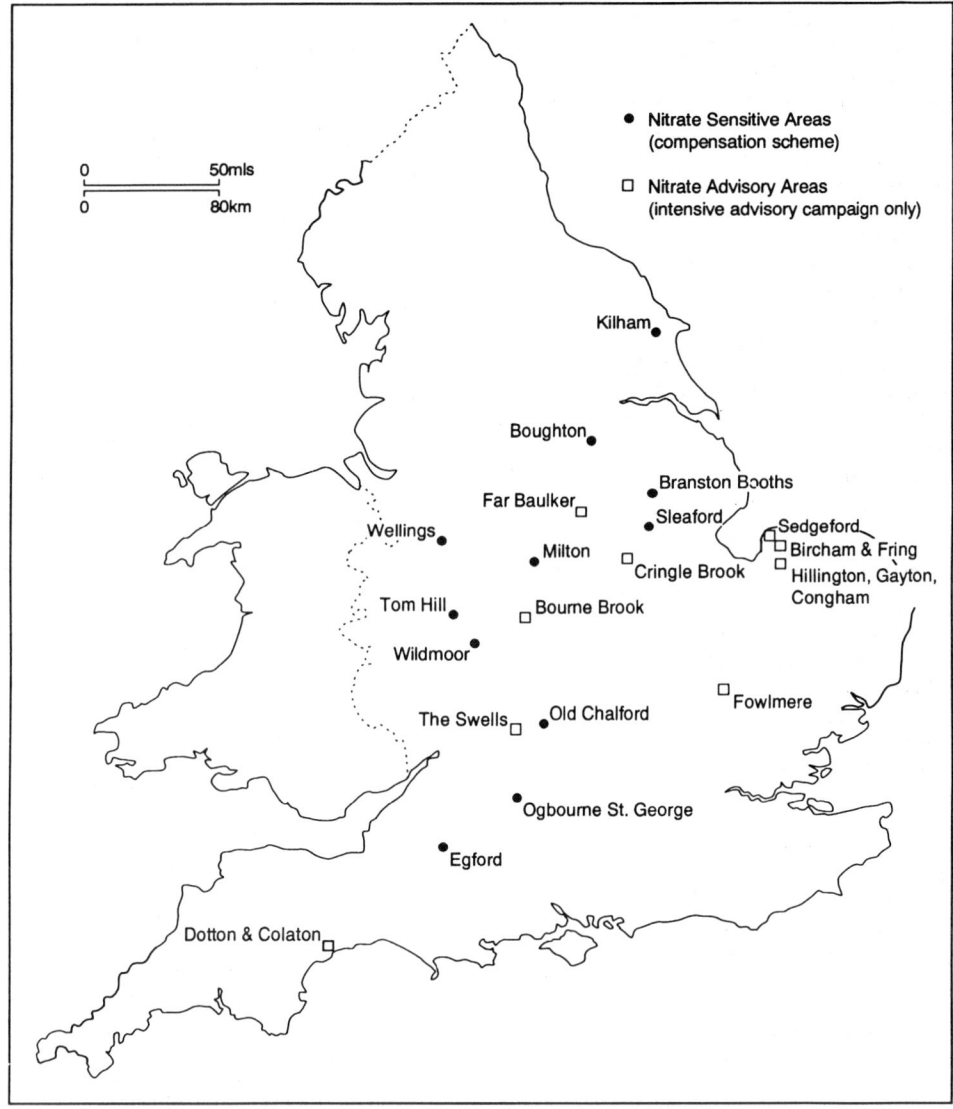

Figure 15.1 Pilot nitrate protection zone scheme (from Burt, 1992, unpublished)

65% of farmers had adopted the Basic Scheme, at hardly any cost to them, only 5% had agreed to join the Premium Scheme. Yet the latter was the principal means whereby nitrate levels would be substantially reduced (see Table 15.3). The additional designation of Nitrate Advisory Areas, where levels of application would be reduced without payment, appears to have been a complete flop. It is worth noting that a similar voluntary scheme in Jersey, which has been in force for 17 years, has also failed to generate any change in fertiliser applications or farmer awareness of nitrogen as a possible environmental nuisance (Foster *et al.*, 1989). One concern over the whole voluntary approach is that land currently under grass, if ploughed at some time in the future, could well release very substantial quantities of stored nitrate. At present there are no safeguards for this.

Table 15.2 Payments for Nitrate Sensitive Areas

| | Basic Scheme | Premium Scheme | |
| | | Unfertilised | Limited fertiliser application |
	£	£	£
Ogbourne St George	55	200−380	90−270
Kilham	55	200−380	90−270
Egford	55	200−380	90−270
Old Chalford	55	200−380	90−270
Wellings	65	200−330	90−220
Tom Hill	70	200−330	90−220
Wildmoor	70	200−280	90−170
Boughton	75	200−280	90−170
Sleaford	85	200−280	90−170
Brinston Booths	95	200−280	90−170

Rates of payment depend on the geology and on the amount of land in a fertiliser reduction scheme. The government was forced to produce a complex array of options, including conversion to woodland, as part of the total offer. In the event, much of this has failed to attract farmers.

Table 15.3 Response to Nitrate Sensitive Area schemes

| | Total area | Applicants to join NSA | | | |
| | (ha) | By area | | By farmers | |
		(ha)	(%)	(no.)	(%)
Ogbourne St George	900	692	77	7	78
Kilham	756	192	25	8	53
Egford	427	202	47	4	33
Old Chalford	629	536	85	9	75
Wellings	529	175	33	12	50
Tom Hill	570	417	73	17	71
Wildmoor	890	584	66	31	84
Boughton	1 784	305	17	9	43
Sleaford	2 375	1 314	46	15	68
Brinston Booths	1 585	1 277	81	17	81
Total	10 945	5 694	52	129	65

The failure of farmer response is not just a problem of policy implementation. It is also a problem for science, as one of the aims was to provide field conditions to see just what would be the effect of measured reductions on fertiliser application and land-use practices or groundwater nitrate concentrations. This important work will be difficult to pursue in the light of the disappointing results so far. Both the NRA and the pressure groups are angry at the slow response: the government is adamant that it will not raise the payment on grounds of principle. By mid-1991, 85% had joined the Basic Scheme, covering 87% of the available area, but only 14% of the land (mostly in three NSAs) had been entered in the Premium Scheme.

It is inevitable that stronger nitrate reduction safeguards will be required. One possibility is to require the use of nitrate filter strips of permanent ungrazed grassland or woodland around all sensitive water courses (see Section 12.4). Another possibility is to increase the price of nitrogenous fertiliser to reflect its environmental burden. This would be in accord with the fashionable principle of environmental pricing advocated by the government in its annexe to the White Paper on Environmental Strategy (UK Government, 1990). The trouble is that

Table 15.4 The effects of a nitrogen tax

Tax rate (%)	Revenue (£m)	Change in farm output (£m)	Reduction in farm profits (£m)	Change in nitrate concentrations (%)
5	15.2	na	15.5	−0.7
10	37.4	−17.7	31.1	−1.4
20	65.8	−37.4	62.5	−2.6
40	142.0	−82.3	126.3	−5.0

Source: *The Economist*, 12 January 1990. Reproduced by permission.

the tax rate would have to be high to cut nitrate levels by any significant extent, such is the payoff on yield increases. A study by the London School of Economics suggested that a tax of 40% would generate £142 million in revenue, cut farm output by £82 million, but would only reduce nitrate concentrations by 5% (reported in *The Economist*, 12 January 1990 — see Table 15.4). Yet this revenue could be deployed in a variety of useful ways if targeted to, say, leguminous biotechnology, or to improved training schemes for organic production, or to provide better incentives for fertiliser application at safe times on safe crops. So far, the Treasury has not seen fit to apply a fertiliser tax, nor has the European Community pushed the idea on a Community-scale. The farming vote is still very powerful.

Eventually, more draconian curbs will be required. Inevitably, this will be coupled to extensification proposals now emanating from the Commission (see Chapter 1). The trouble is that unless price support is all but eradicated, there will be no real incentive to curtail nitrate applications. So we may have to await the ultimate policy shift — a genuinely free market in agricultural produce with a concomitant social contract on farms to tend their farms for public pleasure and nature conservation than for food output, before we begin to address the issue of over-fertilisation really seriously. Meanwhile the nitrate 'time bomb' ticks away, and the expensive programme of blended supplies and bottled water will have to be enhanced. By the turn of the century, the average British consumer in the south and east may be following their French counterpart with most potable water coming from plastic bottles. The nitrate problem will be with us for a long time to come.

REFERENCES

Al-Dabbagh, S., Forman, D., Bryson, D., Stratton, I. and Doll R. (1986) Mortality of nitrate fertilizer workers. *British Journal of Industrial Medicine*, **43**, 507—15.

Armijo, R. and Coulson, A.H. (1975) Epidemiology of stomach cancer in Chile — the role of nitrogen fertilizers. *International Journal of Epidemiology*, **4**, 301—9.

Bartsch, H., Ohshima, H. and Pignatelli, B. (1988) Inhibitors of endogenous nitrosation mechanisms and implications in human cancer prevention. *Mutation Research*, **202**, 307—24.

Beresford, S. (1985) Is nitrate in drinking water associated with the risk of cancer in the urban UK? *International Journal of Epidemiology*, **14**, 57—63.

Britton, M., Fox, A.J., Goldblatt, P., Jones, D.R. and Rosato, M. (1990) The influence of socio-economic and environmental factors on geographic variation in mortality. *In Britton, M. (ed.), Mortality and Geography. OPCS Series DS No. 9*, HMSO, London, pp. 57—78.

Comly, H.H. (1945) Cyanosis in infants caused by nitrates in well water. *Journal of the American Medical Association*, **129**, 112—16.

Conrad, J. (ed.) (1991) *Nitrate Pollution and Politics*, Avebury Press, Aldershot.

Correa, P., Haenszel, W., Cuello, C., Archer, M. and Tannenbaum, S. (1975) A model for gastric cancer epidemiology. *Lancet,* **ii**, 58–60.

Cuello, C., Correa, P., Haenszel, W., Gordillo, G., Brown, C., Archer, M. and Tannenbaum, S. (1976) Gastric cancer in Colombia. 1. Cancer risk and suspect environmental agents. *Journal of the National Cancer Institute,* **57**, 1015–20.

Davies, J.M. (1980) Stomach cancer mortality in Worksop and other Nottinghamshire mining towns. *British Journal of Cancer,* **41**, 438–45.

Department of the Environment (1986) *Nitrate in Water*, Report of the Nitrate Co-ordination Group, HMSO, London.

Food and Agriculture Organisation (1988) *Integration of Environmental Aspects in Agriculture, Forestry and Fisheries Policies in Europe*, FAO, Rome.

Forman, D. (1989) Are nitrates a significant risk factor in human cancer? *Cancer Surveys,* **8**, 443–58.

Forman, D., Al-Dabbagh, S. and Doll, R. (1985) Nitrates, nitrites and gastric cancer in Great Britain. *Nature,* **313**, 620–25.

Foster, I.D.L., Ilberry, B.W. and Hanton, M.A. (1989) Agriculture and water quality: a preliminary examination of the Jersey nitrate problem. *Applied Geography,* **9**, 95–113.

Fraser, P. and Chilvers, C. (1981) Health aspects of nitrate in drinking water. *The Science of the Total Environment,* **18**, 103–16.

Goodman, D. and Redclift, M. (1991) *Refashioning Nature: Food Ecology and Culture*, Routledge, London.

Hartman, P.E. (1983) Nitrate/nitrite ingestion and gastric cancer mortality. *Environmental Mutagenesis,* **5**, 111–21.

Hill, M., Aaronovitch, S. and Baldock, D. (1989) Non decisionmaking in pollution control: nitrate pollution, the EEC Drinking Water Directive and agriculture. *Policy & Politics,* **17**(3), 727–41.

Hill, M.J., Hawksworth, G. and Tatersall, G. (1973) Bacteria, nitrosamines and cancer of the stomach. *British Journal of Cancer,* **28**, 562–7.

House of Lords, Select Committee on the European Community, Session 1988–89, 16th Report (1989) *Nitrates in Water*, HMSO, London.

Lowe, P., Cox, G., McEwen, M., O'Riordan, T. and Winter, M. (1986) *Countryside Conflicts*, Gower, Aldershot.

McCormick, J. (1991) *Environmental Politics in Britain*, Earthscan, London.

National Rivers Authority (1989, 1990) *Annual Reports*, HMSO, London.

Porritt, J. and Winner, D. (1989) *The Coming of the Greens*, Collins/Fontana, London.

Preussman, R. and Stewart, B.W. (1984) N-nitroso carcinogens. In Searle, C.E. (ed.), *Chemical Carcinogens A.C.S. Monographs 182*, American Chemical Society, Washington, DC, pp. 643–828.

Risch, H.A., Jain, M., Choi, N.W., Fodor, J.G., Pfeiffer, G., Howe, G.R., Harrison, L.W., Craib, K.J.P. and Miller, A.B. (1985) Dietary factors and the incidence of cancer of the stomach. *American Journal of Epidemiology,* **122**, 947–59.

Rose, C. (1990) *The Dirty Man of Europe*, Simon and Schuster, London.

Smil, K. (1991) The nitrogen and phosphorus cycle. In Turner, W.B. and Kates, R.W. (eds), *The Earth as Transformed by Human Action*, Cambridge University Press, Cambridge, pp. 419–35.

Timberlake, L. and Thomas, L. (1991) *When the Bough Breaks: Our Children: Our Environment*, Earthscan, London.

UK Government (1990) *This Common Inheritance: Britain's Environmental Strategy*, Cmd 200, HMSO, London.

World Health Organisation (1985) *Health hazards from nitrates in drinking water*, Copenhagen.

Zaldivar, R. (1977) Nitrate fertilizers as environmental pollutants. *Experimentia,* **33**, 264–5.

Zaldivar, R. and Robinson, H. (1973) Epidemiological investigation of stomach cancer mortality in Chileans: associations with nitrate fertilizer. *Zeitschrift Krebsforschung and Kilinische Onkologie,* **80**, 289–95.

16 Nitrate: Future Problems — Future Solutions?

A.L. HEATHWAITE, T.P. BURT AND S.T. TRUDGILL

16.1 AN INTEGRATED APPROACH

In essence, the 'nitrate issue' is concerned with the increasing concentrations of nitrate in surface waters, groundwaters, lakes and the marine environment. How this increase has been brought about is unlikely to be the result of a single environmental factor. Although much of the blame has been directed towards agricultural intensification as the source of nitrogen enrichment, increased nitrogen flux from the atmosphere to the terrestrial environment, together with an increase in the nitrogen loading from human (sewage) sources, are also implicated. Addiscott, Whitmore and Powlson (1991), for agricultural systems, suggest that nitrate *availability* can be equated with nitrate *vulnerability*. That is, *available* nitrogen in soils (ammonium or nitrate) is also *vulnerable* to loss. Catchment management needs to reduce the amount of nitrate present in the soil by, for example, controls on nitrogen fertiliser application, to the point at which plant uptake balances the available nitrogen supply. The key to the nitrate problem in agricultural systems could, therefore, be said to be *untimely nitrate* (i.e. nitrate present in the soil that is excess to crop requirements). Whether 'untimely nitrate' becomes nitrate pollution depends on the size and frequency of the hydrological pathways connecting the land to the stream and the relative importance of transport (hydrological) controls on nitrate delivery to the stream as compared with source or rate controls which govern the *supply* of *available* nitrate in the soil (see Section 16.3).

If the environment is subdivided into different sectors (atmosphere, land, river, ocean) it is clear, on a very simplistic basis, that the flux of nitrogen from the atmospheric and terrestrial sources has increased, the recipients of this nitrogen being the freshwater and marine environment. How the changing flux of nitrogen from one sector to the next is evaluated and therefore, by inference, controlled depends on the 'building block' used for interpreting these processes. In this book, we argue that the fundamental building block for the interpretation of nitrogen processes and for the implementation of nitrogen control must be the catchment or drainage basin. At this scale, it is possible to envisage that a systems basis to solutions to the nitrate issue can be implemented. The key factors to be considered in the catchment system are (1) interdependence of constituent parts and (2) time-dependent behaviour (Newson, 1992). The timescale of the nitrate issue is important as in the case of, for example, nitrate pollution of the chalk aquifer in south-east England (Chapter 8). Furthermore, given the inevitability of climate change, temporal predictability is further lessened, so putting solutions to the nitrate issue within the framework of large-scale catchment planning is essential. Gardiner and Cole (1992) suggest that:

> The fundamental aim of river catchment planning is to conserve, enhance, and, where appropriate,

Nitrate: Processes, Patterns and Management. Edited by T.P. Burt, A.L. Heathwaite and S.T. Trudgill
© 1993 John Wiley & Sons Ltd

restore the total river environment through effective land and resource planning across the total catchment area.

The joint management of land and water has not been UK public policy; nor has it, on the whole, been the focus of research activities. There are still the problems of scale to be resolved in, for example, relating hydrological research — often conducted in research basins which are less than 10 km^2 — to the much larger basins appropriate for catchment planning (Burt, 1989; Heathwaite, Burt and Trudgill, 1989; Newson, 1992).

Taking the catchment as the basic unit for nitrate control, it has been recognised relatively recently that some areas of land offer more scope for nitrate control than others. Décamps et al. (1990) suggest that nutrient retention efficiency is positively related to percentage of catchment composed of terrestrial/aquatic ecotones (see Section 12.4). Such ecotones include wetlands, groundwater exchange zones and riparian forests. The degree of protection or water quality control is thought to be proportional to the length and shape of contact zone. Décamps et al. (1990) suggest that ecotones occurring at the terrestrial/aquatic interface are particularly important in regulating the flow of water and materials across the landscape. Such zones are likely to be critical in nitrate control at a catchment scale (Pinay and Décamps, 1988). For example, the spatial and temporal variations in redox conditions, which are characteristic of ecotones, enhance rates of denitrification. Rates of denitrification will be slower in more 'hydrologically stable' adjacent land where microbial activity will be relatively restricted.

Denitrification is currently viewed as a key process in nitrate control. It is important to note, however, that the value of denitrification in nitrate control at the catchment scale depends on the nature of the gaseous end-product: dinitrogen (N_2) is acceptable, whereas nitrous oxide (N_2O) is regarded as a serious atmospheric pollutant (see Section 16.2). The ratio of N_2:N_2O produced through denitrification is largely determined by soil factors (nitrate concentration, pH, oxygen concentration, organic matter content). In strongly acid soils, characteristic of many upland regions in the UK, N_2O is the main product of denitrification and, as a result, any increase in the concentration of nitrate in upland waters may indirectly alter the local or regional atmospheric nitrogen balance. This has serious implications for the control of nitrate in upland areas (see Section 16.2.2). In alkali or circum-neutral soils, N_2 is normally released through denitrification, except where the soil nitrate concentration is high as, for example, in fertilised soils (Addiscott, Whitmore and Powlson, 1991).

16.2 ATMOSPHERIC NITROGEN: FUTURE ATMOSPHERIC CONCENTRATIONS AND PATTERNS OF DEPOSITION

Land in the UK, whether in agricultural production or not, may receive up to 40 kg ha^{-1} N from the atmosphere each year (Goulding, 1990). This is equivalent to 20% of the average annual application of N fertilisers to arable crops. The range of nitrogen compounds deposited on agricultural land in south-east England from the atmosphere is shown in Figure 16.1. Approximately 10–12 kg N ha^{-1} are deposited by rainfall (Addiscott, Whitmore and Powlson, 1991), the remainder being absorbed by plant material mainly from gaseous or aerosol sources. It is interesting that figures for total N deposition reported by Barrett et al. (1987) are of a similar order of magnitude (see Figure 16.2).

Atmospheric nitrogen inputs to the terrestrial and aquatic environmental sectors have been increasing, largely as a result of human activities (Dollard et al., 1987). The main sources are:

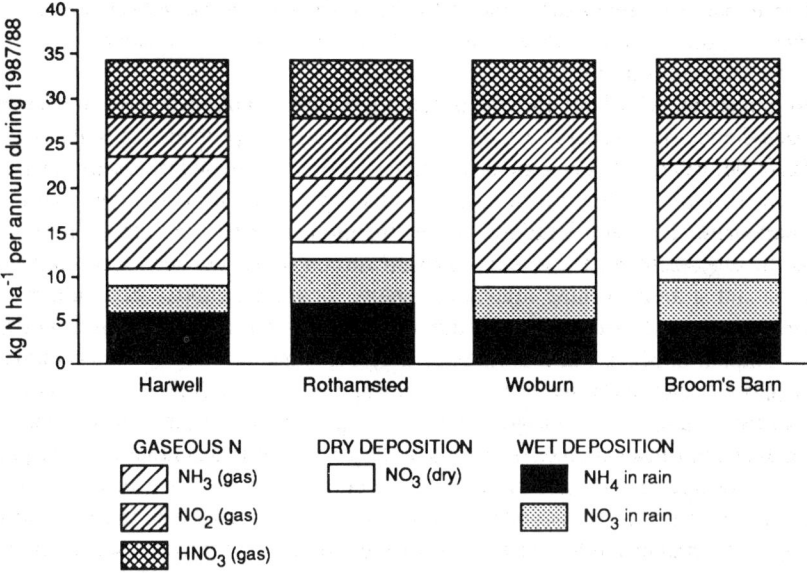

Figure 16.1 Nitrogen deposition from atmospheric sources to agricultural land in the UK (modified from Addiscott, Whitmore and Powlson, 1991)

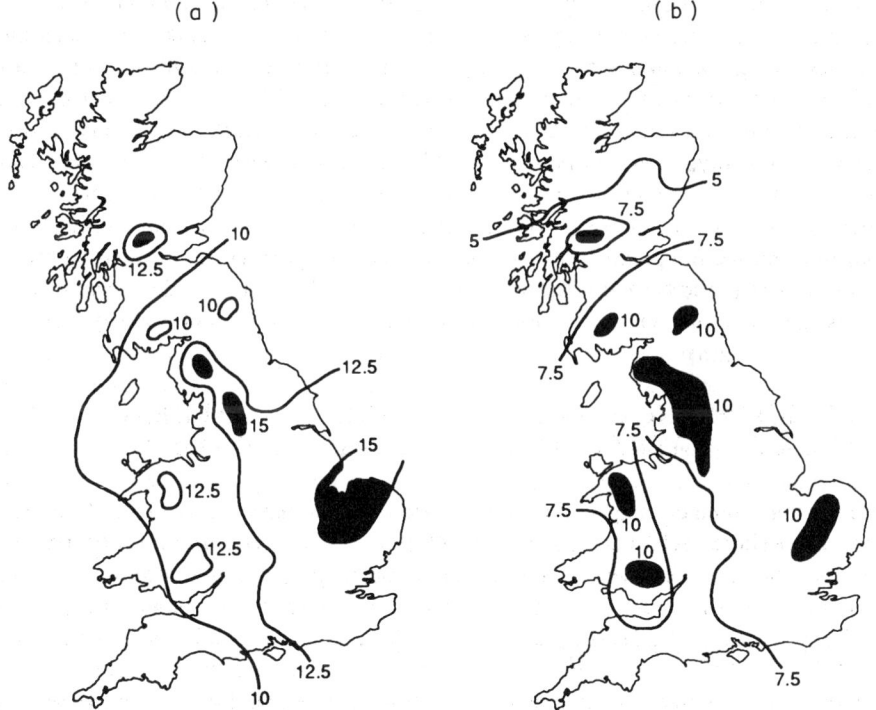

Figure 16.2 Atmospheric nitrogen deposition to land in Britain. (a) Total N (wet + dry deposition); (b) total inorganic N (NH_4 + NO_3) in wet deposition. All units in kg N ha^{-1}. Data compiled from information provided by the Warren Spring Laboratory

(1) high-temperature combustion (motor vehicles and power stations);
(2) ammonia volatilisation from agricultural land, especially from intensive livestock or 'factory farming' enterprises;
(3) stubble burning, which releases NO_x (although current UK legislation will restrict this practice); and
(4) denitrification.

Whereas industrial or motor vehicle emissions of nitrogen affect the atmospheric nitrogen loading on a *regional* scale and form *diffuse* atmospheric nitrogen sources, ammonia volatilisation and stubble burning may have a more *localised* impact where it is possible to identify such emissions as *point* sources of atmospheric nitrogen through wet and dry deposition of NH_3. Ammonia deposited in this manner may contribute locally to soil acidification as the oxidation of NH_3 to NO_3 releases hydrogen ions.

   Assuming that some sort of nitrogen balance exists in soils, any increase in nitrogen input from the atmosphere must be balanced by a nitrogen loss (for example, by leaching to the aquatic system or by denitrification or ammonia volatilisation back to the atmosphere). The partitioning of nitrogen pathways and rates of flux from the atmosphere to the soil (and back again) is complex, and has only recently become the focus of attention now that atmospheric pollution from SO_2 has decreased. The UK emits approximately 1 900 000 tonnes of NO_x per annum (Barnaby, 1988); about 40% comes from power stations and 40% from motor vehicles. Although lower than some European countries (notably Denmark and Germany) and about one-tenth that emitted by the USA, emissions of NO_x are expected to increase in the future, primarily from vehicular sources. In May 1987, the UK government announced a 10-year programme to control NO_x emissions through the installation of low-NO_x burners at the CEGB's twelve largest power stations (Barnaby, 1988). The Draft EC Directive seeks to reduce NO_x emissions by 40% by 1995: current UK policy will not achieve this. This means that any catchment-based nitrate control strategy will have to take into account the effect of increased atmospheric nitrogen loading.

   The atmospheric nitrogen loading from industrial and vehicular sources does not have an even spatial distribution in the UK. A few key areas appear to be susceptible to high atmospheric nitrogen deposition. These areas are shown in Figure 16.2, which illustrates the total atmospheric nitrogen deposition (kg N ha^{-1}) from both wet and dry deposition sources (Figure 16.2(a)) together with the total atmospheric inorganic nitrogen deposition from wet deposition sources alone (Figure 16.2(b)). In England, areas of high N deposition are identified in East Anglia where dry deposition is dominant and the Pennines and Lake District, together with upland regions of north and south Wales, where wet deposition is more important because of high rainfall. In the light of continuing atmospheric pollution and, in particular, the continued increase in NO_x emissions in the future, it is likely that sensitive areas such as the uplands of the Pennines (Section 16.2.1) and the lowlands of East Anglia (Section 16.2.2) may be subject to further modification of the nitrogen cycle.

16.2.1 SENSITIVITY OF UPLAND CATCHMENTS TO N INPUTS

On a national scale, NO_3-N concentrations in rivers range from 0.1 mg l^{-1} in upland areas to in excess of 15 mg l^{-1} in lowlands (Walling and Webb, 1981; Betton, Webb and Walling, 1991). Such figures do not reflect the sensitivity of different areas to nitrogen inputs. Upland soils normally retain nitrogen but such soils may have a limited capacity to buffer excess nutrient inputs. It is conceivable that excess nitrogen entering upland soils from the atmosphere may promote nitrate leaching to watercourses or localised nitrogen pollution as a result of denitrification and the release of N_2O where the soils are acid. The predominance of acid, organic-rich peaty soils in upland regions such as the Pennines may result in extensive N mineralisation by shifting the C:N ratio in the soils. This may, in turn, promote nitrate leaching. Although high annual rainfall in upland regions has been assumed to be capable of diluting the nutrient input to the aquatic system, the extent to which such environmentally sensitive aquatic systems will be able to withstand sustained nitrogen deposition from the atmosphere remains to be seen.

Upland 'eutrophication' or, more correctly, 'enrichment' of watercourses may well become a major nitrogen issue in the future, particularly because headwater catchments are often viewed as the last remaining source of low-nitrogen water for nitrogen-control strategies such as blending (see Chapter 14). In addition, nitrogen enrichment of upland soils may also adversely affect the nutrient balance and ecosystem composition of plant communities found in such areas.

Table 16.1 gives estimates of the rate of increase of nitrate concentrations in UK rivers for the mid-1980s based on data from the Nitrate Co-ordination group (1986). It is interesting that research at that time was focused on major lowland rivers. This is partly because such rivers have long-term nitrate records but also because nitrate pollution of surface waters was perceived as being (and still is) most severe in such areas. Whereas there is some recent evidence to suggest that a reduction in the application of fertiliser nitrogen in the autumn has reduced nitrate concentrations in lowland catchments which are highly responsive to rainfall (Croll and Hayes, 1988; Addiscott, Whitmore and Powlson, 1991), there remains a strong case for the investigation of nitrate enrichment of upland catchments from atmospheric deposition because here agricultural controls are unlikely to offer a major solution to the nitrate issue. It should not be thought, however, that upland soils are immune from the kind of effects noted when lowland soils are cultivated. Cultivation of soils to improve the quality of pasture can result in nitrate concentrations in streamflow which, while unlikely to be of national concern, might cause problems locally. Drainage of upland soils, particularly deep ribbon ploughing, might cause similar problems.

Table 16.1 Estimates of rate of change in long-term trends in nitrate levels (Nitrate Coordination Group, 1986. Reproduced by permission of the Controller of Her Majesty's Stationery Office)

Location	Concentration trend
Scotland Wales NW	$0.1-0.4$ mg l^{-1} yr^{-1} NO_3 $0.02-0.09$ mg l^{-1} yr^{-1} NO_3-N
NE England	$0.1-0.7$ mg l^{-1} yr^{-1} NO_3 $0.02-0.16$ mg l^{-1} yr^{-1} NO_3-N
Yorkshire Severn Trent Water Authority Thames Water Authority South West Water Authority Wessex Water Authority	$0.3-0.8$ mg l^{-1} yr^{-1} NO_3 $0.07-0.19$ mg l^{-1} yr^{-1} NO_3-N
Anglian	$0.7-1.1$ mg l^{-1} yr^{-1} NO_3 $0.16-0.26$ mg l^{-1} yr^{-1} NO_3-N

16.2.2 LOWLAND SENSITIVITY TO DRY DEPOSITION

Goulding (1990) suggested that $35-40$ kg N ha^{-1} are deposited from the atmosphere each year in south-east England. Our figures shown in Figure 16.2, based on data from Warren Spring Laboratory (Barrett et al., 1987), suggest a lower average for total nitrogen (wet + dry) deposited of around 15 kg N ha^{-1} per annum which may, nevertheless, locally approach the levels quoted by Goulding (1990). The atmospheric nitrogen loading derived from wet deposition of nitrate and ammonium (Figure 16.2(b)) is centred over the Breckland area of East Anglia where the altitude is slightly higher. The area receiving high dry deposition of atmospheric nitrogen (Figure 16.2(a) minus Figure 16.2(b)) covers a large area of East Anglia, extending over north and central Norfolk and into Lincolnshire.

These figures suggest that attempts to control nitrate losses from farmland in perceived 'problem' areas such as East Anglia may be limited by the magnitude of atmospheric nitrogen inputs. Goulding (1990) estimates that if the average drainage for the region is 200 mm per year and assuming nitrogen inputs are balanced by nitrogen losses, the concentration of nitrate in drainage water could reach 20 mg NO_3-N l^{-1}, which is almost double the EC limit. Any attempts at nitrate control through, for example, land-use control or fertiliser restrictions could be jeopardised by the anticipated increases in atmospheric nitrogen inputs.

16.3 EVALUATION OF THE CONTROLS ON NITRATE PRODUCTION FROM AGRICULTURAL SOILS

In Section 16.1 the value of ecotones as a means of nutrient control was stressed. Current research (Décamps et al., 1990) suggests that the composition and structure of the biotic community of ecotones may be more sensitive to hydrological changes than to changes in nutrient availability. This implies that hydrological pathways through ecotones are critical in governing the rates and magnitude of processes such as denitrification, which are important in nitrate control. Furthermore, for any given hydrological regime in an ecotone, the rate of nutrient cycling and of net mineralisation in particular is thought to be strongly influenced by the inputs and nutrient concentration of particulate and dissolved organic matter. As the chemical quality of water percolating through an ecotone will vary seasonally, both as a result of fluctuations in physical (hydrological) and biological parameters, so the rate of nutrient cycling will vary. Emphasis on the supply of particulate and dissolved organic matter in the ecotone reflects the importance of microbial processes on rates of nutrient retention and release. The value of ecotones for nitrate control relies on the fact that the concentration of nitrate-rich water entering, for example, a riparian woodland from an intensively grazed grassland hillslope will be reduced as a result of assimilation and/or denitrification by biota within the ecotone (Gilliam, Skages and Weed, 1979; Peverly, 1982; Reddy and Reddy, 1987; Trudgill, Heathwaite and Burt, 1991a). The size and chemical composition of the organic matter pool within an ecotone will influence the rate of nitrate utilisation, as will any change in the hydrological regime as a result of, for example, fluctuation in the water table level.

Our discussion of ecotones for nutrient control did not include consideration of the form in which water enters such zones from the upslope catchment area. We suggest that there are two possible controls on catchment solute production and ultimately stream water quality: *rate* controls or *transport* controls, which are determined by the size of the solute source and the magnitude and rate of water movement from the land to the stream (see, for example, Trudgill et al., 1991b).

In a *transport control* situation the supply of a solute is large relative to the water flow rate and solute (e.g. nitrate) concentrations in the soil and in drainage water percolating through the soil are dependent on the rate of flow of water. This means that, regardless of the size of the solute source (e.g. where it is supplemented by nitrogen (fertilisers)), nitrate will only be transferred from the hillslope to the stream at a rate which is governed by the rate at which water can percolate through the soil. In effect, it does not matter how much nitrogen is added to the soil in the form of, for example, chemical fertilisers or organic manures because their transfer to the stream is hydrologically controlled not supply controlled.

In a *rate control* situation the supply of a solute is small relative to the rate of water flow through the soil and solute concentrations in the soil and in water draining from the soil are dependent on the solute supply. As water passes through the soil (for example, following a storm event) the concentration of a solute in the percolating water will either decrease as the rate of water flow increases (because the rate of solubilisation of the solute is less than the rate of flow of water thus a dilution effect will be observed) or the solute concentration may initially increase as the initial pulse of infiltrating water passes through the soil but it will eventually decrease as the solute supply is exhausted. Thus in a rate controlled situation any additions to the solute store (e.g. fertiliser application) are likely to significantly increase the concentration of the solute in drainage water and hydrological controls are less important than solute source controls.

The concepts of rate control and transport control outlined above provide a simple basis to water quality modelling. Identification of the key solute production factor (rate or transport control) should allow, for example, nitrate control strategies to be tailored to dynamics of the catchment system. In a transport controlled catchment or for a transport controlled solute, water quality modelling must focus on hydrological pathways. By implication, spatial and temporal variability in solute supply, reaction rates and biological solubilisation processes are less important than rainfall−runoff ratios and soil water flow rates. Nitrate control through land-use control would be inappropriate under such circumstances. Conversely, for a rate controlled solute, solute losses will be sensitive to changes in the supply of the solute. Land-management practices and, in particular, biological controls on solute release and cycling are far more important in determining the rate and magnitude of solute release to water percolating through the soil.

16.4 LAND USE AND AGRICULTURAL INTENSIFICATION

16.4.1 ARABLE LAND

Arable cultivation has frequently been cited as the source of nitrogen enrichment in surface waters and lakes, and ultimately for the marine ecosystem. The risk of nitrate leaching from arable land increases considerably post-harvest when plant uptake of nitrogen is zero but N release through mineralisation is high. A number of authors maintain that the source of nitrate is this mineralisation rather than fertiliser nitrogen remaining in the soils (see, for example, Addiscott, Whitmore and Powlson, 1991; Addiscott and Powlson, 1992). Different arable crops leave different amounts of nitrogen residue in the soil; MacDonald, Poulton and Powlson (1990) found the pattern: potatoes > oilseed rape > winter wheat for nitrogen residues.

In controlling the loss of nitrogen from arable land (assuming that nitrate production is rate controlled: see Section 16.3) it is important to establish whether there is any 'memory effect' in the system. Does, for example, extra N fertiliser result in extra nitrate mineralisation?

Such a 'memory effect' has been demonstrated for grassland (Dowdell and Webster, 1980). Addiscott, Whitmore and Powlson (1991) suggest that the memory effect depends on crop type (for example, oilseed rape has a clear memory effect after the first year of crop, whereas winter wheat has no memory effect after the first year) and the number of years that a particular crop has been grown at the same site. There is clear evidence to suggest that the rate of production of nitrate by microbes increases with the amount of fertiliser which has been applied each year. This has important repercussions for nitrate control on arable land because it suggests that soil nutrient amendments take a number of years to pass through the system but, more importantly, atmospheric N inputs could also contribute to a memory effect in the agricultural system. In any event, a solution to the reduction in nitrate losses from arable land will not be quick or easy.

16.4.2 GRASSLAND

Grassland can absorb a higher nitrogen supply. For example, up to 400 kg of fertiliser N per hectare can be applied to cut grass before substantial leakage of N occurs. The fertiliser application rate for grass can exceed that of arable land because it has continuous ground cover, nitrogen is removed in the crop and it has an extensive root system through which more nitrogen is directed to the soil organic matter store. Ultimately, however, the soil N store becomes a major potential source of nitrate so that ploughing of permanent grassland can release up to 4 t ha^{-1} (e.g. Whitmore, Bradbury and Johnson, in press) and ploughed temporary grass, 100–200 kg N ha^{-1} depending on the length of ley (Darby et al., 1988).

However, the nitrate problem associated with grassland is also strongly related to the presence or absence of livestock. In grazed grassland, over 80% all N consumed by grazing animals is returned to soil as urine or dung (Addiscott, Whitmore and Powlson, 1991). This effectively creates point sources of N pollution within the catchment (Heathwaite, Burt and Trudgill, 1990). Grazed grassland can be a major source of nitrate pollution of surface waters where livestock graze land vulnerable to leaching losses (e.g. adjacent to watercourses) and also where they have direct access to the stream. The grazing density is also important. Heathwaite, Burt and Trudgill (1990) showed that the total nitrogen load in surface runoff from heavily grazed permanent grassland was nine times that of lightly grazed grassland. Most of the nitrogen was lost in inorganic (mainly NH_4-N) form.

16.4.3 MODIFICATION OF AGRICULTURAL PRACTICES TO REDUCE NITRATE LOSSES

Bearing in mind the public perception that agriculture is the main source of excess nitrate in the aquatic system, research and policy efforts have been focused on modifying agricultural practice so that nitrate loss can be reduced. These strategies are fully discussed in Chapters 10–12 of this book. Some recent modifications of agricultural practice (Addiscott, Whitmore and Powlson, 1991) include early sowing of winter arable crops to maximise the plant uptake of nitrogen mineralised in autumn. However, this strategy may simply create other environmental problems because early sowing of winter crops increases their vulnerability to disease and hence the need for pesticide application, and autumn-sown fields are vulnerable to soil erosion.

Now that stubble burning is restricted, many farmers are ploughing in straw. As a nitrate control strategy this practice may be valuable as a result of the microbial utilisation of nitrogen to break down the carbon contained in the straw. However, Powlson, Brookes and Christensen (1987) point out that this ultimately increases the soil organic nitrogen store and thus the potential leachable nitrogen source.

Winter-grown catch crops have also been considered as a potential nitrogen control technique as long as the problems associated with disease risk and a suitable crop can be overcome (Addiscott, Whitmore and Powlson, 1991). Furthermore, as denitrification is often carbon limited, the addition of carbon source in form of a catch crop may increase N loss via this pathway, although the suitability of this approach depends on the gaseous form in which N is lost.

The above approaches are mainly related by the control of nitrate loss from arable farmland. On grassland, control strategies largely relate to stocking densities *per se* rather than focusing on the identification and control of livestock numbers in vulnerable zones such as riparian land. Nitrate losses from grassland may also be controlled by modifying the nitrogen fertiliser loading. Addiscott, Whitmore and Powlson (1991) found that halving the amount of nitrogen applied as fertiliser to grassland grazed by cattle halved the nitrogen returned to the soil as excreta. This is because livestock voids nitrogen in a more *concentrated* form in urine or dung from grassland which is supplied with fertiliser N.

16.5 HUMAN PRODUCTION OF NITROGEN

Most of the discussion in this chapter has focused on agricultural or atmospheric sources of nitrogen. It is important to put these in context with a third major source of nitrogen: the human production. The annual production of N in the UK from humans is approximately 6 kg N ha^{-1} per capita (Royal Society, 1983). Keeney (1989) suggests that for an average (in his example, American) family, the average annual nitrogen load if applied to land as sewage sludge is equivalent to around 165 kg N ha^{-1}. In order to dilute the nitrogen waste produced by the family to bring the drainage water quality in line with WHO limits for NO_3-N it would require over 1400 mm of runoff. In the UK, disposal of sewage effluent from eastern and southern England, where the population is high and the runoff is low, is becoming a major 'nitrate issue'.

16.6 TECHNICAL ISSUES FOR NITRATE CONTROL

16.6.1 LONG-TERM MONITORING SITES

In order to understand the rates and magnitude of change in nitrogen cycling and nitrate leaching in particular, information from long-term monitoring sites, which show the extent of any such changes, are essential. In the UK, long-term research basins are limited. They include relatively small-scale studies at, for example, Rothamsted Experimental Station (see Addiscott, Whitmore and Powlson, 1991) and stations maintained by the Institute of Terrestrial Ecology (see Hornung, Roda and Langan, 1990). Long-term monitoring studies at the small catchment-scale are few in number but in the UK they include the Plynlimon catchment in (upland) mid-Wales and the Slapton catchment in (lowland) south-west Devon.

In the UK, the establishment in 1992 of an Environmental Change Network (ECN) is designed to yield data on physical, chemical and biological variables for a number of representative sites over a period of at least 30 years. For nitrate, this should yield long-term data which, at present, are chiefly available only for large rivers. Research at ECN sites will be designed to investigate observed changes in more detail. Early recognition of such changes should allow processes to be investigated, models to be developed and, as a result, policy to be better informed.

On a different scale, a similar study is being undertaken by the Institute of Terrestrial Ecology, who are reviewing the results from small catchment studies in western Europe which are producing hydrochemical budgets (Hornung, Roda and Langan, 1990). The ENCORE project (European Network of Catchments Organised for Research on Ecosystems) consists of a network of catchments covering a range of environmental conditions in Europe. The study is conducted at the catchment scale because 'it is the smallest complete unit of a landscape that combines linked aquatic and terrestrial systems ... [but is] also a large enough unit to comprise all the components which interact in the terrestrial and aquatic system' (Hornung, Roda and Langan, 1990).

There is also a pressing need to begin linking nitrogen processes in the different environmental sectors into some form of comprehensive model which will allow the identification of changes in nitrogen cycling in one sector to be traced to another. The marine environment is a good example of this need (see Chapter 7). Here it is essential that the river: estuary: coastal sea: ocean pattern of nitrogen loading is understood in order to alleviate the risk of marine eutrophication in the future.

16.6.2 NITROGEN SPECIATION

Where long-term monitoring information on nitrogen is available it tends only to be in the form of dissolved inorganic (usually nitrate) concentrations in freshwaters. In some cases the discharge has also been measured so that a measurement of the nitrate loading of the stream or river may be obtained. However, in providing solutions to the 'nitrate issue' it is not enough to simply monitor the inorganic loading of the aquatic system (see Chapter 5). Knowledge of the spatial and temporal variation in all nitrogen species in a drainage basin is essential (Johnes and Heathwaite, 1992; Heathwaite and Johnes, in press). This is because nitrogen fluxes between different species may change during transport from the land to the stream and as a result of in-stream, in-river or in-lake transformations once the 'nitrogen parcel' has entered the aquatic system. Nitrogen control at the catchment scale must start with the measurement of all nitrogen species: total nitrogen, particulate organic nitrogen, dissolved organic nitrogen, particulate inorganic nitrogen and dissolved inorganic nitrogen. For a more complete understanding, the gaseous losses of nitrogen through denitrification should also be measured, but as yet there is no simple, satisfactory technique available to routinely measure rates of denitrification.

16.6.3 NITRATE: RATE CONTROL OR TRANSPORT CONTROL?

The concept of nitrate production and loss from agricultural land on the basis of transport or rate controls has been discussed in Section 16.3. This theoretical development has not yet been widely tested at different catchment scales or on different catchments. This is partly because rigourous testing requires long-term data at a number of different scales (e.g. plot, hillslope, subcatchment, catchment) and there are few sites at which with this level of data is available. However, the concept of rate or transport control does allow a relatively simple approach to the identification of the importance of hydrological, chemical and biological *pathways* of nitrogen cycling and release. By modelling such pathways and determining the processes governing the attenuation of nitrate in leachate, we will be a step further towards solving the nitrate issue.

16.7 CATCHMENT PLANNING FOR NITRATE CONTROL

To a large extent, the nitrate issue depends on current (public) perceptions: is it nitrate losses from arable land or accidental spills from silage clamps and slurry lagoons or even atmospheric nitrogen pollution which is the 'nitrate issue'? The answer is that all sources are important but their emphasis shifts according to public or government attention (see Chapter 15). This results in rather piecemeal attempts at controlling nitrate pollution. Nitrate control is not possible unless all environmental sectors are tackled simultaneously and the basis for control must be the catchment or drainage basin which enables the net effect of a wide number of control strategies (e.g. protection of ecotones, reduction of nitrogen fertiliser application and livestock densities) to be monitored and evaluated. Improved knowledge of the processes of nitrogen cycling and release and the rates of these reactions in different environmental sectors is required in order to allow catchment interlinkages to be made.

The idealised monitoring system of such processes is at the catchment scale (Newson, 1992; Burt and Haycock, in press) but the question remains: who will pay? Where is the burden of proof or burden of payment? As the ultimate recipient for nitrate enrichment is the aquatic system, what is the role of the individual water companies and pollution control authorities (in the UK, the National Rivers Authority)?

It is perhaps too soon to evaluate the success of land-regulation programmes such as the NSA scheme or set-aside. They remain only part of the story but at least they are a start in the recognition of catchment planning as the solution to the nitrate issue. It seems increasingly likely that legislation will focus on catchment planning (Newson, 1991; Chapter 14, this volume) and environmentally targeted set-aside is likely to become an important feature of land-use planning in the 1990s.

REFERENCES

Addiscott, T.M., Whitmore, A.P. and Powlson, D.S. (1991) *Farming, Fertilizers and the Nitrate Problem*, CAB International, Wallingford.

Addiscott, T.M. and Powlson, D.S. (1992) Partitioning losses of nitrogen fertilizer between leaching and denitrification. *Journal of Agricultural Science* (in press).

Barnaby, F. (1988) Acid rain: UK Policies. *Ambio*, **17**(2), 160−2.

Barrett, C.F. *et al.* (1987) Acid deposition in the United Kingdom. Warren Spring Laboratory, Stevenage, UK. Crown Copyright.

Betton, C., Webb, B.W. and Walling, D.E. (1991) Recent trends in NO_3-N concentration and loads in British Rivers. *IAHS* publication 203, 169−80.

Burt, T.P. (1989) Storm runoff generation in small catchments in relation to the flood response of large basins. In Beven, K. and Carling, P. (ed.), *Floods*, Wiley, Chichester, pp. 11−36.

Burt, T.P. and Haycock, N.E. (in press) Catchment planning and the nitrate issue: a UK perspective. *Progress in Physical Geography, 16*.

Croll, B.T. and Hayes, C.R. (1988) Nitrate and water supplies in the United Kingdom. *Environmental Pollution*, **50**, 163−87.

Darby, R.J., Hewitt, M.V., Penny, A., Johnston, A.E. and McEwen, J. (1988) The effects of increasing length of ley on the growth and yield of winter wheat. *Rothamsted report for 1987*, Part I, pp. 101−2.

Department of Environment/Welsh Office (1988) *Integrated Pollution Control: A Consultation Paper*, HMSO, London.

Décamps, H., Fournier, F., Naiman, R.J. and Petersen, R.C. (1990) An international research effort on land/water ecotones in landscape management and restoration. *Ambio*, **19**, 175−6.

Dollard, G.J., Atkins, D.H.F., Davies, T.J. and Healy, C. (1987) Concentrations and dry deposition velocities of nitric acid. *Nature*, **326**, 481−3.

Dowdell, R.J. and Webster, C.P. (1980) A lysimeter study using ^{15}N on the uptake of fertiliser nitrogen in perennial ryegrass swards and losses by leaching. *Journal of Soil Science*, **31**, 65−75.

Gardiner, J.L. and Cole, L. (1992) Catchment Planning: the way forward for river protection in the UK. In Boon, P.J., Calow, P. and Petts, G.E. (eds), *River and Conservation Management*, Wiley, Chichester, pp. 397−406.

Gilliam, J.W., Skaggs, R.W. and Weed, S.B. (1979) Drainage control to diminish nitrate loss from agricultural fields. *Journal of Environmental Quality*,**8**, 137−42.

Goulding, K.W.T. (1990) Nitrogen deposition to arable land from the atmosphere. *Soil Use and Management*, **6**, 61−3.

Heathwaite, A.L., Burt, T.P. and Trudgill, S.T. (1989) Runoff, sediment and solute delivery in agricultural drainage basins — a scale dependent approach. *IAHS*, **182**, 175−91.

Heathwaite, A.L., Burt, T.P. and Trudgill, S.T. (1990) The effect of agricultural land use on nitrogen, phosphorus and suspended sediment delivery to streams in a small catchment in south west England. In Thornes, J.B. (ed.), *Vegetation and Erosion*, Wiley, Chichester, pp. 161−79.

Heathwaite, A.L. and Johnes, P.J. (in press) The chemical composition of nitrogen and phosphorus in runoff from agricultural land. *Journal of Applied Ecology*.

Hornung, M., Roda, F. and Langan, S.J. (1990) *A review of small catchment studies in Western Europe producing hydrochemical budges*. Air Pollution Research Report Number 28, Commission of the European Communities, Brussels.

Johnes, P.J. and Heathwaite, A.L. (1992) The simultaneous determination of total nitrogen and total phosphorus in freshwaters using a microwave digestion procedure. *Water Research*, **26**(10), 1281−7.

José, P. (1989) Long-term nitrate trends in the River Trent and four major tributaries. *Regulated Rivers: Research and Management*, **4**, 43−57.

Keeney, D.R. (1989) Sources of nitrate to groundwater. In Follet, R.F. ed.), Nitrogen management and groundwater protection. *Developments in Agricultural and Managed-forest Ecology*, **21**, 23−34.

MacDonald, A.J., Poulton, P.R. and Powlson, D.S. (1990) Arable crops and farming practices: Effects on nitrate in arable soils. In Scaife, A. (ed.), *Proceedings of the First Congress of the European Society of Agronomy*, ESA, Session 5, p. 28.

Newson, M.D. (1991) Catchment control and planning: emerging patterns of definition, policy and legislation in UK water management. *Land Use Policy*, **8**, 9−15.

Newson, M.D. (1992) River conservation and catchment management: a UK perspective. In Boon, P.J., Calow, P. and Petts, G.E. (eds), *River and Conservation Management*, Wiley, Chichester, pp. 385−96.

Nitrate Co-ordination Group (1986) *Nitrate in Water*, Department of Environment Central Directorate of Environmental Protection, Pollution Paper No. 26, HMSO, London.

Peverly, J.H. (1982) Stream transport of nutrients through a wetland. *J. Environmental Quality*, **11**, 38.

Pinay, G. and Décamps, H. (1988) The role of riparian woods in regulating nitrogen fluxes between the alluvial aquifer and surface water: a conceptual model. *Regulated Rivers: Research and Management*, **2**, 507−16.

Powlson, D.S., Brookes, P.C. and Christensen, B.T. (1987) Measurement of soil microbial biomass provides an early indication of changes in total soil organic matter due to straw incorporation. *Soil Biology and Biochemistry*, **19**, 154−64.

Reddy, G.B. and Reddy, K.R. (1987) Nitrogen transformations in ponds receiving polluted water from nonpoint sources. *Journal of Environmental Quality*, **16**(1), 1−5.

Royal Society (1983) *The Nitrogen Cycle of the United Kingdom: A Study Group Report*, The Royal Society, London.

Trudgill, S.T., Heathwaite, A.L. and Burt, T.P. (1991a) The natural history of Slapton Ley Nature Reserve XIX: A preliminary study of the control of nitrate and phosphate pollution in wetlands. *Field Studies*, **7**, 731−42.

Trudgill, S.T., Burt, T.P., Heathwaite, A.L. and Arkell, B.P. (1991b) Soil nitrate sources and nitrate leaching losses. *Soil Use and Management*, **7**(4), 200–6.

Walling, D.E. and Webb, B.W. (1981) Water quality. In Lewin, J. (ed.), *British Rivers*, Allen and Unwin, London, pp. 126–69.

Whitmore, A.P., Bradbury, N.J. and Johnson, P.A. (in press) The potential contribution of ploughed grassland to nitrate leaching. *Agriculture, Ecosystems and Environment*.

Author Index

Subject Index